T0350945

PUBLIC INFRASTRUCTURE MANAGEMENT

Tracking Assets and Increasing System Resiliency

FREDERICK BLOETSCHER, PH.D., P.E., LEED-AP

J.ROSS
PUBLISHING

Copyright © 2019 by J. Ross Publishing

ISBN-13: 978-1-60427-139-3

Printed and bound in the U.S.A. Printed on acid-free paper.

10 9 8 7 6 5 4 3 2 1

Library of Congress Cataloging-in-Publication Data

Names: Bloetscher, Frederick, author.
Title: Public infrastructure management : tracking assets and increasing
  system resiliency / by Frederick Bloetscher.
Description: Plantation, FL : J. Ross Publishing, [2019] | Includes
  bibliographical references and index.
Identifiers: LCCN 2019010606 (print) | LCCN 2019016385 (ebook) | ISBN
  9781604278125 (e-book) | ISBN 9781604271393 (hardcover : alk. paper)
Subjects: LCSH: Engineering—Management. | Public works—Maintenance and
  repair.
Classification: LCC TA190 (ebook) | LCC TA190 .B63 2019 (print) | DDC
  363.6068—dc23
LC record available at https://lccn.loc.gov/2019010606

Direct all inquiries to J. Ross Publishing, Inc., 300 S. Pine Island Rd., Suite 305, Plantation, FL 33324.

Phone: (954) 727-9333
Fax: (561) 892-0700
Web: www.jrosspub.com

DEDICATION

This book is dedicated to my wife Cheryl Fox,
the love of my life. Thank you for being you!

CONTENTS

SECTION I—Infrastructure: A Comprehensive Overview 1

SECTION III—The Future, Based on Avoiding the Past **429**

Chapter 14: What Could Possibly Go Wrong? **431**

Chapter 15: Sustainability of Infrastructure: Looking at the
Long-Term Future Trends **449**

Chapter 16: Leadership **485**

Chapter 17: Conclusions **491**

PREFACE

Training the next generation of infrastructure engineers, operators, and managers is a national priority. There is, however, a critical shortage of people educated and trained for the workforce required to build, operate, and manage the nation's infrastructure. Too often the tasks are left to those with field knowledge, but who lack the technical skills to assess, budget, manage, and argue to make repairs or replacement of critical infrastructure assets. To offset this shortage, we have seen the development of "STEM," or Science, Technology, Engineering and Mathematics programs at the high school and college level. The long-term growth and economic development of our society will rely on the ability to resolve challenges using these four key components of STEM education as they relate to the infrastructure systems that sustain our way of life. In addition to STEM programs is the continual need to think creatively when solving problems and to plan for problems before they even arise.

The intent of this book is to provide a comprehensive introduction to public sector infrastructure asset management at the local level. The infrastructure focus is on water and wastewater treatment systems, stormwater systems, and roadways, which is what most people use on a daily basis and is generally taken for granted until it fails. Typically, expenses required to operate, repair, maintain, and expand this infrastructure are too often deferred or artificially reduced to meet budget constraints by elected officials and budget managers. Unfortunately, there is a disconnect between the operational needs of the infrastructure system and the perceived budget realities. Both must be adjusted to maintain high quality infrastructure. Operations managers must develop tools to insure high quality assets are constructed, appropriate budgets to maintain them, and a means to assess future needs. Budget managers and elected officials must understand that the needs drive the budget, not vice versa. Otherwise, the deferral of maintenance will increase the risk of premature failure of the system which is always the most expensive to fix.

This text reviews the basic functions and maintenance of our critical infrastructure systems, while providing readers with an understanding of asset management. Chapter 1 is an introduction to infrastructure with a historical perspective. Chapter 2 will outline the economic benefits that investments in infrastructure can provide, based on peer reviewed literature. This chapter makes the case for why society should invest in infrastructure.

Chapters 3 through 6 will cover the infrastructure systems most commonly associated with local and state governments: water, sewer, storm water, and roadways. The chapters are designed to create a basic understanding of the system, how it operates, the major components of the system, and issues that will require maintenance attention (like corrosion, roadway base failure, etc.). Power and communication are not covered in this volume as they are generally owned by private providers which operate under different state and federal regulations.

Chapter 7 initiates a discussion of asset management—what it is and how to develop this type of system. The term "asset" can mean any physical component that makes up the type of infrastructure system being discussed. For example, water clarifiers at a water treatment plant are an asset, as are the pipes in the water distribution system. For asset management, an inventory of all assets within a particular infrastructure system must be developed. Once all the assets are inventoried, then the monetary values of the assets can be assessed, the condition of the assets assessed, the needs of the assets assessed, and then the resources to address those needs must also be assessed. The chapters following Chapter 7 outline how to develop this information. Chapter 8 includes a brief discussion about how to value infrastructure. Knowing valuation techniques will help managers and engineers shed light on financial needs of the system. Chapter 9 focuses on vulnerability assessments and risk management. This is not to say any public works official should make decisions based on the risk alone, but one should be cognizant of where potential vulnerabilities are and prioritize maintenance and repair to those assets that put the system at greatest risk. Chapter 10 outlines a means to evaluate the condition of an infrastructure system, even when parts of it are hard to access. This is a huge issue for water, sewer, and stormwater systems which have miles of buried pipe. Asset management, condition assessment, and risk management should lead to maintenance, repair, and replacement planning and funding, which is the focus in Chapter 11. Resources—money, skills, people, equipment, materials—all must be identified to develop a maintenance program. Chapters 12 and 13 will outline funding for maintaining infrastructure systems and life-cycle cost analysis, respectively, which is a useful tool for identifying when improvements should be planned.

Chapter 14 provides case studies associated with the failure of infrastructure systems (i.e., where things went wrong). Chapter 15 outlines future trends or challenges that infrastructure managers will face, beyond just the deteriorating condition of the assets themselves. These challenges include climate change and its effect on coastal infrastructure, as well as problems associated with actual water supply. Chapter 16 delves into the need for managers to provide leadership when dealing with infrastructure issues.

It is intended that people who read this book will (1) gain a very good understanding of infrastructure management and how important it is to the health, safety, and welfare of a community, (2) use the tools and resources discussed for managing infrastructure, (3) understand the need for infrastructure maintenance to ensure continuous operation of critical components, and (4) gain an introductory knowledge of fiscal constraints related to capital planning and construction of infrastructure systems.

ACKNOWLEDGMENTS

The author wishes to thank Gwen Eyeington of J. Ross Publishing who suggested I write this book; Daniel E. Meeroff, Ph.D., Professor and Associate Chair and Dr. Yan Yong, Professor and Chair, Department of Civil, Environmental and Geomatics Engineering at Florida Atlantic University; Mark G. Lawson, Esq., in Tallahassee, FL; Michael F. Nuechterlein, Esq., retired from Carlton Fields, Tampa, FL; A. John Vogt, Ph.D., C.P.A., retired from the School of Government at the University of North Carolina at Chapel Hill; over 4 dozen students who have taken an infrastructure management class from me and provided great insight; and the hundreds of operators, managers, engineers, and field personnel over the years who have dealt with many of these issues and have allowed me to be a part of their solutions. I also want to thank industry partners: Brad Kaine, Dominic Orlando, and Ronnie Navarro in Dania Beach, FL; Karl Kennedy and Jon Cooper in Pembroke Pines, FL; Dave Stambaugh, Dennis Giordano, and Jim Hart at Calvin, Giordano and Associates; Albert Muniz, VP at Hazen and Sawyer; and my many AWWA and Florida Section AWWA friends who were willing to share. Without your experiences, this would not be possible.

ABOUT THE AUTHOR

Fred Bloetscher, Ph.D., is currently a Professor of Civil, Environmental and Geomatics Engineering and Associate Dean for Undergraduate Studies and Community Outreach at Florida Atlantic University in Boca Raton, Florida. His research focus has been urban infrastructure systems, particularly public water, stormwater, and sewer systems and their sustainability. Dr. Bloetscher teaches the capstone senior design sequence at FAU, plus classes in water/wastewater, construction, environmental engineering and modeling, hydraulics, and infrastructure management.

Dr. Bloetscher received his bachelor's degree in civil engineering from the University of Cincinnati and earned his Master of Public Administration Degree from the University of North Carolina at Chapel Hill. His Ph.D. is in civil, environmental, and architectural engineering from the University of Miami in Coral Gables, FL. His experience prior to academia is helpful to his ongoing research and outreach efforts. He was a city manager for two communities in North Carolina that were undergoing significant infrastructure improvements. In Florida, he was a utility director and deputy director for several large water and sewer systems that included groundwater, aquifer storage, reclaimed water systems, wastewater plants, and miles of piping that needed to be maintained.

Dr. Bloetscher has many years of consulting experience as well. He is the President of Public Utility Management and Planning Services, Inc. (PUMPS), a company dedicated to the evaluation of utility systems, needs assessments, condition assessments, strategic planning, capital

improvement planning, grant and loan acquisition, inter-local agreement recommendations, bond document preparation, consultant coordination, permitting, and implementation of capital improvement construction. In managing both large and small infrastructure systems, and consulting for local governments, he has been involved in the construction of over $500 million in improvements—both new for expansion, and for replacement as "owner," engineer, or project consultant.

Dr. Bloetscher has volunteered his time for many professional associations. He is the former Chair for the Water Resource Division Trustees, a member and 3-time Chair of the Groundwater Resource Committee, Chair of the Aquifer Storage and Recovery Sub-committee, and former Chair of the Education Committee for the American Water Works Association (AWWA). Currently Dr. Bloetscher is the chair of the Distribution Engineering and Construction Committee for AWWA and a former member of AWWA's Technical and Education Council. He is the Vice Chair for the Florida Section of AWWA and has been the Technical Program Chair since 2004. He serves on several local committees. He is an LEED-AP, holds professional engineering licenses in 9 states, and operations licenses in water, wastewater, distribution, and collection systems. He has had a North Carolina General Contractor's license for pipeline and small utility construction for over 20 years.

Dr. Bloetscher has been nominated for the "Distinguished Teacher of the Year" award a number of times by his students and has received three University-wide leadership awards including the prestigious "Presidential Award for Community Engagement" in 2018, received several national awards including the "Distinguished Educator Award" from the National Engineers Council in 2017, and has been recognized by the AWWA, the National Engineer's Council, the Florida Section of AWWA, and the local chapters of ASCE for his contributions to the industry. In 2012, Dr. Bloetscher received the "National Council of Examiners for Engineering and Surveying (NCEES) Award for Connecting Professional Practice and Education" for his work on the Dania Beach Nanofiltration Facility, which is the first LEED-Gold water treatment facility in the world (see photo on page xv). Dr. Bloetscher was the LEED administrator for the project.

Dr. Bloetscher can be reached via e-mail at h2o_man@bellsouth.net. His website is www.h2o-pe.com and his blog is at publicutilitymanagment.com on Wordpress.

At J. Ross Publishing we are committed to providing today's professional with practical, hands-on tools that enhance the learning experience and give readers an opportunity to apply what they have learned. That is why we offer free ancillary materials available for download on this book and all participating Web Added Value™ publications. These online resources may include interactive versions of the material that appears in the book or supplemental templates, worksheets, models, plans, case studies, proposals, spreadsheets and assessment tools, among other things. Whenever you see the WAV™ symbol in any of our publications, it means bonus materials accompany the book and are available from the Web Added Value Download Resource Center at www.jrosspub.com.

Downloads for *Public Infrastructure Management* feature instructor materials for adopting professors, including a proposed syllabus for the course, PowerPoint lectures covering chapters in the book, end-of-chapter homework questions and answers, project ideas for the instructor, GIS data sets for students to do their own condition assessments, and sample final and midterm exams.

SECTION I

Infrastructure:
A Comprehensive Overview

1

HISTORY OF INFRASTRUCTURE

For the purposes of this book, infrastructure is defined as capital projects—things of significant value and long life. They are physical, as opposed to policy-oriented. Hence for the purposes of discussion, what is not considered infrastructure in this text, even though they may include physical components, are: information technology (the Cloud), financial institutions, health institutions, educational institutions, and any nonphysical assets. This is not to diminish any of these necessary portions of society, but the definition of infrastructure within this volume needs to be narrowed to better understand the essential systems.

INFRASTRUCTURE OVERVIEW

Dr. Brian Murray Fagan is a Professor Emeritus at the University of California at Santa Barbara. After graduating from Pembroke College in Cambridge, he used his degrees in anthropology to travel the world to study human prehistory. Dr. Fagan explained, at an American Water Works Association conference several years ago, that because he has studied water and drainage ditches across the globe, he can speak to the rise and demise of nearly every civilization in the past 5,000 years. In over 50 years of field studies, Dr. Fagan has studied more drainage ditches than anyone ever has. Based on those 50 years of study, he notes that the *development of civilization can be predicated, and its fate predicted, based on irrigation practices*. People need clean water to survive. For example, in studying the Middle East—specifically the ancient Babylonian cultures of 4,000 years ago in current day Iraq—he was able to explain the mystery of the sudden disappearance of the ancient city of Ur, one of the largest and most fabulous cities in the ancient world at the time (thought to have had a population of over one million inhabitants). The city is recorded in many ancient books, including the Bible, yet it disappeared almost overnight. Fagan surmised that the Euphrates River moved after a flood, and when it did, the drainage ditches no longer flowed as they had before, so the population rapidly dispersed, and the city fell precipitously into ruin. He suggested that the fate of the Anasazi tribes in the American southwest, a number of ancient Egyptian pharaohs, many island communities, and other civilizations from antiquity were predicated on the ability to construct infrastructure to convey water for irrigation and drinking. He makes the point that civilization relies on reliable water infrastructure to survive and thrive.

With adequate, reliable water supplies and the infrastructure to deliver it, agriculture can prosper, providing food for growing civilizations. Agricultural surpluses are available for trade, creating the need for merchants and commercial centers in the developing villages and towns. But with growth comes the need for additional infrastructure, including infrastructure for waste disposal, transportation for the movement and exchange of those goods and services being traded, public buildings for markets, and communication systems. The advent of the ability of humans to harness energy accelerated development and the growth of civilizations.

Today we live in an energy-based economy, whereby our modes of transportation, communication, and structures are unlike anything envisioned by the ancients. At the root of all modern societies is a series of infrastructure systems that provide clean water, waste removal, flood protection, energy, and transportation to facilitate the economy. Maintenance of these same systems is vital to the continued economic stability of these communities and their ability to compete in an increasingly developed world.

The difference between developed nations and those that are not developed basically comes down to the degree of infrastructure improvements that are constructed and maintained. The rise in Western Europe and the United States in the early 20th century can be traced to infrastructure investments. Many of these infrastructure systems are interconnected. Underground utilities such as natural gas, water, and sewer lines are buried beneath roadways. Roadways provide access to property and connect to other communities, rivers, and ports. Stormwater systems are designed to keep roadways passable and drain flood waters while limiting private property damage. Roadways convey stormwater along the curbs and gutters, while maintaining the continuity of traffic. Power and communication systems are often located on the same power poles along road rights-of-way or buried beneath them. The result is that these systems are interconnected and interdependent on one another.

There is a direct correlation between investments in infrastructure and economic growth (as will be discussed in Chapter 2). In more developed nations, reliable infrastructure leads to a higher quality of life that citizens enjoy and come to expect. Four hours without water a year will be long remembered by residents; yet it means the water was present 99.74% of the time, which is hardly a failing grade—unless 100% is your only acceptable rate of service provision. All it takes is one extended interruption in service for the fabric of society to begin to unravel. Ask any young person or business what happens when the internet is disrupted.

We have had more significant incidents. For example, in 2003 much of the northeast was blacked out due to a failure in a small part of the connected power grid, creating significant loss of productivity. Fortunately, the outage was short-lived, but public confidence was diminished, even being suggested by elected officials in Flint, Michigan, as one of the reasons for their desire to change water supplies and restart their own plant after 50 years, which subsequently led to the lead contamination problems there (as will be discussed in Chapter 14). New Orleans, Louisiana, was devastated from the failure of dikes during Hurricane Katrina. The city of New Orleans has not yet fully recovered; over 10 years after the storm, the population is half what it was in 2005. Billions of dollars were spent on dikes, property damage, utility repairs, etc., yet the city remains vulnerable to the next storm event. In Minneapolis, Minnesota, an interstate highway bridge failed due to lack of proper maintenance, prompting a nationwide study of bridge conditions. In Alamosa, Colorado; Flint, Michigan; and Walkerton, Ontario, public health was compromised by contaminated water systems due to a combination of errors and misjudgments by one or more of the operators, elected officials, managers, and regulatory staff involved in the management of those systems.

When infrastructure systems fail in a major way, the disruption of human life can be significant—not just in loss of life, but property, economic activity, and confidence in the governments that typically provide the infrastructure systems. Billions of dollars have been spent on these infrastructure systems (as will be discussed in Chapter 2), yet more needs to be spent, and many communities are not spending enough to prevent their overall deterioration. Trillions of dollars in economic activity are dependent on these investments working as intended. But failures will occur, and those charged with operating and maintaining our critical infrastructure must deal with pragmatic issues like funding limitations, material availability, changes in technologies, parts availability, personnel skill sets, assigned responsibility, and public expectations. The reality is that, for the most part, our infrastructure systems do function as intended as a result of the combined efforts of those who are dedicated to managing and operating the systems—which is a testament to those employees whose job it is to keep them running despite the challenges. The concern is that while these systems have long lives, they do deteriorate with the passage of time. National entities such as the American Society of Civil Engineers, the U.S. Environmental Protection Agency, and the Department of Defense have been assessing the state of U.S. infrastructure for decades, and their findings indicate that sufficient funds have not been allocated to keep infrastructure systems running optimally. The warning bells have been ringing for some time that this lack of attention will eventually impact economic growth and quality of life. The question is how to *fix* it?

INFRASTRUCTURE INVESTMENTS

It can easily be shown that progress in society depends on the presence and maintenance of physical infrastructure (as will be discussed in Chapter 2). Societies that invest in infrastructure are able to produce goods, develop trade markets, and even improve the life expectancy of their society. Improvements in sanitation and public health have allowed people to be more productive and live longer. The greatest health improvement in the 20th century was the disinfection of water where annual mortality to waterborne illness has been reduced from 1:100,000 to less than 1 in 15 million in developed countries. This combination has also increased life expectancy from just over 47 years, to well over 80 years (Center for Disease Control, 2012 via commons.wikipedia.org). Safer drinking water, better sewage disposal, and upgraded roads demanded by the automobile have also led to a general economic expansion throughout the 20th century and into the 21st century (the Great Depression and Recession aside). As a result of 20th-century investments in infrastructure, economic activity in the United States has increased from $400 billion/yr, to over $17 trillion/yr since 1900 (Isites.Harvard.edu).

While development of infrastructure systems has improved public health and facilitated economic growth, it should be noted that most infrastructure systems are related; in part because they are often co-located and work together, which is why they are designated as infrastructure *systems*. Infrastructure systems include:

- Water
- Sewer
- Stormwater
- Roads
- Other transportation modes (rail, air, ship, transit)
- Energy
- Communications

The first four infrastructure systems listed, plus rail, will be discussed in this volume. All are costly to install and require constant maintenance, but it is the failure to maintain, upgrade, improve, and rebuild them that leads to disruptions in economic growth. Thus, as infrastructure systems grow, the local economy grows because investments in infrastructure tend to benefit the local economy (see Chapter 2). Areas with more developed infrastructure systems find that their communities attract more people because the job opportunities are greater.

More precisely, at present, over 80% of people in the United States live in urban areas (Berg 2012), a figure typical of most industrialized nations. Energy usage grows as people move to urban areas because there is more industrial growth and more wealth created by industry. Thus individual income is higher and synergistic opportunities are greater due to a more diverse population when there is more modern infrastructure present. A recent U.S. Department of Agriculture study (USDA 2014) makes this point when it reports that:

- Rural areas grew 0.5% versus 1.6% in urban areas from mid-2011 to mid-2012
- Rural incomes are 17% lower than urban incomes
- The highest income rural workers (95th percentile) earn 27% less than their urban counterparts
- 17.7% of rural constituents live in poverty versus 14.5% in urban areas
- 80% of high poverty rate counties surveyed were rural
- All high income counties surveyed were urban

Social structure grows as industrialization increases because there are more people. The opposite is true when the state of the infrastructure is poor to nonexistent.

Need proof? Look no further than China, where investments in physical infrastructure have created large migrations of rural Chinese to the urban areas, which are among the fastest growing cities in the world. Major advances in technology and human development tend to occur in population centers (think Detroit for cars in the first half of the 20th century, Pittsburgh and Cleveland for steel during the same time period, and Silicon Valley for technology in the last quarter of the 20th century, etc.). People with ideas tend to migrate to urban areas, increasing the number of people and the proximity to each other. Universities, research institutions, and the like tend to grow up around these industries, further increasing the draw of talent to urban areas. However, when the infrastructure ages or is not maintained, the potential for the opposite occurs—an active area of study is Rust Belt cities of the Midwest like Pittsburgh, Cleveland, Detroit, Buffalo, Toledo, Gary, and Flint.

No one should buy a house, refuse to maintain it, and expect to live in it without problems for 50 years. Roofs leak, pipes and mechanical equipment need replacing, and appliances fail. It is part of the cost of home ownership. Home assets need routine and periodic updates. Often the new equipment is more efficient than the old equipment, saving homeowners money. So why should anyone expect anything different with critical infrastructure systems? Why defer maintenance of critical infrastructure when it is the lifeblood of the community? Even in a small community infrastructure systems may be worth hundreds of millions of dollars. That is a big asset, and it needs big maintenance and repair budgets.

THE DEVELOPMENT OF INFRASTRUCTURE OVER TIME

The development of infrastructure has occurred in spurts over time based on needs, funding, or visionary leadership. Each time, huge changes in population and/or development followed the implementation of large infrastructure projects. Let's go back to Dr. Fagan. He will tell you that 10,000 years ago people were basically hunter-gatherers. Then the only infrastructure needed was a little shelter, but that was personal, not societal. It is only when agriculture was first established that the need for infrastructure to the community was understood. Agriculture meant that people could become stationary because their needs were met in the local community. As long as crops continued to grow, people would not be hungry—a second basic personal need met. But more consistent availability of food meant more people to feed, which increased the demand for crops, which could only be met through a more concerted (societal) effort to extend irrigation systems from local farms to larger areas servicing multiple farms. Dr. Fagan (2012) will tell you that the first irrigation ditches for this purpose were in Babylonia and Egypt. The rivers supplied the needed water, and the ditches were extended as far as gravity would carry the water. Drought, river course changes, and upstream disruptions could have catastrophic impacts on the communities, but these were relatively rare. Societies flourished—until a calamity stopped the flow of water or the community outgrew its ability to feed people. Then society would collapse. Water meant food; clean water meant health and commerce.

Water was also useful for removing waste. Communities with opportunities to use floods or topography to remove the wastes benefited, although their downstream neighbors were less fortunate. The Celtic and Teutonic cultures realized that drinking water from streams would make you sick. Beer and tea were a solution, but efforts continued to improve waters supplies. The Romans built aqueducts, tunnels, and other water conveyance structures to bring in clean water from long distances to Rome, which improved urban health, fostered more commerce, and removed more waste. The entire empire benefited from these improvements, allowing commerce to flourish and create additional demands for trade and the need for defense. Both led the Romans to extensive road building. Many Roman roads are still in use today (see Figure 1.1) and we still use the Roman chariot axle dimension of 4 feet 8.5 inches in railroads, automobiles, trucks, and other vehicles.

Populations decreased after the fall of Rome, in part because the infrastructure was no longer maintained. Following the Dark Ages, populations increased slowly until the Industrial

Figure 1.1   Old Roman road still in use today leading to the ancient city of Assos in Turkey. (*Source*: image courtesy of www.HolyLandPhotos.org)

Revolution, which increased the ability to efficiently create products and develop commerce, and with it, the opportunity to increase the reach of goods and services. Both increased the attraction to urban areas and subsequently, increased the need for additional infrastructure. Safe, clean water supplies and adequate waste disposal were important community functions. It is a bit of a vicious cycle.

However, infrastructure normally is constructed as a result of a real or perceived need. A cholera epidemic from well contamination in London in 1848 caused more than 14,600 deaths. Cholera again erupted in 1854 causing 10,675 deaths (Burian et al. 2000). Many large cities in Europe started to treat water with filtration and constructed sewer lines to reduce disease as a result of these outbreaks.

In 1820 less than 5% of all Americans lived in urban areas (cities with a population larger than 8,000), but by 1860 the percentage increased to 16% and by 1880 had risen to 22.5% (Burian et al. 2000). All those people living in urban areas required more robust means to deal with the demands for water, sewer, transportation, etc.

The demand for transportation increased in the trade centers, leading to better and more numerous roads and canal systems to convey large quantities of goods from ports to inland communities. In the northeast United States, the Erie Canal was one such improvement—allowing for a more rapid and safe connection between the east coast and the Great Lakes region. Canal boats were towed by horses on paths along the canal (see Figure 1.2). But boats were slow, and so were horse- or oxen-drawn wagons.

The next generation of major infrastructure improvements were railroads (see Figure 1.3), which were developed as a faster means to transport goods and freight, but the tracks had limited geographic access. Governments facilitated the construction and expansion of rail after the Civil War, opening the western United States and many small communities to trade opportunities that did not exist before. Tunnels (see Figure 1.4) were constructed to allow trains to move more easily across the Rockies. Communities sprung up wherever trains stopped or tracks crossed. But these stops were likely to be places where there was land and water, both of which

Figure 1.2   A stone aqueduct of the Erie Canal crosses the Mohawk River at Rexford, New York. (*Source*: Clifton Park Collection)

Figure 1.3   A railroad bridge built in Georgetown, Colorado, in the 1880s.

Figure 1.4   Railroad tunnel located in Winter Park, Colorado—constructed in 1927.

were needed by the locomotives and would facilitate development of the new communities. The desert and high mountains had few stops.

As cities grew, the limitations of railroads became more obvious. They could not be constructed to get people and goods everywhere. Horses and horse-drawn carts were used to address this need, but as better rail transportation increased economic activity and attracted more people to cities, the limitations of horse-drawn carriages manifested. Horses created health concerns in densely populated communities as a result of copious amounts of droppings. New York City alone employed 15,000 people to clean up horse droppings and dead animals on their streets each day (see Figure 1.5). Most major cities had similar employees. The health issues associated with this magnitude of waste in urban streets were obvious. In conjunction with pipes to remove waste, pipes to remove rainfall were constructed. Burian et al. (2000) note that in 1909, in the largest cities (populations over 100,000) there were 17,068 miles of sewers, of which 14,240 miles were combined sewers (where both sanitary sewage and stormwater were conveyed together to a water body for disposal because stormwater was assumed to flush both the streets and the sanitary sewers), 2,194 miles were separate sanitary sewers, and 634 miles were storm sewers.

The next investments came with the advent of the automobile, which was a boon for urban areas as well as individual freedom of movement. The automobile eventually eliminated the

MORTON STREET, CORNER OF BEDFORD, LOOKING TOWARD BLEEKER STREET,
MARCH 17, 1893

Figure 1.5   Prior to the advent of the automobile, huge amounts of horse droppings had to be cleaned every night in large cities. This photo is from New York City during the 1890s.

horse manure problem, along with the jobs for some 15,000 people in New York alone. Despite this massive job loss from obsolescence, the feces pollution problem went away, but it was replaced by congestion, which soon demanded that urban roads have better pavement, traffic signals, signs, street lights, and wider, multi-lane roads.

The 1920s and 1930s saw a boom in the construction of dams and water conveyance systems. The Hoover Dam (see Figure 1.6) and related canal systems to move water to Los Angeles are examples of infrastructure projects that were constructed by far-sighted people. They looked at the needs of southern California and developed a canal system that enabled it to grow and develop as they hoped. Los Angeles could not grow without expansion of secure water supplies. The urban Los Angeles area now includes over 20 million people (up from under a million in 1920) that is served by these extensive canals (see Figure 1.7), some of which start at the Hoover Dam. Visionary leaders in many communities realized that with water, growth was possible—without it, the communities would stagnate. The Las Vegas area had a population of 4,800 in 1940. By 2015, the population had grown to over 2 million, not including the 42 million annual tourists.

The Great Depression of the 1930s led to the Work Progress Administration (WPA) program where thousands of Americans were hired to build water and sewer systems. Many rural communities had their original utility systems, along with recreation sites and paved roads built as a part of WPA programs. The WPA built or improved 651,000 miles of roads, 19,700 miles of water mains, 500 water treatment plants, 122,000 bridges, 1,000 tunnels, 1,050 airfields, 4,000 airport buildings, 800 pumping stations, 1,500 sewage treatment plants, 24,000 miles of sewers

Figure 1.6 Hoover Dam is a concrete arch-gravity dam in the Black Canyon of the Colorado River, on the border between the states of Nevada and Arizona. It was constructed during the Great Depression and was dedicated on September 30, 1935 by President Franklin D. Roosevelt.

Figure 1.7 A portion of the canal system that brings raw water to Los Angeles from the Colorado River.

and storm drains, 36,900 schools, 2,552 hospitals, 2,700 firehouses, and nearly 20,000 county, state, and local government buildings (The Economist 2011; Roosevelt Institute 2011). Much of this infrastructure was in the south where there were far fewer developed areas, fewer water and sewer systems, and fewer paved roadways, all factors that had delayed growth in southern communities. Most of those assets remain in service today.

Electricity was not delivered by power companies to rural areas because of the general belief that the infrastructure costs could not be recouped, so federal efforts of rural electrification began in small communities in 1936 (state efforts were even earlier in the Midwest and northeast), leading to hundreds of small dams for water supply and power production purposes throughout the country, particularly in the rural south. Prior to these improvements, industrialization of the south was limited due to the lack of infrastructure—water, sewer, roads, and power. But with these improvements, communities began to grow and diversify from farming. The Tennessee Valley Authority was one such rural electrification entity—chartered in 1933 as a result of severe economic impacts in the southeast (see Figure 1.8). Much of the dam building in the U.S. ceased in the 1950s due to increasing demands and an inability to find effective dam sites. New power plants were constructed to burn coal; then nuclear power followed in the 1960s and natural gas after 2000.

Henry Flagler built the railroad along the east coast of Florida on the high ground from Jacksonville to Miami to encourage development. However, the high ground was 2 miles inland from the Atlantic coast, so much of the early development was concentrated in this narrow corridor. Storm and water impacts from hurricanes in 1926 and 1928, and especially 1947, disrupted this plan for growth in south Florida triggering the need to address flooding in the state (see Figure 1.9). As California was running water pipes from the mountains to Los Angeles, engineers in Florida were moving water from the southern half of the state to the ocean, draining much of

Figure 1.8   A Tennessee Valley Authority dam completed in 1930 in Blairsville, Georgia.

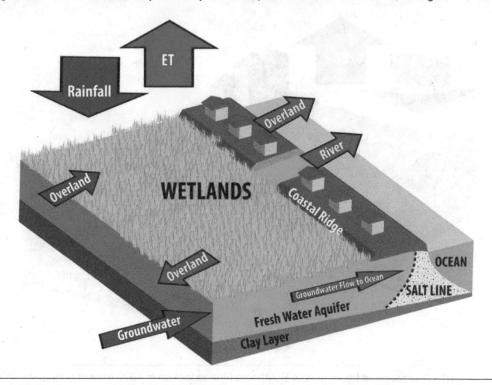

Figure 1.9   A conceptual figure of how Southeast Florida looked in 1900. (*Source: Journal of Environmental Science and Engineering*, 5 (2011) 1507–1525, Bloetscher et al. 2011. David Publishing.)

the historical Everglades. As a result, 1,800 miles of canals, a series of pump stations, etc., were constructed over the next 30 years to provide land for development in Florida—consisting of 6.6 million people, $7 trillion in property values, and $400 billion in annual economic activity as of 2015 (see Figure 1.10). Today, Federal and state funding trickles into Florida to address the enormous negative impacts from draining much of the Everglades ecosystem, including the loss of groundwater sources.

During World War II, General Eisenhower was impressed by the Autobahn system in Germany, so when he became President, he chose to invest in the Interstate highway system by commissioning the first 42,500 miles of the Dwight D. Eisenhower System of Interstate and Defense Highways in 1956 (FHWA 2018). Its estimated cost, in 1996 dollars, was over $329 billion. In 1960 there were 10,000 miles of Interstate opened. By 1970 there were 30,000 miles, and by 1980 over 80,000 miles were completed (McNichol 2006). These new roadways allowed people to travel farther and faster than before; and as a result, facilitated the ability of urban residents to move away from the inner city. This helped establish suburbs and urban sprawl, which necessitated a new wave of water, sewer, stormwater, roadway, and power infrastructure construction in the 1950s and 1960s. However, with more people driving today, the long commute times from the suburbs has started a return to urban environments for the younger generation. More vehicle miles traveled per person, with less reliance on public transportation, has put a congestion pressure on the interstate systems, while at the same time, more fuel efficient vehicles have reduced the revenues from the gas tax that funds maintenance and improvement programs. Hence, funding for roads has decreased since 1959.

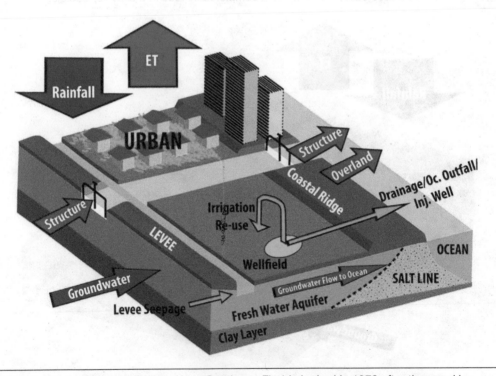

Figure 1.10   A conceptual figure of how Southeast Florida looked in 1970 after the canal improvements. (*Source: Journal of Environmental Science and Engineering*, 5 (2011) 1507–1525, Bloetscher et al., 2011. David Publishing.)

From the late 1960s to the early 1970s, a revolution in environmental awareness and environmental protection legislation occurred to address pollution impacts to water and air. The Clean Water Act, Safe Drinking Water Act, and Clean Air Act all required investments in infrastructure to reduce pollution to water bodies, air, and drinking water across the country. Grants were provided through 1988 for these improvements with low interest loans provided thereafter. However, the urge to deal with water and sewer infrastructure seems isolated to areas of growth, while the older, developed urban and rural systems have seen less investment. Rural systems are particularly at risk since economic growth appears to trail urban areas.

Post World War II saw the development of jet engines, which in turn increased the demand for air traffic and the need for airports. As air travel continues to increase, it facilitates economic opportunities across the world, necessitating an ongoing construction boom for both domestic and international airports. Shipping by boat remains the lowest cost option for moving goods, leading to a re-visitation of older infrastructure like the Panama Canal. The size of new ships for transport created the need for investments to expand the Panama Canal, which has also spurred changes to the east coast ports—Miami and Fort Lauderdale, for example, are expecting to spend billions to upgrade and deepen their ports to handle the larger cargo ships. Technology and commerce spurs infrastructure in these areas.

IMPACT ON THE PUBLIC

Infrastructure improvements yield major changes for the population. While the discussion about economic development is contained in Chapter 2, other impacts are just as significant. With respect to the provision of drinking water, it was previously noted that waterborne illness deaths plummeted from 100,000 per year in 1900 (of 100 million) to virtually one in 15 million people in 2010. Why? Because water treatment technology, design, and construction improved the disinfection—and improved distribution piping systems serve the majority of people in all urban and most rural areas in the U.S.

Sewer systems also permit economic development. Since the Clean Water Act was passed and that money was spent, the water quality in many rivers has improved. Likewise, the huge migration of people to the suburbs and to otherwise dry areas—like Phoenix, Denver, Las Vegas, and Tucson—was made available via roadways constructed by federal, state, and local governments, supplemented with gas tax revenues. It is easy to travel on good roads to virtually any place in the U.S. as a result.

Power is also available everywhere in the U.S., which permits not only an improved way of life via technology, but access to news, internet, social media, and other communication systems. Today, infrastructure connects us around the world and as a result, this has created expectations from the public that water will flow when they turn on the faucet, that cell phones will charge when plugged in to the wall socket, that sinks will drain, toilets will flush, and stormwater systems will reduce the potential for flooding of roads and property. The public expects that all transportation modes will work to move goods and people from one point to another. From an operational perspective we call this resiliency and reliability. Operations managers may also call it sustainability; the public calls it service.

Infrastructure is public works, which are physical structures and facilities developed or acquired to provide water, sewer, power, waste disposal, transportation, etc. The value of infrastructure in the U.S. was estimated at $37 trillion by the consulting firm ARCADIS (China was $47 trillion). However, the start of the decline in investment in the U.S. began in 1968. In part, the perception may have been—at the federal level—that the major investment needs for roads, bridges, and dam projects were complete. However, local funding for capital improvements slowed as well. But because the maintenance component of infrastructure was not budgeted, the results of deterioration started to manifest 10 years later. Today much of our infrastructure has issues associated with:

- Age
- Neglect
- Overuse
- Excess wear
- Exposure
- Mismanagement
- Obsolescence

These issues have not been dealt with due to various reasons. In an effort to control rates and taxes, along with the lack of apparent problems, elected officials have underinvested in repair, rehabilitation, and replacement of infrastructure systems. Cutbacks in budgets and deferral of expenses have been the result. Expenses have been based on available revenues, not needs. There has been a change by some officials who look more at a bottom line for investing as opposed to the need for insuring levels of service are maintained, making it more difficult to support added monies. The result is a lack of management of infrastructure systems and the failure to recognize the importance of these systems to the future economy. Developers have pushed new projects to local officials without the recognition of the impact of their projects on the level of service to existing residents. A fresh look at these policies is needed.

What is at risk? From a public service prospective, local officials risk a number of impacts from catastrophic failure of the system and deaths caused by that failure, to a disruption of way of life/economy that limit growth opportunities for the community. As an example, over 10% of the economy of the U.S. is related to transportation and shipping of goods; yet we are spending hundreds of billions less for repair and replacement than needed, which impacts our ability to grow or even sustain our current economy. For example, repair and replacement on the multi-trillion dollar transportation system was only 20–40% of the value of monies spent for new transportation construction in 1995. So what is needed to manage the infrastructure? The five major needs for infrastructure management are: (1) regulations to require assessment and investment strategies, (2) identification of capital projects, (3) more robust maintenance and repair programs, (4) people with skills, and (5) money.

The cost is high, which means there is a legitimate need for a more holistic evolution of managing infrastructure; one that is not piecemeal, not reactive, not forensic (post failure), and not without consideration of interrelationships.

CURRENT INFRASTRUCTURE IS BROKEN

The message is that infrastructure systems need to be constructed and maintained. The U.S. Army Corps of Engineers reported that the need to reinvest in infrastructure is estimated to be about 3.6% of the infrastructure value per year, but that the U.S. is spending about 2.4% (CBO 2015). The best condition of American infrastructure was in the 1980s (National Council on Public Works 1988), but decreases in reinvestment due to funding limitations have caused an ongoing decline in infrastructure value, which is why the ASCE report cards show most of American infrastructure at the D or D– level (ASCE 2001, 2009, 2013, 2017). The average score is a D.

The television program *60 Minutes* did a piece about the deterioration of bridges. The magazine *American City and County* has published articles about the risks of aging infrastructure. The discussion in Chapter 11 of this book outlines the long litany of needs. We have many issues with infrastructure in the U.S. Part of the problem is that the infrastructure operations industry does a poor job of communicating its importance. At the same time, it is hard to get into the news when the crisis is not apparent. People just are not interested when there is nothing wrong (and when there is something wrong, people will likely create issues that obfuscate the reason that something is wrong—listen to almost any election). Because communication and marketing has been so poor by the infrastructure community, the industry has been out-competed by cable television, cell phones, and Netflix for consumer dollars. Despite what the youth of today might believe, one can live without a cellphone and cable TV, but you can only live three days without water. In large part, the industry has lost its place in the minds of its customers due to its failure to market its product. The failure to market is often a time, leadership, and political issue. In addition, it is a tribute to those operating the systems—despite the needs or problems, the operators make sure the systems work. Hence—out of sight, out of mind—as long as the system works!

UNDERSTANDING THE PROBLEM

Local officials do not convey an understanding of these complex systems to the public very well. In part this may be because understanding the maintenance needs is difficult and highly variable. And many do not fully comprehend the assets they have, their condition, life expectancy, or technological needs. No one knows when things will fail, so maintenance or replacement of some equipment or pipeline is always cut in the budget with no real understanding of the consequences.

The public does not see many of the assets (or fully understand the ones they do see), assumes they will have a long life, and thus is unconcerned until they are affected; then it gets personal. When the public does not understand the impact or value that these assets have to society, they tend to be personally focused, not societally. That is a leadership issue. That leadership starts with vision and communication from those who understand the issue to the elected officials who need to advocate for their infrastructure.

Despite the infrastructure crisis, the good news is that construction of piping is increasing—both new and replacement. Every so many months, the magazine *Utility Contractor* will note current trends—and piping seems to be going up. That's good, but there is a long way to go. Better news—the construction of buildings is increasing. That could lead to more revenue. In Florida, finding experienced construction workers has recently become a problem (in 2015—more work than workers). Things are definitely better economically, but are we taking advantage of opportunities to improve the local infrastructure? And what happens when the next economic downturn hits?

UNDERSTANDING THE SOLUTION

The infrastructure crisis is a political and business leadership crisis. Asset management is practiced by few governments. The public doesn't want to foot the bill and some politicians want taxes and fees cut further. Where does it end? The reality is that local officials need to make their infrastructure systems self-sustaining and operating like a utility business whereby revenues are generated to cover needed maintenance and long-term system reliability. For example, for water and sewer utilities, the adage that *we can't afford it* simply ignores the fact that most communities cannot afford *not* to maintain their utility system since the economic and social health of the community relies on safe potable water and wastewater systems that operate 24/7; likewise with roadway, stormwater, and other infrastructure systems. Too often decisions are made by an official whose vision is limited by short-term gains as opposed to long-term viability and reliability of the utility system and community. This is why boom communities fall precipitously, often never recovering—the boom is simply not sustainable. Long-term planning is a minimum of 20 years—well beyond the next election and often beyond the reign of current managers. Decisions today absolutely affect tomorrow's operators. Dependency on water rates or taxes that fund public sector infrastructure systems may be perceived to be a barrier, but this ignores the fact that private power, telephone, cable television, gas, and internet access industries are generally more expensive than either water or sewer in virtually all communities. Better education of the public about the importance of improved infrastructure is needed for public sector agencies to obtain the funding needed. We need improved water, sewer, roads, and stormwater systems for our health and economic sustainability.

REFERENCES

American Society of Civil Engineers (ASCE). 2001. *2001 Report Card for America's Infrastructure.* ASCE, Alexandria, VA. http://www.infrastructurereportcard.org/making-the-grade/report-card -history/2001-report-card/. Accessed 4/3/16.

____. 2003. *Report Card for America's Infrastructure, 2003 Progress Report.* ASCE, Alexandria, VA. http:// www.asce.org/reportcard/index.cfm. Accessed 4/3/16.

____. 2009. *2009 Report Card for America's Infrastructure.* ASCE, Alexandria, VA. http://www.infra structurereportcard.org/2009/. Accessed 4/3/16.

____. 2013. *2013 Report Card for America's Infrastructure.* ASCE, Alexandria, VA. http://www.infra structurereportcard.org/tag/2013-report-card/. Accessed 4/3/16.

____. 2017. *2017 Report Card for America's Infrastructure.* ASCE, Alexandria, VA. http://www.infra structurereportcard.org/. Accessed 3/3/17.

Berg, N. 2012. *U.S. Urban Population Is Up . . . But What Does 'Urban' Really Mean?* Citylab .com. http://www.citylab.com/housing/2010/03/us-urban-population-what-does-urban-really-mean/ 1589/. Accessed 4/20/17.

Bloetscher, F., Heimlich, B. N., and Romah, T. 2011. "Counteracting the effects of sea level rise in Southeast Florida." *Journal of Environmental Science and Engineering,* 5 (2011) 1507–1525.

Burian, Steven J., Nix, Stephan J., Pitt, Robert E., and Durrans, S. Rocky. 2000. "Urban wastewater management in the United States: Past, present, and future." *Journal of Urban Technology.* Vol. 7, N. 3, pp. 33–62.

Congressional Budget Office. 2015. *Public Spending on Transportation and Water Infrastructure.* 1956 to 2014, CBO, Washington, D.C.

CDC. 2012. Life expectancy in the U.S. 1900–2011. http://www.cdc.gov/nchs/data/nvsr64/nsvr64_11 .pdf, accessed via commons.wikipedia.org 4/20/17.

The Economist. 2011. *America's transport infrastructure: Life in the slow lane,* April 28, 2011. http://www. economist.com/node/18620944.

Fagan, B. 2012. *Elixir: A History of Water and Humankind.* Bloomsbury Press: New York, NY.

FHWA. 2018. Interstate System, Dwight D. Eisenhower National System of Interstate and Defense Highways. https://www.fhwa.dot.gov/programadmin/interstate.cfm.

Isites. Harvard, 2016. http://isites.harvard.edu/fs/docs/icb.topic247603.files/Lecture3.pdf. Accessed 4/20/17.

McNichol, Dan. 2006. *The Roads That Built America: The Incredible Story of the U.S. Interstate System.* Sterling: New York, NY. p. 87.

National Council on Public Works Improvement (U.S.). 1988. *Fragile foundations: a report on America's public works: final report to the President and the Congress/National Council on Public Works Improvement.* Washington, D.C. The Council: For sale by the Supt. of Docs., U.S. G.P.O., (1988).

Roosevelt Institute. 2011. *The WPA that Built America is Needed Once Again*, 050611. http://roosevelt institute.org/wpa-built-america-needed-once-again/.

USDA. 2014. *Rural America at a Glance*, United States Department of Agriculture Economic Research Service, Economic Brief Number 26, November 2014. https://www.ers.usda.gov/webdocs/publica tions/eb26/49474_eb26.pdf.

2

ECONOMIC IMPACT OF INFRASTRUCTURE

At a presentation at the Florida Section of the American Water Works Association (AWWA) annual conference opening general session on November 28, 2016, DC Water CEO George Hawkins noted that he was often asked how many jobs DC Water created in the DC area. His answer was "all of them." He was then asked how much DC Water contributed to the economic viability of the area and his answer was "all of it." The reality is that infrastructure systems are necessary for any society to function, let alone at the level that is expected in the United States.

Prud'homme (2004) defines infrastructure as capital goods. *Infrastructure* thus includes roads, tunnels, bridges, railways, airports, harbors, canals, subways and tramways, dams, irrigation networks, water pipes, water purification plants, sewers, water treatment plants, sanitary landfills, and incinerators; power plants, power lines, and distribution networks; oil and gas pipelines; telephone exchanges and networks; district heating equipment, etc. (Prud'omme 2004). As noted in Chapter 1, the focus of this book will be water, sewer, roads, and stormwater—with some lesser discussion of rail and bridges. The reason? Virtually every community has three or more of these asset systems and as a result, they are primarily local and are affected by local decision makers and maintained by local personnel.

Infrastructure provision results from the efforts of individuals and communities to modify their physical surroundings to improve their comfort, economic development and productivity, and protection from the elements (World Bank 1994). This means the construction of infrastructure to improve conditions and develop the local economy: access to roads, water, sewer, communication technologies, and electricity—which are all essential to the economy (Kamps 2005). The characteristics of modern infrastructure systems include the delivery of a given service through a networked delivery system designed to serve multiple users, such as piped potable and irrigation water, electric power, gas, telecommunications, sewer, roadways (including mass transit and other modes that use roadways), and rail. However, each of these infrastructure systems tends to be capable of delivering only one service and, in most cases, cannot be repurposed for other uses or moved elsewhere (World Bank 1994).

The benefits of infrastructure systems are broad based and within the public interest. Due to the fact that the costs are high and the payback periods are long, most infrastructure systems are constructed with public funds. Public finance theory stresses that a basic rationale for government provision of goods and services is that the private market economy is unable or unwilling to accomplish the task—the payback for these types of projects is normally too long for private equity interests and the risk of losses from the investment may be too great. Governments, however, can wait for longer payback periods and have opportunities with taxes, fees, credits, and other tools to deal with the potential that the economic advances may not occur immediately, which is why governments, as opposed to the private sector, generally tend to take the economic risk. Grimsey and Lewis (2002) noted that since World War II, local, state/province, and federal governments have been the primary constructors of infrastructure projects (albeit less so in the United States with respect to communications and power).

Public or private, the reality is that infrastructure expenditures are an investment decision—these investments are made to solve a current problem, with an eye to addressing a future concern or in anticipation of future growth. For most, the hope is that if infrastructure is built, development and/or an overall improved quality of life will follow, which may continue the cycle—growth and age of the assets will require more investments, in both new infrastructure and older systems, to maintain the quality of life. The belief has also been that good *core* infrastructure will increase productivity and economic activity, including capital (Barro 1990), labor and total factor productivity (Fedderke and Bogetic 2006), or a combination of factors (Barro 1998). *Core* infrastructure includes streets and highways, airports, electrical and gas facilities, mass transit, water systems, and sewers. However, defining the amount of economic growth created by infrastructure investment is more elusive.

MEASURING ECONOMIC GROWTH DUE TO INFRASTRUCTURE INVESTMENT

Most of the economic theory and practice has suggested that government can stimulate the economy in difficult times largely through the construction of infrastructure projects. The most notable examples are the Works Progress Administration (WPA) projects of the Great Depression and the American Recovery and Reinvestment Act of 2009 (ARRA) Public Law 111-5 legislation of the Great Recession. Both are credited with putting people to work and preventing further economic hardship on both people and the economy (Krugman 2014; Kavoussi 2011). By putting people to work, they had income that could be spent and therefore, the greater potential for economic growth (Krugman 2014). In both cases, after the investments were made, the economy stabilized and then began to expand; the argument that Krugman makes for more governmental investment in challenging economic times. One of the problems however, is that the response is not immediate, which is why some politicians argue that infrastructure benefits are limited—the delay factor. Also, if the government does not possess adequate reserves, the money must be borrowed, increasing debt and possibly weakening their financial condition.

Most of the research examples on infrastructure leading to economic growth are based on the concept of *elasticity* as defined in the field of economics. In economics, elasticity is the measurement of how responsive an economic variable is to a change in another variable. In this case, the variable is the change in infrastructure investments and looking for a causal change in economic value, generally defined by gross domestic product (GDP). In this case, the elasticity is defined as the percentage of change in GDP divided by the percentage of change in infrastructure investment. However, GDP is normally developed on a national, not a local basis. A challenge is downscaling to the local level as local results may not address externalities associated with issues that occur outside the locale, especially for certain types of infrastructure such as transportation. There is some evidence to prove that the investment in public infrastructure will foster economic growth beyond just the rational thought process. These results appear to be similar worldwide. Thus, the data appears to support the theory that an increase in public infrastructure spending/investment should increase the output of the economy as a whole, directly inducing economic growth (Fedderke and Garlik 2008).

Arrow and Kurz (1970) were the first to develop theoretical work on the contribution of infrastructure to output, productivity, and welfare—finding a correlation between infrastructure development and economic growth. Borcherding and Deacon (1972) found large and statistically significant income elasticities for highway and water-sewer expenditures. Aschauer (1989c) advanced the concept that public investment will induce an increase in the rate of return on private capital and, as a result, stimulate private investments. He asserted that infrastructure expenditures in the 1930s and 1940s may well have been a key ingredient to the robust performance of the economy in the 1950s and 1960s (Aschauer 1989b), and suggested that public infrastructure investments are the primary factors in fostering economic growth and productivity improvement (Aschauer 1990). Both Aschauer (1989a) and Munnell (1992) found a strong positive relationship between infrastructure and growth. Barro (1990) noted that production is driven by its flow of productive expenditures toward infrastructure. Duggal, Saltzman, and Klein (1999), Munnell (1992), and Nadiri and Mamuneas (1994) also found that increases in public capital raise output growth. Empirically, these results have been confirmed by Duffy-Deno and Eberts (1989), Garcia-Mila, McGuire, and Porter (1996), Carlino and Voith (1992), and Morrison and Schwartz (1996). High returns were also found with respect to infrastructure investment by Canning and Pedroni (2004), the World Bank (1994), Sanchez-Robles (1998), Canning (1999), Bougheas, Demetriades, and Mamuneas (2000), Röller and Waverman (2001), Esfahani and Ramirez (2003), Estache (2006), Calderón and Servén (2008), and Sahoo and Dash (2008, 2009). Other literature highlights the distributive impact of infrastructure provision and reform (Estache, Foster, and Wodon 2002; Calderón and Chong 2004). Calderón and Servén (2008) found robust evidence that infrastructure development—as measured by an increased volume of infrastructure investments and an improved quality of infrastructure services—has a positive impact on long-run growth and a negative impact on income inequality. Esfahani and Ramirez (2003) found the impact of power and communications infrastructure on GDP growth to be substantial. Fan and Zhang (2004) noted a high correlation between roadway, education, and power infrastructure development and economic development. Bougheas, Demetriades, and Mamuneas (2000) and Moomaw, Mullen, and Williams (1995) also found a positive correlation between public infrastructure and economic output in almost all cases, although they suggested that states get greater returns from investing in water and sewer systems than from investing in highways. Pereira (2001) found that public investment in water and sewer infrastructure has lower long-term elasticities than power or communications, but higher than highways and streets.

With respect to the concept of elasticity specifically, over 40 studies have been performed. Aschauer (1989a) found that the capital elasticity for economic growth (the ratio of capital compared to nonmilitary capital spending) was robust (0.39). Munnell (1992) estimated that the elasticity was 0.34, which she felt was larger than expected (Fedderke, Perkins, and Luiz 2006). She later estimated the elasticity at 0.15, still creating a positive relationship between growth and infrastructure investments (Munnell 1992). Easterly and Rebelo (1993) found similar high elasticity. Chandra and Thompson (2000) found that for the post-1969 sample there did not appear to be diminishing returns for infrastructure investments. Bom and Ligthart (2008) and Ligthart and Suárez (2011) found that the elasticity with respect to infrastructure capital lies around 0.15 for developed countries. During the early 1990s, research evaluated the impact of infrastructure as a determinant of economic production functions, with a view to estimate its contribution to economic growth. Prud'homme (2004) noted elasticities varying from 0 to 0.50 or 0.60 in the literature.

The elasticity argument holds outside the United States as well. Abedian and Van Seventer (1995) found output elasticities between 0.17 and 0.33 and economic rates of return between 0.2

and 0.23 (depending on the definition of the infrastructure stock). A 1998 Development Bank of South Africa (DBSA) Development Report also focused on public sector infrastructure stock, and using the same statistical techniques employed by earlier studies, found output elasticities between 0.15 and 0.3 and economic rates of return between 0.11 and 0.9 (Heymans and Thome-Erasmus 1998). Using a more appropriate statistical technique that took explicit account of stochastic time trends, the calculated elasticities were between 0.25 and 0.3, with economic rates of return between 0.17 and 0.33 (Heymans and Thome-Erasmus 1998). Perkins, Fedderke, and Luiz (2005) denoted increases in South African GDP based on public investment; road mileage; and rail, power, and communications technology. Kularatne (2006) also found a positive infrastructure/growth relationship. The relationship is positive for both economic and social infrastructure for all values of infrastructure investment. Fedderke and Bogetic (2006) found that aggregate infrastructure stock and investment impact positively on labor productivity, with elasticities of 0.19 and 0.2, respectively. Table 2.1 outlines the results of these many efforts. Of importance, the elasticity is always positive.

Table 2.1　Elasticity from 1990s and 2000s literature

Author(s)	Sectors/Location of Output	Elasticity to Sector Investment
Calderón and Servén (2009)	Mixed	0.08
Estache et al. (2005)	Transports	0.34
Estache et al. (2005)	Water	0.45
Hurlin (2006)	Transports	0.07
Ratner (1983)	USA	0.057
Aschauer (1989)	USA	0.400
Ram and Rasmey (1989)	USA	0.240
Merriman (1990)	USA and Japan	0.418
Munnell and Cook (1990)	USA	0.360
Eisner (1991)	USA (national and 4 regions)	0.027
Ford and Poret (1991)	10 OECD countries	0.378
Tatom (1991)	USA	0.087
Berndt and Hansson (1992)	Sweden	0.687
Garcia-Milà and McGuire (1992)	USA (48 states)	0.105
Bajo-Rubio and Sosvilla-Rivero (1993)	Spain	0.190
Finn (1993)	USA	0.010
Munnell (1992)	USA (national and 4 regions)	0.155
Eisner (1994)	USA	0.270
Evans and Karras (1994)	7 OECD countries	0.005
Holtz-Eakin (1994)	USA (48 states and 8 regions)	0.009
Ai and Cassou (1995)	USA	0.308
Baltagi and Pinnoi (1995)	USA (48 states)	0.073
Dalamagas (1995)	Greece	0.532
Holtz-Eakin and Schwartz (1995a)	USA	0.010

Continued

Author(s)	Sectors/Location of Output	Elasticity to Sector Investment
Holtz-Eakin and Schwartz (1995b)	USA (48 states)	0.004
Sturm and De Haan (1995)	The Netherlands and USA	0.635
Garcia-Milà et al. (1996)	USA (48 states)	0.023
Hulten (1996)	USA	0.317
Mas et al. (1996)	Spain (17 regions)	0.050
Otto and Voss (1996)	Australia	0.232
Crowder and Himarios (1997)	USA	0.291
Kavanagh (1997)	Ireland	0.495
Vijverberg et al. (1997)	USA	0.119
Batina (1998)	USA	0.110
Boarnet (1998)	USA (State of California)	0.083
Flores de Frutos et al. (1998)	Spain	0.210
Ramirez (1998)	Mexico	0.315
Delorme et al. (1999)	USA	0.276
Canning and Bennathan (2000)	97 countries	0.084
Charlot and Schmitt (2000)	France (22 regions)	0.229
Nourzad (2000)	24 countries	0.469
Vanhoudt et al. (2000)	15 EU countries	0.042
Yamano and Ohkawara (2000)	Japan (47 regions)	0.034
Yamarik (2000)	USA (48 states)	0.087
Stephan (2001)	France and Germany	0.100
Yilmaz et al. (2001)	USA	0.032
Kemmerling and Stephan (2002)	87 German cities	0.169
Ligthart (2002)	Portugal	0.189
Dodonov et al. (2002)	13 Eastern European countries	0.525
Song (2002)	Australia	0.005
Stephan (2003)	Germany (11 states)	0.659
Kamps (2005)	22 OECD countries	0.452
La Ferrara and Marcellino (2005)	Italy (4 regions)	0.017

Source: Ligthart and Suárez, 2011

ECONOMIC BENEFITS DUE TO INFRASTRUCTURE INVESTMENT

How does this relate to actual economic growth? Aschauer (1989a) estimated that the productivity-raising power of infrastructure investment to be huge, as much as quadrupling that of private investment. The World Bank (1994) notes that infrastructure capacity grows step for step with economic output—a 1% increase in the stock of infrastructure is associated with a 1% increase in GDP across all countries and an estimated average rate of return on infrastructure investments between 7% and 29% per year.

Cohen et al. (2011) estimated that a dollar spent on infrastructure construction produces roughly double the initial spending in ultimate economic output in the short term, primarily in the manufacturing and business services sectors, which is down from Aschauer's ratio, but still positive. At the same time, the more robust the economic climate, the more spending on infrastructure construction generates a larger return (Cohen et al. 2011). During poor economic climates, Cohen et al. (2011) determined that the overall effect of initial spending are still double output primarily by providing money for people to spend (realizing that spending improves the economy and pulls communities out of the economic doldrums—a key Krugman argument). Krop, Hernick, and Frantz (2008) estimate that one dollar of water and sewer infrastructure investment increases private output (GDP) in the long term by $6.35 (a 6.35:1 ratio). The boost to GDP from a dollar spent on building new bridges and schools is estimated to be $1.59 (Zandi 2008). Straub (2008) argues a positive impact of energy infrastructure on output/growth to be positive. Garsous (2012) and Estache, Speciale, and Veredas (2005) find the contribution of water and sanitary infrastructure to be positive.

Cohen et al. (2011) estimated that aggregate public investment in these five types of infrastructure is estimated to result in a marginal product of $3.21 economic output (GDP) over a twenty-year period. Transportation and power provide the largest economic gain, where spending $1.00 results in over $14.00 of output for a twenty-year period (but confined primarily to urban areas). The 2014 Draft Report to Congress identified aggregate annual benefits of between $217 billion and $863 billion and costs between $57 billion and $84 billion, both in 2001 dollars and averaged over the period of 2003 to 2013 (Office of Management and Budget 2014), a ratio of return between 4:1 and 10:1. Five years later, Cohen et al. (2016) calculated the ultimate economic impacts over twenty years to be 4.24:1 for transportation and power investments, and $2.03 in revenue per $1.00 spent for water and sewer investment.

However, during the period of 1973 to 1985, net investment in water, sewer, stormwater, power, and transportation infrastructure in the United States and Japan averaged 0.3% and 5.1% of GDP, while their respective growth rates of real gross domestic output per employed person were 0.6% and 3.1% per annum (Organization for Economic Cooperation and Development National Accounts and Historical Statistics). This decrease in expenditures coincides with the point where U.S. infrastructure started to see more obvious deterioration. Duffy-Deno and Eberts (1991) noted then that a lack of attention given to the deterioration of the nation's public infrastructure raised the question of whether the lack of public capital expenditures significantly affects economic development. That was 25 years ago—so the issue of infrastructure condition and deterioration is not new. Outside the U.S., Kumaraswamy and Zhang (2001) note that in order to achieve growth, countries must develop a means to construct and finance infrastructure; and Coetzee and Le Roux (1998) suggest a focus on financial measures in public-sector infrastructure when analyzing the relationship between infrastructure and growth. D'emurger (2001) explored the growth in Chinese provinces and noted that transportation infrastructure was the key differentiating factor in the development of economic activity. Sahoo, Dash, and Nataraj (2010) note that the results reveal that infrastructure investment plays an important role in economic growth in China. Fedderke, Perkins, and Luiz (2006) note that infrastructure investment has led to robust economic growth in South Africa. Mitra, Varoudakis, and Véganzonès-Varoudakis (2002) found a similar response in India. Banerjee, Duflo, and Qian (2012) note that proximity to transportation networks has a moderate positive causal effect on per capita GDP levels across sectors. They further note that richer countries have dramatically better transportation infrastructure than poorer ones (Bannerjee, Duflo, and Qian 2012).

Few empirical studies have tackled directly the inequality impact of infrastructure at the microeconomic level. Among them are those of López (2004) and Calderón and Servén (2008). Eberts (1990) concluded with an overall assessment of a positive relationship between public infrastructure and regional growth. Röller and Waverman (2011) noted that telecommunications infrastructure investment can lead to an increase in economic growth. Rehabilitation of rural roads raised male agricultural wages and aggregate crop indices in poor villages of Bangladesh (Khandker, Bahkt, and Koolwal 2006). Likewise, in Vietnam the result of improvements in rural roadways was an increase in the availability of food, the completion rates of primary school, and the wages of agricultural workers (van de Walle and Mu 2007). In the same vein, Escobal and Ponce (2002) found that access to new and improved roads in rural areas enhances opportunities in nonagricultural activities in Peru; and Lokshin and Yemtsov (2005) found a similar correlation with nonfarm activities among women in the Republic of Georgia. The overall conclusion that emerges from this is that infrastructure seems to have a relatively high rate of return, so increased infrastructure spending appears to be a means to stimulate the economy, not just in the U.S., but also in Bangladesh, Vietnam, India, China, South America, Europe, and South Africa.

Direct investment in infrastructure creates increases in production of goods, reduces transaction and trade costs, improves competitiveness, and provides economic opportunities—especially for those at lower income levels (Sahoo, Dash, and Nataraj 2010). Dalenberg and Partridge (1997) studied the relationship between public infrastructure and wages, and noted that highways raise wages in manufacturing sectors, but may cause a decline in the size of the overall private sector. Duffy-Deno and Eberts (1989) defined public capital stock to be a *rival good* in the sense that local public services—such as transportation and highways, and water treatment and distribution systems—are subject to competition for the same dollars, which is an obvious fact, but an economic issue that seems to be largely lost on local officials.

And finally, noneconomic benefits may accrue. The construction grants program associated with the Clean Water Act of 1972 spurred the expenditure of over $40 billion on the building and updating of sewage treatment facilities, which has had a "significant positive impact on the nation's water quality" (Urrea 2002). In addition, the measure of economic performance should include not only the GDP increase, but also benefits and costs not yet accounted for, according to a note by Musgrave in Aschauer (1990).

ACCOUNTING FOR OTHER FACTORS IN ECONOMIC GROWTH

There are some findings that differ from the ones previously mentioned or the robustness of the economic returns (Munnell 1992; Tatom 1993; Gramlich 1994). Bom and Ligthart (2008) suggest that estimates of growth may be optimistic, but they were looking at the 2009 financial crisis. Other studies use different statistical methodologies, such as those of Calderón and Servén (2008) and Estache, Speciale, and Veredas (2005) that demonstrate growth. Romp and de Haan (2007) conclude that "there is more consensus than in the past that public capital positively affects economic growth, but the impact seems to be lower than previously thought."

As noted previously, one reason some variation occurs is that downscaling may not work. Local-level investments may not demonstrate the GDP growth curves because the impacts may manifest both locally and regionally, and the regional component is not captured in the local economy. Stephanades (1990) and Stephanades and Eagle (1986) found that highway spending increases economic growth in urban counties; less in rural areas. Chandra and Thompson

(2000) found that opening new interstate highways will not increase net economic activity in nonmetropolitan regions. The dual effects of economic activity moving away from adjacent counties toward highway counties, and the intra-county reallocation of industrial structure lead to a rearrangement of economies. Hence downscaling may be an issue, especially in rural communities. Janeski and Whitacre (2014) suggested that rural Oklahoma housing values do show a statistically significant increase in communities receiving water or sewer infrastructure funding over the long term. However, increasing the price of housing may not translate to economic development in rural communities. Recent studies that were focused on the impact of highway improvements on county economies found limited evidence that highway investments increase the size of local economies. Given that data on output are not generally available for rural counties, these studies have focused on changes in income or employment. The outcome from Holtz-Eakin (1994), Holtz-Eakin and Schwartz (1995), Garcia-Mila, McGuire, and Porter (1996), and Canning and Pedroni (1999) is that infrastructure is very heterogeneous with respect to its relationship to economic development at the large scale, but the amount of infrastructure investment may not be a good predictor of economic activity.

Further, the contribution of infrastructure to economic local development will differ based on whether it is free or priced, or if it is overutilized (i.e., congested) or underutilized. The magnitude of infrastructure contributions to local economies varies from one study to another. However, in a study that ties to prior benefits of infrastructure investment, Duffy-Deno and Eberts (1989), in estimating a profit function, found a positive relationship between public capital and manufacturing output.

Results are more mixed among the growth studies using measures of infrastructure spending (Straub 2008). The argument is that while infrastructure may affect productivity and output, economic growth can also shape the demand and supply of infrastructure services, which is likely to cause an upward bias in the estimated returns to infrastructure. That means that economic activity demands more infrastructure, which permits more economic activity, which creates a greater infrastructure demand—it's a cycle. However, the cycle does not start without initial infrastructure investments, as indicated by Borcherding and Deacon (1972), Aschauer (1989a,b,c), and Munnell and Cook (1990).

Not studied with the same veracity is the impact of the lack of infrastructure—a question raised by Eberts (1990). Duffy-Deno and Eberts (1989) indicate that decaying public capital appears to be one factor that can retard regional economic development, as measured by per capita personal income. The results derived from annual data for 28 metropolitan areas in the U.S. from 1980 through 1984 revealed that public infrastructure has positive and statistically significant effects on per capita personal income (Duffy-Deno and Eberts 1989). In contrast, Estache and Vagliasindi (2007) argue that an insufficient power generation capacity limits growth in Ghana. Lumbila (2005) found that deficient infrastructure may hinder the growth impact of direct foreign investment in Africa. In addition, Bhattacharyay (2009) noted that infrastructure development was deemed to be essential to the realization of the Association of Southeast Asian Nations' goal of economic integration and their ability to survive the 2008 global economic crisis. And we know the lack of infrastructure in the South limited its economic growth prior to the 1930s. Hence, Bhattacharyay (2009) believed the development of infrastructure needed to be accelerated to enhance physical connectivity as well as encourage resource sharing.

The American Society of Civil Engineers (ASCE 2013a) opined that unless the infrastructure deficit is addressed by 2040, the cumulative loss in business sales is projected to be $734 billion, the cumulative loss to the nation's economy $416 billion in GDP, and 1.4 million jobs will be at risk. The World Bank (1994) noted that closely related to operating inefficiencies is a lack of maintenance: roads deteriorate, irrigation canals leak, water pumps break down, sanitation systems overflow, installed phone lines fail, and power generators are not available when needed. Infrastructure reliability is impacted, and as a result, economic output declines and economic risk increases. With such situations, substantial additional investment is required to bring the infrastructure system back online. The World Bank (1994) noted that when times are hard, capital spending on infrastructure is often the first item to be removed from budgets; operations and maintenance are often close behind. According to Krugman (2014) and Keynes (1936), it should be just the opposite. They opine that investing in infrastructure during difficult economic times provides two benefits—income for displaced people and often a reduced cost for that infrastructure since competition is lessened.

INVESTING IN INFRASTRUCTURE

Urban communities appear to have more invested in infrastructure. The thought is that the critical mass of urban development permits ongoing investment and maintenance of same. In rural communities the cost to extend water, sewer, roads, etc., are often more per person than for urban settings. For example, the cost for central sanitary sewers usually exceeds the cost of septic tanks, so septic tanks are used instead. Likewise, rural roads do not have curbs, sidewalks, and hard stormwater systems—the cost is too great and the use too low. The financial investment decisions are made based on use and cost in many places. But does this hurt the private sector investment or economic growth? Does deteriorating infrastructure do the same? And while deterioration is easy to note in some instances, what about a buried pipe? These are questions that must be considered. As noted previously, much of this infrastructure is distinctly local. At the same time, the amount spent on infrastructure varies with time, economic activity, and population density. Since infrastructure is not consumed directly, is not particularly incremental, and is usually very long lasting, its life is often measured in decades, if not in centuries. Unlike most goods, it is generally immobile.

What makes the situation more difficult to analyze is that most communities also have a limited understanding of their assets. The collection of asset data began in 1958 by federal agencies (not local), so the estimates of the number and needs for the many older assets (especially all those WPA investments) may be underestimated (Javetz 2013). In the 1980s the U.S. government added notes to its basic accounting statements to include roads, bridges, curbs, gutters, streets, sidewalks, drainage systems, lighting, and other immovable objects that are of value mainly to the public. The National Council on Public Works concluded their first assessment grade for infrastructure in the 1980s—note that piping was not discussed in this report. ASCE's first report card in 1998 looked at much of the visible infrastructure (in poor condition), but did not express concern about piping systems. GASB34—the accounting mechanism for tracking infrastructure asset value—did not appear until after 2000. The result is that for much of the last 120 years, a knowledge-based system of the assets of any specific community was missing, and as a result, the infrastructure needs, beyond those for growth, were likewise unclear.

The Technology Innovation Program (TIP) of the National Institute of Standards and Technology (NIST) (TIP 2009) noted that as the economy grows, people become even more dependent on the interdependent infrastructure systems that provide services and support social and economic activities—transportation systems (e.g., highways, roads, rail systems, ports), utilities (e.g., water, power, communications), and public facilities (e.g., schools, recreation, prisons, postal facilities). More complex networks of civil infrastructure require ever increasing expenditures to maintain their safety and security. TIP (2009) noted that research is required to overcome civil infrastructure challenges because the incremental improvements typically employed by local governments will not meet the challenges of providing cost-effective, widely-deployable solutions. Unfortunately, for the most part, this research is lacking because local governments do not have research budgets and often lack access to university research groups.

The 2008 fiscal crisis and subsequent recession had major effects on the economy and on infrastructure investments. While Congress battled to save the financial system, funds were designed for infrastructure investment and job creation. Blinder and Zandi (2010) noted that the effects of the fiscal stimulus alone appear very substantial, raising 2010 real GDP by about 3.4%, and holding the unemployment rate about 1.5 percentage points lower than otherwise projected by adding almost 2.7 million jobs to U.S. payrolls. Romer (2010) also estimated that the ARRA saved or created roughly 2.5 million jobs. While achieving many of its goals, Krugman (2009) noted that the proposed stimulus was insufficient due to the sheer size of the Great Recession. He suggested that the $2.9 trillion economic shortfall should have been replaced with federal dollars over the subsequent three years (Krugman 2009). Food and Water Watch (2010) recommended using more federal dollars for water infrastructure, including green infrastructure.

Paulais (2009) noted that local governments grappling with the crisis face a number of constraints. Given that three-quarters of local government spending is for current operations, that leaves a small amount for capital. In 2012, only 13% of budgets went for capital outlays (Barnett and Vidal 2013). That creates an accumulated, deferred effect (Kramer and Sobel 2014). In addition, communities often struggle to pay for the capital costs of infrastructure they have already built, and debt payments can lead to difficult budget cuts in other areas and/or local tax increases (Walsh-Sarnecki 2012). During the Great Recession many states shored up their financial positions by cutting shared revenues and aid to local governments (Pagano 2012)—further weakening local financial pictures. Some states had active legislatures that deliberately sought to limit local government access to revenue streams by eroding home rule powers. Yet these same governments had to take responsibility to mitigate, counteract, and overcome the impact of the economic crisis of their residents (Davey 2012).

After 2012, infrastructure spending began to rebound from the global financial crisis and was projected to grow significantly over the coming decade. However, the grades for infrastructure continue to stagnate. Gregory (2013) suggested that any method of making infrastructure investments that are deficit neutral reduces their impact on near-term activity and employment, but every method except for cuts to (state or federal) government transfers still leaves a net-positive impact (Gregory 2013). In 2015 the Congressional Budget Office (CBO) reported a persistent decline in actual public spending on transportation and water infrastructure since 2003, and that both construction and rehabilitation of highways has declined since 1959. Although all levels of government have spent less, the greatest reduction was reported at the federal level (about 19% since 2003). Public spending on transportation and water infrastructure only accounted for 2.4% of GDP in 2014, down from 3.0% in 1959 (CBO 2015). According to the Government Accountability Office (GAO), because constructing roadway improvements represents a significant investment in future economic activity, the lack of funding suggests a longer-term decrease in economic efficiency (GAO 2015). Since the 1960s, total public spending on transport and water infrastructure in the U.S. has fallen to 2.4% of GDP in 2011 from a high of over 3.6% (CBO 2015). In addition, funds for both capital investments and operations and maintenance have steadily dropped (The Economist 2011; The Roosevelt Institute 2011). In contrast, worldwide, the World Bank (1994) noted that public infrastructure investment ranges from 2% to 8% (and averages 4%) of GDP, nearly twice what the U.S. spends. Europe invests 5% of GDP in its infrastructure, while China spends 9%. Munnell and

Cook (1990) suggested that underinvestment in public capital may ultimately retard our economic growth. Perhaps the slow growth experienced by the U.S. economy since 2012 is partially due to decreased infrastructure funding and accompanying reliability concerns within the economy—an area worth further research.

RISK

Many local policymakers and researchers who are concerned with regional issues have claimed for years that public infrastructure investment is one of the primary means to implement a strategy of regional growth. One of the ways local governments compete for new firms is through investing in various types of public facilities. The robust infrastructure installed prior to 1950 has been a significant factor in economic development for many communities. But today, the effects of age, material choices, obsolescence, and the lack of maintenance has surfaced. Piping continues to age and exposes communities to risk. In many communities greater than 60% of their assets are buried pipes (Bloetscher 2017). Bridges (both concrete and steel) show the effects of weather deterioration, just as roadway surfaces show the effects of weather and wear. The result is that public infrastructure has been poorly rated by the ASCE (2001, 2005, 2009, 2013, 2017), while the GAO and most public officials acknowledge the deterioration of the infrastructure we rely on daily (ASCE 2017). At the same time, local budget growth rate has decreased as local officials try to rein in costs. According to the U.S. Conference of Mayors (USCM), the growth in local public water spending from the 1970s was 7.6% year-over-year (USCM 2015). However, spending growth decreased to 6% and decreased further to 5% after 2010 (USCM 2015). Many utilities went negative on growth rates from 2009–2011 due to the Great Recession. Larger utilities can often weather such storms, but the majority of water utilities are small—serving populations of 10,000 or less.

Tools for assessing the condition of buried pipes, especially water distribution pipes, are limited. As a result, the true risk to the community of pipe damage was likely underestimated and the potential for economic disruption likely increased. This led to Bondo (2008) suggesting that the status of the nation's and the states' water and wastewater infrastructure was *problematic*. The argument—the age of water and wastewater pipes—is one important consideration for the quality of water and wastewater infrastructure. Those with older infrastructure may face increased costs due to increased maintenance and leakage.

Wastewater piping (for sewers) is in similar condition. Inflow and infiltration (I/I), water that gets into the piping from rainfall or groundwater (as will be discussed in Chapter 4), may increase as sewer lines age. The ASCE (1999) reported that with respect to sewer systems, of the 1.2 million miles of sewer lines, 57.9% of the pipes were reported to be between 21 and 100 years old with 41.1% reported as between 21 and 50 years old and 16.8% greater than 51 years old. According to the U.S. Environmental Protection Agency (EPA), these data suggest that by 2020, up to half of the assets in these systems may be beyond the midpoint of their useful lives (EPA 2009).

The nation's highway infrastructure is relatively robust and redundant (Ham and Lockwood 2002). However, the facilities that are most vulnerable to disruptions are those playing important regional and strategic roles. Since transportation constitutes 10% of the U.S. economy, the loss of critical assets would be highly disruptive in terms of reliability, safety, and punctuality, and would likely involve greater and more expensive replacement challenges. Certain rail and interstate corridors are of particular concern due to the amount of goods and services that are transported on these assets. Bertram (2010) suggests that pavement will dominate highway construction spending over the next 20 years. Maintenance spending includes different types of quality enhancement—such as local repair, winter maintenance, renewal, or the addition of new functionalities (bridge, tunnel, etc.), as well as prolongation of the lifetime of existing infrastructures.

So how large is the potential backlog? The answer depends on the point in time when the estimate was made. The numbers have increased since the 1980s. For example, in 1989, Duffy-Deno and Eberts (1989) estimated that the shortfall between the amount of investment needed to provide *adequate* public water, sewer, stormwater, and transportation infrastructure and the available revenues to fund these projects ranged from $17.4 billion to $71.7 billion annually by 2000. As time and research continues, the costs continue to rise as the investments are delayed. The following are examples:

- Ham and Lockwood (2002) estimated a transportation system shortfall of $1.5 billion, plus $880 million for operations and maintenance (O&M) for bridges alone
- Bertram (2010) estimated the backlog on roadway projects was $32.1 billion
- Anderson (2010) estimated the needs at $500 billion by 2030 for drinking water and wastewater systems alone
- The AWWA estimated the needs over 20 years at $1 trillion for water systems (AWWA 2012)
- The USCM (2015) argued that a $111 billion a year investment will not satisfy future water demand
- The EPA estimates that more than $655 billion may be needed to repair and replace drinking water and wastewater infrastructure nationwide over the next 20 years (GAO 2017)
- The EPA estimates that small water utilities may need an estimated $110 billion for drinking water infrastructure and $33 billion for wastewater infrastructure (GAO 2017)
- The GAO (2107) estimated that large water utilities—those serving populations of 100,000 or more—account for an estimated $145 billion to repair and replace drinking water infrastructure and an estimated $219 billion for wastewater infrastructure (for a total of $364 billion GAO 2017)
- The United States Department of Agriculture (USDA) is planning to invest over $300 million in rural water infrastructure (USDA 2016)
- The ASCE claimed that additional spending of $1.6 trillion, in 2010 dollars, is needed by 2020 to bring the quality of the country's infrastructure up from *poor* to *good*.

The ASCE (2013) predicted that the deficit in water and sewer investments may lead to $206 billion in increased costs for businesses and households, along with putting nearly 700,000 jobs at risk by 2020.

A fear is that the impacts of these infrastructure-related job losses will be spread throughout the economy. Lower-wage workers will be most at risk of job loss and will have the least ability to pay. The impacts on jobs are a result of costs to businesses and households managing unreliable water delivery and wastewater treatment services. Bivens (2014) notes that an infrastructure investment by $250 billion annually over seven years would likely increase productivity growth by 0.3% annually, and Walker (2016) says that infrastructure development was the best way to avoid austerity and economic contraction. The ASCE (2017) suggested that the U.S. could lose $18 trillion in GDP in the next 10 years due to infrastructure deficiencies. That is why this is a congressional topic.

The European Union (2014) estimated considerable investment will be needed in energy infrastructure; but such investment decisions are largely in the hands of the private sector and need to take place in well-designed markets. The World Economic Forum's Global Competitiveness Report 2014–2015 ranked the U.S. 16th in *quality of overall infrastructure*, 15th in the quality of its rail system, and 16th in the quality of its roads (Holmes 2015, PWC Global (Oxford) 2015). This rating for the largest economy in the world that also has the most resources to redevelop that infrastructure should not be deemed to be acceptable.

SOLUTIONS

Despite decreasing appropriations, federal spending on transportation and water infrastructure benefits society on an ongoing basis, often by improving economic productivity, although those benefits are generally difficult to quantify. All funding helps economic activity based on the literature devoted to analyzing these impacts. From a purely economic perspective, the availability of transportation infrastructure lowers the costs to private firms obtaining input for the goods and services they produce. It also lowers the costs of delivering those products to their customers, increasing the productivity of the labor and capital at work in those firms, and therefore improving the rates of return on (private) investment. That increases the local economy. Federal spending on infrastructure can also provide noneconomic benefits to society—for example, clean drinking water and adequate sanitation improve public health, which can lead to noneconomic benefits that have economic advantages: improved public health and quality of life can lead to more productive workers. The 2016 Presidential election featured ongoing discussions about the need for infrastructure funding, and much of the ARRA was focused on funding *shovel-ready* infrastructure projects in order to put people to work. Congress recently passed the Water Infrastructure Finance and Innovation Act (WIFIA) at $1 billion. However, none of these sources significantly addresses the backlog of needs—they barely address the current needs. So much of the backlog will be left to the locals.

Canning and Pedroni (2004) note that infrastructure must be paid for. Since the 1950s, local communities have taken on trillions of dollars of debt to build the infrastructure that facilitated spread-out, automobile-oriented development. This growth came with infrastructure for transportation and utilities that initially had low maintenance costs—and the revenues it generated helped fund the maintenance of older developments. But times have changed. In too many cases, there is insufficient base to raise the funds to maintain and upgrade those assets. The automobile-oriented development rarely can generate enough revenue to cover the long-term maintenance and eventual replacement costs of the infrastructure that serves it (StrongTowns 2011). Suburbanization has left many inner cities without a viable economic model to maintain their infrastructure as people fled the inner cities and reestablished themselves in suburban communities (Glaeser and Gottlieb 2009).

Getting funding relies on economic strength, a problem if you are in a depressed area (Detroit) or a boom that could crash at any time (North Dakota). Both create a situation where there are insufficient resources to maintain the assets. One option is that growth in costs can lead to mergers where an entity cannot afford to go it alone, as the economy-of-scale of larger operations continues to play out in communities. Several small water plants cannot operate at the same cost as one larger plant. This option could include the establishment of a utility authority, whereby several communities work together to develop how to structure costs and authority. However, many smaller communities resist this based on a loss of local control of the system. Many also pull monies to the general fund from their infrastructure systems.

The fact that quality infrastructure impacts growth is now relatively well recognized and widely understood among practitioners and policy makers (Estache and Garsons 2012), which means money may be available as the financial market view infrastructure debt positively and a *safe* risk. The reason is that even during severe economic times, debts are paid, and therefore infrastructure credit ratings are generally strong. For example, during the Great Recession, there were only three rated water-sewer bond defaults (Breckenridge Capital Advisors 2014). In part, this is because of local governments' unilateral rate-setting authority, which provides a distinct credit strength (Moody's 2015). Yet, despite this power, local government debt climbed through 2010, while revenues decreased 2007–2010 (USCM 2012; Moody's 2015)—an issue that will concern the financial market if the trend continues.

So, while long-term financing alternatives are being explored by various government agencies, the question is: how do we lead our customers to investing in their/our future? That is the question as the next 20 years play out. Many risk factors will be exposed. The fact is that the true risk to the community of pipe damage is underestimated and the potential for economic disruption increases daily.

Going forward, the U.S. Census estimates the U.S. population will increase to 400 million by 2050, which will trigger an 8% increase in service requirements. The growing demand will create the need for rate increases to pay for system expansion and to sustain and renew the physical plants that serve the current population. Yet, in a nationwide survey of stormwater utilities, only 40% of stormwater managers are planning stormwater capital projects in the next five years (Landis 2015). A 2017 survey of water and wastewater utilities indicated that over 25% had no capital projects in their budgets (Bloetscher 2017), and sustainability plans are being developed by less than half the utilities surveyed by Landis (2015). This means there is a disconnect between the needs to the system and the understanding of those needs by the public. Better education is needed. Hopefully, the coming chapters will help readers understand the breadth and complexity of these infrastructure systems and the need to develop plans to address inevitable deterioration of the assets.

REFERENCES

Abedian, I. and Van Seventer, D. 1995. *Productivity and Multiplier Analysis of Infrastructure Investment in South Africa: An Econometric Investigation and Preliminary Policy Implications*. Mimeo. Pretoria, South Africa: Ministry of Finance.

American Radio Networks. 2017. http://americanradioworks.publicradio.org/features/infrastructure/b1.html.

———. 2005. *Report Card for America's Infrastructure, 2003 Progress Report*. ASCE, Alexandria, VA. http://www.infrastructurereportcard.org/making-the-grade/report-card-history/2005-report-card/. Accessed 12/15/16.

———. 2009. *2009 Report Card for America's Infrastructure*. ASCE, Alexandria, VA. http://www.infrastructurereportcard.org/making-the-grade/report-card-history/2009-report-card/. Accessed 12/15/16.

———. 2011. *Failure to Act: The Economic impact of current investment trends in water and wastewater infrastructure*. ASCE, Alexandria, VA.

———. 2013. *2013 Report Card for America's Infrastructure*. ASCE, Alexandria, VA. http://www.infrastructurereportcard.org/a/#p/home.

———. 2013a. *Failure to Act Report*. http://www.asce.org/uploadedFiles/Infrastructure/Failure_to_Act/Failure_to_Act_Report.pdf.

———. 2017. *2017 Report Card for America's Infrastructure*. ASCE, Alexandria, VA. https://www.infrastructurereportcard.org/. Accessed 3/15/17.

Amitabh, C. and Thompson, E. 1999. "Does public infrastructure affect economic activity? Evidence from the rural interstate highway system." *Regional Science and Urban Economics* 30 (2000) 457–490.

Anderson, R. F. 2010. "Trends in local government expenditures on public water and wastewater services and infrastructure: Past, present and future." *The U.S. Conference of Mayors—Mayors Water Council. 2010.* http://www.usmayors.org/publications/201002-mwc-trends.pdf.

Anderson, R. F., Gatton, D. and Sheahan, J. 2013. "Growth in local government spending on public water and wastewater—But how much progress can American households afford? *The United States Conference Of Mayors,*" April 2013, Washington, D.C.

American Society of Civil Engineers (ASCE). 2001. *2001 Report Card for America's Infrastructure.* ASCE, Alexandria, VA. http://www.infrastructurereportcard.org/making-the-grade/report-card -history/2001-report-card/. Accessed 12/15/16.

Arrow, K. and Kurz, M. 1970. *Public Investment, the Rate of Return and Optimal Fiscal Policy.* Johns Hopkins University.

Aschauer, D. 1989a. "Is public expenditure productive?" *Journal of Monetary Economics.* Vol. 23, pp. 177–200.

———. 1989b. "Public investment and productivity growth in the group of seven?" *Economic Perspectives.* Vol. 13, pp. 17–25.

———. 1989c. "Does public capital crowd out private capital?" *Journal of Monetary Economics.* Vol. 24, pp. 171–188.

———. 1990. *Why Is Infrastructure Important?* https://www.bostonfed.org/-/media/Documents/confer ence/34/conf34b.pdf. Accessed 12/15/16.

AWWA. 2012. *Buried No Longer: Confronting America's Water Infrastructure Challenge.* AWWA, Denver, CO.

Banerjee, A., Duflo, E., and Qian, N. 2012. *On the Road: Access to Transportation Infrastructure and Economic Growth in China.* MIT Department of Economics Working Paper Series, MIT, Cambridge, MA. file:///E:/Book%20CHapters%20Draft/Banerjee12-06.pdf. Accessed 3/11/17.

Barnett, J. L. and Vidal, P. M. 2013. *State and Local Government Finances Summary: 2011.* U.S. Census Bureau. 2013. http://www.census.gov/govs/local.

Barro, R. 1990. "Government spending in a simple model of endogenous growth." *Journal of Political Economy.* Vol. 98, No. 5, pp. 102–125.

———. 1998. *Notes on Growth Accounting.* National Bureau of Economic Research Working Paper No. 6654. NBER, Cambridge, MA.

Bee, C.A. 2013. "The geographic concentration of high-income households: 2007–2011," *American Community Survey Briefs.* file:///E:/CGN%206506%20Infra%20Mgmt/9968%20acsbr11-23%20 (2015_03_08%2017_53_48%20UTC).pdf.

Bertram, C. 2010. U.S. Congress, House Committee on Transportation and Infrastructure, Subcommittee on Highways and Transit, Testimony of Christopher Bertram, Assistant Secretary for Budget and Programs and Chief Financial Officer, U.S. Department of Transportation, Hearing on Using Innovative Financing to Deliver Highway and Transit Projects, 111th Cong., 2nd sess., April 14, 2010. http://republicans.transportation.house.gov/Media/file/TestimonyHighways/ 2010-04-14-Bertram.pdf.

Bhattacharyay, B. N. 2009. *Infrastructure Development for ASEAN Economic Integration.* ADBI Working Paper 138. Tokyo: Asian Development Bank Institute. Available: http://www.adbi.org/working -paper/2009/05/27/3011.infrastructure.dev.asean.economic/.

Bivens, J. 2014. *The Short- and Long-Term Impact of Infrastructure Investments on Employment and Economic Activity in the U.S. Economy.* Economic Policy Institute, July 1, 2014, Briefing Paper #374.

Blinder, A. S. and Zandi, M. 2010. *How the Great Recession Was Brought to an End.* https://www .economy.com/mark-zandi/documents/End-of-Great-Recession.pdf. accessed 2/10/17.

Bloetscher, F. 2017. "Risk and economic development in the provision of public infrastructure," *Florida Section of the American Water Works Association Annual Conference Proceedings—November 30, 2017, Championsgate, Orlando, FL.* FSAWWA, St. Cloud, FL.

Bom, P. and Ligthart, J. E. 2008. *How Productive Is Public Capital? A Meta-Analysis.* CESifo Working Paper Series No. 2206; Center Discussion Paper No. 2008-10. Available at: SSRN:https://ssrn .com/abstract=1088651. Accessed 2/15/17.

Bondo, M. 2008. *An Analysis of Public Water and Sewer Providers Rates and Practices,* University of South Carolina Institute for Public Service and Policy research, Columbia, SC.

Borcherding, T. E. and Deacon, R. T. 1972. "The demand for the services of non-federal governments," *American Economic Review*, 62, pp. 842–853.

Bottini, N., Coelho, M. I., and Kao, J. 2013. *Infrastructure and Growth; Launch Version*. Institute for Government/Center for Economic Performance. http://www.lse.ac.uk/researchAndExpertise/units/growthCommission/documents/pdf/SecretariatPapers/Infrastructure.pdf. Accessed 2/15/17.

Bougheas, S., Demetriades, P. O., and Mamuneas, T. P. 2000. "Infrastructure, specialization, and economic growth." *Canadian Journal of Economics*. Vol. 33, No. 2.

Breckenridge Capital Advisors. 2014. *Water and Sewer Utilities: Risks not yet a drain on credit quality*. April 2014, Breckinridge, Boston, MA.

Burian, S. J., Nix, S. J., Pitt, R. E., and Durrans, S. R. 2000. "Urban wastewater management in the United States: past, present, and future," *Journal of Urban Technology*. Vol. 7, No. 3, pp. 33–62.

Burton, K., Kates, R., and White, G. 1993. *The Environment as Hazard*, 2nd edition. Guilford Press, New York, NY.

Calderón, C. and Chong, A. 2004. "Volume and quality of infrastructure and the distribution of income: An empirical investigation." *Review of Income and Wealth*. 50, 87–105.

Calderón, C. and Servén, L. 2008. *Infrastructure and Economic Development in Sub-Saharan Africa*. The World Bank Development Research Group Macroeconomics and Growth Team. file:///E:/Book%20CHapters%20Draft/Infrastructure_and_Economic_Development_in_Sub-Sah.pdf. Accessed 2/11/2017.

Canning, D. 1999. *The Contribution of Infrastructure to Aggregate Output*. The World Bank Policy Research Working Paper 2246. The World Bank, Washington, D.C.

Canning, D. and Pedroni, P. 1999. *Infrastructure and Long Run Economic Growth*. Center for Analytical Economics Working Paper No. 99-09. Cornell University, Ithaca, NY.

———. 2004. *The Effect of Infrastructure on Long Run Economic Growth*. Rice University, 6100 Main St., Houston, TX. http://web.williams.edu/Economics/wp/pedroniinfrastructure.pdf. Accessed 2/17/17.

Carlino, G. A. and Voith, R. 1992. Accounting for differences in aggregate state productivity. *Regional Science and Urban Economics*. 22, 597–617.

Chandra, A. and Thompson, E. 2000. *Does Public Infrastructure affect economic activity? Evidence from the rural interstate system, Regional Science and Urban Economics*. 30, 457–490.

Coetzee, Z. and Le Roux, E. 1998. "Does public infrastructure affect economic growth?" Presented at the *Annual EBM Conference of the National Productivity Institute*.

Cohen, I., Freiling, T., and Robinson, E. 2011. *The Economic Impact and Financing of Infrastructure Spending*, Thomas Jefferson Program in Public Policy at the College of William & Mary, Williamsburg, VA. file:///E:/CGN%206506%20Infra%20Mgmt/9951%20aed%20(2015_03_08%2017_53_48%20UTC).pdf.

Cohen, J., Moeltner, K., Reichl, J., and Schmidthaler, M. 2016. Linking the value of energy reliability to the acceptance of energy infrastructure: Evidence from the EU. *Resource and Energy Economics*. 45, 124–143.

Congressional Budget Office. 2015. *Public Spending on Transportation and Water Infrastructure, 1956 to 2014*. CBO, Washington, D.C.

———. 2013. *The Distribution of Household Income and Federal Taxes, 2010*. CBO, Washington, D.C.

Dalenberg, D. R. and Partridge, M. D. 1997. "Public infrastructure and wages: Public capital's role as a productive input and household amenity." *Land Economics*. 73, 268–284.

Davey, K. 2012. *Local Government in Critical Times: Policies for Crisis, Recovery and a Sustainable Future*. Council of Europe Strasbourg, France, 2012.

D'emurger, S. 2001. "Infrastructure development and economic growth: an explanation for regional disparities in China?" *Journal of Comparative Economics*. 29, 95–117. doi:10.1006/jcec.2000.1693, available online at http://www.idealibrary.com.

Deno, K. T. 1988. "The effect of public capital on U.S. manufacturing activity: 1970 to 1978." *Southern Economic Journal*. 53, 400–411.

Development Bank of Southern Africa. 1998. "Infrastructure: A foundation for development." Development Report 1998. Pretoria, South Africa: DBSA.

Drake, S. 2016. "WPA 2.0 beauty, economics, politics, and the creation of new public infrastructure." *Land Lines*, October 2016. https://www.lincolninst.edu/sites/default/files/pubfiles/wpa-2-0-1016ll.pdf. Accessed 2/15/17.

Duffy-Deno, K. T. and Eberts, R. W. 1989. *Public Infrastructure and Regional Economic Development: A Simultaneous Equations Approach*, Working Paper 8909 National Science Foundation under Grant *SES-8414262-01, Federal Reserve Bank of Cleveland, Cleveland, OH. http://clevelandfed .org/research/workpaper/index.cfm. Best available copy.

———. 1991. "Public infrastructure and regional economic development: A simultaneous equations approach." *Journal of Urban Economics* 30, 329–343.

Duggal, Vijaya, Saltzman, Cynthia R., and Klein, Lawrence. 1999. "Infrastructure and productivity: A nonlinear approach." *Journal of Econometrics*. 92, 47–74. 10.1016/S0304-4076(98)00085-2.

Easterly, W. and Rebelo, S. 1993. "Fiscal policy and economic growth: An empirical investigation." *Journal of Monetary Economics*. 32, 417–458.

Easterly, W. and Servén, L. 2003. *The Limits of Stabilization: Infrastructure, Public Deficits, and Growth in Latin America*. Stanford University Press, Redwood City, CA.

Eberts, R. W. 1990. *Public Infrastructure and Regional Economic Activity*. file:///E:/Book%20CHapters %20Draft/6230055.pdf. Accessed 2/15/17.

The Economist. 2016. *Buy Local*. May 15, 2016. http://www.economist.com/news/finance-and-economics/ 21698301-america-increasingly-relies-state-governments-fund-public-investment-buy.

———. 2011. *America's transport infrastructure—Life in the slow lane*. April 28, 2011. http://www .economist.com/node/18620944.

EPA, 2009. Drinking Water Infrastructure Needs Survey and Assessment: Fourth Report to Congress, EPA 816-R-09-001. Washington, D.C.

Escobal, J. and Ponce, C. 2002. *The benefits of rural roads: Enhancing income opportunities for the rural poor*. Group of Analysis for Development (GRADE) Working paper No. 40-I, Lima, p. 52. Retrieved from http://www.grade.edu.pe/upload/publicaciones/archivo/download/pubs/ddt/ddt40ES.pdf. Accessed 2/17/17.

Esfahani, H. S. and Ramirez, M. T. 2002. "Institutions, infrastructure, and economic growth." *Journal of Development Economics*. 70, (2003) 443–477.

Estache, A. 2006. *Infrastructure: A survey of recent and upcoming issues*. Washington, D.C.: The World Bank.

Estache, A. and Garsons, G. 2012. "The impact of infrastructure on growth in developing countries." *IFC Economic Notes*. Note 1.

Estache, A. and Vagliasindi, M. 2007. *Infrastructure for accelerated growth for Ghana: Needs and challenges*. Powerpoint presentation from an unpublished manuscript. http://siteresources.world bank.org/INTGHANA/Resources/CEM_infrastructure_presentation.pdf. Accessed 2/5/17.

Estache, A., Foster, V., and Wodon, Q. 2002. *Accounting for Poverty in Infrastructure Reform: Learning from Latin America's Experience*. WBI Development Studies, Washington, D.C.: The World Bank.

Estache, A., Speciale, B., and Veredas, D. 2005. *How much does infrastructure matter to growth in Sub-Saharan Africa?* http://citeseerx.ist.psu.edu/viewdoc/download?doi=10.1.1.508.7016&rep=rep1&typ e=pdf. Accessed 2/5/17.

European Union. 2014. *Infrastructure in the EU: Developments and Impact on Growth*. Occasional Papers 203 European Union, European Commission Directorate-General for Economic and Financial Affairs Unit Communication and interinstitutional relations B-1049. Brussels, Belgium.

Fan, S. and X. Zhang. 2004. "Infrastructure and Regional Economic Development in Rural China." *China Economic Review*. 15, 203–214.

Fedderke, J. 2005. "Technology, human capital and growth in economic growth." *Proceedings of a G20 Seminar*, Pretoria, South Africa, August 2005.

Fedderke, J. and Bogetic, Z. 2006. *Infrastructure and Growth in South Africa: Direct and Indirect Productivity Impacts of 19 Infrastructure Measures*. Economic Research Southern Africa Working Paper No. 39, forthcoming World Development. World Bank, Washington, D.C. https://www .csae.ox.ac.uk/conferences/2007-EDiA-LaWBiDC/papers/020-Fedderke.pdf. Accessed 2/5/17.

Fedderke, J. and Garlick, R. 2008. *Infrastructure Development and Economic Growth in South Africa: A review of the accumulated evidence*, Policy Paper Number 12, School of Economics, University of Cape Town and Economic Research Southern Africa

Fedderke, J., De Kadt, R., and Luiz, J. 2001. "Indicators of political liberty, property rights and political instability in South Africa." *International Review of Law and Economics*. Vol. 21, No. 1, pp. 103–134.

Fedderke, J., Perkins, P., and Luiz, J. 2006. "Infrastructure investment and long-run economic growth: South Africa 1875–2001." *World Development*, Vol. 34, No. 6, pp. 1037–1059.

Fitch Ratings, 2008. *Revenue Criteria Report—Water and Sewer Revenue Bond Rating Guidelines*, Fitch Ratings, New York, NY.

Food and Water Watch. 2010. *Water and Sewer Privatization Contributes to Sprawl*, Fact Sheet, January 2010, Food and Water Watch, Washington, D.C.

GAO. 2015. *Collaboration among Federal Agencies Would Be Helpful as Governments Explore New Financing Mechanism, Report to the Chairman, Committee on the Budget*, U.S. Senate, GAO-15-646 GAO, Washington, D.C.

———. 2017. *Drinking Water and Wastewater Infrastructure: Information on Identified Needs, Planning for Future Conditions, and Coordination of Project Funding* GAO-17-559 GAO, Washington, D.C.

Garcia-Mila, T. and McGuire, T. J. 1992. "The contribution of publicly provided inputs to states' economies." *Regional Science and Urban Economics*. Vol. 22, Issue 2, pp. 229–241.

Garcia-Mila, T., McGuire, T. J., and Porter, R. H. 1996. "The effect of public capital in state level production functions reconsidered." *Review of Economics and Statistics* 78, 177–180.

Garsous, G. 2012. *How Productive Is Infrastructure? A Quantitative Survey*. ECARES Working Paper, Université libre de Bruxelles.

Glaeser, E. and Gottlieb, J. 2009. *The Economics of Place Making Policies, Brookins Institutes*. https://www.brookings.edu/wp-content/uploads/2008/03/2008a_bpea_glaeser.pdf.

Gramlich, E. 1994. "Infrastructure investment: a review essay." *Journal of Economic Literature*, Vol. XXXII, September, pp. 1176–1196.

Gregory, P. R. 2013. "Infrastructure gap? Look at the facts. We spend more than europe." *Forbes*. https://www.forbes.com/sites/paulroderickgregory/2013/04/01/infrastructure-gap-look-at-the-facts-we-spend-more-than-europe/#7c1a25ff5bf6. Accessed 2/17/17.

Grimsey, D. and Lewis, M. K. 2002. "Evaluating the risks of public private partnerships for infrastructure projects." *International Journal of Project Management*. 20, pp. 107–118.

Ham, D. B. and Lockwood, S. 2002. *National Needs Assessment for Ensuring Transportation Infrastructure Security*. American Association of State Highway and Transportation Officials (AASHTO) Transportation Security Task Force. http://freight.transportation.org/doc/NCHRP_B.pdf

Heymans, C. and Thome-Erasmus, J. 1998. "Infrastructure: A foundation for development—key points from the DBSA Development Report 1998." *Development Southern Africa*. Vol. 15, Issue 4.

Hibbard, P. J. 2006. *U.S. Energy Infrastructure Vulnerability Lessons from the Gulf Coast Hurricanes*. Analysis Group Boston, Boston, MA. http://bipartisanpolicy.org/wp-content/uploads/sites/default/files/Infrastructure%20Vulnerability%20Hibbard_44873b7081ec6.pdf. Accessed 2/17/17.

Hideki, T. and Skidmore, M. 2005. *Economic Development and the Impacts of Natural Disasters*. Working Paper 05—04 University of Wisconsin—Whitewater. Whitewater, WI. 53538 file:///E:/Book%20CHapters%20Draft/10.1.1.660.1931.pdf. Accessed 3/15/2017.

Holmes, F. 2015. "What China can teach us about infrastructure investing." *Forbes*. https://www.forbes.com/sites/greatspeculations/2015/05/29/what-china-can-teach-u-s-about-infrastructure-investing/#496d11e460ed. Accessed 2/15/17.

Holtz-Eakin, D., 1994. "Public sector capital and the productivity puzzle." *Review of Economics and Statistics*. 76, 12–21.

Holtz-Eakin, D. and Schwartz, A. E. 1995. "Infrastructure in a structural model of economic growth." *Regional Science and Urban Economics*. 25, 131–151.

Janeski, I. and Whitacre, B. E. 2014. "Long-term economic impacts of USDA water and sewer infrastructure investments in Oklahoma." *Journal of Agricultural and Applied Economics*. 46, 1 (February 2014), 21–39.

Javetz, E. 2013. *The US water sector on the verge of transformation*. Global Cleantech Center white paper Simul Consulting Corp. http://www.ey.com/Publication/vwLUAssets/Cleantech_Water_Whitepaper/$FILE/Cleantech-Water-Whitepaper.pdf. Accessed 2/17/17.

Kamps, C. 2005. "The dynamic effects of public capital: VAR evidence for 22 OECD countries." *International Tax and Public Finance*. Vol. 12 (August), pp. 533–58.

Kavoussi, B. 2011. *Paul Krugman: U.S. Economy Needs 'The Financial Equivalent Of War'* 09/28/2011 01:03 pm ET. Updated Mar 02, 2012. http://www.huffingtonpost.com/2011/09/28/paul-krugman-spending_n_984921.html. Accessed 3/15/17.

Keynes, J. M. 1936. *The General Theory of Employment, Interest and Money.* Palgrave Macmillan, London, UK.

Khandker, S. R., Bahkt, Z., and Koolwal, G. B. 2006. *The Poverty Impact of Rural Roads: Evidence from Bangladesh.* World Bank Policy Research Working Paper 3875, April, World Bank, Washington, D.C.

Khramov, V. and Lee, J. R. 2013. "The Economic Performance Index (EPI): An intuitive indicator for assessing a country's economic performance dynamics in an historical perspective." IMF Working Paper No. 13/214.

Kramer, M. and Sobel, L. 2014. *Smart Growth and Economic Success: Strategies for Local Governments.* EPA, Washington, D.C.

Krop, R. A., Hernick, C., and Frantz, C. 2008. *Local Government Investment in Municipal Water and Sewer Infrastructure: Adding Value to the National Economy.* The Cadmus Group, Inc. 57 Water Street Watertown, MA.

Krugman, P. 2009. "The true fiscal cost of stimulus." Op-ed appears in print on September 29, 2009, on Page A25 of the New York Times, New York, NY.

———. 2014. "The stimulus tragedy." Op-ed appears in print on February 21, 2014, on Page A25 of the New York Times, New York, NY.

Kularatne, C. 2006. *Social and Economic Infrastructure Impacts on Economic Growth in South Africa.* Presented at the UCT School of Economics Sta/Seminar Series, October 2006.

Kumaraswamy, M. M. and Zhang, X. O. 2001. "Government role in BOT led infrastructure development." *International Journal of Project Management.* 19, 195–2005.

Kuznets, S. 1945. *National Product in Wartime.* National Bureau of Economic Research, Washington, D.C. http://www.nber.org/chapters/c5465.pdf.

Kuznets, S., Epstein, L., and Jenks, E. 1941. *National Income and Its Composition, 1919–1938,* National Bureau of Economic Research, 1941, Washington, D.C. http://www.nber.org/chapters/c4225.pdf. Accessed 2/15/17.

Landis, A. E. 2015. "The state of water/wastewater utility sustainability: A North American Survey. Journal." *American Water Works Association.* Vol. 107, Issue 9.

Ligthart, J. E. and Suárez, R. M. 2011. "Chapter 2: The productivity of public capital: A meta-analysis." W. Manshanden and W. Jonkhoff (eds.), *Infrastructure Productivity Evaluation.* SpringerBriefs in Economics 1, doi: 10.1007/978-1-4419-8101-1_2, © TNO (Dutch Organization for Applied Scientific Research).

Lokshin, M. and Yemtsov, R. 2005. "Has rural infrastructure rehabilitation in Georgia helped the poor?" *The World Bank Economic Review.* 19(2), 311–333.

López, H., 2004. "Macroeconomics and inequality." *The World Bank Research Workshop, Macroeconomic Challenges in Low Income Countries, October, 2004.* World Bank, Washington, D.C.

Lumbila, K. 2005. *What Makes FDI Work? A Panel Analysis of the Growth Effect of FDI in Africa.* Africa Region Working Paper Series No. 80, The World Bank, Washington, D.C.

McFarland, C. and Pagano, M. A. 2012. *City Fiscal Conditions in 2012.* National League of Cities. 2012. http://www.nlc.org/sites/default/files/2016-12/City%20Fiscal%20Conditions%202016_1.pdf. Accessed 2/15/17.

Mitra, A., Varoudakis, A., Véganzonès-Varoudakis, M-A. 2002. "Productivity and technical efficiency in Indian States's manufacturing: The role of infrastructure." *Economic Development and Cultural Change.* 50(2), 395–426.

Moody's. 2015. U.S. Water and Sewer Utilities Outlook Is Stable for 2016. Dec. 2015. Moody's, New York, NY. https://www.moodys.com/research/Moodys-US-water-and-sewer-utilities-outlook-is-stable-for—PR_340720. Accessed 3/17/2017.

Moomaw, R. L., Mullen, J. K., and Williams, M. 1995. The interregional impact of infrastructure capital. *Southern Economic Journal.* 61, 830–845.

Morrison, C. J. and Schwartz, A. E. 1996. "State infrastructure and productive performance." *American Economic Review.* 86(5), 1095–1111.

Munnell, A. H. 1992. "Infrastructure investment and economic growth." *Journal of Economic Perspectives.* 6(4), 189–198.

Munnell, A. H. and Cook, L. M. 1990. *How Does Public Infrastructure Affect Regional Economic Performance?* Federal Reserve Bank of Boston, Boston, MA. https://www.bostonfed.org/-/media/Documents/conference/34/conf34c.pdf. Accessed 2/2/17.

Nadiri, I. M. and Mamuneas, T. P. 1994. "The effects of public sector infrastructure and R&D capital on the cost structure and performance of U.S. manufacturing industries." *Review of Economics and Statistics* LXXVI(1), February: 22–37.

National Council on Public Works Improvement (U.S.). 1988. *Fragile foundations: a report on America's public works: final report to the President and the Congress.* Washington, D.C. The Council; For sale by the Supt. of Docs., U.S. G.P.O., [1988].

Noss, A. 2013. *Household Income: 2012 American Community Survey* Briefs, mfile:///E:/CGN%206506%20 Infra%20Mgmt/9969%20acsbr12-02%20(2015_03_08%2017_53_48%20UTC).pdf. Accessed 2/15/17.

Office of Management and Budget, 2014. Draft Report to Congress on the Benefits and Costs of Federal Regulations and Unfunded Mandates on State, Local, and Tribal Entities, Office of Information and Regulatory Affairs, Washington, D.C.

Pagano, Michael A. 2012. "The typical municipal budgeting process is rigged against infrastructure investments." *The Atlantic Cities.* Dec. 18, 2012. http://www.theatlanticcities.com/politics/2012/12/how-typical-municipal-budgeting-process-riggedagainst-infrastructure-investments/4190. Accessed 2/15/17.

Paulais, Thierry 2009. "Local governments and the financial crisis: An analysis." *The Cities Alliance.* 1818 H Street, NW, Washington, D.C.

Pereira, A. M. 2001. "Public investment and private sector performance: International evidence." *Public Finance and Management.* Vol. 1, No. 2, pp. 261–277.

Perkins, P., Fedderke, J., and Luiz, J. 2005. "An analysis of economic infrastructure investment in South Africa." *South African Journal of Economics.* Vol. 73, 2.

Piketty, T. E., Hess, P., and Saez, E. 2004. "Income inequality in the United States, 1913–2002*" UC Berkeley and NBER, November 2004. file:///E:/CGN%206506%20Infra%20Mgmt/9964%20piketty-saezOUP04US%20(2015_03_08%2017_53_48%20UTC).pdf. Accessed 2/17/17.

Prud'homme, R. 2004. "Infrastructure and development." In: Bourguignon, François and Boris Pleskovic, eds. 2005. *Lessons of Experience* (Proceedings of the 2004 Annual Bank conference on Development Economics). 2005. Washington: The World Bank and Oxford University Press, pp. 153–181. file:///E:/Book%20CHapters%20Draft/Prud$27homme+2005a.pdf. Accessed 3/15/2017.

PWC Global (Oxford). 2015. *Capital Project and Infrastructure Spending, Outlook to 2025.* PWC Global. http://www.pwc.com/gx/en/industries/capital-projects-infrastructure/publications/cpi-spending-outlook.html. Accessed 2/15/17.

Röller, L-H. and Waverman, L. 2001. *Telecommunications infrastructure and economic development: a simultaneous approach.* WZB Discussion Paper, No. FS IV 96-16. https://www.jstor.org/stable/2677818?seq=1#page_scan_tab_contents. Accessed 2/14/17.

Romer, Christina D. 2010. "Treatment and prevention: Ending the Great Recession and ensuring that it doesn't happen again." Council of Economic Advisers, City Club of Cleveland, Cleveland, OH, May 3, 2010.

Romp, W. and De Haan, J. 2007. "Public capital and economic growth: A critical survey." *Perspektiven der Wirtschaftspolitik.* 8, 6–52.

Roosevelt Institute. 2011. *The WPA that Built America is Needed Once Again,* 05/06/11. http://roosevelt institute.org/wpa-built-america-needed-once-again/.

Sahoo, P. and Dash, R. K. 2008. Economic Growth in South Asia: The Role of Infrastructure, Institute of Economic Growth. Working paper No. 288.

Sahoo, P. and Dash, R. K. 2009. "Infrastructure development and economic growth in India." *Journal of the Asia Pacific Economy.* 14, 351–365.

Sahoo, P., Dash, R. K., and Nataraj, G. 2010. *Infrastructure Development and Economic Growth in China,* IDE Discussion Paper No 261, IDE-JETRO, Chiba, Japan.

Sanchez-Robles, B. 1998. "Infrastructure investment and growth: Some empirical evidence." *Contemporary Economic Policy.* 16, 98–108.

Stephanades, Y. J. 1990. "Distributional effects of state highway investment on local and regional development." *Transportation Research Record.* 1274, 156–164.

Stephanades, Y. J. Eagle, D. M. 1986. "Time-series analysis of interactions between transportation and manufacturing and retail employment." *Transportation Research Record.* 1074, 16–24.

Stonecypher, Lamar. 2011. Updated: 9/29/2011. *A New Works Progress Administration for Renewing American Infrastructure.* http://www.brighthubengineering.com/structural-engineering/125164-new-works-progress-administration-for-renewing-american-infrastructure/ accessed 2/5/17.

Straub, S. 2008. "Infrastructure: Recent advances and research challenges", unpublished manuscript. file:///C:/Users/Frederick/Downloads/SSRN-id1080475.pdf. Accessed 2/15/2017.

Straub, S. 2011. "Infrastructure and development: A critical appraisal of the macro-level literature." *The Journal of Development Studies,* Vol. 47, 5, 683–708.

Strong Towns. Curbside Chat. 2011. http://www.strongtowns.org/companion-booklet. Accessed 2/17/17.

Tatom, J. A. 1993. "The spurious effect of public capital formation on private sector productivity." *Policy Studies Journal.* 21, 391–395.

Taylor, Nick. 2017. *The WPA: Antidote to the Great Depression?* https://www.gilderlehrman.org/history-by-era/new-deal/essays/wpa-antidote-great-depression. Accessed 2/17/17.

Technology Innovation Program. 2009. *Advanced Sensing Technologies and Advanced Repair Materials for the Infrastructure: Water Systems, Dams, Levees, Bridges, Roads, and Highways.* National Institute of Standards and Technology, Gaithersburg, MD. https://www.nist.gov/sites/default/files/documents/tip/prev_competitions/ci_wp_031909.pdf. Accessed 2/17/17.

Tol, R. and Leek, F. 1999. "Economic analysis of natural disasters." In T. Downing, A. Olsthoorn, and R. Tol, eds, *Climate Change and Risk,* London, UK: Routledge.

Urrea, Jorge. 2002. "Regional development and the action of public investment: The FNDR and the ERDF, a comparative analysis." PhD thesis. http://theses.gla.ac.uk/2739/1/2002urreaphd.pdf.

USDA. 2016. "USDA is planning to invest over $300 million in rural water infrastructure." Press release #0235.116. https://www.usda.gov/media/press-releases/2016/11/01/usda-announces-331-million-investment-clean-water-infrastructure.

U.S. Conference of Mayors. 2012. "USCM staff report on water and wastewater affordability." November 16, 2012. Mayor's Water Council, Washington, D.C.

———. 2015. "Struggling local government finances and decelerating public water investment." U.S. Conference of Mayors, Washington, D.C.

U.S. Department of Commerce. 2000. *Bureau of Economic Analysis. National Income and Product Accounts of the United States, 1929–97.* U.S. Department of Commerce, Washington D.C. www.bea.doc.gov/bea/dn/nipaweb/. Accessed 2/17/17.

van de Walle, D. and Mu, R. 2007. "Fungibility and the flypaper effect of project aid: Micro-evidence for Vietnam." *Journal of Development Economics.* 84(2), pp. 667–685.

Walker, A. 2016. "Infrastructure investment is best way to avoid austerity says World Bank." file:///D:/Book%20CHapters%20Draft/infra%20Econ/Infrastructure%20investment%20is%20best%20way%20to%20avoid%20austerity%20says%20World%20Bank%20_%20Infrastructure%20Intelligence.pdf. Accessed 8/15/17.

Walsh-Sarnecki, P. 2012. "Housing development slowdown leaves suburbs to pay for new infrastructure." *Detroit Free Press.* Sep. 18, 2012. http://www.freep.com/article/20120918/NEWS05/309180071/Housing-development-slowdown-leavessuburbs-to-pay-for-new-infrastructure. Accessed 2/17/17.

Wildavsky, A. 1988. *Searching for Safety,* New Brunswick, N.J.: Transaction Books.

World Bank. 1994. *World Development Report 1994—Infrastructure for Development.* The International Bank for Reconstruction and Development. The World Bank, Washington, D.C. https://openknowledge.worldbank.org/bitstream/handle/10986/5977/WDR%201994%20-%20English.pdf?sequence=2&isAllowed=y. Accessed 2/1/17.

Yan, W. 2008. *The Impact of Revenue Diversification and Economic Base on Revenue Stability: An Empirical Analysis of County and State Governments.* University of Kentucky Doctoral Dissertations. Paper 619. http://uknowledge.uky.edu/gradschool_diss/619.

Zandi, M. 2008. *Assessing the Macro Economic Impact of Fiscal Stimulus 2008.* Moody's Economy.com • www.economy.com • help@economy.com • January 2008.

3

PUBLIC WATER SYSTEMS

OPERATIONAL GOALS OF WATER TREATMENT SYSTEMS

Most water systems in the United States and Canada are operated by local governments. It should be the goal of local officials to develop water systems that will meet the water demands for quality and quantity expected by the customers of the areas served by the utility, yet ensure that existing and future utility systems are constructed, operated, and managed at a reasonable or fair cost to the users while developing a system that is compatible with the area's future growth. Meeting this desire involves understanding the utility operating environment, performing planning activities, and making financial decisions that will provide appropriate funding to achieve the utility's needs. Water systems provide public health protection, fire protection, economic prosperity, and a high quality of life to their customers at a minimal cost (under $3.75 for every 1,000 gallons of safe water delivered to their taps according to the American Water Works Association) (AWWA 2012). That is 0.5% of the cost of bottled water. The operational model for functioning water systems appears to work since, for the most part, publicly owned water systems meet this goal in developed countries.

Water utilities are also highly regulated entities because human health and welfare are affected by drinking water quality (Bloetscher 2009). The 1974 Safe Drinking Water Act (SDWA) and its associated amendments are focused on protecting the public health from various contaminants in potable water supplies (Bloetscher 2009). Cost is not a consideration in the law—public health is the only consideration. Why? Because the public *expects* that the water is safe to drink so laws have been passed to ensure that their expectations are met. Whether surface waters, groundwaters, or a combination, the SDWA sets basic water quality parameters that must be met, which the vast majority of systems do on an ongoing basis.

Most local officials have little concept of the extent and value of the buried portion of the water infrastructure. Bloetscher (2017) found that about 60% of the value of a water system is in piping; meaning that even in systems with exotic and expensive treatment plants, the value of the buried infrastructure often exceeds treatment plant components. For the 40,000 small systems that withdraw groundwater and only chlorinate it, nearly 100% of the value of their infrastructure is in buried pipe. Hence the piping system is a very expensive asset that is located underground. The age of many of these pipe assets exceed 70 years; in some jurisdictions there are assets over 100 years old. For many communities that saw their major water system and sewer system components installed via the Works Progress Administration (WPA) during the Great Depression, their systems are 80 years old. The fact that they are still in service should be a concern for planning the future infrastructure needs of the community. It is also a testament to their ongoing operations and the engineering that went into their original construction.

SOURCES OF SUPPLY FOR WATER SYSTEMS

There are over 54,000 community water systems in the United States, up from 16,000 in 1960. The major growth is in small, rural utilities. The 500 largest water utilities serve half the nation's population, just as they did in 1960. Figure 3.1 outlines how most utility systems are configured, from raw water supply to the home.

Surface Water Sources

Less than 25% of the 54,000 water utilities in the United States use surface waters for their water supplies, but they serve nearly ⅔ of the population since the catchment areas for water and the associated impoundments can be very large. Most are older systems, first developed during the early 20th century. Surface water sources include any river, stream, lake, reservoir, ocean, or other body of water on the ground surface. However, surface water sources are limited

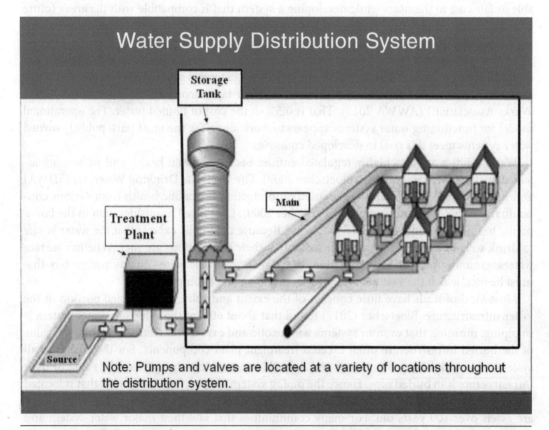

Figure 3.1 Basic water utility system. Water comes from a source (surface or groundwater), then is treated and stored in an elevated (or ground storage) tank before being pumped into a grid that is the distribution system that serves homes and businesses. *Source*: EPA https://www.epa.gov/dwsixyearreview/drinking-water-distribution-systems

in location—albeit modern technology has permitted the construction of artificial reservoirs to hold water. Surface water bodies are recharged by rainfall that runs off the land, reaching streams, rivers, or lakes. Springs are areas where the groundwater table intersects a surface water body, thereby contributing groundwater to the surface water body.

Regulatory permits are normally required to develop new surface water sources. The situation is more challenging in the western 18 states where water is regulated under the Prior Appropriation Doctrine, commonly deemed the water rights doctrine. In these states, water *rights* for new users must be obtained from existing users to create a new surface water source. Even in non-water-rights states, permitting is normally required because of the potential impacts on downstream users (communities, agriculture, ecosystems, etc.). Extensive permitting may be required if the use of those new surface water sources requires the construction of a dam because of the challenges posed as a result of potential adverse environmental impacts the dam creates due to the subsequent flooding of property and disruption of ecosystems—to say nothing of the displacement of residents and acquisition of their property for the reservoir. Communities like New York, Philadelphia, Cleveland, Cincinnati, Detroit, Chicago, San Francisco, and Los Angeles are examples of large surface water systems. Surface water systems are simply connections to rivers, streams, and reservoirs (see Figure 3.2). Often, they include a second artificial reservoir, or off-stream reservoir, that can be used in the event there is a spill of contaminants that would complicate treatment in the primary surface water (such as an oil spill). Many Ohio River utilities use off-stream reservoirs because extensive river traffic of goods increases the potential of an accident that creates an unintended release. The Ohio River Valley Water Sanitation Commission is a combined effort of these utilities that monitors water quality in the Ohio River and issues notifications to downstream utilities to divert to off-stream reservoirs if they detect a spill.

Surface water systems require precipitation. Precipitation requires evapotranspiration (ET) from the surface. Figure 3.3 is a map of ET rates in the United States from a National Oceanic

Figure 3.2   Large lake behind a dam in North Carolina used as a raw surface water source

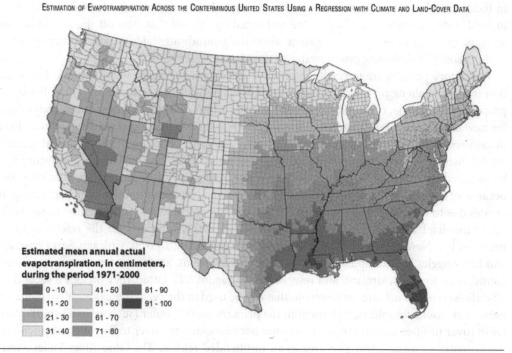

ESTIMATION OF EVAPOTRANSPIRATION ACROSS THE CONTERMINOUS UNITED STATES USING A REGRESSION WITH CLIMATE AND LAND-COVER DATA

Estimated mean annual actual
evapotranspiration, in centimeters,
during the period 1971-2000

0 - 10	41 - 50	81 - 90
11 - 20	51 - 60	91 - 100
21 - 30	61 - 70	
31 - 40	71 - 80	

Figure 3.3   Average annual evapotranspiration rate in the United States. *Source*: NOAA 2013

and Atmospheric Administration website (NOAA.gov)—there are few reliable, public access maps that include Canada and Mexico, but similar mapping may be available, at least locally, for both countries. Keep in mind these rates assume native groundcover, not altered terrain that results from pavement, buildings, deforestation, and other human impacts to the natural cover that will intensify ET yet reduce available water (due to increased runoff from asphalt, concrete, buildings, etc.). The map shows that in subtropical areas during the wet season or during summer months in northern latitudes, large bodies of water, including wetlands and estuarine areas, have high evaporation rates—meaning that surface reservoirs will lose water quickly during those months. The highest evaporation rates are associated with shallow, open water bodies. Groundwater, as much as four feet below the surface, may be subject to evaporation to some degree. This rising moisture forms clouds that condense and return the water to the land surface or oceans in the form of precipitation.

Precipitation occurs in several forms, including rain, snow, sleet, and hail. Figure 3.4 shows average annual precipitation in the United States. The intensity of the rainfall is a major area of hydrologic study by academics and scientists as it seems rainfall intensity is increasing with time. That is of concern because high intensity storms create a lot of runoff, but limited infiltration for groundwater recharge. Rainfall intensity can vary up to five inches per hour in subtropical areas, but is commonly two to three inches per hour across North America. High intensity storms increase runoff—the precipitation that runs off of the land, reaching streams, rivers, or lakes, contributes to the surface water. Surface water flow is essentially controlled by the topography because water on the surface flows in a downward direction, which means it generally flows toward the oceans (or the Great Lakes in the Midwest). Surface water bodies that mix with saltwater bodies along the coast are called estuaries; and where a river meets the ocean, a delta will form. These coastal systems are rarely used for water supplies due to the high salinity content in the mixed water.

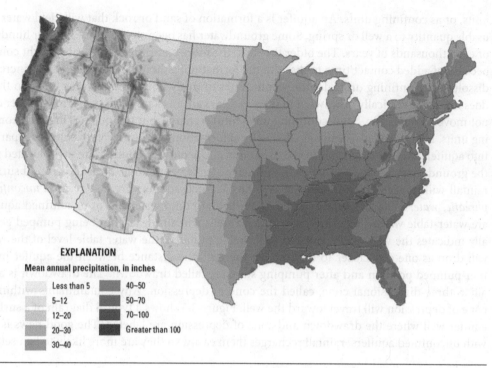

Figure 3.4   Map of average annual precipitation in the United States. *Source*: Reilly et al. 2008

Groundwater Sources

There are over 41,000 utilities that use groundwater as their source in the United States. Precipitation that infiltrates, or percolates, downward through porous surface soils is the primary source of water for groundwater. Areas that have geology that allows downward flow of water into the ground through infiltration are called *recharge areas*. The characteristics of soil depend on the soil-forming parent material, the climate, soil chemistry, types of organisms in and on the soil, the topography of the land, and the amount of time these factors have acted on the material. Aside from soil types, infiltration rates vary widely depending on land cover (forests versus grasslands for example), land use and development (impervious area creates more runoff and less infiltration), the character and moisture content of the soil, and the intensity and duration of the precipitation event. In developed areas, stormwater systems can become overloaded; the excess runoff means that infiltration lessens, so groundwater does not rebound—a major concern for groundwater users.

Infiltration rates can vary from as much as one inch per hour (inch/hr) or 25 millimeters per hour (mm/hr) in mature forests on sandy soils, to almost zero in clay soils and paved areas. When the rate of precipitation exceeds the rate of infiltration, runoff occurs. This is why development and deforestation are critical concerns in watersheds since both diminish infiltration and increase runoff which increases the loss of soil. Because vegetative types differ in their nutrient requirements and in their ability to live in water-saturated or saline areas, soil types also play a role in determining plant distribution and can be indicative of certain groundwater table conditions (AWWA 2014).

Below the unsaturated soil zone are a series of layers of sediments and rocks (including unconsolidated sediments like sand) that can be classified either as aquifers, semi-confining units, or as confining units. An aquifer is a formation of sand or rock that will yield water in a usable quantity to a well or spring. Some groundwater has been stored in aquifers for hundreds or even thousands of years. The older the rock, the more constituents the water might contain because of added contact time to dissolve the formation, although flow velocity also increases dissolution. Confining units do not permit water to freely flow—they *confine* water so that it does not move vertically downward in the soil. Clays are an example of such a soil—water does not move through clay. Semi-confining units limit recharge, but not quite as much as confining units. Hydrogeologists can help define local formations. Groundwater is further separated into aquifers that are either *confined* or *unconfined*. Unconfined aquifers are those located near the ground surface—and rainfall is the primary source of recharge. Near the ground surface, rainfall will impact the groundwater by filling it. Such aquifers are also termed *unconfined, phreatic, water table*, or *surficial aquifers*. Wells that pump water out of unconfined aquifers are water-table wells. The water level in these wells when they are not being pumped generally indicates the water table level. In wells being pumped, the water table level of the water will drop as one gets closer to the well. At the well, the distance between the aquifer in the non-pumped position and after pumping starts, is called drawdown. The drawdown is actually a three-dimensional cone, called the cone of depression. Any contaminants within the cone of depression will travel toward the well. Figure 3.5 shows a typical diagram of a surficial aquifer well where the drawdown and cone of depression are labeled. The good news is that with unconfined aquifers, rainfall recharges them easily, so they are more likely to be useful as

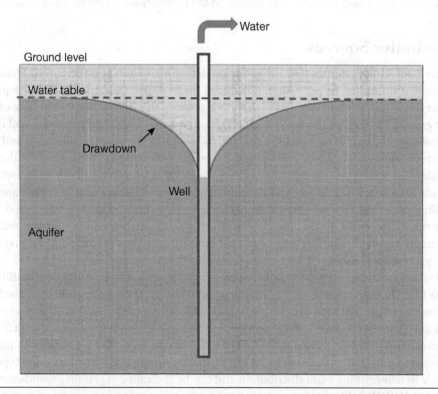

Figure 3.5   In an unconfined aquifer, the well penetrates the surficial or water table aquifer. When the well pumps, water will move toward the well (drawdown) creating a cone of depression around the well casing. The area in the cone of depression has been *dewatered*.

sustainable supplies as long as operators have a good understanding of recharge and withdrawals (withdrawals must be less than recharge rates).

Unconfined aquifers are normally characterized by water quality issues like organic constituents, nutrients, microbial activity (related to nutrients and organics in the soils), dissolved metals and impacts of surface activities like septic tanks, gas stations, and dry cleaners. As a result, while water quality may be more stable than surface waters that vary with rainfall, there are constituents to be removed. The presence of microbial activity, which thrives on the organics and nutrients in the water, increases the amount of degradation to well casings and pumps. Ongoing maintenance of wells is important for continuing operational reliability (more on this shortly).

If the aquifer is overlain by a layer of clay or other impermeable formation, it is termed a *confined* aquifer. In confined aquifers, water completely fills the formation, and the distance to the recharge area often causes it to be under pressure. The pressure can cause the water to rise in a well above the top of the aquifer formation. Such aquifers are referred to as artesian aquifers. If the water level in an artesian well stands above the land surface, the well is a flowing artesian well—no pump is required. Artesian wells can reduce the costs to pump water to the surface. Submersible pumps are designed for confined aquifers because they push the water to the surface, as opposed to sucking it up. By pushing the water, the water can be pumped up hundreds of feet to the surface. However, because confined aquifers are not exposed to the surface, they do not recharge with local rainfall. That means the aquifer may be drawn down gradually, which is a sustainability concern. The water is usually older than water found in unconfined aquifers, so metals are often prevalent, hardness is increased, and microbes should be assumed to be present. But, confined aquifers are less likely to be impacted by nutrients and surface activities.

Figure 3.6 shows these hydrogeologic formations. As can be seen in that figure, confining layers or confining units are geologic formations that water cannot flow through easily. The ability of water to flow through the porous media is referred to as hydraulic conductivity (K) or transmissivity (T). Clay layers (and some rock formations such as shale), although having high porosity, have *low hydraulic conductivity* and can be functionally impermeable. If the aquifer has low hydraulic conductivity, drawdown will be significant and the cone of depression will be steep. The higher the rate of flow, the steeper the drawdown and the larger the diameter for the cone of depression. Therefore, the types of rock strata that underlie the area have significant bearing on the quality and quantity of groundwater available and the ability of the rocks to absorb and store water (AWWA 2014). Engineers and hydrogeologists typically look for limestone, sandstone, and alluvial formations when siting wells because they tend to have high T. They avoid clay, shale, and similar zones which have low T (Bloetscher 2011).

Note that the amount of water available in the surficial/unconfined aquifer will vary as the water table changes. The impacts are less obvious in confined aquifers where pressure changes will need to be monitored in order to determine how much water is available. Too many utilities simply lower the pumps in confined aquifers, creating a long-term decrease in available water and the potential for depleting or mining the aquifer over time. This happened in the eastern Carolinas in the mid-1990s, and ultimately, the aquifer was no longer productive. Those utilities converted to surface waters, an option not available everywhere.

Under natural conditions, groundwater in surficial aquifers moves downward and then laterally along a confining unit until it reaches the land surface at a spring or through a seep along the side or bottom of a stream channel or estuary. In deeper aquifers, water flows toward the ocean. In many areas, the direction of groundwater movement can be derived from observations of land topography when the land slopes toward water bodies. Thus, the water table

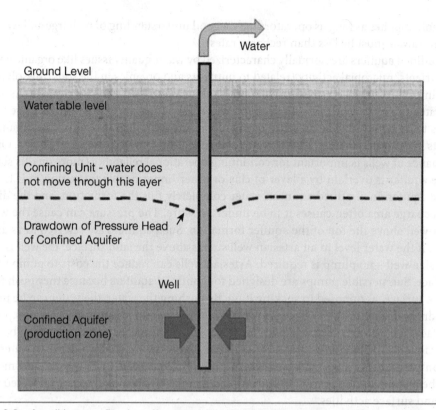

Figure 3.6 A well in a confined aquifer penetrates the confining unit and draws water from the lower aquifer. This has little effect on the water table aquifer. When the well pumps, water will move toward it horizontally, creating a cone of depression around the well casing, but water in the surface aquifer within the cone of depression will not be *dewatered*, only the pressure in the lower aquifer will be decreased (piezometric head).

usually follows the land surface. In confined aquifers, the recharge area is located some distance away and defines the pressure head. Changes upstream to land use can definitely alter confined aquifer yields. This pressure is termed the *potentiometric* surface.

Groundwater Issues

Kenny et al. (2009) note that water use has increased across all sectors (including agriculture which uses 40% of all water) through the 1980s. Pumpage has been relatively constant since 1980 due to better water-use technologies and water-conservation practices, but in many places groundwater use is significant. Unfortunately, most places do not have the luxury of abundant water supplies that, for example, Florida has. So, while water is available, it is unsustainable in most places. The west is a prime example—surface water supplies are limited, and until rural electrification in the 1930s and 1940s, pumping groundwater was not feasible. Electrification made wells and thereby extensive agriculture possible, so groundwater has been pursued both as a small utility solution and as a large-scale solution for agriculture. This seemed like an ideal solution to develop and farm the west since many thought that groundwater was a drought-proofing solution where surface waters varied too much or required too much treatment. But aquifers require recharge, and areas with limited surface water and/or rainfall do not have recharge. To address the issue, utilities and the U.S. Geological Survey (USGS) use small wells to monitor water levels (called monitoring wells). The continued lowering of groundwater levels that the USGS and state agencies often see in monitoring wells is an indication that recharge is often overestimated—giving a false picture of water availability. If an aquifer declines year after year, it is not from drought—it is because the aquifer is being *mined*.

Several important issues arise with aquifer systems: the safe yield of the aquifer, water quality, and potential contamination. One problem with pumping groundwater from confined aquifers is that our means to assess recharge are not very good. Reilly et al. (2008) outlined the condition of groundwater in the United States and found that the loss of groundwater supplies is due to overpumping in many areas. They identified areas where the ET rate is higher than the rainfall, meaning net rainfall (rainfall minus ET) for crops and other purposes is not available. Most of these regions have used groundwater since the advent of rural electrification. But the surface activities that require significant water from confined formations that do not recharge quickly does not constitute a sustainable practice. In many of these areas, streamflows are variable and limited, so groundwater is used to insure water supplies for crops and people. Thus, what USGS showed (Reilly et al. 2008) was that while water is currently available, it is being used at an unsustainable rate in most places. The result is an even bigger problem—overpumping. Throughout the West/Southwest, the Plains states, the upper Midwest (WI, MN, IA), and the Southeast (SC, NC), the crop circles resulting from rotating irrigation systems that pump groundwater can be seen from the air (see Figure 3.7).

Unfortunately, there is a lack of willingness to confront the overpumping issue in many areas. There are states with a lack of regulations on groundwater pumping, water laws that inhibit responsible use, and groundwater modeling that is limited to larger utilities, while smaller, rural systems may be most in need of such modeling due to competing interests. Meanwhile, the groundwater is extracted, used, and often discharged to waterways as a result of runoff or utility use. The amount is significant enough that retired USGS scientist Leonard Konikow (2015) suggested that rising sea levels had a small contribution from groundwater extraction.

Figure 3.7 Crop circles in the Midwestern United States.

A significant consequence of groundwater development from unconsolidated or highly friable rock formations can be a collapse of the land surface, called subsidence. Subsidence can occur when the groundwater, which exerts pressure on the adjacent soil and rock, is removed, thereby relieving the pressure. The formation then collapses, causing the surface topography to be altered. The collapse is caused by the reorientation of aquifer grains as a result of the loss of pore pressure exerted by the water in the aquifer. Recharging an aquifer that has collapsed will not restore the land surface to its prior state because the collapsed formation has less void space after the collapse. These collapses in limestone can create sinkholes—a phenomenon found in central Florida (among other states). Hence, development of groundwater sources needs to include consideration of possible land-surface subsidence, especially from overdevelopment. The magnitude of subsidence in areas subject to flooding either by tidal inundation or alteration of surface drainage should be estimated, especially in aquifers with high clay or sand content, or where prior subsidence has been noted. Faults may be an indicator as well. Where subsidence has been well documented, subsidence data may be coupled with the amount of compressible material to determine compressibility. Unfortunately, the information needed is not available in sufficient detail in most areas. Hence, part of the operation and maintenance of wells is observing the specific capacity, tracking aquifer levels, and monitoring water quality to ensure that the groundwater is not being mined.

A secondary issue is the water quality of groundwater. For many years, groundwater was assumed to be pristine, unaffected by surface activities and therefore safe to drink without treatment or disinfection. Undisinfected groundwater is delivered to customers in some places. Alamosa, Colorado found out about the downside of this situation in 2008 when *Salmonella* sickened many customers. Disinfection would have resolved the issue; they now disinfect (see Chapter 14 for more discussion). While some groundwater sources are safe to drink without significant treatment, the public's perception of purity is misplaced since there are few *pure* groundwater sources. As many as one-quarter of the wells associated with an Environmental Protection Agency (EPA) survey in the 1960s were found to be contaminated. Contaminants, like arsenic, solvents, pesticides, and hydrocarbons are sometimes present in sufficient quantities to render the water supplies unsuitable for drinking water. In addition, in stark contrast to public perception that aquifer environments are *pristine*, bacteria exist naturally in most aquifer systems; almost any aquifer with an organic content will have bacteriological activity to some degree. Some of these bacteria will be pathogenic to humans. Bloetscher, Witt, and Dodd (1997) note that they have never found an aquifer that is free of bacteria.

Still, groundwater often requires less treatment than surface water treatment and groundwater is considered more aerially extensive (extends over large areas). Since most groundwater systems serve under 500 people, the perception is that their impact of groundwater availability is small. But while aerially extensive, cumulative withdrawals over a larger area can lead to aerially extensive drawdowns (like the 1990s in the Carolinas or southwestern Utah in the 2000s). A few exceptions exist that have the fortune of being located above aquifers that have plentiful water supplies which are recharged constantly by rainfall (southeast Florida being one such example).

Groundwater Wells

Groundwater supplies are developed by constructing wells into an aquifer and pumping water out of the aquifer. There are thousands of wells drilled for potable water supplies by community water systems in the United States. The means to drill those wells varies by region, well needs, and aquifer formation. Table 3.1 outlines well-drilling methods that are commonly practiced. As part of the proper operation of any well, the gathering, compiling, and recording of a wide variety

Table 3.1 Common techniques used for drilling wells

Drilling Technique	Advantages	Disadvantages	Speed
Cable tool	Accurate samples Stable borehole Fluids not required No disposal Not complex Lost circulation can be identified Depth known Casing installed as you go	Loose material caves around bit Slow Limited depth Limited diameter	Slow
Jetting	Used in sand Water needed under pressure	Small diameter wells only Shallow Disposal of material Cannot obtain samples	Fast
Reverse air	Accurate samples Stable borehole Fluids not required No disposal Yield can be determined in place Lost circulation can be identified Depth known Casing installed as you go	Small holes Shallow	Fast
Reverse rotary with fluids	Large wells Use in a variety of formations (including sand) Aquifer protected Economical Screens and casing not required during drilling but can easily be set	Drill fluids need to be disposed of Cannot get good water samples Water/mud pits required Space needed Need experience Lost circulation possible	Fast
Direct rotary	No water required Borehole stable during drilling No lost circulation Water samples possible Unconsolidated materials ok	Operational noise Cost	Fast
Direct rotary with fluids	Large wells Variety of formations (including sand) Aquifer protected Economical Screens and casing not required during drilling but can easily be set	Drill fluids need to be disposed of Cannot get good water samples Water/mud pits required Space needed Need experience Lost circulation possible Cost	Fast

of data must be performed to develop an operating history that can be useful for long-term decision making and as a means to detect changes in the aquifer, well equipment, or water quality. Maintaining wells and gathering data will help utilities identify where there may be concerns. Monitoring water levels on an annual basis, tracked over the years, should be required practice for all groundwater users. That way one can determine if the aquifer is being mined. A loss of production efficiency indicates a need to schedule maintenance. Knowing trends and baseline information allows operations personnel to schedule maintenance at opportune times to avoid emergency shutdowns, to evaluate the cost of water production, and to schedule capital improvements. The forms used for record-keeping are not critically important—the key is that the records must be collected and maintained in a logical fashion regardless of the form that is used. AWWA's *Groundwater Manual* (M-21) contains information recommended for collection, and includes design, construction, and operational data. Included in the data collection should be:

- Detailed individual well (geologic) logs
- Well diameter
- Proposed total depth
- Position of the screens (or portion of the open hole if constructed in rock)
- Method of construction and materials to be used for the screen and casing
- Pump design
- Water-quality analyses
- Static (non-pumping) water levels in the aquifer
- Design pump discharge pressures
- Other data developed during the design phase

After a production well has been constructed, *as-built* records of the well should be recorded. These records should include:

- The method of construction used to drill the well
- The driller's log of the materials encountered during drilling
- Geophysical logs
- Diameters (and materials of construction) of well casing and screens
- Slot sizes of the screen (if a screen is used)
- The depths (settings) of the casing and screen
- Cementing information
- The total depth of the well

Figure 3.8 is an example of an installed well where the motor is above ground while the pump is down the well casing. Pumps should always be within the groundwater to prevent cavitation. Other motors may be connected to the pump below ground in what is referred to as a submersible pump, which is becoming increasingly popular. With submersible pumps, the pump and motor must remain submerged so the water cools the motor. With any well, pump data should include:

- The type (and make) of the pump installed
- The type and horsepower of the motor (driver)
- The pump setting (depth to the pump intake)
- The setting of the air line or other device for measuring the water level in the well
- Notation for the point (and reference elevation) used for measurement of the water level

Figure 3.8   This is an example of vertical turbine pumps in an unconfined aquifer in Florida that pumps to a water treatment plant. This is a typical well installation.

- All information provided by the pump and motor manufacturer, such as capacity and efficiency data

A newly constructed well with a new pump should be evaluated to create a baseline against which future data will be analyzed. This data should be used to determine well losses that relate to well design and construction. Comparative data pertaining to the physical condition of the pumped unit should also be collected, including:

- Water level measurements made before, during, and after the (drawdown) pumping test
- A record of the pumping rate
- Hydrographs generated during the test
- Any raw data collected (manual or computer generated)
- A copy of the hydrogeologist's report on the procedures and test results

The well production rate is usually determined by a hydrogeologist, based on aquifer tests performed at the time of drilling. Overpumping can damage the well by reducing the water level and production capacity of the groundwater system. In granular formations, the water-bearing formation may consolidate.

Records of water levels in the well during periods of non-use (static) and during pumping should be recorded to provide a baseline for determining the amount of drawdown. The static levels can identify changes in the amount of water that may be available in the aquifer with time or at any given time. The total pumpage for each well should be recorded daily and reported monthly on monthly operating reports (most regulatory agencies require some kind of report like this anyway). These numbers can be graphed to illustrate the seasonal and yearly production rates and for trend analysis to project future water withdrawal rates, and to monitor the actual volume of water produced from each well to predict periods between maintenance. Data can usually be recorded from a totalizer on flow meters that have been installed in the discharge piping for each well.

Since temperature is often indicative of changes in flow regimes in aquifers, the groundwater temperature should be recorded and plotted. As the temperature of the groundwater varies, the capacity of the well fluctuates due to the viscosity of the water. In projects where recharge to the aquifer may come from infiltration of surface water, the temperature of the adjacent surface water body should also be recorded.

Operations personnel should evaluate any well failure or long-term decline in performance to determine if physical or mechanical problems are causing the decline. Specific capacity is the pumping rate per foot of drawdown, which is one of the best ways to evaluate performance. The specific capacity, or ratio of the yield of each well to its drawdown, should be calculated each year in order to identify the potential need for maintenance, plugging problems, or water supply concerns, although specific capacity depends not only on the T of the aquifer but also on well construction factors such as screen type, well diameter, degree of aquifer penetration, and degree of well development. When plugging problems occur, the drawdown increases, and therefore, the specific capacity decreases despite the fact that the total yield of the well may not decrease significantly.

The annual specific capacity analysis should occur only after the well is allowed to fully recover. Then, the well should be run for an hour to determine specific capacity. Because injection wells may clog more easily than withdrawal wells, this calculation should be performed more often so trends will be clear.

If the specific capacity decreases, it may be the result of a drop in the water levels or a reduction in pumping yield due to microbiological fouling, chemical precipitation, formation, well screen or gravel pack plugging, pump corrosion, and/or biofouling. Water level declines can be caused by regional water level declines or reduced hydraulic efficiency in the well, most commonly caused by plugging or incrustation of the borehole, screen, or gravel pack. Other specific yield problems may relate to:

- Changes in the water-bearing zone
- Insufficient development of the well at time of drilling
- Pump wear
- Impeller detachment from shaft

Other operational conditions are suggested, including design problems like overpumping (which results in lowering of the water table), clogging or collapse of a screen or perforation of a screen section, corrosion caused by poor selection of well materials, incrustation by biological agents or silt, and wear aggravated by excessive intake velocities. Wear due to the entrance velocity occurs as water passes through the well screen (or the edge of the formation depending on the type of well). As the entrance velocity increases, sand, silt, and colloidal matter can enter the flow stream. Generally, entrance velocities are designed to be less than 0.1 foot per second on the assumption that flow at this low velocity remains in the laminar condition, thereby minimizing turbulence around the well screen and precipitation of iron, manganese, calcium, and particulates at the well screen.

All wells can plug or foul due to hydrogeologic, geologic, engineering, or construction factors. The problems are usually physical, mechanical, or environmental. Several performance problems are caused by fouling or sand and silt production in wells. These problems and their likely causes include (Bloetscher, Witt, and Dodd 1997):

- *Water level decline in the well*
  - Reduced hydraulic efficiency in the well, most commonly plugging or incrustation of the borehole, screen, or gravel pack

□ Regional water level declines
□ Well interference or plugging of a gravel pack by sand, silt, or clay

- *Lower specific capacity*
  □ A drop in pumping water level
  □ Reduction in pumping yield due to incrustation, formation plugging, pump corrosion, or biofouling

- *Lower yield*
  □ Dewatering or caving in of a major fracture or other water-bearing zone
  □ Insufficient development of the well
  □ Lack of connection to water-bearing fractures
  □ Pump wear
  □ Impeller detachment from shaft
  □ Incrustation, plugging, or corrosion and perforation of column pipe

- *Sand/silt pumping*
  □ Presence of sand or silt in fractures intercepted by a well-completed open hole
  □ Leakage around casing bottom
  □ Inadequate screen and filter-pack selection or installation
  □ Screen corrosion
  □ Collapse of filter pack due to excessive vertical velocity and wash-out

- *Silt/clay infiltration*
  □ Inadequate seal around the well casing or casing bottom
  □ Infiltration through filter pack
  □ *Mud seams* in rock

Many of these performance problems indicate an issue associated with the design and construction of the well, for example:

- Poor selection of materials that leads to significant corrosion or collapse of the well casing or screen
- Poor construction—casing cracks or leaks, leaking or missing grout, misplacement of screens and gravel pack, or misalignment (enhanced corrosion or collapse can result)
- Lack of well development—poor well yield, turbidity and sand pumping, biofouling, incrustation, and excessive drawdown can result

Other problems include incorrect specification of pumps, suction breaks, and electrical surges. No pump should operate at a rate at which it breaks suction because that will cause severe damage to both the pump and the aquifer as a result of a cavitation. Surging in the well may collapse the well if it was not properly stabilized and can stimulate sand, silt, and colloidal activity or dislodge corrosion and precipitates. Air bubbles may be entrained into the wells, which can damage the distribution system piping by causing air pockets. The solution is not to close the valves to reduce pumping, but to remove a bowl or slow the motor speed. Lightning strikes and poor grounding may cause electrical surges that damage motors and pumps. Appropriate lightning attenuation should be installed where required.

Understanding these issues requires field investigations to confirm site-specific characteristics, usually by a registered professional hydrogeologist. Finding a location favorable to the utility and its customers is also important. Proper study and comparison of such data enables the operator (or consultant) to anticipate maintenance and repair needs. Well maintenance activities should also be recorded. This data can be used to predict times when maintenance needs to be performed, identify possible causes of well decline, and plan for annual budgets for well-field management. Therefore, these records should include:

- The dates that maintenance was performed
- Results of pre- and post-maintenance pumping tests
- Methods (and materials) used in the maintenance procedures
- Other factors such as the coloration of the pumped water, amounts of sand removed, odors, and water quality analyses

POTABLE WATER SYSTEM TREATMENT

Once the raw water supplies have been secured, water treatment normally follows. Drinking water will normally receive some form of disinfection (chlorine gas, liquid hypochlorite, or onsite chlorine generation) (see Figures 3.9–3.11). There may be more extensive treatment such as lime softening that is used for groundwater to remove hardness caused by calcium and magnesium in the raw water (Figure 3.12 shows a close-up of the center of a lime softening unit showing the mixing zone and the clarifier zone where the particles settle).

Many surface water systems have variable water quality (due to rainfall and runoff), and thus will require chemical use that will facilitate the particles settling. The means to do this includes

Figure 3.9   Installation with chlorine gas cylinders, which is one procedure to disinfect raw water in the water treatment process.

Figure 3.10 A hypochlorite system is another method to disinfect raw water in the water treatment process.

Figure 3.11 Onsite chlorine generators can also be used to disinfect raw water during the water treatment process.

the addition of coagulants like ferric chloride and alum in a flocculation basin, which causes the particle to conglomerate. Larger particles settle faster. The water then enters a settling basin where the particles are allowed to settle, prior to filtration and disinfection. The process is called flocculation and sedimentation (Figure 3.13).

Surface waters must be filtered per the Surface Water Treatment rules prior to disinfection (see Figure 3.14). Gravity filters will need to be backwashed to prevent plugging. Other options for treatment include ion exchange for metals and softening (Figure 3.15) and membrane filtration/reverse osmosis for brackish waters (Figure 3.16). All of these systems involve chemicals, pumps, and significant equipment that requires ongoing maintenance.

Figure 3.12 Example of lime softening treatment system in an accelator that shows the mixing area (lighter, yellow color) and the area where particles settle—the clarifier. Note that in lime softening the mixing creates a chemical reaction that is very fast, so the clarifier may be very small. These units remove hardness (dissolved, polyvalent positive ions).

Figure 3.13 Example of flocculation and sedimentation system at a surface water treatment plant. Note the flocculators move slowly beneath the water to prevent obvious mixing that might break up the floc. Flocculation is short; the majority of the time this system is undergoing sedimentation—the process where the suspended solids settle (sedimentation).

Figure 3.14   Example of a gravity sand filter with backwash troughs shown.

Figure 3.15   Example of magnetic ion exchange system (MIEX). These mixing systems are usually in stainless steel tanks and mix an ion exchange resin with the water to remove hardness and other metals.

Figure 3.16   Example of membrane system—these are 3 MGD skids. The white pressure vessels contain seven 40″ in length and 8″ in diameter membranes that remove dissolved solids such as salt.

Fortunately, most water treatment plants and chemical feed equipment have manuals that include manufacturer literature, parts lists, operation parameters, and other useful information. It is critical that water plant operators have access to these operations and maintenance (O&M) manuals and that copies are available in archives and at the plants (as opposed to a file cabinet or storage that is not accessible). The manuals will often outline common problems that may be encountered and their corresponding solutions. They will normally include suggested maintenance schedules: for example, bearings should be greased at regular intervals, packing and O-rings may require replacement or maintenance if pumps leak, and parameters should be used to check if motors are operating within tolerances. The O&M manuals can be invaluable to prevent problems or diagnose them if they occur. Also, all warranties for equipment and the parameters by which the warranties will be honored are explained: for example, failure to grease bearings may invalidate the warranty. Certified operators and engineers have significant information on how to operate plants. The plants must submit monthly reports, perform water quality and process testing, and track chemical use on an ongoing basis. Those discussions are outlined in numerous texts and will not be discussed further here but note that maintenance schedules that match O&M manuals, documentation of the parts and maintenance needs, tools and training, and work orders to track maintenance are all items that should be in place in all water utilities. This is a critical preventative maintenance function at water plants!

Once the water is treated and disinfected, the water is generally stored in ground storage tanks (Figure 3.17) and pumped into the water distribution system. A series of high-service pumps (Figure 3.18) send the treated water to customers through an extensive series of underground pipes or into elevated storage tanks.

Figure 3.17   Example of concrete ground storage tank where treated, disinfected water awaits distribution. Ground storage requires high-service pumps to pressurize the system—here they are located in the center left.

Figure 3.18   An example of high-service pumps (blue), with the motor control center in the white building. These pumps will send the treated water to homes and businesses through an extensive series of underground pipes or into elevated storage tanks.

WATER DISTRIBUTION PIPING SYSTEMS

Forkman (2012) estimates that there are over 1.2 million miles of water pipe in the United States and Canada, ⅔ of which are eight inches or less in diameter. Over 50% of these pipes are cast iron (CI) or ductile iron (DI). One quarter are polyvinyl chloride (PVC), although the percentages vary widely by region (the northeast for example is 90% CI or DI pipe). Forkman (2012) reported that 43% of all water mains are between 20 and 50 years old, and 22% of all mains are over 50 years old. He also reported that the average age of water mains that fail is 47 years old, although he does not indicate materials for these mains, and the average age of pipe can vary widely between communities (Forkman 2012). Noncorrosive materials like PVC have an estimated life of over 100 years (Burn 2006), but these pipes have been only installed in the last 30–40 years. The WPA pipes from the 1930s are far closer to 80 years old, and pipes may be even older in more established communities in the northeast and Great Lakes states. Forkman (2012) and the EPA (2002) suggest that over 8% of installed water mains are beyond their useful life.

Water distribution systems are constructed with a variety of materials: older systems are often CI, but a few wood water mains exist. Forkman (2012) reported that the newer water mains are primarily DI and C900 PVC, although with the advances in directional drilling, high-density polyethylene (HDPE) pipe has increased in usage. All of these commonly installed pipes are shown in Figures 3.19–3.21. DI is the primary pipe for larger mains and may be preferred regionally—for example, virtually all of the New England states use only DI. Materials that are no longer used include asbestos concrete because of shear breaks, brittleness, and concern about exposure to asbestos fibers when maintaining the pipe, and galvanized iron (zinc coated iron) because of galvanic and soil-induced corrosion—but that does not mean that many miles of these pipes are not still in use. Newer pipes are generally more durable and less likely to corrode than older pipe, but corrosion, tubercles, breaks, and other failure modes periodically occur in DI pipe and various breaks can occur with PVC.

Figure 3.19 DI pipe is used as water mains in water distribution systems. DI pipes will have a black coating of epoxy paint on them. This is the bell end—DI used in water systems is typically a push-together pipe.

Figure 3.20   Blue, C900 PVC pipes that are used as water mains in water distribution systems. These pipes push together and require no glue.

Figure 3.21   More recently, HDPE pipes are being used as water mains in water distribution systems. These are a black plastic that must be butt welded together.

Water distribution systems are generally constructed in grids (see Figure 3.22) with strategically located valves that can turn off small portions of the system for maintenance purposes. The grid means that water is provided from multiple directions and minimizes disruption to customers when emergencies occur. The number of valves is usually the number of *legs* minus one. Hence a standard grid intersection has four *legs*, and requires three valves at a minimum (see Figure 3.22), although some utilities may require additional valves. Valves also should be located no more than every 1,000 feet. The standard for pipes under 16 inches in diameter is a gate valve (see Figure 3.23). Looping of lines in a grid to eliminate dead ends is also important for fire flow and water quality purposes.

The larger pipes are transmission mains and are designed to serve a region. Transmission mains are installed to move large quantities of water from one point to another and are often connected to storage tanks, reservoirs, the treatment plant, or pumping stations. Typically, utilities will define transmission versus distribution pipelines. In many cases, pipes above 12 inches

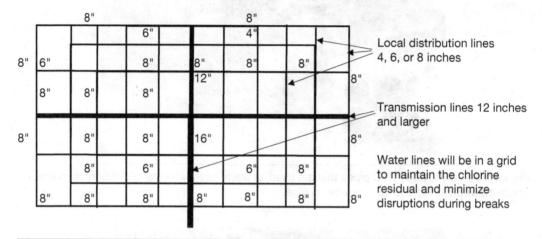

Figure 3.22   Example of water distribution grid showing transmission and local distribution lines with their respective diameters in inches.

Figure 3.23   A gate valve that can be used to stop the flow of water in a water distribution system—as the nut turns, a disk rises or falls to open or close the valve.

are deemed transmission, but this may have more to do with the size of the utility than a strict definition. Transmission mains normally do not have services connected to them.

Local water distribution mains are the lines normally found in front of houses. The piping in the local distribution system should be large enough to meet the maximum domestic and industrial use by customers, provide ample flow for fire protection, and allow for future expansion. Since fire flow is almost always the largest demand, it usually determines the pipe sizes required in the system. Six-inch and eight-inch diameter mains are installed in most residential areas because they are considered the minimum size that will provide adequate fire flow. Occasionally, fire flow capacity cannot be provided due to economic reasons. This situation exists in some rural areas where homes are far apart and can only be served practically by small-diameter pipes that furnish only domestic needs. Such lines should not have fire hydrants attached to them because the fire pumps can collapse the lines if pumped beyond the capacity of the pipelines to deliver water. There are also some rural subdivisions that do not have a water distribution system and rely on lakes or ponds for fire protection. Loxahatchee Groves, Florida, is an example of one such development.

Fire hydrants and flows in pipes are a major part of the fire insurance evaluations for communities. Most communities require fire hydrants 300–500 feet from every household and require a flow of 500–1000 gallons per minute (gpm) available. Hydrants are normally spaced closer together and provide higher flows in commercial areas. Many local jurisdictions have standards for fire hydrant spacing. Fire hydrants attach to six inch lines and include a gate valve to isolate the fire hydrant to facilitate maintenance. Fire hydrants must be accessible (see Figure 3.24).

Figure 3.24   Fire hydrants must be easily accessible for maintenance and fire crews; although, there are sometimes other challenges regarding access (like this alligator).

Pipe Materials for Water Distribution Systems

Failures of these pipelines, especially large ones, potentially will cause road and property damage as well as service disruptions, so a proactive approach is needed. CI pipe has been used for water mains for many years—some CI pipe is more than 100 years old and still provides good service. CI pipe use diminished in the 1950s when it was replaced with DI pipe. Many water systems built with WPA funds during the Great Depression were made of cement lined CI. An ongoing issue with these older CI pipes is that the joints were often made using lead that was either driven or poured into the joint to seal it. Periodically the lead pops out and the pipes leak. To repair these lines, the lead must be driven back into the joints and bell-clamps installed. Lead joints in high traffic areas are particularly susceptible to leaks due to truck vibrations on the pipe causing the joints to flex. CI is brittle under certain loads or freezing.

Some older CI pipe had no lining and under some water conditions, moderate to severe *tuberculation* occurs in the pipe. Tuberculation involves the build-up of rust/bacteria in mounds on the pipe interior. Tuberculation reduces the capacity of the pipe both because of constriction of the opening and the added roughness of the walls. When tuberculation seriously restricts flow, the main must be mechanically cleaned or replaced. Note that tubercles often include a biological component that facilitates transfer of electrons during the corrosion process. Biofilms will form, and bacteria will periodically be shed from such pipelines. For health reasons, chlorination of pipes is a continual process to combat bacteria in pipes. This is discussed in more detail later in this chapter.

DI is the newer version of CI that is now in general use. The chief difference between DI and CI is the form of carbon within the metal matrix. Rather than the graphite flakes found in CI, carbon in DI is formed into round nodules which have less of a tendency to propagate cracks, making DI less brittle and less prone to longitudinal or circumferential cracking. The material appears to be about the same as CI except the walls are normally thinner. Most DI pipe is delivered with a cement coating to prevent interaction of water with the iron. Ductile joints can be mechanical, but are usually push-on joints with rubber gaskets. Mechanical joints have typically featured DI or CI flanges and bolts. To deal with the potential for corrosion, some utilities have replaced the old DI and CI bolts with stainless steel, figuring the stainless steel is less likely to deteriorate. This is not true. Stainless steel contains a relatively high amount of chromium which reacts with oxygen to form a protective coating that resists corrosion. Soils tend to be oxygen deficient, so the coating will be compromised, creating a dissimilar metals issue, and thereby a corrosion cell. Chlorides worsen the situation. Hence stainless steel may make the situation worse. Likewise, if not properly handled during delivery or construction, the pipe's exterior coating can be damaged, which will also create a galvanic cell that accelerates corrosion from the outside of the pipe (Ellison et al. 2014). The lining reduces the tubercle problem and rubber gaskets replaced the lead in the old CI pipe. AWWA standard C104 addresses DI pipe coatings.

Asbestos-cement (AC) or transite pipe is made of cement mixed with asbestos fibers for reinforcement (AWWA Standard C400). It was initially used extensively in the 1960s because of its light weight, ease of handling, competitive price, and relative resistance to corrosion. AC pipe constituted 14% of the U.S. water main inventory when its use was discontinued around 1980 (Logsdon and Millette 1981). However, when disturbed, AC pipe can become brittle and shear, especially when digging beneath it. Also, concern about asbestos fibers when sawing the pipe has dictated special procedures for workers. Many water systems are planning long-term replacement programs to remove it from service. Life expectancy for this pipe seems to be about 50 years, which means much of this pipe is ripe for replacement. Asbestos pipe is often a larger maintenance concern than older CI pipe.

PVC water main pipe has been used extensively in many water systems since the 1980s. PVC is light, flexible, easy to cut and work with, color coded, competitive cost-wise with DI, and does not appear to deteriorate in the ground (except in cases where there are volatiles in the soil from solvent or petroleum spills). The external diameter is the same as DI pipe, making it possible to connect with DI directly. AWWA has created standard C900 that should be used for PVC water distribution piping to ensure pipe quality. Thinner walled PVC pipe often does not provide the same life as thicker C900 PVC. C900 PVC is available in 4, 6, 8, 10, and 12 inch sizes and at least three pressure classes. C905 is the AWWA standard for PVC pipe larger than 12 inches. Ellison et al. (2014) noted that a greater percentage of PVC pipes fail in the first 20 years of use than in the next 20 years. This was confirmed in a previous survey funded by the AWWA Research Foundation (Moser and Kellogg 1994). Construction issues are indicated by early failure of the pipe. PVC needs proper bedding. Too often, contractors just dump backfill material onto the pipe. That causes the pipe to deflect into an oval shape as opposed to retaining its stronger circular shape. Note that DI pipe will also deform under high loads so proper backfill applies to both types of pipe and should be evaluated during the design phase.

Many water systems specify blue pipe for water, green for sewer, and pantene purple #51 for reclaimed wastewater. These colors have been established by the underground industry as standard colors. Location tape, if used, will reflect these colors as well. From a durability perspective, PVC pipe should not be stored outside for any period of time—sunlight quickly degrades the pipe's strength. Discolored or pipe that has not retained its original color should be rejected on construction sites for new installations and discarded from utility parts yards. Well-bedded PVC should provide a life similar to DI.

Galvanized iron was a common material for small diameter pipelines (three-inch and smaller) and service lines in the 1950s and 1960s. Galvanized iron is simply iron pipe with a metallic coating of zinc. If damaged, the zinc coating provides a point for corrosion to start. Experience throughout the southeast indicates that the acidic soil conditions do not promote long life of galvanized pipelines, so these small pipelines may be significant sources of leakage. They also may contribute to water quality problems as a result of deterioration of the iron from aggressive waters. Replacement of these pipelines should be a priority because sandy or clay soils are not conducive to indicating small leaks. As a result, these galvanized lines could be a source of significant leakage on a water system.

Larger size water mains are also constructed of reinforced concrete, steel, or fiberglass—depending on their intended use, type of installation, and soil conditions. AWWA publishes installation and material standards that should be followed for each type of pipe and fitting to ensure long, low maintenance life for field crews. Prestressed concrete pipe (C303) and steel are used for large transmission piping, but not for local water mains. In reality, the term *concrete pipe* is misleading because except for AC pipe, the remaining concrete pipes get their strength from the steel within the pipe. Hence the focus of design and construction is on protecting this steel from corrosion. Prestressed concrete cylinder pipe (PCCP, ANSI/AWWA C301) uses high-tensile, high-carbon, small-diameter prestressing wire. Failure of the external concrete makes the steel much more vulnerable to corrosion and embrittlement.

Steel pipe has the same issues as any steel in a water system might—the risk/likelihood of corrosion. Most steel, including stainless steel, is not buried as a result. Fiberglass is used in low pressure situations and in plants. While unlike steel, concrete, PVC, or DI, it can be easily constructed into any needed configuration; it is brittle and intolerant of a water hammer.

Table 3.2 outlines the common types of pipes used in water systems. Wood is not used any longer, so it is not included.

Table 3.2 Common types of pipe materials used in water and sewer systems

Pipe	Years Used	Tapping/Connections	Standard Lengths	Standard Sizes	Joint	Uses
Cast iron	1845–1950s	Direct, saddle	12–18 ft.	3–49 in.	Lead, gasket	Water, sewer primarily
Ductile iron	1948 to date	Direct, saddle	18–20 ft.	3–64 in.	Gasket	Water, sewer primarily
Steel	1850, with numerous upgrades to methods	Special fittings, welded into pipe	40–40 ft.	4"–12 ft.	Special rubber gasketed connections, welds	High pressure water
Polyvinyl chloride C900	1950s to date	Direct, saddle (preferred)	20 ft.	4–48 in.	Embedded gaskets, fused	Water, sewer, stormwater
Polyethylene	1970s to date	Special fittings, saddle	1,000 ft. depends on diameter	3/4–36 in.	Fused	Water, sewer, stormwater
Asbestos cement	1918–1980	Saddle	13 ft.	4 to 24 in.	Gasket	Water, sewer
Concrete cylinder pipe (non-prestressed)	Ancient	Manhole	8–40 ft.	0.5–20 ft.	Gaskets, none	Low pressure, stormwater, gravity sewers
Concrete cylinder pipe (prestressed)	Mid-20th century	Saddle	8–40 ft.	Up to 12 ft.	Embedded steel connection	Pressure water, sewer pipes with lining
Vitrified clay	Mid-19th century	Fitting only	4, 6, or 8 ft.	4–42 in.	Gaskets now, burlap or mastic in the past	Gravity sewer

Source: Beieler, R.W. (2013). Pipelines for water conveyance and drainage. ASCE: Reston, VA.

Depth of Pipe

The critical concerns in burying pipe are the prevention of freezing, the minimizing of temperature fluctuations (hot or cold) and protection of the pipe. As a result, the depth at which water mains are buried varies greatly throughout the U.S. Water mains can be buried at shallow depths in warm southern states (a minimum of 30 inches in most cases) because the only concern is physical damage. Mains are buried deeper where there is moderate ground frost and may be buried up to eight feet (2.5 meters) deep in northern states where temperature changes and frost may expand and contract the pipe. Depth of bury is an important consideration when specifying pipe. The weight of materials and potential vibrations may indicate a preference for DI pipe in such situations. The formula for load on a pipe is:

$$P_v = \gamma H = 0.33\gamma h + \gamma_w h$$

Where P_v is the earth load on the pipe (psi)
γ is the weight of the backfill (lbs./cubic inch (ci))
H is the depth of backfill (inches)
h is the water height above the pipe (inches)
γ_w is the unit weight of water (lbs./ci.)

Note that Marston's research indicates that very deep pipes also have added loads from the side of the trench that must be taken into account (Marston 1930). In addition, sewer and drainage pipes have no inside pressure like water mains, so the pipe wall and bedding must be able to withstand the full load.

Depth of bury is also an important consideration for construction and maintenance. Deep pipelines require significant costs for excavation to reach them. Figure 3.25 shows a comparison of a trench for a shallow and deep pipeline. Note much more excavation is needed for deep lines. More material removal means more time and money. Trench boxes will help reduce excavation and protect workers, but the trench box will slow things down.

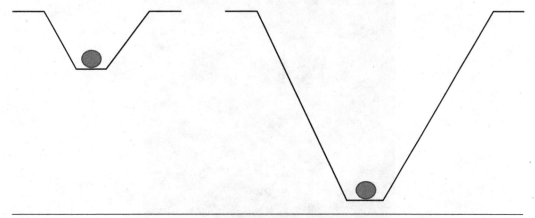

Figure 3.25 Example of excavation associated with shallow and deep trenches for the installation of water mains (pipes). Deeper pipes require far more excavation than shallow pipes and are therefore more expensive to install, maintain, repair, and replace.

CONSTRUCTION OF WATER DISTRIBUTION SYSTEM

Construction is a key to long pipe life, so trenches and pipe laying should be performed in accordance with engineering specifications (see Figure 3.26). Aside from poor construction, perhaps the most common cause of pipe failures involves five areas: (1) breaks due to freezing and differential settling (all piping), (2) pipes that burst (typically from a water hammer), (3) tubercles that restrict water (normally metal pipe), (4) corrosion (iron and galvanized), and (5) excessive age. Large pipe breaks may require contractors with specialized equipment, but the utility's water crews should be expected to be able to fix small pipe breaks. To accomplish this, an appropriate inventory of materials must be kept on hand—clamps, sleeves, and extra pieces of pipe.

Water main break rates should be calculated for all pipe materials used in distribution systems to create a means to judge pipe performance and durability. The location, type of pipe, dates, and other information are important. Water main break rates can vary year to year and by utility (Forkman 2012). A Water Research Foundation study of 20 cities found that 0.1 to 0.2 breaks per mile per year was considered to be an *acceptable benchmark* (AWWA 2001)—but did not answer the question about which pipes may fail. Pipes fail when the forces on the pipe exceed its remaining structural capacity (Marston equation). In addition, materials deteriorate through corrosion and other aging processes (see Table 3.3). The result is that designed-in safety factors are compromised until a failure occurs (see Figure 3.27). Unfortunately, most pipes fail from either bending or slow crack growth, which are often construction related and not related to design factors. The most common type of pipe failure is a circumferential crack or break and the cause is most commonly bending of the pipe like a beam (see Figure 3.28). Buried pipes are

Figure 3.26 Installation of C900 PVC water main in accordance with engineering specifications. It is being installed under an old galvanized main that has been abandoned.

Table 3.3 Summary of pipe failure modes (applies to all pipe)

Type of Deterioration	How Deterioration Occurs				Interaction	Other actions
	Load/use	Environment	Material	Construction		
Surface defects	Secondary	Primary	Primary	Secondary	Material and environment, extended by loading	Man-made defects
Deformation	Primary	Secondary	Primary	Secondary	Material, load, and environment	n/a
Cracking/breaks disintegration	Primary	Primary	Secondary	Secondary	Load, environmental degradation	Corrosion damage, accidents, lack of maintenance
Failure (aging/inadequate structural capacity)	Structural deficiency caused by surface defects, deformation due to construction, cracking/load exceedance, obsolescent	Structural deficiency caused by surface defects, deformation due to construction, cracking/load exceedance, obsolescent	Structural deficiency caused by surface defects, deformation due to construction, cracking/load exceedance, obsolescent	Structural deficiency caused by surface defects, deformation due to construction, cracking/load exceedance, obsolescent	Natural causes	Capacity and safety, failure to maintain
Catastrophic failures	Natural disasters, accidents, hazardous materials, man-made actions	Natural disasters, accidents, hazardous materials, man-made actions	Natural disasters, accidents, hazardous materials, man-made actions	Natural disasters, accidents, hazardous materials, man-made actions	Poor construction quality, design deficiency, fires	Fire, accidents, crashes

Figure 3.27   Example of corrosion and breaks in CI pipe. The inside of this pipe is in good shape but there is corrosion damage at the bells. The pipe shattered when the break occurred.

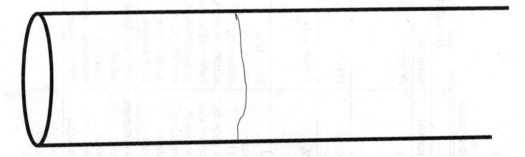

Figure 3.28   Example of a break caused by bending of the pipe like a beam (beam break).

generally designed to be fully supported, and many can handle only a small amount of bending deflection before breaking or leaking. Fatigue likely plays an overlooked role in the initiation and propagation of cracks in brittle pipe materials creating longitudinal cracks (see Figure 3.29). Pressure cycles and traffic loadings can be sources of fatigue loadings. While pump starts and stops are obvious causes of pressure cycles, diurnal pressure fluctuations throughout the system may also lead to failures (Bardet et al. 2010). When rocks are not removed from the backfill, they will migrate to the pipe and create pinholes and breaks as a result of surface vibrations (see Figure 3.30). Table 3.4 outlines the basic causes of pipe breaks and appropriate corrective actions. Table 3.5 addresses challenges and solutions for water distribution systems. Note that this information applies to all pipe; including the sewer and drainage pipes covered in Chapters 4 and 5. Gaining useful information on pipe requires some form of work order tracking tool to be in place, especially with respect to breaks.

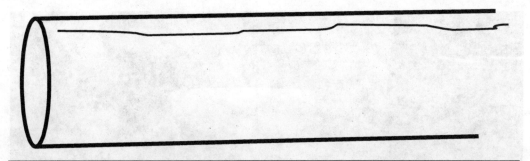

Figure 3.29   Example of longitudinal break in a water main—a water hammer can cause this type of break and explode the pipe.

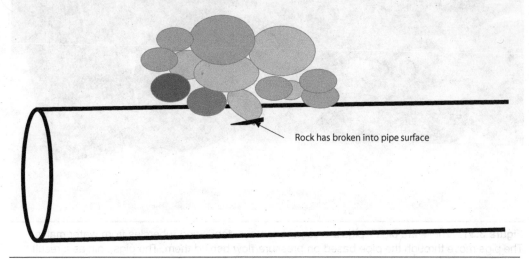

Rock has broken into pipe surface

Figure 3.30   Rock migration to the pipe can cause pitting failure from the outside of the pipe.

Table 3.4   Causes of breaks in water distribution pipe

Break Type	Cause
Circumferential	Thermal contracting, bending
Longitudinal	Excessive crush, internal, damage to surface
Hole	Internal pressure, biofilm corrosion
Bell crack	Thermal contracting, bending, joint material
Shear	Void under pipe

Table 3.5   Water distribution challenges and solutions

Challenge	Repair and Replacement Action
Water quality affected by tubercles	Main cleaning, pigging
Low pressure	Main cleaning (PM)
Valves not maintained, poor condition	Develop valve exercise program, replace non-working valves
Leakage in pipes	Leak survey, assess condition, service line replacement
Main breaks due to deterioration	New piping, CIP

Figure 3.31   These are pigs—pigs are used to remove debris and tubercles from water mains. The pigs move through the pipe based on pressure/flow behind them. The pigs spin as a result of their surface (see the black areas that are hard plastic or steel). The pigs spin and scrape the inside of the pipe.

Flushing programs remove sediments from pipes, reducing the risk of water discoloration, taste and odor problems, biofilm regrowth, and regulatory non-compliance. Pigging (Figure 3.31) will help remove tubercles and biofilms from sections of pipe. Care should be taken to track the pig because they can get stuck and create water hammer conditions. Ellison (2014) discussed measures for determining pipe cleanliness, before and after flushing projects (and similar interventions), including customer complaints; turbidity, color, bulk iron from sampling stations (not from hydrants), and heterotrophic plate counts (HPC) that measure non-coliform bacteria in the water lines.

Water Hammer

A water hammer is one of the most destructive forces in a closed (pressure) piping system (see Figure 3.32). A water hammer results from the start-up of pumps or shutdown of valves or pumps in a system without attenuation. To address this, many utility systems use some form of attenuation to dampen the water hammer. However, accidents, breaks, sudden failures, etc., can create a water hammer on a pressure pipe system. Even with attenuation, waves of high pressure fluids can impact pipe. Figure 3.33 shows the pressure wave of an attenuated pump, where the

Figure 3.32   This is what happens when a water hammer—a high pressure wave—occurs in a pipe. This 36-inch fiberglass pipe exploded.

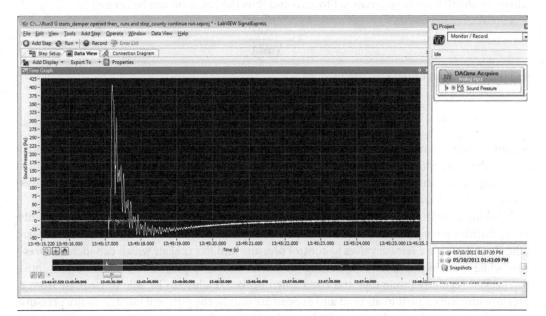

Figure 3.33   A water hammer pressure wave—this is a pressure wave that occurred after a pump with pressure attenuation started up. This instantaneous pressure exceeded 1000 psi. It was not measurable with the instrumentation on the pipe. The wave lasted about one second.

attenuation is designed to reduce the pressure wave from pump start-up—pressures still exceed 1000 psi instantaneously.

A water hammer is an unsteady flow condition. The fundamental equation for a water hammer is (Ghidaoui et al. 2005):

$$\Delta P = +/- \rho a \Delta V \quad or$$

$$\Delta H = +/- \frac{a \Delta V}{g}$$

where a is the wave speed, V is the average velocity of the pipe, and P is the piezometric pressure in the pipe defined by:

$$P = \rho g (H - Z)$$

where H is the piezometric head and Z is the elevation of the center of the pipe. A pipe stress solution is obtained with setting Bernoulli's equation to zero:

$$\frac{dV}{dt} + \frac{gdH}{dv} + \frac{4\tau_{auw}}{rD} = 0$$

where D is the pipe diameter and τ_{auw} is the shear stress at the pipe wall. When this value exceeds the design shear stress for the pipe material, a shear break will occur, permitting large volumes of water to escape.

The speed of the wave is defined as:

$$\frac{1}{a^2} = \frac{dr}{dP} + \frac{r}{A}\frac{dA}{dP}$$

where the dr/dP term is generally set to zero and thus the equation can be revised to:

$$A = \sqrt{\frac{AdP}{rdA}}$$

and the dissipation of the wave defined by

$$\tau_{auw} = \tau_{ws} + \frac{krD}{4}\frac{dV}{dt} - a\frac{dV}{dx}$$

Because a water hammer can be destructive, utilities normally attempt to include weighted check valves, pressure release valves, and vacuum valves in the system to address the potential for damage from water hammer waves. As a result, some effort is required to insure these valves operate as intended and that the appropriate relief systems work properly.

In addition, high points on lines, open water, canal or roadway crossings, and other points in the distribution system may require pressure relief. All pressurized piping systems need periodic air and pressure release valves to control pressure and to expel air which can increase pressure while reducing fluid flow (including pressurized sewer pipes). The control devices must be checked regularly (annually) and fixed when broken. Failure to be able to clear air from a pressure line will cause it to air bind and as pressure builds, the pipe will burst. A faulty pressure relief valve is as useless as having no valve; in both cases, the impact on the pipe from a water hammer is significant.

Water Main Upgrade Programs

Water main upgrade programs are designed to replace current pipelines—those that are small, galvanized, or provide insufficient service and/or no fire protection. Typically, the water utility replaces these small lines with six-inch or eight-inch pipelines made of PVC C900 or DI to provide fire protection. Leaky pipes and salt-immersed pipes should be a priority for replacement with appropriate pipe materials. PVC C900 and DI pipelines generally cost under $100 per foot (2016 U.S. dollars) to install for small pipelines (six inches to eight inches in diameter). Many can be installed by water system crews. Pipeline maintenance and upgrade programs should address low flow or low pressure problems in the system—existing, developed areas are the first priority. Other priorities should be focused on completing loops that will address water distribution system pressure and disinfection residual issues. Both of these programs have the benefit of increasing water sales to the areas where the loops are made or the lines upsized, since flows are no longer restricted.

Along the way are a series of fire hydrants, elevated tanks, and other appurtenances generally constructed of steel, DI, or brass. Valves and fire hydrants are mechanical devices that contain numerous moving parts and rubber gaskets that require periodic maintenance and will wear with age. They are the next issues to discuss.

Valves (AWWA Standard C500)

The AWWA creates the standards for water treatment plants in the United States. The types of valves permitted in water treatment plants and water distribution systems are AWWA Standard C500 Series. Valves are installed at intervals in water main piping so that segments of the distribution system can be shut off for maintenance or repair and as mentioned earlier, many utilities have their own criteria for valve placement. Gate valves are the standard on water systems. Valves should be located close enough so that only a few homes or businesses will be without water while a leak or break is being repaired. More valves mean that smaller areas can be isolated, but those valves must be operable. Water distribution system valves should be periodically operated as this provides an opportunity to ensure that valve boxes are exposed and have not been filled accidentally with dirt or damaged by paving or snow removal activities. The AWWA recommends performing this procedure annually. A valve exercise program ensures that the valves are open and work properly, and it loosens up the valves so that they will operate more easily. Systems with a large number of valves often purchase power valve-turning equipment to speed the job of *exercising* their valves.

Valve exercising programs may be the most ignored program for water systems—very few systems actually exercise valves because it is not seen as important. When valves are operated, the number of turns should be counted to make sure they are fully operated in both directions. However, many utilities, especially smaller systems, never perform this function. It is time intensive and as a result there are many other things that take priority. But when a major break occurs, the need for the valve exercise program gains traction since often only a few of the necessary valves work properly and large sections of the system must be shut down to make repairs.

Each valve should be installed with a valve box that extends to the ground surface and has a cap that can be removed so that a valve key can be used to operate the valve. Valves should, if possible, be located where the box is easily accessed and where damage by snowplows and other equipment is least likely. Cleaning the valve boxes of debris as a part of an exercise program insures that the valve is accessible (see Figure 3.34). Valves that do not operate properly or leak

Figure 3.34   Example of debris in valve box blocking access to valve nut (square).

should be dug up and repaired as soon as possible. Valves that cannot be repaired should be replaced when discovered to minimize outages.

The size of gate valves, the most common distribution system valves, is defined by the relationship:

Size of water main in inches = 3 × full rotations of the valve stem (turns) + 3

For example, a six-inch valve will have 21 rotations. Note that some valves turn clockwise and some turn counterclockwise. Operations personnel and engineers must insure consistency in the distribution system.

Fire Hydrants (AWWA Standards C502 and C503)

Fire hydrants are of two general types. The AWWA has a standard for each permitted in water distribution systems (C502 and C503). A wet-barrel hydrant (AWWA standard C503) is full of water at all times. Because much of the country is subject to freezing temperatures, there is the potential for the water to freeze in the hydrants (and subsequently shatter the CI pipe barrel). As a result, wet-barrel hydrants are less common than dry-barrel hydrants (AWWA standard C502). Figure 3.35 shows a cutaway of a dry-barrel hydrant—the long shaft connects the nut at the top of the hydrant with the valve that is at the bottom of the barrel. Figure 3.35 also shows the breakaway flange that is designed to permit the hydrant to break off and close the valve if hit by a vehicle (to eliminate the water spraying from a damaged hydrant as is so often seen in old comedy films). Also, near the bottom are a series of weep holes to permit the hydrant to drain after use. Figure 3.36 shows a cutaway of the top of a dry-barrel hydrant showing the

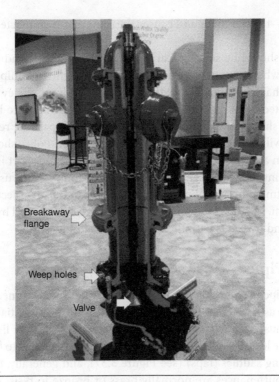

Breakaway flange

Weep holes

Valve

Figure 3.35 A display of a dry-barrel fire hydrant valve, with arrows pointing to the barrel, weep holes, and breakaway flange.

Figure 3.36 Cutaway of the top of a dry-barrel hydrant showing the thread mechanism below the nut (see arrow). The nut does not rise, but the threads open downward to open the valve at the base of the hydrant.

valve nut mechanism—the nut turns and causes the shaft that is connected to the valve to open downward.

Hydrant locations should be selected carefully. They should be readily visible and located near a paved surface where they will be accessible to fire-fighting equipment. Many jurisdictions have a Fire Marshal who will approve the location of hydrants. They should also be placed where they are protected from damage by vehicles and are less likely to be covered by plowed snow. Public officials should always insist that police enforce parking restrictions adjacent to fire hydrants, so they will not be blocked if needed to fight a fire. Police should also be reminded to watch for vandalism and unauthorized use of hydrants and to report incidents to the utility manager. Frequent painting with bright paint protects hydrants from rusting and makes them easy for the fire department to find. Well-maintained hydrants also project a positive public image of the water system. In some systems, fire hydrants are color coded by flow volume to help the fire department find hydrants with the greatest flows to fight fires.

Water Service Pipes

The small-diameter pipe used to carry water from the water main connection to an individual building is referred to as a water service line or pipe. For most residential uses, these pipes are small, although one must keep in mind the length of run of the service line as that may dictate the size due to pressure loss. Service lines deliver water from the mains to homes and businesses through surface tapping saddles (taps) (see Figure 3.37), and generally, through water meters (see Figure 3.38). Appurtenances are normally brass or bronze to resist corrosion (see Figure 3.39). Older service lines tend to be copper, galvanized iron, and in some cases lead. Newer service lines are often polyethylene (PE). In the late 1980s, the Lead and Copper Rule required utilities to test water inside customers' homes to determine the potential for metals leaching

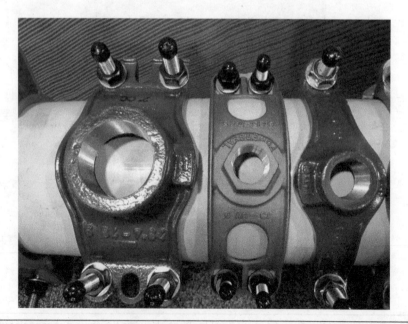

Figure 3.37   Examples of service tapping saddles; these are used to connect service lines to the main lines.

Figure 3.38 Displacement water meter that can be read electronically.

Figure 3.39 Most service line appurtenances are made of brass. This includes valves, couplings, backflow devices, and corporation stops, which thread into the saddles shown in Figure 3.37.

from corrosion. Utilities conducted these tests and made adjustments to their treatment practices to minimize metal leaching.

At some point on the service line there should be a meter. The meter may sit in a meter setter (Figure 3.40—includes a backflow device) or the piping (Figure 3.41). A water service pipe may range from ⅝ or ¾ inch (20 mm) in diameter for a small home to 6 inches (150 mm) for an

Figure 3.40   Example of a water service line and meter setter that will hold a water meter—with a shutoff valve (top left) and a backflow device (top right). The setter is included at a point along a water service line where the meter will be installed.

apartment building. An example is included in Figure 3.41. Large buildings and industries often have service and fire sprinkler pipes that are even larger. AWWA standard C800 was designed for service lines (C901 and C903 apply to plastic pipe).

Each water service pipe usually has a buried valve called a curb stop inserted in the line at a point at the edge of the public street or alley right-of-way or an easement. Where curbs and sidewalks exist, water system policies generally standardize the curb stop location at a set distance between the curb and sidewalk or at the lot line. The buried valve is fitted with an adjustable service box (curb box) that extends to the surface and has a removable cap so a valve key may be inserted to operate the valve (note meter setters have the valve included). The curb stop is primarily used to shut off the service if the building being served is vacant or repairs are needed. It is also a way of discontinuing service for nonpayment of the water bill. Local ordinances are usually enacted to address turn-off of services for nonpayment.

Water system policy varies on responsibility for maintenance and replacement of water services. Some systems require that all of the service, beginning at the main, be maintained by the property owner. Other systems require property owners to maintain only the portion beyond the curb stop, lot line, or meter box. The utility system policy should be made clear by ordinance or approved policy resolution.

Water service pipes are generally made of lead, galvanized iron, copper, PVC, or PE. Lead was the material most easily worked and available for small pipes when the first water systems were developed. Lead water service pipes are still in use in many older systems. Lead service lines are relatively flexible and resist corrosion but gradually become more likely to leak or break as they get older. Lead leaching is a serious health issue for residents. Flint, Michigan (2014) and DC Water (early 2000s) had issues with lead, which translated to children with high lead levels in their blood. Lead is no longer used for service lines, but millions may still exist in water systems. In part this is because service lines are repaired rather than replaced when they leak.

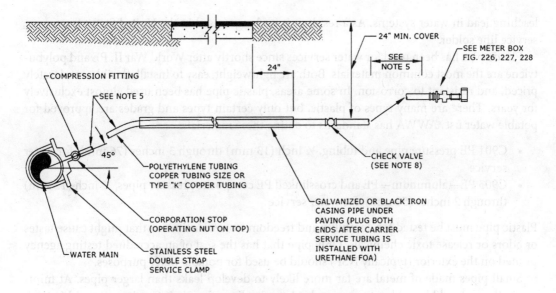

1. SUCCESSIVE TAPS INTO THE WATER MAIN SHALL BE SPACED NOT LESS THAN 18" ON CENTER. NO TAPS SHALL BE CLOSER THAN 18" TO A JOINT.
2. 1" SERVICE REQUIRE A 2" MINIMUM INSIDE DIAMETER CASING PIPE.
3. 2" SERVICE REQUIRE A 3" MINIMUM INSIDE DIAMETER CASING PIPE.
4. ALL CASING PIPE SHALL EXTEND A MINIMUM OF 2' BEYOND THE EDGE OF PAVING STREETS.
5. FOR 1" SERVICE LINES THE MINIMUM RADIUS SHALL BE 14".
   FOR 2" SERVICE LINES THE MINIMUM RADIUS SHALL BE 21".
6. ALL CASING PIPE ENDS SHALL BE FILED SMOOTH WITH NO BURRS AND SEALED WITH URETHANE FOAM.
7. THE POLYETHYLENE OR COPPER TUBING SHALL BE ONE CONTINUOUS PIECE FROM THE CORPORATION STOP TO THE CHECK VALVE. NO JOINTS WILL BE PERMITTED BETWEEN THESE POINTS.
8. THE CHECK VALVE IS TO BE INSTALLED 5 FEET BEFORE THE METER VALVE.

THIS DETAIL APPLIES ONLY TO RESIDENTIAL ROADS WITH LESS THAN 70' R.O.W. (NO MEDIAN) OR WITHIN EASEMENTS

Figure 3.41 Engineering design of a service line installation to a house. *Source*: Broward County, FL

New federal regulations, designed to protect the public from the danger of lead in drinking water, require systems to ensure that leaching of lead from water services is minimized. Systems with aggressive water that tends to dissolve lead may have to install additional chemical treatment to meet the requirements. Systems that cannot adequately control the leaching of lead may be forced to remove existing lead service pipes and replace them with other material. Galvanized iron pipe, used for water service piping for many years, corrodes very quickly in some types of soil. This material should be replaced since it is a source of leakage in the system. The same issues apply for all galvanized pipe, although the service lines are smaller than galvanized mains, and tend to deteriorate more quickly.

Copper pipe came into use in the early 1900s and gradually became the preferred material in many parts of the country. Copper is flexible, fairly easy to install, resistant to corrosion, and lasts almost indefinitely under most water and soil conditions. However, copper can also leach from the service to water supplies where the potable water is aggressive. Lead solder was also used until the mid-1980s for connecting copper pipe. This lead solder was noted as a source of leaching lead in water systems. As a result, it is no longer used in the United States for copper service line solder.

Plastic pipe has been used for water services since shortly after World War II. PE and polybutylene are the most common materials. Both are lightweight, easy to install, flexible, moderately priced, and resistant to corrosion. In some areas, plastic pipe has been used almost exclusively for years. There are many types of plastic, but only certain types and grades are approved for potable water use. AWWA has standards to define these materials:

- C901 PE pressure pipe and tubing, ½ inch (13 mm) through 3 inches (76 mm), for water service
- C903 PE—aluminum—PE and crosslinked PE composite pressure pipes, ½ inch (12 mm) through 2 inches (50 mm), for water service

Plastic pipe must be tested for durability and freedom from constituents that might cause tastes or odors or release toxic chemicals. Only pipe that has the seal of an accredited testing agency printed on the exterior (typically NSF) should be used for potable purposes.

Small pipes made of metal are far more likely to develop leaks than larger pipes. At minimum, there should be a plan to remove lead service lines. Lansing, Cincinnati, and Madison are examples of water systems that have programs to accomplish this replacement. In reality, because of the cost for repairs versus replacement, in most cases service lines should not be repaired—it is much easier just to remove them and replace them. Utilities spent thousands of man-hours to repair these lines. Many times, utilities will find service lines have been fixed previously. Repairs are often more costly than replacement.

The costs to repair service lines will vary across the United States depending on material, bury, and location of the break. However, an estimate can be made and then be adjusted according to local conditions. Installing a new service line using traditional methods would involve digging across the road (half of the time, anyway) and laying a new service line (a conduit is suggested to make it easier to replace in the future). Assuming a shallow service line in the street that is 30 inches in depth, the equipment required would be a tool to break through the asphalt (saw), a small excavator, backhoe, or vac-truck to dig to the leak, a couple of people, and tools. If a clamp were put on the leak (minimal cost), the full time for the repair would be about three hours and involve two or three men. Table 3.6 outlines the estimate for such a repair. A deeper repair would be more costly, as noted in Table 3.7.

Table 3.6 Estimated cost for typical shallow service line repair (2015 U.S. dollars, nonunion, south Florida)

Item	# of Units	Units	Cost/unit	Total Cost
Man 1 ($40k/yr + 30% Ben)	3	hr.	$ 27.00	$ 81.00
Man 2 ($34k/yr + 30% Ben)	3	hr.	$ 24.00	$ 72.00
Man 3 ($28k/yr + 30% Ben)	3	hr.	$ 21.00	$ 63.00
Backhoe	3	hr.	$ 125.00	$ 375.00
Clamp	1	ea.	$ 10.00	$ 10.00
Asphalt repair	2	sy.	$ 10.00	$ 20.00
Fill	1	cy.	$ 30.00	$ 30.00
				$ 651.00

sy. = square yard; cy. = cubic yard

Table 3.7 Estimated cost for typical deep service line repair (2015 U.S. dollars, nonunion, south Florida)

Item	# of Units	Units	Cost/unit	Total Cost
Man 1 ($40k/yr + 30% Ben)	5	hr.	$ 27.00	$ 135.00
Man 2 ($34k/yr + 30% Ben)	5	hr.	$ 24.00	$ 120.00
Man 3 ($28k/yr + 30% Ben)	5	hr.	$ 21.00	$ 105.00
Backhoe	5	hr.	$ 125.00	$ 625.00
Clamp	1	ea.	$ 10.00	$ 10.00
Asphalt repair	4	sy.	$ 10.00	$ 40.00
Fill	2	cy.	$ 30.00	$ 60.00
				$ 1,095.00

sy. = square yard; cy. = cubic yard

New service lines can be laid with PE as noted in AWWA standards, or copper. Both require bronze and/or brass fittings to connect to existing corporation stops and valves. Both have a long life although PE is preferred in corrosive soils or with corrosive water. Materials like galvanized iron, lead or lead solder, PVC, and others should be completely eliminated from the system. The costs for the traditional dig-and-replace option with new PE tubing connected to existing fixtures involves equipment similar to that noted in Tables 3.6 and 3.7. The patch and repair would take longer, and traffic maintenance can be an issue when cutting across a street.

Trenchless technologies include directional drilling, pneumatic moles, and pulling. All require digging a hole at the corporation stop on the main and at the meter or valve box. The direction drill uses augers to drill under the road. They can drill in directions other than straight lines. The directional drill would enter one pit and drill to the other. The process usually takes under 30 minutes. The service line is either fed through the borehole or pulled back with the drill head. PE is the preferred material to use due to its flexibility. Then the new service line is connected to the existing corporation stop and meter/valve. A two- or three-man crew, equipment to excavate, and a directional boring machine are required.

A pneumatic mole is similarly installed. A pneumatic mole is a cylindrical device connected to an air compressor that uses air to punch under pavement. The result is an opening that will allow the service line to be fed or pulled through by backing up the mole. Moles go in a straight line. Two pits, one over the corporation stop and one at the meter/valve, are dug. The pneumatic mole is attached to a standard air compressor. The mole is launched from one pit and compacts its way to the other. The service line is then attached to the mole and is backed up, putting the service line with it. PE is the preferred material to use in this application. The time to mole is 15–30 minutes. Then the new service line is connected to the existing corporation stop and meter/valve. A two- or three-man crew, equipment to excavate, an air compressor, and a mole are required.

A backhoe can sometimes connect to an existing service line (that is disconnected from the main line, but connected to the new service line), and using the backhoe bucket, *pull* the old service line, thereby laying the new line. This works if soils are soft and the pipe has remaining integrity. With this method, a pit is dug at the main and the meter/valve. The existing service line is disconnected. The new service line is connected to the old one. A backhoe bucket pulls one end by slowly curling the bucket. If the old line does not have too many clamps or too many weak points, if can literally be *pulled* through the soil. PE is the only replacement material applicable for this solution. The pull time can vary depending on the soils and compaction, condition of the existing pipe, etc. A two- or three-man crew and equipment to excavate are required.

In all cases, as a means to reduce water losses, reduce the potential for metal leaching, and reduce maintenance costs, it is preferable to replace the lines as opposed to fixing them. The costs to replace a service line may be little more than the actual repair, but the likelihood of a return visit is minimized.

Table 3.8 outlines the costs of all four methods. Note they are similar, but of real interest is the fact that none are much more costly than the repairs in Tables 3.6 and 3.7. The suggestion is that replacement of service lines is a better use of scarce utility resources than repair of service lines.

Water Metering

The hugely overlooked parts of water distribution systems are the meters. The principal uses of water meters on a water system are to record the amount of water that is treated and delivered to the water system, and to measure the amount of water used by customers. They are the source of water billing and, therefore, neglecting them costs the utility money directly. The two general types of meters used for individual services are velocity meters and displacement meters. Velocity meters are used when large quantities of water must be measured. These are large meters used to measure the amount of water admitted to and/or pumped from reservoirs and may also be used for customers that use large quantities of water. They lose calibration quickly and need ongoing maintenance to restore veracity. Displacement-type meters measure the number of times a container of known volume is filled and emptied. This method is more accurate than a velocity meter but is only practical for measuring relatively low flow rates. The nutating disk meter (Figure 3.42) is a type of displacement meter that is commonly used for water services because it is durable, relatively trouble-free, moderately priced, and quite accurate over the normal flow range of most water customers. Note that they never over-read.

Figure 3.42   A common nutating disk displacement meter—these are normally installed at residential service locations or small commercial locations.

Table 3.8 Comparative costs for service line replacement options

Cost Item	Cost/unit	Units	Option 1 Excavation to Repair a Service Line # of Units	Option 1 Excavation to Make a Repair on Service Line—Total Cost	Option 2 Directional Drill # of Units	Option 2 Cost to Replace Service Line with a Directional Drill	Option 3 Replacement of Service Line Using a Pneumatic Mole # of Units	Option 3 Total Cost for Pneumatic Mole	Option 4 Replacement of Service Line Using a Pull (if Possible) # of Units	Option 4 Total Cost for Pulling New Line
Man 1 ($40k/yr +30% Ben)	$ 27.00	hr.	5	$ 135.00	3	$ 81.00	3	$ 81.00	3	$ 81.00
Man 2 ($34k/yr +30% Ben)	$ 24.00	hr.	5	$ 120.00	3	$ 72.00	3	$ 72.00	3	$ 72.00
Man 3 ($28k/yr +30% Ben)	$ 21.00	hr.	5	$ 105.00	3	$ 63.00	3	$ 63.00	3	$ 63.00
Backhoe	$125.00	hr.	5	$ 625.00	3	$ 375.00	3	$ 375.00	4	$ 500.00
Boring machine	$ 75.00	hr.			3	$ 225.00				
Mole	$ 50.00	hr.					3	$ 150.00		
Clamp	$ 10.00	ea.	1	$ 10.00						
PE tubing	$ 2.00	LF	0	$ 0.00	40	$ 80.00	40	$ 80.00	40	$ 80.00
Asphalt repair	$ 10.00	sy.	4	$ 40.00	2	$ 20.00	2	$ 20.00	2	$ 20.00
Fill	$ 30.00	cy.	2	$ 60.00	2	$ 60.00	2	$ 60.00	2	$ 60.00
TOTAL COST				$1,095.00		$ 976.00		$ 901.00		$ 876.00

LF = linear foot; sy. = square yard; cy. = cubic yard

Table 3.9 Appropriate sizing and type of water meters—depending on the number of plumbing fixtures or anticipated usage

Water Meter Size (inches) for a Given Building Based on Fixture Units	Meter Type	Maximum Fixture Units	Peak Maximum Volume (gpm)—AWWA
5/8–3/4	Displacement	25	20
1	Displacement	45	50
1.5	Displacement	100	100
2	Compound	225	160
3	Compound	500	320
4	Compound	750	500
6	Compound		1000
8	Compound		1600
10	Turbine		2900
12	Turbine		4300

Source: AWWA

Meters should be sized properly according to their use. Oversized meters will under-register usage due to wear, while large meters will tend to lose calibration within 24 months of installation. The economics of disposing of worn meters and replacing them with new meters should also be considered. Small meters are rarely worth fixing. Table 3.9 shows the appropriate sizing for water meters depending on the units, plumbing fixtures, or usage anticipated.

The following AWWA Standards are available for meters:

- ANSI/AWWA C700-02 AWWA Standard for cold-water meters—displacement type, bronze main case
- ANSI/AWWA C702-01 cold-water meters—compound type
- ANSI/AWWA C704-02 propeller-type meters for waterworks applications
- ANSI/AWWA C710-02 cold-water meters—displacement type, plastic main case

The total quantity of water pumped to the distribution system should be accurately measured and recorded so that it can be compared with the water sold. The difference between the water pumped and water sold is the amount of water unaccounted for—that is the amount that is lost to leaks or otherwise lost and not paid for in the distribution system. Water systems should strive for an unaccounted-for fraction of less than 15%; less than 10% is good and often easily achieved if the meters are kept up-to-date. Analysis of production meter records should also be made to check such things as well productivity, trends in customer use, and the need for system expansion (Bloetscher 2011). Old, poorly registering meters—as opposed to leaks—are a common source of unaccounted-for water.

WATER STORAGE FACILITIES

The primary reason for providing storage of treated water is to allow the water plant to run at a constant speed. This allows the distribution system to have a reserve supply readily available for peak hour demands (mornings and evenings), firefighting, emergencies such as repairs to treatment facilities or pumps, and loss of water supplies due to pipe breaks or when flushing out contamination. The stored water can be used to maintain pressure in the distribution system if a well or pressure pump should fail or lose power. The storage tanks are designed to permit the plan to operate at one speed, while the demands may float higher or lower as needed. Operating at one speed (or flow rate) makes operations far more efficient than variable flow rates. Water enters and exits the plant throughout the day as needed: when customers are using water at a low rate, excess water can be stored; when use is high, stored water is used to meet the demand without having to alter the operation of the treatment plant (Bloetscher 2011).

The volume of water needed in storage differs by system, but there are standards such as the 10 States Standards (see https://www.broward.org/WaterServices/Documents/states_standards _water.pdf). Typically, the storage is up to half the plant capacity. Too much storage will impact water quality—essentially, the chlorine residual will drop because the water cannot be used fast enough.

There are several means of storage, although they may be limited by regulatory mandates. Water storage facilities may be elevated tanks, standpipes, hydropneumatic systems, and ground reservoirs. Elevated tanks are the most familiar because they are visible in prominent locations in most communities. Elevated tanks are generally constructed of steel, with the tank portion supported on legs or a pedestal (see Figure 3.43). Prestressed concrete elevated tanks do exist (Boynton Beach, FL). The height of the tank is determined based on the pressure desired in the distribution system. Most tanks are about 135 feet high, which translates to pressures of 55–65 psi. The tank volume is in the air to permit changes in demands to have limited impact on the system pressure. As a result, an elevated tank normally fills and empties in response to demands on the water system (referred to by operators as *riding the demand curve*—see Figure 3.44). A signal indicating the water level in the elevated tank is commonly used to vary the operation of the pressure pumps that are supplying the system. However, a means to turn off the pump that fills the tank is needed; otherwise, when the water level is near the top of the tank, the tank overflows.

Figure 3.43   Elevated storage tanks are generally constructed of steel.

A standpipe tank generally refers to an aboveground tank that is the same size from the ground to the top. Standpipes are primarily used where they can be located on a high point of land so that all or most of the stored water will furnish usable pressure to the water system (see Figure 3.45). They may be located on high ground to provide more volume at pressure. However, the lower part of the tank provides much lower pressure.

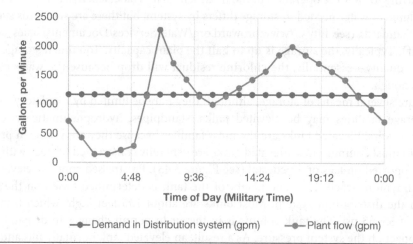

Figure 3.44   Example of a typical demand curve for a water system. The orange line shows how the typical demand is high in the morning, lower mid-day, with an increase in the evening. At night there is normally very little flow since most of the people go to bed. The blue line shows that the actual production level of the plant is constant. At night the plant pumps water that goes to storage; during the day water comes out of storage. As a result, all water systems need storage to draw from in order to operate at a constant production level, adjusting to actual demands. This is the most efficient way to operate the plant.

Figure 3.45   A standpipe tank is typically an aboveground tank that is the same size from the ground to the top of the tank.

Figure 3.46 Hydropneumatic tanks are used to help deliver pressure in smaller water distribution systems.

Smaller systems that cannot turn over the tank but need the pressure balance may use hydropneumatic tanks. In such systems, a large pressure tank is buried or located above ground and kept partially filled with water and partly with compressed air (see Figure 3.46). The balance of compressed air against the water maintains the desired pressure in the system and forces water out of the tank when needed. An air compressor is required to maintain the proper air-to-water ratio.

At water plants, it is not uncommon for the water to be stored at ground level. These ground storage tanks will have walls 30–50 feet high but they cannot generate the pressure required by the system. As a result, a high service pumping station is required to pressurize the system. Ground storage tanks may be prestressed or reinforced concrete, steel, or steel with glass lining, depending on the location of the utility. Ground storage tanks have covers for the purposes of this chapter, but local preferences may refer to them as reservoirs. Geographical preferences drive tank construction methods. The prime advantages of ground storage tanks are that they can be constructed to store relatively large quantities of water and can be completely buried where an aboveground structure would be objectionable to residents. When a large tank is completely buried, the land above it is sometimes used for a park or recreational area. The prime disadvantage is the cost of power to operate the pumping equipment and the potential for contamination.

An open reservoir has no cover. As a result, open reservoirs are discouraged by regulatory agencies because of the potential for bird droppings contaminating the water. However, they are used in some jurisdictions for raw water, or potable water that will be re-treated.

Pumps

Pumps are a major area of expense for both water and wastewater systems. For water systems, high service pumps may constitute 80% of all electrical costs on the water system due to the pressure needed on the system. Hence when trying to save money, pumps are an area where much improvement can often be made. To start, many older pumps do not have the efficiency that newer pumps and motors do. A 15-year-old pump can probably be replaced with a newer pump and save money through improved efficiency when pumping the same amount of water. The motors are more efficient as well. Since the efficiencies are multiplied with one another, small changes can facilitate large savings.

Water plants have high service pumps that pressurize the system or pump to the elevated storage tanks. Most major pumps are centrifugal pumps. Step and piston pumps and motors, and other pumps may be used for chemical feed systems, but are normally much smaller. The basis of centrifugal pumping lies in the capacity of the pumps, the revolutions per minute, and the diameter of the pumps. These come in two basic configurations: vertical turbine and horizontal split case pumps. Both types are common in application and both provide good service. The choice has more to do with cost, space, and maintenance than any other factors.

Horizontal pumps are generally all above ground. The casing can be opened on the surface and repairs made. Vertical turbine pumps have vertical pump shafts and sit in pump cans. The pump impeller is below the surface. The footprint for vertical pumps is much smaller than horizontal pumps, but ceiling height is needed to remove them (and a crane). Both are high efficiency, and neither is very good for suction. Ludwigson, Rago, and Greenfield (2016) outlines advantages of vertical turbine pumps, including better pressure control, lower horsepower (efficiency), and wider operating range. Horizontal pumps have the advantage of less need for cranes, easier maintenance, less corrosion, and often less capital cost. But service life may be different and operating costs are higher with horizontal pumps.

Pump efficiency is defined by the brake horsepower (BHP), which is hydraulic horsepower (HHP) divided by the motor input power; a.k.a. effective horsepower (EHp). Total efficiency is defined by:

$$Total\ Efficiency\ (\%) = HHP/EHp$$

Where HPP is hydraulic horsepower. The hydraulic horsepower is defined by:

$$HHP = pressure\ (ft) \times flow\ (gpm)/3{,}960$$

Motor horsepower is defined as $1.341 \times EHp$
The brake horsepower would be:

$$BHP = Eff(\%) \times EHp$$

The power factor would be:

$$Pfactor = Power\ input \times 1{,}000/(current\ (amps) \times voltage \times phases)$$

The cost per year is related to the motor input power, which includes both the motor and the pump efficiency. At \$0.07 kW/hr., for example, if a utility operated a pump 6 hours per day using a 220-volt, 10-amp/3-phase service (more efficient than single phase), 70% efficient motor, a pump efficiency of 70%, a 4 kW demand, with a head of 22 feet and a volume of 200 gpm, the total hydraulic horsepower would be:

$$HHP = pressure\ (ft) \times flow\ (gpm)/3{,}960 = 22 \times 200/3{,}960 = 1.1\ HP$$

The power demand would be 4 kW × 1,000/(10 × 220 × 3) = 0.6 W
The motor input would be 1.341 × 4 kW = 5.36 HP
The break (motor) horse power would be 0.737 × 5.36 = 3.95 HP
The total efficiency would be 1.1 HP/5.36 HP = 20.5%, very poor
The pump efficiency is 1.1 HP/3.95 HP = 27.8%, also poor.

Both efficiencies should be well above 50% for most pumping situations. The pump and motor in this case need to be changed since the cost savings may be substantial.

MAINTENANCE NEEDS

Maintenance needs differ between buried pipe and aboveground facilities. Both the inside and the outside of the tanks and vessels are apt to corrode, depending on materials. The above-ground facilities are primarily steel and concrete, with a little stainless steel, plastic, fiberglass, and aluminum mixed in. Weather, moisture, soil chemistry (resistivity, pH, etc.), and materials of construction all affect condition. Metals, which are often the most common materials used in water systems due to their tensile strength, are the first item to discuss.

Metal Corrosion

The selection of the best material for a specific application can be a complicated engineering analysis—yet very important to the success of the project. Metals, such as bronze, copper, and iron have been used for thousands of years. Steel has many uses, and therefore is one of the largest industries worldwide today. Steel was an advancement made when molybdenum was added to harden wrought iron. Otherwise steel is iron with a small amount of carbon (usually less than 0.5%), and other elements in small quantities. Grades of steel are determined by their amounts of non-ferrous components such as carbon, chromium, nickel, molybdenum, copper, and titanium. Depending on the specific content and manufacturing process, steel mixes can be designed for a range of properties, including (Bloetscher et al. 2003):

- Overcome brittleness
- Cohesive strength
- Corrosion resistance
- Tensile strength
- Fatigue protection
- Ductility
- Hardness
- Malleability
- Shear strength
- Torsional strength
- Electrical conductivity
- Thermal conductivity
- Thermal expansion (or resistance)
- Magnetic properties
- Heat treatability

Advantages of steel include (Bloetscher et al. 2003):

- High strength
- Uniformity
- Elasticity under high stress
- Ductility
- Adaptations for connections and manufacture
- Fatigue strength and toughness

- Resistance to heat deformation
- No ultraviolet light degradation

Disadvantages of steel use include maintenance costs due to susceptibility to corrosion by air, water, and microbiological elements and fatigue if subjected to stress reversals (pressure on, then off).

The most critical drawback in using steel is corrosion. Common types of corrosion in steel include galvanic, dezincification, pitting, crevice, intergranular, stress, cracking, erosion, and microbiologically induced corrosion (MIC). There are a series of things that will accelerate corrosion, and multiple corrosion activities are often present. In water, exposure to a periodic wet/dry cycle, saltwater, or bacteria accelerate corrosion. Connections between dissimilar metals likewise accelerate corrosion. The farther apart the metals are on Figure 3.47, the more quickly corrosion will occur. High dissolved oxygen in the water is a problem because it feeds bacteria in the water and can encourage oxidation. Chlorides (salt) likewise increase the potential and development of corrosion. High alkalinity in water can increase corrosion potential because it creates an unstable water condition that can deposit materials on the pipe and create dissimilar points on the pipe surface that can develop into corrosion cells. Figure 3.48 shows excess deposition in the galvanized pipe. A high or low pH, high temperatures, and low flow worsen corrosion as well. This is why most utilities try to have a pH above eight and maintain ongoing flushing programs. Corrosive water will dissolve (leach) metals from pipe and fittings into the system. Lead, copper, zinc, and iron are common leaching metals. It should be noted that asbestos cement pipe is *not* immune to deterioration.

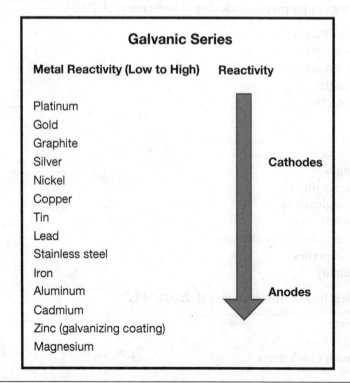

Figure 3.47 Galvanic series chart. The concept is that the lower metal will be sacrificed for the higher metal. Hence, iron will deteriorate faster in the presence of copper—dissimilar metals make a galvanic (battery) cell.

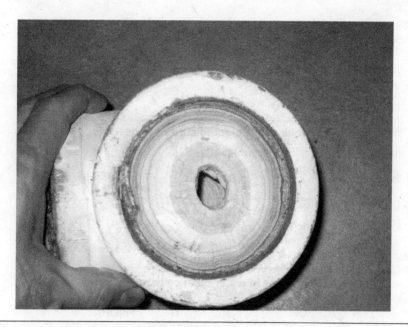

Figure 3.48 Excess deposition in a galvanized pipe.

Corrosion is the deterioration of a metal or alloy as the result of exposure to and reaction with its environment. How corrosion occurs is useful to understand because it affects all metal piping and metal surfaces (tanks, ladders, equipment, etc.). It is an electrochemical process, like a battery, with an anode and a cathode reaction, and ions migrating through an external circuit, even in single metal steel. Positive electrolytes migrate through the liquid electrolyte toward the cathode, whereas negative ions are attracted to the anode. The corrosion reaction cannot occur without a simultaneous cathodic reaction. The typical reactions are:

$$Fe^0 \rightarrow Fe^{2+} + 2e^-$$

$$2H^+ + 2e^- \rightarrow H_2$$

$$2H^+ + \tfrac{1}{2}O_2 + 2e^- \rightarrow H_2O$$

$$H_2O + \tfrac{1}{2}O_2 + 2e^- \rightarrow 2OH^-$$

where e^- is a free electron. The first equation is anodic and the rest are cathodic. Figure 3.49 outlines the manner in which tubercles and pitting is formed. Bacteria are normally a part of this process.

There are several corrosion types:

- Crevice corrosion—which is most likely to occur on the outside of the pipes. The tightness of the crevice will affect the speed at which crevice corrosion occurs (Dillon 1995). The dominant mechanism is the presence of bacteria, which will accelerate initiation of localized attack (Videla 1996).
- Chloride pitting—where the chlorides exceed 50 mg/L.
- Pinhole leaks—often with microbiological components, pinhole leaks will occur as a result of the field welding process in stainless steel pipe or from rocks when the heat-affected zone is not properly removed or from other objects vibrating against buried pipe walls from traffic. Backfill issues (rocks) can contribute to the latter issue.

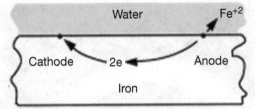

A. Minor variations cause electric current to develop

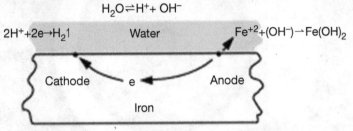

B. Chemical reactions in water balance those in iron

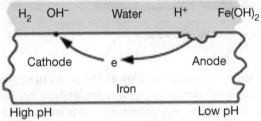

C. Rate of corrosion is accelerated

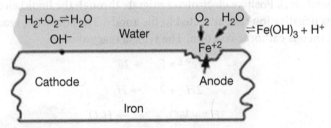

D. Rust forms

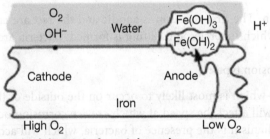

E. Rust precipitates to form tubercules

Figure 3.49   Example of how a tubercle forms in a water line. Even with a pipe that is all one material (avoiding dissimilar metal corrosion), any defects, dings, scratches, or other imperfections will create a galvanic cell that will lead to tubercles. *Source*: AWWA

- External corrosion—in the form of rust stains is often found in multiple locations on the exterior of the pipe. Bands of rust streaks on the exterior of the pipes generally result from the inappropriate use of an unpadded sling on the crane to lift the pipes into place. They are generally cosmetic in nature, but create a dissimilar metal cell condition in the ground.
- Stagnant water—always presents a risk for the occurrence of corrosion and microbiological activity. The ability to completely drain the pipeline during periods of non-use is important because static water encourages microbial activity even in stainless steel pipe (Bloetscher et al. 2002, 2002a). A minimum velocity of 3 fps is required to prevent corrosion (Dillon 1995).

Rajani and Makar (2000) estimated the time for full penetration of a 6 inch (150 mm) diameter pipe in moderately corrosive environments to be 11–14 years for pressure class 350 and class 50 (the thinnest pipe walls), 23–52 years for class 52, and 335–415 years for class 56, which is the thickest wall option. For 8 inch (200 mm) pipes under the same conditions, Rajani and Makar (2000) estimated times for penetration were 11–14, 13–24, 46–83, and 395–485 years for pressure class 350, and thickness classes 50, 52, and 56 pipes, respectively. However, conditions can change quickly in the ground.

To address the corrosion problem with iron and steel pipes, a number of solutions have been pursued. The development of stainless steel is one. Stainless steel has provided good service in many applications; however, use of the appropriate grade is important to the ultimate success of the steel. There are many stainless steel options and an engineer who understands the metallurgical properties should be consulted to get the optimal solution. Grades 304, 304L, 316, and 316L are commonly used but may not provide the protection needed, especially since stainless steel is normally thinner than the comparative DI or standard black steel. The appropriate choice of stainless steel grade and accompanying appurtenances, plus the appropriate construction and treatment techniques should allow the metal to provide a reasonably long life by resolving most of these corrosion issues. However, corrosion should be an expected by-product of any steel use, including stainless steel. Operations issues, such as scaling, over chlorination, poor fabrication, and MIC can initiate corrosion in stainless steel (Dillon 1995).

Other options include using resistant materials such as aluminum or plastic pipe (noting that plastic pipe in the sun presents other complications). A protective coating (epoxy, other coatings) can help. Paint is usually used. However, paint sprayed on a metal surface can create edge corrosion issues. When spray systems are used, there are volatiles in the paint. As the volatiles disperse as a part of normal paint drying, the paint pulls away from edges (it shrinks—as does the paint thickness—which is why you cannot test paint thickness immediately after application (see Figure 3.50). Edge painting is particularly difficult to deal with unless a roller is used and copious amounts of paint are applied (a secondary issue). Hence painting is needed regularly to keep corrosion at bay. In lime softening systems, some calcium carbonate (lime sludge) on the walls is desirable to protect the paint (see Figure 3.51).

Wrapping pipe in plastic (PE wrap) is common for DI pipe. Cathodic protection applies a small voltage to the metal with a sacrificial (magnesium) plate that reduces the electrolytic issues with corrosion. Cathodic systems must be kept up-to-date or they will fail. Most distribution systems apply orthophosphates and other corrosion inhibitors to reduce the potential for corrosion. Much of this came after the lead and copper testing programs in 1990/1991.

There are models to create statistical likelihoods, but results need to be supplemented by inspection due to the great natural variability that is inherent in the conditions under which buried pipes perform (Rajani and Makar 2000). Nondestructive testing would help, but is not readily available

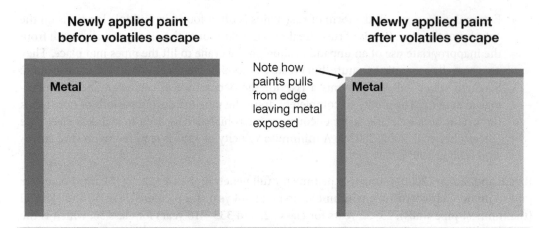

Newly applied paint before volatiles escape

Metal

Newly applied paint after volatiles escape

Note how paints pulls from edge leaving metal exposed

Metal

Figure 3.50 Paint on metal will exhibit *shrinkage* once the paint dries. (a) Newly applied paint has volatile organic compounds in it. These compounds are released after the paint is applied and starts to dry. Once the paint dries and the volatile organic compounds are gone, the paint layer is thinner. (b) An associated problem is that while the paint dries, it also shrinks and pulls away from the corner edges exposing them to corrosion.

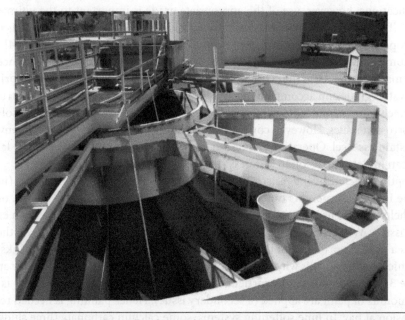

Figure 3.51 Lime sludge on the walls of a steel accelator at a lime softening water treatment plant. An added benefit of lime softening is that a buildup of calcium carbonate develops which protects the paint and the metal. You see this at most lime softening plants.

for DI pipe and requires digging up the pipe to perform (very expensive). Thomson, Flamberg, and Condit (2013) attempted to create structural/physical condition curves of pipelines based on modeling data. However, they noted that only large water mains with costly consequences of failure currently justify the cost of accumulation of data that are required for physical model application.

Marlow et al. (2009) attempted to assess the remaining asset life by determining the threshold above which risk is deemed to be unacceptable and then using this assessment as a means to determine the current state of an asset, expressed in terms of failure likelihood. Factors that Marlow et al. (2009) and Thomson, Flamberg, and Condit (2013) found to be helpful were leakage, hydraulic capacity, water quality, and the capability of a system of assets to deliver a required level of service to customers. Grigg (2007) discussed the link between main break models and contributing risk factors (predictor variables) in a Water Research Foundation project. Examples of these predictor variables include pipe age (minor factor in later stages of failure), material, diameter, soil corrosivity, and operational pressures. Ultimately, Thomson, Flamberg, and Condit (2013) determined that the condition curves had potential application for smaller diameter pipelines and were useful for long-term water asset planning. However, to date such models have not been used much by utilities for predicting failure of pipelines.

Concrete Corrosion

Concrete is far more resistant to corrosion than steel, except when it is in gravity pipes with the potential for hydrogen sulfide (see Chapter 4) or when the pipes are reinforced concrete. Understanding concrete corrosion requires thorough knowledge of the concrete itself. Concrete is a mix of water, cements, aggregates, and other products specified by the design engineer. When poured, concrete heats up. The heat helps the curing process; the lime in the cement will boil the water away—hence the reason to keep the surface wet and cool. The heating and cooling process causes micro-cracking in the concrete (without reinforcing steel the micro-cracks become visible). The steel is in the concrete to help pull the cracks tight via tension (concrete has very limited tensile ability). As a result, the surface of the concrete may appear to be a solid smooth surface, but in reality it is full of micro-cracks. Water can seep into the micro-cracks, and since some of these cracks lead down to the reinforcing steel, there is the potential for the steel to corrode. When reinforcing steel corrodes, it expands and loses some of its tensile capabilities. Hence the cracks open wider and the steel corrodes faster. All the issues discussed previously that apply to steel will also apply to reinforcing steel.

Since concrete has micro-cracks, all concrete structures will have cracks—the goal is to limit the structurally important ones. To distinguish important from merely cosmetic cracks, a structural engineer must often be consulted. But even nonstructural cracks can be a concern if they can expand to structural cracks in the future. Knowing the cause of cracking is important when it comes to selecting appropriate mitigation or repairs. There are several types of cracks to be aware of for maintenance purposes:

- *Structural cracks*—structural cracks are the greatest concern and occur when the loading exceeds the capacity of the structure. Differential settlement is a common example, but design errors, construction errors, and unexpected loadings can also be causes.
- *Shrinkage cracks*—shrinkage of the concrete as it dries is the most common source of cracking. Expansion and construction joints are designed to minimize the cracking. They are not structural concerns, most of the time. However, if the shrinkage cracks result from a lack of reinforcing steel, structures like tanks may leak, and the leakage will accelerate corrosion.
- *Cold joints*—partial cold joints result from poor consolidation between two layers of concrete. While these cracks are generally not structural concerns, leakage from water retainment structures may result.

Fixing concrete requires some form of sealer. Epoxy injection seals cracks and can restore a portion of the strength that has been lost, but is only appropriate for cracks where continued movement is not expected. Membrane liners have also been used to stop leakage through working cracks and from leaky basins and reservoirs. Repair of the concrete itself can be accomplished by using calcium aluminate, silanes, and siloxanes, which are sealers that penetrate into the concrete pores, then harden into solid silica-based polymers, thereby blocking the pores. Calcium aluminate is among the better options for concrete repair as it creates a chemical reaction that seals the surface. Calcium aluminate is preferred in corrosive environments. Silicates are densifying sealers that react with calcium hydroxide within the concrete, filling the pores with a crystalline product that hardens the concrete surface. These products can be sprayed or brushed onto the surface. Epoxy or other polymer coatings are commonly applied as final surface coats to prevent water intrusion. For nonstructural concrete cracks, routing and then sealing with an elastomeric caulk such as PE is an option.

Concrete also corrodes quickly in the presence of hydrogen sulfide (from bacteria in groundwater, which will corrode the surface or wastewater, as discussed in Chapter 4). Epoxy and other coatings are used to reduce the potential for hydrogen sulfide damage. However, delamination of coatings with time is an issue that must be monitored.

PLANT OPERATIONAL ISSUES

There are many plant issues that can arise; far too numerous to discuss in this document. However, there are several issues that are useful to consider that affect maintenance by causing accelerated needs to refurbish the plant, clean the plant, or replace the media in filters. First is that the equipment and surfaces must be kept clean and that ongoing maintenance for pumps, motors, and painting be kept up-to-date. Every plant has a series of operations and maintenance manuals that were compiled by the design engineers and manufacturers that outline how often equipment should be greased, have packing replaced, be tested for alignments, etc. Many utilities have rarely dusted these documents off—that needs to change and a listing of all activities should be compiled along with hiring the needed staff to perform those tasks. Otherwise, preventive measures will not be undertaken, which suggests that failure will be accelerated and the life of equipment reduced.

Next, skimping on chemicals to reduce costs is not a good means to cut costs. The world saw the impact when the city of Flint, Michigan, failed to use polyphosphates (among other things). Failure to use enough coagulant will prevent settling of the particles, meaning more sludge to dispose of and more difficulty in dewatering it. That will put a strain on sludge pumps, sludge equipment, trucks, and other resources that are involved with sludge removal. It will increase sludge disposal costs significantly and accelerate wear on dewatering equipment.

Failure to use enough chlorine will cause the residual to diminish too quickly, causing boil water orders to be issued by the public health agency. Failure to monitor the water quality and maintain the chlorine residual has the impact of permitting biofilms to grow in the distribution system. Biofilms can manifest in a number of ways—nitrification in the distribution system is one. Failure to monitor pH may also encourage biofilms and nitrification.

Likewise, changing polyphosphates can alter the corrosion potential of water. Collier County (FL) changed polyphosphates in 1990 based on a low bid response between their first and second lead/copper test and found very different responses (all non-detects for lead and copper the first round, many detects the second time). Reverting back to the initial polyphosphate created all non-detects again. The polyphosphate mattered. Every system should test results to verify the impact because all waters are different.

More costly are major plant renovations needed due to the way plants are operated. Water coming into membrane plants must be free of sand, silt, and most bacteria. Otherwise the membranes will foul and potentially be damaged, requiring accelerated replacement. Sand separators are useful and have a low cost, but are often not installed (see Figure 3.52).

Lime softening plants are another issue where problems occur when trying to save money. Lime softening is a fast chemical reaction, normally taking less than a minute in practice. Keeping the pH around 10.3 is critical to good settling. Cascading lime sludge is often visible in lime softening units when this occurs. Polymers can be used to improve settling, but two things are important to make lime softening work well: a good mixing system and high pH. Figure 3.53 shows mixing at a plant. Swirls of clear water can be noted, which indicates a poor mix. It can also indicate insufficient lime or poor mixing at the slaker (all of the above at this plant). Using less expensive lime, a smaller dose to save money, or both will create a cascade of issues: poor settling, carry-over into the filters, calcium carbonate coating the filter media or cementing it, coating ion exchange beads, and higher chlorine demand because of higher turbidity. It may save money, but it will create major repair costs later. When the pH does not reach 10.3, the reaction is incomplete (Hammer and Hammer 2011). The sludge also settles poorly (compare Figures 3.54 and 3.55). Filters have sand, which is a catalyst for calcium carbonate. The calcium carbonate attaches to the media, making particles larger and lessening filter effectiveness. Eventually the filters may clog up, requiring expensive rehabilitation. Ion exchange media will need to be changed. Operators will see that the filters do not plug as fast, which seems good, but the larger particles filter water less effectively, meaning turbidity will not be decreased as much as desired, which risks pushing material into the next process (disinfection). That means more chlorine demand. And worse, the reduced pH can encourage biofilm growth. Hence, one decision affects the entire plant and accelerates the need for maintenance.

Figure 3.52   Sand separator

Figure 3.53   Poor mixing at a lime softening water treatment plant can occur if not enough lime is used for proper softening. Here the liquid is not fully mixed (dark and light areas).

Figure 3.54   The failure to use enough lime will cause an incomplete reaction. The incomplete reaction means that calcium carbonate will not settle properly and will carry over to the next process affecting the filters. These turbid filters follow the reactor in Figure 3.53. Ultimately, this utility had to replace all the filter media and rehabilitate the filters due to the lime buildup.

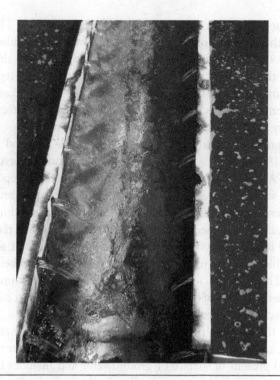

Figure 3.55   The water above is from a lime softening water plant that used the proper amount of lime; the water is noticeably clearer. Compare this to the filter in Figure 3.54—note edge corrosion (see Figure 3.48 for explanation).

Water Distribution System Operational Challenges

Table 3.5 outlines the challenges facing water distribution systems. Ellison et al. (2014) notes that the risks associated with poor infrastructure include public health risks associated with contaminants entering during breaks, biofilm formation, poor water quality, loss of disinfection residuals, nitrification in the distribution system, and accelerated corrosion of the pipe and fittings. Oxenford et al. (2012) outlined 22 different metrics for water system performance, including: break rate, pressure, leak rate, service interruption frequency, regulatory compliance, water quality, water quality complaints, disinfectant residual concentration, and lead and copper contaminant levels at the tap. Maintaining chlorine residuals, controlling disinfection byproducts, and controlling biofilms and nitrification are ongoing challenges for many water systems—especially in warm areas because nitrification is an indicator of bacteria, and bacteria prefer warm conditions.

Disinfection By-Products and Nitrification

Disinfection via chlorine was the biggest public health improvement of the 20th century. Small concentrations led to huge decreases in waterborne illness and death each year. As a result, most water supplies are required to be disinfected and a chlorine residual must be present at all times. However, concern exists over the by-products of chlorination—trihalomethanes (THMs) and haloacetic acids. Residual chlorine molecules react with harmless organic material to form a group of chlorinated chemical compounds called THMs. They are tasteless and odorless. The four THMs are chloroform ($CHCl_3$), bromodichloromethane ($CHBrCl_2$), chlorodibromomethane ($CHClBr_2$), and bromoform ($CHBr_3$). In 1974, THMs were discovered to be formed during the disinfection step of drinking water if free chlorine was used as the disinfectant (EPA 1979). This, coupled with the perceived hazard to the consumer's health, led the EPA to amend the National Interim Primary Drinking Water Regulations to include a maximum contaminant level of 0.10 mg/L for total THMs (TTHMs). Lowering the THM precursor (organic) concentrations has the effect of reducing overall disinfectant demand, thereby reducing the possibility of the formation of all disinfection by-products.

Under its Disinfectants and Disinfection By-Products Rule (Federal Register, 1998 63 FR 69390: December 16, 1998), the EPA set the maximum contaminant level (MCL) for TTHMs at 0.080 mg/L. The EPA did not set an MCL goal (MCLG) for the group of TTHMs, although there are MCLGs for some of the individual constituents, including bromodichloromethane (zero), bromoform (zero), dibromochloromethane (0.06 mg/L), and chloroform (0.07 mg/L). The MCL of 0.080 mg/L was set based on the potential for an increased risk of cancer and other health effects (Federal Register 1998). The EPA considers that the MCL for TTHMs is feasible and achievable for a chlorinated drinking water supply. The EPA also believes that by meeting MCLs for TTHMs and haloacetic acids (HAA5) (other disinfection by-products), water suppliers will also control the formation of other disinfection by-products not currently regulated that may also adversely affect human health (EPA 2002a).

THMs result from the reaction of chlorine with natural organic material and bromine in source waters (Miller and Uden 1983; Coleman et al. 1984). Natural organic matter (NOM) is the organic material present in surface or groundwater and includes both humic and non-humic fractions. Organic compounds are generally found in very small quantities, unless humic substances are involved. Humic substances are terrestrially based organics; they are the organic portion of soil that remains after prolonged microbial decomposition from the decay of leaves, wood, and other vegetable matter. Song (1999) documented the impact of NOM on chloramine decay by altering NOM concentrations through activated carbon adsorption. Bone et al. (1999) hypothesized that the NOM oxidation mechanism is the dominant pathway for chloramine decay early in the decay process (i.e., within 24 hours), and that auto-decomposition is the dominant cause of chloramine decay later. Ødegaard et al. (2010) found that granulated active carbon (GAC) was not a suitable treatment of NOM in water supplies due to clogging.

Potential health effects from ingestion of TTHMs in water include liver, kidney, and central nervous system problems, as well as an increased risk of cancer. The MCL of 0.08 mg/L was set based on the potential for an increased risk of these health effects (Federal Register 1998). It should be noted that the EPA has *not* conducted a cancer assessment for the TTHMs. However, the individual TTHM constituents have been evaluated and qualitative descriptors of their carcinogenicity include:

- *Bromodichloromethane*—is likely to be carcinogenic to humans by all routes of exposure.
- *Bromoform*—is likely to be carcinogenic to humans by all routes of exposure.
- *Dibromochloromethane*—there is suggestive evidence of carcinogenicity but not sufficient to assess human carcinogenic potential.
- *Chloroform*—is likely to be carcinogenic to humans by all routes of exposure under high-dose conditions that lead to cytotoxicity and regenerative hyperplasia in susceptible tissues. Chloroform is not likely to be carcinogenic to humans by all routes of exposures at a dose level that does not cause cytotoxicity and cell regeneration (EPA 2004).

A number of researchers have raised questions about their health hazards (Jolley et al. 1990; IARC 1991). All four THMs, when administered by food or drugs through a tube leading down the throat to the stomach, are carcinogenic in the livers of female B6C3F1 mice (NCI 1976; NTP 1985, 1987, 1989). The carcinogenic activity of the THMs has been proposed to be mediated through a nongenotoxic mechanism based on their weak activity in mutagenicity and genotoxicity assays (Le Cureiux et al. 1995; Reitz et al. 1989; Rosenthal 1987). However, only one THM—bromodichloromethane—has demonstrated mutagenic activity, apparently though a glutathione (GSH) metabolite (Pegram et al. 1997).

The route of administration appears to play a significant role in the metabolism of the THMs and consequently in their carcinogenic and toxic activity. Although chloroform was carcinogenic in mice when administered via a feeding tube, it was not carcinogenic and did not promote liver cancer when administered in the drinking water (Jorgenson et al. 1985; Klaunig, Ruch, and Pereira 1986; Pereira, Knutsen, and Herren-Freund 1985; Ammann, Laethem, and Kedderis 1998; Lilly, Ross, and Pegram 1997; Pankow, Damme, and Wünscher 1997; Ilett 1973; Pohl et al. 1977). Furthermore, cell proliferation was enhanced by chloroform administered by feeding tube, but not in drinking water (Larson, Wolf, and Butterworth 1994; Pereira 1994; Pereira and Grothaus 1997). Hence the real risk is less clear when drinking the water from the tap.

Neither chlorine dioxide nor ozone nor chloramines produce THMs at significant concentrations when used alone as disinfectants. Furthermore, the cost of any of these unit processes is very low. The major disadvantage of using alternate disinfectants for THM control relates to the lack of any precursor removal. Chloramines are an option where ammonia and organics naturally persist.

Chloramination is commonly used for secondary disinfection purposes to control microbial growth in finished water. This is particularly at issue in water systems that have warm water (>60°F) and organics. Chloramines include monochloramine, dichloramine, trichloramine, and organochloramines. For disinfection purposes, monochloramine and dichloramine are the preferential forms. Nitrification can occur in water systems that contain chloramines, as organic content and warm temperatures limit the ability to maintain a chlorine residual. Chloramines last longer in the distribution system, which is helpful in large systems and those with warm waters—both of which tend to degrade chlorine residuals. In these systems, the concentration of free ammonia present in the distributed water will be a function of the chlorine to ammonia-N ($Cl_2:NH_3$-N) ratio. Free ammonia is almost completely eliminated when a 5:1 weight ratio of $Cl_2:NH_3$-N is used (Kirmeyer et al. 1993, 1995, 2000). The downside is that chloramination appears to increase the potential for biofilms, and specifically nitrification in the distribution system.

Biofilms in Water Distribution Systems

One reason that disinfection is so important is *biofilms*. Biofilms are biological matrices that form on the pipe walls under certain conditions. Chapelle (1993) notes that bacteria are known to inhabit and thrive in the presence of moisture and nutrients, both of which exist in plentiful supplies in water distribution networks, where they form biofilms (Videla 1996). Biofilm problems persist in water distribution systems and as a result, have been the focus of remediation strategy development. Biofilms are complex aggregates of microorganisms embedded in a highly hydrated extracellular matrix that show structural heterogeneity resulting from a diverse and complex microcosm. Biofilms are often observed as an unwanted accumulation attached to a surface, such as the inner wall of water distribution lines or well casings. During biofilm growth, microorganisms excrete a matrix of extracellular polysaccharides (EPSs), which lead to the formation of a slime layer that connects cells and anchors them to the surface and to each other. From the microbial perspective, biofilms provide an ideal habitat as a source of nutrients, oxygen stratification, resistance to velocity currents, and protection from grazers and biocides (Videla 1996). From the utility perspective, the undesirable accumulation of biofilms with actively growing slime layers can lead to sloughing, which can increase HPCs (bacteria counts in water). If left uncontrolled, biofilms can become a considerable issue for water distribution systems, particularly with regard to hygienic, operational, and economic consequences.

According to the World Health Organization, diseases associated with unsafe water distribution, sanitation, and hygiene cause approximately 1.7 million deaths per year (Prentice 2002). Mature biofilms in drinking water distribution systems are a highly diverse potential source of human pathogens. A wide range of primary pathogens (i.e., ones that cause disease in healthy individuals) and opportunistic pathogens (i.e., ones that cause disease in individuals with underlying conditions that may facilitate infection) have demonstrated the ability to survive and thrive in biofilms. Primary pathogens, opportunistic pathogens, and indicator organisms including *Clostridium* (Emde, Smith, and Facey 1992), *E. coli* (Emde, Smith, and Facey 1992; Geldreich 1996; Sartory and Holmes 1997), *Enterobacter* (LeChevallier, Babcock, and Lee 1987; Emde, Smith, and Facey 1992; Geldreich 1996; Sartory and Holmes 1997; Lee and Kim 2003), *Legionella* (Murga et al. 2001), *Pseudomonas* (LeChevallier, Babcock, and Lee 1987; Emde, Smith, and Facey 1992; Geldreich 1996; Norton and Le Chevallier 2000; Lee and Kim 2003), and *Staphylococcus* (Geldreich 1996; Lee and Kim 2003), among others, have been reported in biofilms collected from water distribution networks. An important point to consider is that only coliforms are routinely analyzed for in drinking water, as mandated under the Total Coliform Rule (TCR) and the Groundwater Rule of the SDWA; other pathogens and opportunistic pathogens are not.

Other pathogenic microorganisms have also been isolated from biofilms including viruses, fungi, yeast, protozoa such as amoebae and ciliates, diatoms and other algae, invertebrates, and microbial toxins (Eboigbodin, Seth, and Biggs 2008; Bachmann and Edyvean 2006; and references therein). The presence of these microorganisms in potable water distribution systems represents a potential public health threat. Specifically, release of pathogens harbored in biofilms can lead to an increase in the incidence of gastrointestinal symptoms from waterborne infections caused by bacterial, viral, and parasitic microorganisms. The Centers for Disease Control and Prevention (CDC) identified biofilms as the source for 65% of human bacterial infections from community water supply associated outbreaks (EPA 2007). Gofti et al. (1999) reported epidemiological evidence that children showed close to four digestive problems per person per year and one episode of diarrhea per person per year, attributable to pathogens that developed in the water transmission network after centralized disinfection. Table 3.10 lists the typical organisms in a biofilm.

In addition to health effects resulting from pathogens, biofilms can also contribute to taste, odor, and color issues, which may lead to operational changes at the treatment plant or for the transmission network. Biofilms may also compromise the proper enumeration of indicator organisms and weaken pipe integrity by *microbial-influenced corrosion* (MIC), which is defined as a microbial-mediated electrochemical process that permits the onset and acceleration of corrosion (Videla 1996). Once mature colonies are established, the effects of MIC are often seen. Microbes cause corrosion directly through metabolic processes that form corrosive chemical species such as ammonia, hydrogen sulfide, sulfate, ferric, or manganic chlorides (Dillon 1995). Within the community structure of the biofilm, sulfur can be reduced by anaerobic bacteria to release hydrogen sulfide, which can significantly increase the susceptibility of the pipe to pitting. At the same time, any aerobic bacteria present in the biofilm can corrode metals directly via oxidation. The heterotrophic biomass typically found in a biofilm is supported by the

Table 3.10   Bacteria commonly found in water distribution system biofilms

Genus	Source
Gram-Positive Bacteria	
Bacillus spp.	LeChevallier et al. 1987; Emde et al. 1992; Norton and LeChevallier 2000; Lee and Kim 2003
Clostridium spp.	Emde et al. 1992
Total Coliform Group	
Aeromonas spp.	Emde et al. 1992; Geldreich 1996; Lee and Kim 2003
Citrobacter spp.	Emde et al. 1992; Geldreich 1996; Sartory and Holmes 1997
E. coli	Emde et al. 1992; Geldreich 1996; Sartory and Holmes 1997
Enterobacter spp.	LeChevallier et al. 1987; Emde et al. 1992; Geldreich 1996; Sartory and Holmes 1997; Lee and Kim 2003
Klebsiella spp.	LeChevallier et al. 1987; Emde et al. 1992; Geldreich 1996; Sartory and Holmes 1997
Coliform Antagonists	
Actinomycetes spp.	LeChevallier et al. 1987; Geldreich 1996
Flavobacterium spp.	LeChevallier et al. 1987; Emde et al. 1992; Geldreich 1996; Lee and Kim 2003
Pseudomonas spp.	LeChevallier et al. 1987; Emde et al. 1992; Geldreich 1996; Norton and Le Chevallier 2000; Lee and Kim 2003
Opportunistic Pathogens	
Legionella spp.	Murga et al. 2001
Moraxella spp.	LeChevallier et al. 1987; Geldreich 1996; Lee and Kim 2003
Mycobacterium spp.	LeChevallier et al. 1987; Geldreich 1996
Serratia spp.	Emde et al. 1992; Geldreich 1996; Sartory and Holmes 1997
Staphylococcus spp.	Geldreich 1996; Lee and Kim 2003
Iron Bacteria	
Gallionella spp.	Emde et al. 1992
Leptothrix spp.	Emde et al. 1992
Thiobacillus spp.	Emde et al. 1992

synergistic effects of mixed growth rates, mixed metabolisms, and high surface area to volume ratios, which allow the biofilm to remain active within the relatively hostile conditions of a pipe environment. For water distribution systems constructed with metallic materials, corrosion and corrosion control are ongoing issues that, if not addressed, can result in pipe damage/failure, premature aging/replacement, clogging, and increased maintenance requirements. Proper corrosion control has also been shown to increase disinfection effectiveness on biofilms in iron pipes (Le Chevallier, Lowry, and Lee 1990).

Biofilm growth in water distribution systems occurs when microorganisms attach themselves to the pipe walls. The steps in the life cycle of a biofilm include attachment, slime formation, growth, and detachment or sloughing (Videla 1996). Key requirements for biofilm development include an active microbial community and interaction with pipe materials (Videla 1996). The first microorganisms to attach are called *pioneers*; they are most commonly facultative anaerobes that form the foundation for colonization and further growth. Flemming and Wingender (2001) estimated that 50–90% of the biofilm matrix was extracellular material, while only 25% was cell matter. Through microbial-mediated redox mechanisms, micronutrients are released within this growing layer, which attracts other species. As the biofilm continues to coat the pipe surface, acid-formers can reduce the pH near the pipe wall and accelerate corrosion (Videla, 1996). This phenomenon can create localized anodes and, in conjunction with abiotic reactions caused by dissimilar metals or pipe defects, can lead to a steady cathodic deterioration over time.

As the biofilm matures, it grows thicker from the diverse community of microorganisms attracted to the biofilm and from the accumulation of particles that stick to the extracellular polymer matrix. A depleted oxygen layer forms near the wall and an anaerobic environment in which sulfate-reducing bacteria (SRB) proliferate because the transport of oxygen into the anaerobic layer of the biofilm is limited by the biological activity in the upper layers. The appearance of SRB is indicative of mature biofilm growth. The biofilm cannot continue to increase in thickness indefinitely. The bulk fluid velocity will act to reduce the thickness of the biofilm on the pipe wall as a result of friction and shear. Periodically, portions of the growing mass become detached (sloughing), releasing bacteria into the bulk fluid. Sloughing can also occur if the system is disturbed by changes in velocity of the water or water hammer.

A variety of factors affect biofilm development in distribution networks. These include environmental factors (pH, temperature, dissolved oxygen, etc.), water quality (nutrients, inorganics, dissolved organic carbon, etc.), pipe materials, system hydraulic regime (stagnant conditions), corrosion control measures, presence of a disinfectant residual, and sediment accumulation. Most of the data on the factors that influence biofilm development are based upon changes in total viable counts (e.g., heterotrophic plate count) or on changes in the growth of specific microorganisms (e.g., total coliforms). Although a number of comprehensive review articles have been published (Geldreich 1996; Geldreich and LeChevallier 1999; Batté et al. 2003; Bachmann and Edyvean 2006), the interaction among these factors is complex and variable, making predictions difficult.

Currently, researchers lack techniques for effective detection, diagnosis, and control of biofilms. Low pH, low hardness, high chlorides, high sulfates, and a ratio of chlorides to bicarbonate above 0.3, all indicate a greater potential for corrosion and biofouling (Geldreich 1996). One effective diagnostic tool involves the HPC test. Samples that yield more than 500 colony-forming units (CFU)/100 mL and have chlorine residuals less than 0.2 mg/L typically indicate stagnant water or conditions that promote biofilm growth (Geldreich 1996). Since HPC is non-selective, biofilm material can be collected from a pipe and cultured to isolate specific organisms. For instance, persistent coliform levels may indicate extensive biofilm shedding (Crozes and Cushing 2000). Biofilms can become a point source of coliforms, leading to TCR violations. Therefore, mechanisms for controlling biofilms may be of benefit to reducing coliform levels as well as other opportunistic pathogens.

Nitrification is a specific biofilm issue that is not unusual. Nitrification involves ammonia, which can be present in drinking water through either naturally occurring processes or through ammonia addition during secondary disinfection to form chloramines. The treatment protocol will impact nitrification. This includes pH, chloramination, temperature, and water age. The water distribution issue with nitrification is significant and symptomatic of other issues.

According to the EPA (2002), the microbial process for nitrification involves two steps. Step one involves the reduction of nitrogen compounds (primarily ammonia) that are sequentially oxidized to nitrite and nitrate. The nitrification process is primarily accomplished by two groups of autotrophic nitrifying bacteria that can build organic molecules using energy obtained from inorganic sources—in this case, ammonia or nitrite. In the first step of nitrification, ammonia-oxidizing bacteria oxidize ammonia to nitrite according to:

$$NH_3 + O_2 \rightarrow NO_2^- + 3H^+ + 2e^-$$

Nitrosomonas is the most frequently identified genus associated with this step, although other genera include *Nitrosococcus* and *Nitrosospira*. Some subgenera, *Nitrosolobus* and *Nitrosovibrio*, can also autotrophically oxidize ammonia (Watson, Valos, and Waterbury 1981). In the second step of the process, nitrite-oxidizing bacteria oxidize nitrite to nitrate (EPA 2002).

$$NO_2^- + H_2O \rightarrow NO_3^- + 2H^+ + 2e^-$$

Nitrite and nitrate are produced during nitrification through ammonia utilization by nitrifying bacteria. According to the prior equations, for every mole of ammonia-N produced, a one-mole equivalent of nitrite-N is produced. Subsequently, for every mole of nitrite-N produced, a one-mole equivalent of nitrate-N is produced (EPA 2002). Ammonia can also be released from chloramines through a series of complex reactions (Woolschlager et al. 2001). According to Valentine, Ozekin, and Vikesland (1998), the overall net stoichiometries can be used to examine the relationship between chloramine decay and ammonia production.

Nitrobacter is the most frequently identified genus associated with this second step, although other genera, including *Nitrospina*, *Nitrococcus*, and *Nitrospira* can also autotrophically oxidize nitrite (Watson, Valos, and Waterbury 1981). Various groups of heterotrophic bacteria and fungi can also carry out nitrification, although at a slower rate than autotrophic organisms (Verstraete and Alexander 1973; Watson, Valos, and Waterbury 1981). Speciation of nitrifying bacteria in drinking water systems (Wolfe 1990 and 2001) suggest that the number of heterotrophic nitrifiers in drinking water systems may be negligible compared to autotrophic nitrifiers, but analyses are finding this may not be true if conditions are ripe to encourage them.

It should be noted that nitrifying bacteria are obligate aerobic organisms commonly found in terrestrial and aquatic environments (Holt et al. 1995; Watson, Valos, and Waterbury 1981). Their growth rates are controlled by: substrate (ammonia-N) concentration, temperature, pH, light, oxygen concentration, and microbial community composition (EPA 2002).

Of interest is that the groundwater rule applies when water sources have no ammonia in them. This rule permits reduced monitoring if four logs of virus removal (reduction to 1/10,000th of the initial concentration) can be demonstrated. Regulatory agencies encourage this compliance. However, where utilities have ammonia in the raw water, the 4-log guidelines do not apply. As a result, the argument from the regulatory agencies has been to chlorinate the ammonia *out of the water* and then free chlorinate the water—then add ammonia back in to create chloramines (monochloramine). The nitrogen does not disappear—it just reacts with chlorine. As a nutrient, the nitrogen is still present and would potentially be available to be used as a nutrient by bacteria. As a result, this interpretation of the 4-log groundwater rule may be adding unwanted nutrients to the water and encouraging biofilms in the process.

Water pH value is an important factor in nitrification activity for two reasons. First, a reduction of total alkalinity may accompany nitrification because a significant amount of bicarbonate is consumed in the conversion of ammonia to nitrite. A model developed by Gujer and Jenkins (1974) indicates that 8.64 mg/L of bicarbonate (HCO_3^-) will be utilized for each mg/L of ammonia-nitrogen oxidized. While a reduction in alkalinity does not impose a direct public health impact, reductions in alkalinity can cause reductions in buffering capacity, which can impact pH, water stability, and corrosivity of the water.

Second, nitrifying bacteria are very sensitive to pH. *Nitrosomonas* has an optimal pH between approximately 7.0 and 8.0, and the optimum pH range for *Nitrobacter* is approximately 7.5 to 8.6. Some utilities have reported that an increase in pH (to greater than 9) can be used to reduce the occurrence of nitrification (Skadsen and Sanford 1996). According to Wilczak (2001) pH appears to be the most important factor controlling the rate of chloramine auto-decomposition. Thomas (1987) stated that the rate of chloramine decay approximately doubles for a drop in pH of 0.7 units. In Florida, Cates and Lavinder (1999) noted that raising pH reduced nitrification—a pH above 8.7 limits nitrification activity.

Hence, nitrification can be controlled. The problem is greatest when temperatures are warm, water usage is low, and the pH is below 8.6. According to Wilczak et al. (1996), nitrification is often indirectly identified via monitoring the chlorine residual, ammonia, nitrate and nitrite in the water, changes in dissolved oxygen, pH, alkalinity, and temperature (EPA 2002). Frequently, but not always, systems that have nitrification occurring may also have increases in HPC, coliform-positive test results, or both. If these issues occur, nitrification is suggested and a more active control strategy must be put in place:

- *Replace aging infrastructure*—corroding pipes and equipment provide plenty of crevices for nitrifying bacteria to escape residual disinfectant. If a system requires excessive maintenance to keep the infrastructure clean, consider replacing the problematic components with newer, less corrodible equipment.
- *Reduce water age*—nitrification will usually show up first in areas where residence time (or *water age*) is highest—for example, dead-end mains, storage tanks, and areas where pressure planes overlap. Watch these areas carefully. Disinfectant levels drop when water stands still in the system. If water usage drops, a temporary solution is to flush mains to keep new water moving in the system (TCEQ 2017). Pipe looping helps. Water conservation does not help with water age since the pipes are oversized for potable water service in most cases.
- *Routine flushing*—although Schrempp et al. (1994) noted that mechanically cleaning pipelines; draining and cleaning reservoirs; and dead-end, unidirectional, and continuous flushing were not sufficient to control nitrification at one midwestern utility.
- *Starving the nitrifying bacteria of nitrogen*—is the key to stopping nitrification (Odell et al. 1996, Wilczak et al. 1996). Hence, temporarily converting to free chlorine (disinfectant) as a part of their periodic preventive maintenance routine will be beneficial. A common control strategy involves either the maintenance of high distribution system disinfectant residuals (greater than 2 mg/L) or periodic breakpoint chlorination. Analytical survey results of 10 U.S. utilities showed that greater than 90% of distribution system samples with increased nitrite and nitrate levels, indicative of nitrification, occurred in water with disinfectant residuals less than 2 mg/L (Wilczak et al. 1996). Many utilities have found that increasing disinfectant residuals by increasing chemical doses or managing water age has helped to control nitrification. Free chlorine is more effective at inactivating ammonia-oxidizing bacteria colonies than chloramines (Wolfe et al. 1990).
- *Increase lime dosage at the water plant*—and ensure finished water enters the distribution system at a pH above 8.7; for lime softening systems, 9.

Not recommended is boosting the level of chloramination in the distribution systems because chloramines are normally less effective than free chlorine (Woolschlager et al. 2001; Valentine,

Ozekin, and Vikesland 1998). In addition, uncontrolled blending of chlorinated and chloraminated water could occur near a chlorine booster station; in some cases uncontrolled blending
has been shown to cause unintended breakpoint chlorination, increases in DBP levels, or decreases in disinfectant residuals (Mahmood et al. 1999; Muylwyk, Smith, and MacDonald 1999).

Pipes are not the only places where biofilms are an issue. Wells are perfect environments for
bacteria. Bacteria in wells tend to have an impact on the life of the wells, their operation, and the
integrity of the materials used to construct the wells as a result of biofouling degradation. The
typical agents for microbiological fouling include iron-reducing, sulfur-reducing, and slime-
producing bacteria, although many others exist. Iron bacteria, like *Gallionella*, are common in
aerobic environments where iron and oxygen are present in the groundwater and where ferrous
materials exist in the aquifer formation (such as steel-cased wells). These bacteria attach themselves to the steel and create differentially charged points on the surface, which in turn create
cathodic corrosion problems. The iron bacteria then metabolize the iron that becomes more
soluble in the process. Iron bacteria tend to be rust-colored, or cause rust-colored colonies on
the pipe surfaces.

Sulfur-reducing bacteria are often responsible for the release of hydrogen sulfide when raw
water is aerated. These bacteria are common where sulfur naturally exists in the aquifer formation, and will tend to form black colonies on pipe surfaces. While anaerobic, they will exist
in environments where aerobic conditions exist that can lead to symbiotic relationships with
aerobic organisms.

Supporting both are slime-forming bacteria. Slime-producing bacteria are found in surface
waters and in soil. The *Pseudomonas* genera are an example. *Pseudomonas* sp. are adhering
bacteria and produce a polysaccharide matrix (biofilm) that acts to protect the bacteria from
the shearing effect of turbulent flow; they resist disinfectants and provide an environment for
other species to thrive. Microbiological accumulations/biofilms pose two significant concerns.
First, the accumulations on the metallic surfaces create anodes and in conjunction with reactions caused by dissimilar metals, can lead to a steady cathodic deterioration of the well casing
and column pipe with time (with or without iron bacteria). Because the *Pseudomonads* are acid
formers, ferrous materials are particularly vulnerable to this sort of deterioration, especially in
the presence of iron bacteria (which is indicated by iron staining). Removing these bacteria is
impossible, so monitoring and periodic shock chlorination of the wells to control populations
is required when present.

Lead Service Lines and Treatment—a Connection

Lead service lines will be a big issue over the next several years. An Associated Press analysis
of EPA data found that nearly 1,400 water systems serving 3.6 million Americans exceeded the
federal lead standard at least once between January 1, 2013 and September 30, 2015 (Foley and
Hoyer 2016). The affected systems were large and small, public and private, and included 278
systems that are owned and operated by schools and day care centers in 41 states.

As noted previously, lead service lines are present in water distribution systems across Canada and the United States. In some cities, they were installed as late as 1980. In Canada, depending on the utility, up to 22% of service lines are still made out of lead (Nour et al. 2015). When
municipalities replace a lead service line, they often only replace the public section because of
private ownership of service lines past the meter. Therefore, utilities replace the public section
with PE or copper pipe and leave the private section as is, resulting in partial lead service line
replacements (Muylwyk et al. 2011). Depending on water chemistry, such practices may result

in galvanic and deposition corrosion, resulting in possible higher particulate and dissolved lead concentrations (Schock and Lytle 2011).

Physical and chemical factors control the lead release from lead service lines. Alkalinity, pH, dissolved inorganic carbon (DIC) and the oxidation-reduction potential (ORP) are some of the chemical factors influencing the lead release. Hydraulic and physical disturbances can impact the lead release from lead service lines. The lead release and corrosion rate are controlled by the scales present in the lead service lines (Schock and Lytle 2011).

Changes in water quality can increase the lead release from lead service lines, as observed in Washington, D.C., and Flint, Michigan. In Washington, D.C., the disinfectant was switched from chlorine to monochloramine, therefore reducing the ORP of the water (Edwards, Triantafyllidou, and Best 2009). The major phases of lead present in the scales on the inside of the lead service lines switched from stable lead (IV) to lead (II), therefore increasing drastically the lead release from lead service lines (Boyd et al. 2008; Lytle and Schock 2005). In Flint, the utility stopped distributing water from the City of Detroit, which is treated with orthophosphates and has a different alkalinity. Instead, water from the Flint River was distributed without adding ortho-phosphates. This drastic change in water quality caused massive lead release from lead service lines throughout the city. As a result, high blood lead levels were measured in children (Hanna-Attisha et al. 2016). These important changes in water quality are reflected in the mineralogy of the scales. More discussion on Flint, Michigan, can be found in Chapter 14.

Significant changes in scale formation were also observed at the lead-copper galvanic connection of partially replaced lead service lines, induced by local changes in pH (Triantafyllidou and Edwards 2011; Nguyen et al. 2010; DeSantis, Welch, and Schock 2009). Field and pilot studies have shown similar results (Doré et al. 2014). Hence consideration of partial change out of service lines should be carefully considered. Plastics are perhaps the best choice.

Any water quality changes can influence scales and lead release: seasonal water quality changes, treatment changes, or corrosion control optimization. There is a lag phase before new and stable scales become established. Lead release can only be quantified by water sampling to assess the exposure of vulnerable population to lead.

REFERENCES

Ammann, P., Laethem, C. L., and Kedderis, G. L. 1998. Chloroform-induced cytolethality in freshly isolated male B6C3F1 mouse and F-344 rat hepatocytes. *Toxicol. Appl. Pharmacol.* 149, 217–225.

APHA. 2017. *Standard Methods for the Examination of Water and Wastewater, 23rd Edition.* American Water Works Association/American Public Works Association/Water Environment Federation, Washington, D.C.

AWWA (American Water Works Association). 2001. *Dawn of the replacement era: Reinvesting in drinking-water infrastructure.* AWWA, Denver, CO.

AWWA. 2012. "Buried no longer: Confronting America's water infrastructure challenge." AWWA, Denver, CO.

AWWA. 1980. *Standard C400.* AWWA, Denver, CO.

AWWA. 2014. *Groundwater, M-21,* AWWA, Denver, CO.

Bachmann, R. T. and Edyvean, R. G. J. 2006. "Biofouling: An historic and contemporary review of its causes, consequences and control in drinking water distribution systems." *Biofilms.* 2(03), 197–227.

Bardet, J. P. et al. 2010. "Expert review of water system pipeline breaks in the city of Los Angeles during Summer 2009." LADWP. Los Angeles, CA.

Barrett, S. E., Davis, M. K., and McGuire, M. J. 1985. "Blending chloraminated and chlorinated waters." *Journal of the American Water Works Association.* (1), 50–61.

Batté, M. et al. 2003. "Influence of phosphate and disinfection on the composition of biofilms produced from drinking water, as measured by fluorescence in situ hybridization. "*Can. J. Microbiol.* 49, 741.

Bloetscher, F. 2009. *Water Basics for Decision Makers: What Local Officials Need to Know about Water and Wastewater Systems*. AWWA, Denver, CO.

Bloetscher, F. 2011. *Utility Management for Water and Wastewater Operators*. AWWA, Denver CO.

Bloetscher, F. 2017. "Risk and economic development in the provision of public infrastructure." *Florida Section of the American Water Works Association Annual Conference Proceedings—November 30, 2017, Championsgate, Orlando, FL*. FSAWWA, St. Cloud, FL.

Bloetscher, F., Bullock, R. J., and Witt, G. M. 2001. "Brackish water supply corrosion control issues using 316L Stainless Steel." *ASCE-EWRI Annual Conference Proceedings—Orlando*. ASCE, Reston, VA.

Bloetscher, F. et al. 2002. "Revisiting the selection of stainless steel grades in water and wastewater treatment environments—Part 1." *Water Engineering and Management*, Vol. 149, No. 5, pp. 36–44.

Bloetscher, F. et al. 2002a. "Revisiting the selection of stainless steel grades in water and wastewater treatment environments—Part 2." *Water Engineering and Management*, Vol. 149, No. 6, pp. 12–15.

Bloetscher, F. et al. 2003. "Revisiting the selection of stainless steel grades in water and wastewater treatment environments—Part 3." *Water Engineering and Management*, Vol. 149, No. 7, pp. 12–26.

Bloetscher, F., Witt, G. M., and Dodd, A. E. 1997. "Bacterial issues in raw water treatment." *Florida Water Resource Conference Proceedings—Orlando, FL*. April, 1997.

Bone C. C.et al., 1999. "Ammonia release from chloramine decay: Implications for the prevention of nitrification episodes." Proceedings of AWWA Annual Conference, Chicago, IL.

Boyd, G. R. et al. 2008. "Effects of changing disinfectants on lead and copper release." *Journal of the American Water Works Association* 100(11), 75–86.

Burn, S. et al. 2006. "Long-term performance prediction for PVC pipes." AWWARF Report 91092F, May 2006.

Cates, J. and Lavinder, S. 1999. "Improving Chloramine Residuals and Minimizing Nitrification." *Florida Water Resources Journal* v :2. 26–28.

Chapelle, F. H. 1993. *Groundwater Microbiology and Geochemistry*. John Wiley and Sons, New York, NY.

Cloete, T. E., Westaard, D., and Van Vuuren, S. J. 2002. "Biofilm formation in hot water systems." *Water Sci. and Technol.*, 95–101.

Coleman, W. E. et al. 1984. "Gas chromatography/mass spectroscopy analysis of mutagenic extracts of aqueous chlorinated humic acid. A comparison of the byproducts to drinking water contaminants." *Environ. Sci. Technol.* 18, 674–678.

Crozes, G. F. and Cushing, R. S. 2000. "*Evaluating biological regrowth in distribution systems*." AWWARF. Denver, CO.

DeSantis, M. K., Welch, M., and Schock, M. R. 2009. "Mineralogical evidence of galvanic corrosion in domestic drinking water pipes." *2009 WQTC Conference, Seattle, Washington*. AWWA, Denver, CO.

Dillon, C. P. 1995. *Corrosion Resistance of Stainless Steels*. Marcel Dekker, Inc., New York, NY.

Doré, E. et al. 2014. "Impact of stagnation patterns on particulate and dissolved lead release from full and partial LSLs." *WQTC Conference 2014—New Orleans, LA*. AWWA, Denver, CO.

Eboigbodin, Kevin E., Seth, A., and Biggs, C. A. 2009. "A review of biofilms in domestic plumbing." *JAWWA*, 100:10, pp. 131–138.

Edwards, M., Triantafyllidou, S., and Best, D. 2009. "Elevated blood lead levels in young children due to lead-contaminated drinking water, Washington, D.C., 2001–2004." *Environmental Science and Technology*. 43(5), 1618–1623.

Ellison, D. et al. 2014. "Answers to challenging infrastructure management questions." Report #4367 WRF, Denver, CO.

Emde, K. M. E., Smith, D. W., and Facey, R. 1992. "Initial investigation of microbially influenced corrosion (MIC) in low temperature water distribution systems." *Water Research*, 26(2), 169–175.

Fass, S. et al. 2003. "Release of organic matter in a discontinuously chlorinated drinking water network." *Water Research*. 37(3), 493–500.

FDEP. 2014. "Clearances, inspections drinking water permitting," November 19. 2014 https://www.dep.state.fl.us/central/Home/DrinkingWater/Workshops/Clearances%20and%20Inspections.pdf.

Federal Register. December 16, 1998. "National primary drinking water regulations. Disinfectants and disinfection byproducts." Final Rule. *Fed. Regist.* (63 FR 69390).

Flamberg, S. and Condit, W. 2013. *Primer on Condition Curves for Water Mains*. National Risk Management Research Laboratory Office of Research and Development, U.S. Environmental Protection Agency. Cincinnati, OH.

Flemming, H. C. and Wingender, J. 2001. "Relevance of microbial extracellular polymeric substances (EPSs)—Part 1: Structural and ecological aspects." *Water Sci. and Technol.* 1–8.

Foley, R. J. and Hoyer, M. 2016. "U.S. water systems repeatedly exceed federal standard for lead." *AP The Big Story*, April 9, 2016. http://bigstory.ap.org/article/5aff8cb852c94585a85c9dc5fa32e9d8/us-water -systems-repeatedly-exceed-federal-standard-lead. Accessed 4/22/17.

Forkman, S. 2012. *Water Main Break Rates in the USA and Canada, A Comprehensive Study.* Utah State University, Logan, UT.

Galli, E., Silver, S., and Withot, B. 1992. *Pseudomonas: Molecular Biology and Biotechnology.* American Society of Microbiology, Washington, D.C.

Geldreich, E. E. 1996. *Microbial Quality of Water Supply in Distribution Systems.* CRC Press, Boca Raton, FL.

Geldreich, E. E. and LeChevallier, M. 1999. "Microbiological quality control in distribution systems." Chapter 18, pp. 18.1–18.49. In: *Water Quality and Treatment (5th ed.).* Letterman, R. D. (ed.), McGraw-Hill, Inc., New York, NY.

Ghidaoui, M. S. et al. 2005. "A review of water hammer theory and practice." *Applied Mechanics Reviews* Vol. 58, No. 1, pp. 49–76. doi: 10.1115/1.1828050.

Gofti, L. et al. 1999. "Waterborne microbiological risk assessment, epidemiological validation of dose-response functions for viruses and protozoans." Epidemiology 4, S56 (abstract).

Grady, C. P. L, Jr. and Lim, H. C. 1980. *Biological Wastewater Treatment.* Marcel Dekker, New York, NY.

Grigg, N. S. 2004. "Assessment and renewal of water distribution systems." AWWA Research Foundation, Denver, CO.

Grigg, N. 2007. "Water sector structure, size and demographics." *J. Water Resource Planning and Mgmt.* Vol. 133, Issue 1, pp. 60–66.

Gujer, W. and Jenkins, D. 1974. "A nitrification model for contact stabilization activated sludge process." *Water Res.*, 9(5), 5.

Hack, D. J. 1984. "State regulation of chloramine." *Journal of the American Water Works Association.* 77(1), 4.

Hammer, M. J. and Hammer, M. J. 2011. *Water and Wastewater Technology*, 7th ed. Pearson, London, UK.

Hanna-Attisha. et al. 2016. "Elevated blood lead levels in children associated with the Flint drinking water crisis: A spatial analysis of risk and public health response." *American Journal of Public Health* 106(2), 283–290.

Holt, David. et al. 1995. "A study of nitrite formation and control in chloraminated distribution systems." In Proc. 1995 AWWA Water Quality Technology Conference; Part II. AWWA, Denver, CO.

IARC. 1991. "Chlorinated drinking-water; Chlorination by-products." In: IARC Monograph on the Evaluation of the Carcinogenic Risk of Chemicals to Humans, Vol. 52, 45–268. International Agency for Research on Cancer, Lyon, France.

Ilett, K. F. et al. 1973. "Chloroform toxicity in mice: Correlation of renal and hepatic necrosis with covalent binding of metabolites to tissue macromolecules." *Exp. Mol. Pathol.* 19, 215–229.

Jolley, R. L., Condie, L. W., Johnson, I. J., Katz, S., Minear, R. A., Mattice, J., and Jacobs, V. A. (eds). 1990. *Water Chlorination: Chemistry, Environmental Impact, and Health Effects,* Vol. 6. Lewis Publishing, Chelsea, MI.

Jorgenson, T. A. et al. 1985. "Carcinogenicity of chloroform in drinking water to male Osborne-Mendel rats and female B6C3F1 mice." *Fundam. Appl. Toxicol.* 5, 760–769.

Kenny, J. F. et al. 2009. "Estimated use of water in the United States in 2005." *U.S. Geological Survey Circular* 1344. USGS, Washington, D.C.

Kim, P. H-S. and Symons, J. M. 1991. "Using anion exchange resins to remove THM precursors." *JAWWA,* 83.

Kirmeyer, G. J. et al. 1993. "Optimizing chloramine treatment." AWWARF and AWWA, Denver CO.

Kirmeyer, G. J. et al. 1995. "Nitrification occurrence and control in chloraminated water systems." AWWARF and AWWA, Denver, CO.

Kirmeyer, G. J. et al. 2000. *Guidance Manual for Maintaining Distribution System Water Quality.* AWWARF and AWWA, Denver, CO.

Kirmeyer, G. J., Richards, W., and Smith, C. D. 1994. "An assessment of water distribution systems and associated research needs." AWWARF, Denver, CO.

Klaunig, J. E., Ruch, R. J., and Pereira, M. A. 1986. "Carcinogenicity of chlorinated methane and ethane compounds administered in drinking water to mice." *Environ. Health Perspect.* 69, 89–95.

Koch, B.et al. 1991. "Predicting formation of DBPs by simulated distribution system." *Journal of the American Water Works Association.* 83, 62–70.

Konikow, L. F. 2015. "Long-term groundwater depletion in the United States." *Ground Water.* 2015 Jan.–Feb. 53(1), 2–9. doi: 10.1111/gwat.12306. Epub. 2014 Dec. 15.

Krasner, S. W., Sclimenti, M. J., and Means, E. G. 1994. "Quality degradation: Implications for DBP formation." *Journal of the American Water Works Foundation.* 86, 34–47.

Kurtz-Crooks, J. et al. 1986. Technical Note: Biological Removal of Ammonia at Roxana, Illinois. *Journal of the American Water Works Association.* 78(5), 94–95. M/DBP Stage 2 Federal Advisory Committee (FACA2). June 27–28, 2000. Washington, D.C.

Larson, J. L., Wolf, D. C., and Butterworth, B. E. 1994. "Induced cytotoxicity and cell proliferation in the hepatocarcinogenicity of chloroform in female B6C3F1 mice: Comparison of administration by gavage in corn oil vs. *ad libitum* in drinking water." *Fundam. Appl. Toxicol.* 22, 90–102.

LeChevallier, M. W., Babcock, T. M., and Lee, R. G. 1987. "Examination and characterization of distribution system Biofilms." *Applied Environmental Microbiology.* 53:12, 2714–2724.

LeChevallier, M. W., Lowry, C. D., and Lee, R. G. 1990. "Disinfection of biofilms in a model distribution system." *Journal of the American Water Works Association.* 82(7), 87–99.

Le Curieux, F. et al. 1995. "Use of the SOS chromotest, the Ames-fluctuation test, and the newt micronucleus test to study the genotoxicity of the four trihalomethanes." *Mutagenesis.* 10, 333–341.

Lee, D.-G. and Kim, S.-J. 2003. "Bacterial species in biofilm cultivated from the end of the Seoul water distribution system." *Journal of Applied Microbiology.* 95(2), 317–324.

Lilly, P. D., Ross, T. M., and Pegram, R. A. 1997. Trihalomethane comparative toxicity: Acute renal hepatic toxicity of chloroform and bromodichloromethane following aqueous gavage. *Fundam. Appl. Toxicol.* 40, 101–110.

Logsdon, G. S. and Millette, J. R. 1981. "Monitoring for corrosion of asbestos-cement pipe." *Proceedings of the Ninth Annual AWWA Water Quality Technology Conference.* AWWA, Denver, CO.

Ludwigson, M., Rago, L., and Greenfield, J. 2016. "Horizontal of vertical high service pump selection." *FWRJ,* Vol. 67, No. 12, p 54.

Lytle, D. and Schock, M. 2005. "Formation of Pb(IV) oxides in chlorinated water." *J. Am. Water Works Assoc.* 97(11), 102–114.

Mahmood, F. et al. 1999. "Combining multiple water sources and disinfectants: Options for water quality compatibility in distribution systems." In: Proc. 1999 AWWA Water Quality Technology Conference, Tampa, Fla., AWWA. Prepared by AWWA with assistance from Economic and Engineering Services, Inc. 15.

Marlow, D. et al. 2009. "Remaining asset life: A state of the art review. Strategic asset management and communication." WERF, Alexandria, VA.

Marston, A. 1930. "The theory of external loads on closed conduits in the light of the latest experiments." *Bulletin 96.* Ames: Iowa Engineering Experiment Station.

Miller, J. W. and Uden, P. C. 1983. Characterization of nonvolatile aqueous chlorination products of humic substances. *Environ. Sci. Technol.* 17, 150–157. doi:10.1021/es00109a006.

Moser, A. P. and Kellogg, K. 1994. "Evaluation of Polyvinyl Chloride (PVC) pipe performance." AWWA Research Foundation, Project #708, Order #90644, February, 1994. AWWA Research Foundation, Denver, CO.

Munch, D. J., Munch, J. W., and Pawlecki, A. M. 1995. "Method 552.2 Determination of haloacetic acids and dalapon in drinking water by liquid-liquid extraction, derivatization and gas chromatography with electron capture detection." Rev 1, Office of Research and Development, U.S. Environmental Protection Agency, Cincinnati, OH.

Munch, D. J. and Hautman, D. P. 1995. "Method 551.1. Determination of chlorination disinfection by-products, chlorinated solvents, and halogenated pesticides/herbicides in drinking water by liquid-liquid extraction and gas chromatography with electron-capture detection." Office of Research and Development, U.S. Environmental Protection Agency, Cincinnati, OH.

Munch, J. W. 1995. "Method 524.2. Measurement of purgeable organic compounds in water by capillary column gas chromatography/mass spectrometry." Office of Research and Development, U.S. Environmental Protection Agency, Cincinnati, OH.

Murga, R. et al. 2001. "Role of biofilms in the survival of *Legionella pneumophila* in a model potable-water system." *Microbiology.* 147(11), 3121–3126.

Muylwyk, Q., Smith, A. L., and MacDonald, J. A. 1999. "Implications on disinfection regime when joining water systems: A case study of blending chlorinated and chloraminated water." In: Proc. 1999 AWWA Water Quality Technology Conference. AWWA, Tampa, FL.

Muylwyk, Q. et al. 2011. "Full versus partial lead service line replacement and lead release in a well buffered groundwater." Paper presented at the *American Water Works Association, 2011 Water Quality Technology Conference, Phoenix, AZ*. AWWA, Denver, CO.

NCI. 1976. "Carcinogenesis bioassay of chloroform." National Tech. Inform. Service No. PB264018/AS. National Cancer Institute, Bethesda, MD.

Niquitte, P., Servais, P., and Savoir, R. 2000. "Impacts of pipe materials on densities of fixed bacterial biomass in drinking water distribution system." *Water Res.* 34:6, 1952.

Norton, C. D. and Le Chevallier, M. W. 2000. "A pilot study of bacteriological population changes through potable water treatment and distribution." *Applied and Environmental Microbiology.* 66(1), 268–276.

Nour, S. et al. 2015. "Lessons and experience on the management of lead service lines by utilities and public perception." *WQTC conference 2015—Salt Lake City, UT.* AWWA, Denver, CO.

NTP. 1985. "Toxicology and carcinogenesis studies of chlorodibromomethane in F344/N rats and B6C3F1 mice." National Toxicology Program TR 282, DHHS Publ. No. (NIH) 85–2538. NIH, Bethesda, MD.

NTP. 1987. "Toxicology and carcinogenesis studies of bromodichloromethane in F344/N rats and B6C3F1 mice." National Toxicology Program TR 321, DHHS Publ. No. (NIH) 88–2537. NIH, Research Triangle Park, NC.

NTP. 1989. "Toxicology and carcinogenesis studies of bromoform in F344/N rats and B6C3F1 mice." National Toxicology Program TR 350, DHHS Publ. No. (NIH) 85–2850. NIH, Bethesda, MD.

Nguyen, C. et al. 2010. "Impact of Chloride: Sulfate Mass Ratio (CSMR) changes on lead leaching in potable water." 4088; Water Research Foundation: Denver, CO.

Ødegaard, H. et al. 2010. "NOM removal technologies—Norwegian experiences." *Drink. Water Eng. Sci.,* 3, 1–9.

Odell, L. H. et al. 1996. "Controlling nitrification in chloraminated systems." *Journal of the American Water Works Association,* 88(7), 86–98.

Oxenford, J. L. et al. 2012. "Key asset data for drinking water and wastewater utilities" [Project #4187]. Water RF, Denver, CO. http://www.waterrf.org/ExecutiveSummaryLibrary/4187_ProjectSummary.pdf, accessed 111516.

Pankow, D., Damme, B., and Wünscher, U. 1997. "Chlorodibromomethane metabolism to bromide and carbon dioxide in rats." *Arch. Toxicol.* 71, 203–210.

Pegram, R. A. et al. 1997. "Glutathione *S*-transferase-mediated mutagenicity of trihalomethanes in *Salmonella typhimurium*: Contrasting results with bromodichloromethane and chloroform." *Toxicol. Appl. Pharmacol.* 144, 183–188.

Pereira, M. A. 1994. "Route of administration determines whether chloroform enhances or inhibits cell proliferation in the liver of B6C3F1 mice." *Fundam. Appl. Toxicol.* 23, 87–92.

Pereira, M. A. and Grothaus, M. 1997. "Chloroform in the drinking water prevents hepatic cell proliferation induced by chloroform administered by gavage in corn oil to mice." *Fundam. Appl. Toxicol.* 37, 82–87.

Pereira, M. A., Knutsen, G. L., and Herren-Freund, S. L. 1985. "Effect of subsequent treatment of chloroform or phenobarbital on the incidence of liver and lung tumors initiated by ethylnitrosourea in 15-day-old mice." *Carcinogenesis.* 6, 203–207.

Pohl, L. R. et al. 1977. Phosgene: A metabolite of chloroform. *Biochem. Biophys. Res. Commun.* 79, 684–691.

Potts, D. E., Williams, W. G., Hitz, C. G. 2001. "A satellite chloramine booster station: design and water chemistry." In: Proc. 2001 AWWA. Distribution System Symposium, San Diego, California: AWWA.

Prentice, T. 2002. Overview. In: Murray, C., Lopez, A. (eds.), "The World Health Report 2002: Reducing risks, promoting healthy life." Geneva, Switzerland: World Health Organization. pp. 7–14.

Rajani, B. and Makar, J. 2000. "A methodology to estimate remaining service life of grey cast iron water mains." *Can. J. Civ. Eng.* 27, 1259–1272.

Reckhow, D. A. 2012. "Analysis of trihalomethanes and related pentane-extractable organic halides." University of Massachusetts, Environmental Engineering Research Laboratory. University of Massachusetts, Amherst, MA.

Reckhow, D. A. and Edzwald, J. K. 1991. "Bromoform and iodoform formation potential tests as surrogates for THM formation potential." *Journal of the American Water Works Association.* 67–73.

Reilly, T. E. et al. 2008. *Ground-Water Availability in the United States*, USGS Circular 1323, USGS, Reston, VA.

Reitz, R. H. et al. 1989. "Estimating the risk of liver cancer associated with human exposures to chloroform using physiologically based pharmacokinetic modeling." *Toxicol. Appl. Pharmacol.* 105, 443–459.

Rosenthal, S. L. 1987. "A review of mutagenicity of chloroform." *Environ. Mol. Mutagen.* 10, 211–226.

Sartory, D. P. and Holmes, P. 1997. "Chlorine sensitivity of environmental, distribution system, and biofilm coliforms." *Water Science and Technology.* 35(11–12), 289–292.

Sawyer, C. N. and McCarty, P. L. 1978. "Chemistry for Environmental Engineering, 3rd edition." McGraw-Hill, New York, NY. *Stage I DBP Rule FACA Meeting, 2000.

Schock, M. R. and Lytle, D. A. 2011. *Water Quality and Treatment: A Handbook on Drinking Water.* Edzwald, J. K. (ed), pp. 20–103, American Water Works Association and McGraw Hill.

Schrempp, Tom. et al. 1994. "Effect of mechanical cleaning and free residual chlorine on the chlorine demand and nitrification process in a chloraminated system." In: Proc. 1994 AWWA Water Quality Technology Conference; Part II. AWWA, San Francisco, CA.

Selbes, M. et al. 2017. "Removal of selected C- and N-DBP precursors in biologically active filters." *Journal of the American Water Works Association.* 109, 3. https://dx.doi.org/10.5942/jawwa.2017.109.0014.

64E Division of Environmental Health and Statewide Programs Chapter 64e-1 Certification of Environmental Testing Laboratories, Florida Statutes. Accessed 6/1/2017.

Skadsen, J. and Sanford, L. 1996. "The effectiveness of high pH for control of nitrification and the impact of ozone on nitrification control." In: Proc. 1996 AWWA Water Quality Technology Conference. AWWA, Boston, MA.

Song, Daniel J. et al. 1999. "Improvement of chloramine stability through pH control, TOC reduction and blending at EBMUD." In: 1999 AWWA Annual Conference Proceedings. AWWA, Chicago, IL.

Stewart, M. H. and Nancy, I. 1997. "Nitrification in chloraminated drinking water and its association with biofilms." In: 1997 AWWA Water Quality Technology Conference. AWWA, Denver, CO. Prepared by AWWA with assistance from Economic and Engineering Services, Inc. 16.

Symons, J. et al. 2002. "Removing trihalomethanes from drinking water—An overview of treatment techniques." U.S. Environmental Protection Agency, Washington, D.C., EPA/600/J-80/360 (NTIS PB82132572).

TCEQ. 2017. "Controlling nitrification in public water systems with chloramines." TCEQ, Austin, TX. https://www.tceq.texas.gov/drinkingwater/disinfection/nitrification.html.

Thomas, P. M. 1987. "Formation and decay of monochloramine in South Australian water supply systems", 12th Federal Convention, Australian Water and Wastewater Association, Adelaide, Australia, pp. 268–276.

Thompson, J. D. et al. 1997. "Enhanced softening: Factors influencing DBP precursor removal." *Journal of the American Water Works Association.* Vol. 89, No. 6, 94–105.

Thomson, J., Flamberg, S., and Condit, W. 2013. "Final report primer on condition curves for water mains." Battelle, Columbus, OH. Contract No. EP-C-05-057, Task Order No. 0062, Water Supply and Water Resources Division Urban Watershed Management Branch, EPA, Edison, NJ. https://nepis.epa.gov/Exe/ZyPDF.cgi/P100H8W0.PDF?Dockey=P100H8W0.PDF.

Triantafyllidou, S. and Edwards, M. 2011. "Galvanic corrosion after simulated small-scale partial lead service line replacements." *Journal of the American Water Works Association.* 103(9), 85–99.

Uden, P. C. and Miller, J. W. 1983. "Chlorinated acids and chloral in drinking water." *Journal of the American Water Works Association.* 75, 524–527.

U.S. Environmental Protection Agency (EPA). 1979. "National interim primary drinking water regulations; control of the trihalomethanes in drinking water; final rule." *Fed. Regist.* 44, 68624–68707.

———. 1989. "Drinking water; National primary drinking water regulations; Filtration, disinfection; Turbidity, Giardia lamblia, viruses, Legionella, and heterotrophic bacteria; Final rule." *Fed. Regist.* 54(124), 27486–27541.

———. 1994. "National primary drinking water regulations; disinfectants and disinfection byproducts; proposed rule." *Fed. Regist.* 59, 38668–38829.

———. 2001. "National Primary Drinking Water Regulations." *Fed. Regist.* 19(141).

———. 2002. "Nitrification." Office of Water. EPA, Cincinnati, OH.

———. 2002a. "Chemical Health Effects Tables." http://www.epa.gov/safewater/tcrdsr.html. May 7, 2002.

————. 2002b. "Microbial Health Effects Tables." http://www.epa.gov/safewater/tcrdsr.html. May 7, 2002.

————. 2002c. "The clean water and drinking water infrastructure gap analysis." United States Environmental Protection Agency, Office of Water, EPA-816-R-02-020, EPA, Washington, D.C.

————. July 2002. "List of Contaminants and their MCLs." EPA-816-F-02-013.

————. Winter 2004. "2004 Edition of the Drinking Water Standards and Health Advisories." United States Environmental Protection Agency, Office of Water, EPA-822-R-04-005, EPA, Washington, D.C.

————. 2007. "Health Risks from Microbial Growth and Biofilms in Drinking Water Distribution Systems." Office of Ground Water and Drinking Water. Washington D.C.

————. 2010. "Comprehensive disinfectants and disinfection byproducts rules (Stage 1 and Stage 2): Quick reference guide." United States Environmental Protection Agency, Office of Water (4606M), EPA-816-F-10-080, EPA, Washington, D.C. http://water.epa.gov/drink.

————. 2011. "Analytical methods approved for drinking water compliance monitoring under the disinfection byproduct rules." EPA//nepis.epa.gov/Exe/ZyNET.exe/P100PHKC.TXT?ZyActionD=ZyDocument&Client=EPA&Index=2011+Thru+2015&Docs=&Query=&Time=&EndTime=&SearchMethod=1&TocRestrict=n&Toc=&TocEntry=&QField=&QFieldYear=&QFieldMonth=&QFieldDay=&IntQFieldOp=0&ExtQFieldOp=0&XmlQuery=&File=D%3A%5Czyfiles%5CIndex%20Data%5C11thru15%5CTxt%5C00000022%5CP100PHKC.txt&User=ANONYMOUS&Password=anonymous&SortMethod=h%7C-&MaximumDocuments=1&FuzzyDegree=0&ImageQuality=r75g8/r75g8/x150y150g16/i425&Display=hpfr&DefSeekPage=x&SearchBack=ZyActionL&Back=ZyActionS&BackDesc=Results%20page&MaximumPages=1&ZyEntry=1&SeekPage=x&ZyPURL.

————. 2016. *Quick Guide to Drinking Water Sample Collection*, EPA, Region 8, Golden, CO.

————. 2016a. "Analytical methods approved for drinking water compliance monitoring under the disinfection byproduct rules." EPA. Washington, D.C. https://www.epa.gov/dwanalyticalmethods/approved-drinking-water-analytical-methods. Accessed 5/11/17.

Valentine R. L., Ozekin, K., and Vikesland, P. J. 1998. "Chloramine decomposition in distribution system and model waters." AWWA Research Foundation, Denver, CO.

Verstraete, W. and Alexander, M. 1973. "Heterotrophic nitrification in samples of natural ecosystems." *Envir. Sci. Technol.*, 7(39).

Videla, Hector A. 1996. *Manual of Biocorrosion*. CRC Press, Boca Raton, FL.

Washington DOH. 2003. "Total trihalomethane sampling procedure." DOH Pub #331-226 Division of Environmental Health, Office of Drinking Water, Washington State Dept. of Health, 2003, DOH PUB #331-226.

Watson, S. W., Valos, F. W., and Waterbury, J. B. 1981. "The family nitrobacteraceae. In the prokaryotes." Edited by M. P. Starr et al. Springer-Verlag, Berlin, Germany.

Wilczak, A. et al. 1996. "Occurrence of nitrification in chloraminated distribution systems." *Journal of the American Water Works Association*. 88(7), 74–85.

Wilczak, A. 2001. "Chloramine decay rate: Factors and research needs." In: 2001 AWWA Annual Conference Proceedings. AWWA, Washington, D.C.

Wolfe, Roy L. et al. 1990. "Ammonia oxidizing bacteria in a chloraminated distribution system: Seasonal occurrence, distribution, and disinfection resistance." *Appl. Environ. Microbiol.* 56(2), 451–462.

Wolfe, R. L. and Lieu, N. I. 2001. Nitrifying bacteria in drinking water." In Encyclopedia of Environmental Microbiology, edited by G. Bitton. John Wiley and Sons, New York, NY.

Woolschlager, J. E. et al. 2001. "Developing an effective strategy to control nitrifier growth using the comprehensive disinfection and water quality model." In: Proc. World Water and Environmental Resources Congress. ASCE, Renton, VA.

4

SANITARY SEWER SYSTEMS

OPERATIONAL GOALS OF SANITARY SEWER SYSTEMS

There are over 19,000 publicly owned sewer systems in the United States. Many large urban areas are served by a *regional* wastewater system—typically an older central city started the sewer system and kept expanding it to serve the suburbs. Cleveland, Detroit, Chicago, and most other large communities are configured in this manner. Figure 4.1 outlines how most utility systems are configured—from homes and businesses to the wastewater plant. Unlike water systems, sanitary sewers are gravity driven and not configured in grids. The wastewater travels by gravity until a point where the line is too deep to maintain. Then a lift or pump station is installed to lift the sewage back to the surface. Force mains, or sewer pipes under pressure, carry the wastewater from this point to another lift station or the wastewater treatment plant. Note that pumping to gravity systems and then repumping requires larger pumps and higher costs for downstream stations. Problems occur where there is grease or inappropriate materials in the sewer line that can cause blockages (more on this later). Blockages can affect large numbers of people by flooding homes and basements, spilling raw sewage on the ground, etc. Broken force mains can impact even more people. Hence, while sanitary sewer systems are designed to protect the public health by removing the waste from houses and treating it before disposal, keeping the system operating can be a challenge. According to the United States Environmental Protection Agency (EPA 2005), the goals of a sanitary sewer management program should include:

- Protection of public health
- Prevention of unnecessary property damage
- Minimization of infiltration, inflow and exfiltration, and maximum conveyance of wastewater to the wastewater treatment plant
- Provision of prompt response to the number of people and service interruptions
- Efficient use of allocated funds
- Development of a sewer use ordinance
- Identification of and remedy solutions to design, construction, and operational deficiencies
- Performance of all activities in a safe manner to avoid injuries

Sanitary sewer systems were initially developed as trenches by ancient people and served to remove both waste and excess rainfall—the first combined sewer systems. The first modern sewers were constructed in London after Dr. Snow identified waste contaminating the Broad St. well as the cause of the typhoid outbreak in 1845. The focus on sanitary sewers began in many cities before the end of the 19th century, but began in earnest in most medium and larger communities in the early 20th century. These early systems, developed in Europe and the United States, were constructed as combined sewer systems with the idea that periodic large flushes

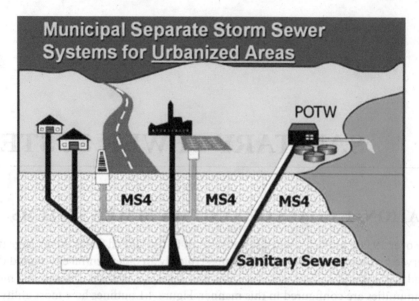

Figure 4.1   General configuration of sanitary sewer systems (black) as compared to storm-water sewer systems (orange). Sewage is moved from homes and businesses to a treatment plant mostly by gravity, and then discharged to a nearby waterway. *Source:* EPA https://cfpub .epa.gov/watertrain/moduleFrame.cfm?parent_object_id=2758

from rainfall would remove the buildups of solids and grease in the pipe from the sanitary flows, decreasing the need for maintenance.

The concept of combined sewer systems worked fine as long as populations were small, industry was limited, and treatment was not required because the receiving waters (primarily rivers and streams) could assimilate the pollution load without adverse impacts. However, as both population and population density increased and industry expanded, adverse conditions from sewage occurred—such as fish kills, increased disease outbreaks in the summer months, smells, obvious degradation, and even rivers catching fire (the Cuyahoga River in Cleveland being the best-documented occurrences). These pipes also had the propensity to overflow (sanitary sewer overflows or SSOs) during rain events due to hydraulic limitations of the piping systems used as combined sewers.

Large cities were the first to pursue sanitary sewers. But rural areas were less likely to install sewers. That changed during the Great Depression. The Great Depression-era Works Progress Administration (WPA) was responsible for the construction of sanitary sewers in many communities, especially rural and southern communities. A focus on the construction of treatment plants began shortly after World War II with the Federal Water Pollution Control Act in 1948 and its updates. Its most famous update is the Clean Water Act of 1972. The new treatment plants were constructed for average daily flows, which meant problems occurred when it rained. The biological community of organisms in a wastewater treatment plant performs well in removing contaminants from the wastewater when flows are constant and the hydraulics are consistent. Rainfall disrupts the hydraulics and the associated equilibrium by washing much of the biological community out of the plant. Good treatment of the wastewater ceases for a period of time thereafter, which impacts permit compliance. Operators and administrators were concerned that when it rained, the peak flows caused by the rainfall could not be handled by combined system treatment plants, necessitating routine discharges of untreated wastewater during high rainfall. This practice continues today in some older, Rust Belt and northeastern cities, although much effort has been undertaken to separate these older systems to prevent the co-mingling waste streams. Boston is an example of a community that has spent billions to separate these systems, an effort that continues today. The EPA has mapped 772 systems that still have some combined features (Evans 2015; see Figure 4.2).

The initial goal of the Clean Water Act of 1972 was oriented to clean up the nation's rivers and streams through the removal of untreated industrial and domestic wastewaters and restore them to their highest and best use, whether that was for fishing, drinking water, transportation, or other. The initial focus was on publicly owned treatment works/plants (POTWs). Hence, a top priority was to provide a level of wastewater service that met state and federal regulatory requirements. With the passage of the Clean Water Act, the priority shifted to separating combined systems— most of which were located in the northeast and Midwest. Newer systems are never designed for the inflow of stormwater to the wastewater system. This is why the concentration of combined systems is in older communities, as opposed to the south, west, or Rocky Mountain states where sewers were not constructed until the 1930s (exceptions like Atlanta and Memphis exist).

Today, collection system management activities form the backbone for operation and effective maintenance activities for sanitary sewer systems. Maintenance and repair of the gravity collection system includes the cleaning of pipes and manholes, the inspection of flow in the manholes, the monitoring of lift station flows, the televised inspection of gravity lines and manholes to look for leaks and breaks, and responding to complaints about blockages. Repairs include excavation and repair of manholes, gravity piping, and service connections. Field crews face many of the same issues when working in the field as other public works crews—safety from traffic, confined spaces, and chemicals with the added requirement for compliance with the federal CMOM program. CMOM stands for *capacity, management, operations, and maintenance*. The CMOM program is a federal regulatory requirement that was intended to ensure that sewer collection system piping, pumps, and wet wells are properly maintained in an effort to eliminate sanitary sewer overflows from plugged pipes or lack of pumping capacity in lift stations. Pipe is inventoried, and cleaning and repair work is tracked. Maintenance logs are

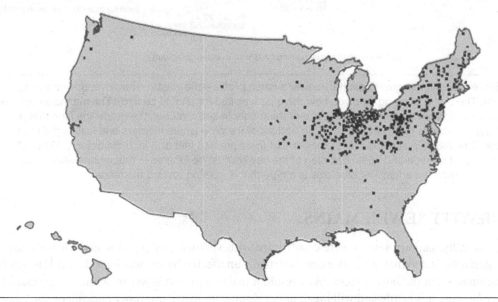

Figure 4.2   The majority of combined sewer systems in the United States are located in the Northeast and Great Lakes regions. *Source:* EPA

required for lift stations. Many wastewater collection systems may not be fully in compliance with the intent of the CMOM regulations because they lack many of the needed record details—few utilities track their work orders to identify that work has been done. The failure to track this information means that the collection system operators may lack historical information that can help improve operations, address problem areas, invest in capital assets, and maintain compliance with federal rules.

SANITARY SEWER SYSTEM COMPONENTS

Sanitary Sewer Gravity Network

The sanitary sewer gravity network is made up of gravity sewer mains, manholes, sewer laterals, and cleanouts. Figure 4.3 is a typical profile of a sanitary sewer. Laterals connect the private customer's system to the public utility network (WERF 2004). The U.S. sewer network totals approximately 800,000 miles, with force mains comprising approximately 7.5% of that total (WERF 2004). There are an estimated 12 million manholes in the United States, while the number of other appurtenances is undocumented. Over 70% of the population and most industrial and commercial enterprises are serviced by sewers. An estimated 25% of the gravity sewer network is more than 40 years old, 77% is 12 inches in diameter or smaller, and 44% is clay or concrete pipe (WERF 2004). Gravity is the primary means of flow.

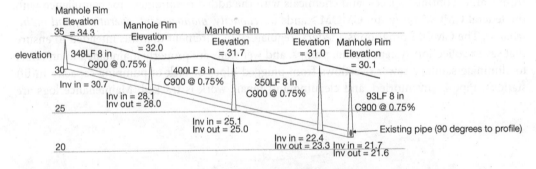

Inv means invert (either coming into the manhole or going out)

Figure 4.3 Typical gravity sanitary sewer system profile—the wastewater flows downhill to the right. The profile drawing tells you how the pipe was laid (or should be laid). The manholes are the triangles. *Top* is the elevation of the top of the manhole and matches the grade line (line that is wavy). *Inv* stands for invert. Invert is the bottom of the pipe where it enters and leaves the manhole. The *348 LF 8 in C900 0.75%* means that there are 348 feet of 8 inch diameter, C900 plastic pipe laid at a grade of 0.75%. Each leg of the sewer must be labeled in this manner. At the far right, the pipe run ends. The blue oval is a pipe that is coming toward the reader.

GRAVITY SEWER MAINS

Historically, sanitary sewers were constructed with vitrified clay pipe because vitrified clay, a material similar to that of a coffee cup, is virtually unaffected by any waste product that might be encountered in the sewer system. As a result, it makes a perfect sewer material, except that like a coffee cup, it is brittle and will break under significant loads, improper backfill, or soil heave. Vitrified clay remains in use today, although subsequent materials have been introduced, each trying to address the leakage created by breakage from pipe brittleness. Prior to World War II, the length of vitrified clay sections were as little as two feet, making the joints two feet apart. The reason for the short lengths was because vitrified clay pipe was heavy, and machination was nonexistent. The pipe had no gaskets; pipes joints were constructed using cement mortar and burlap, commonly referred to as burlap diapers, or oakum (Evans and Spence 1985). The joints were neither waterproof nor capable of dealing with differential settling, so most leaked quickly. Once machination became more common, the joints extended to four, six, and then eight feet. Today, clay pipe has a series of gaskets, compression fittings, and O-ring methods to address the potential for leaks. Vitrified clay pipe today meets ASTM C700, with ASTM C425 O-ring joints and a seal that meets ASTM C12. Table 4.1 shows typical sizes for vitrified clay pipe.

One of the less expensive pipes that was installed beginning in 1893 by the Fibre Conduit Company in Orangeburg, New York was called *Orangeburg pipe*. While its use for sanitary sewers did not occur until the 1940s, the pipe was extensively used for electrical conduit. The walls of Orangeburg pipe were made of cellulose (wood) fibers, impregnated with coal-tar pitch. The first known use of Orangeburg pipe was for a 1.5 mile long water transmission main in the Boston area, which stayed in service for over 60 years—1865 to 1927 (http://www.sewerhistory.org/chronos/convey.htm). The name of the Fibre Conduit Company was changed to the *Orangeburg Manufacturing Company* in 1948. With demand for new housing after World War II, Orangeburg pipe found a following for its small diameter sewer service laterals. Demand for fiber sewer pipe increased dramatically in the 1950s and 1960s until it was replaced by the cheaper polyvinyl chloride (PVC) options in the early 1970s. For sewer laterals, Orangeburg Manufacturing created a heavier walled version of the fiber conduit, manufactured in conformance with ASTM D 1861-73 and ASTM D 1862-73, and sold in sizes ranging from a three to eight inch inside diameter for sewer and drain applications as *Orangeburg Pipe*. Sewerhistory.org (2004) notes that: "The joints were made with couplings of similar material, utilizing no gaskets, joint sealant, etc., just simple compression, thus making the pipe potentially susceptible to inflow and infiltration or root intrusion." The pipe was lightweight (but brittle), and could be easily cut and deformed. It was noted early on that Orangeburg pipe had a tendency to deform when subjected to concentrated pressures over long periods of time. Orangeburg pipe is no longer available and should be replaced if found still in use.

Ductile iron (DI) and asbestos cement are options for sewer pipes that were installed after 1950, but neither were significant options in most jurisdictions due to corrosion potential.

Table 4.1 Typical sizes for vitrified clay pipe

Nominal I.D. (Inches)	Nominal Laying Length (Feet)	Approximate Weight (Lbs. per lineal foot)	Minimum ASTM 3-Edge Bearing (Lbs. per lineal foot)	Minimum SSPWC 3-Edge Bearing (Lbs. per lineal foot)	Nominal Wall Thickness (Inches)	Nominal O.D. Barrel (Inches)
4	4 to 6	10	2000	2000	11/16	5.45
6	4 ½ to 6	20	2000	2000	7/8	7.75
8	4 ½ to 6	30	2200	2200	1 11/16	9.9
10	4 ½ to 6	45	2400	2400	1 1/8	12.4
12	4 ½ to 6	60	2600	2600	1 5/16	14.6

Source: http://missionclay.com/band-seal-dimensions-and-specifications

These are basically the same types of pipes used for water lines (see Chapter 3 for more discussion). However, both suffer from issues with hydrogen sulfide (to be discussed shortly).

To address the brittleness and leakage problems associated with vitrified clay pipe, the *new* options were acrylonitrile butadiene styrene (ABS), and later PVC truss pipe in the 1970s and 1980s. Truss pipes had an inner and outer thin layer of ABS or PVC, and a webbing filled with Mearlcrete®. Mearlcrete is a lightweight cement, so the pipe was lighter than clay pipe. Joints were commonly 18 feet. The pipes were initially glued together but later, gaskets were introduced. Truss pipe is still manufactured as of 2016, and remains an option of choice where deflection is an issue, but it is mostly used for storm sewers, as opposed to sanitary sewers, today. It has the benefit that it deflects less than standard dimension ratio (SDR) 35 PVC pipe (truss pipe has four times the stiffness of SDR 35 PVC) but is more difficult to tap without leaks as the Mearlcrete falls out of the truss upon tapping in many instances (Ferry 2002). The ABS truss pipe had no tolerance to dry cleaning fluid and other solvents, which was an issue with integrity for the long term (Ferry 2002). ABS service laterals (not truss pipe) were brittle.

In the late 1960s and early 1970s, PVC pipe was introduced for sewer applications. It was cheap and almost immediately put the Orangeburg pipe out of business. Initially, thin walled SDR 35 PVC was used, but this was later altered by most utilities to C900 to provide better resistance to fatigue due to loading, given that deflection was a concern with the SDR 35 PVC. PVC—because it is one solid material—provides a better tapping opportunity. The standard color for PVC sewer pipes is green. Today, vitrified clay and C900 PVC are the pipe choices for most utilities.

Chin (2007) identified the flow concepts for a laminar, gravity system. Uniform pipe flow in a sanitary or storm sewer can be defined by (note that units need to be consistent—English or metric):

$$Q = a\sqrt{\frac{8g}{f}}\sqrt{RS_0}$$

Where Q is flow (cfs or m³/s), a (sf or m²) is the cross-sectional area of the pipe, f is the pipe friction factors, R is the hydraulic radius, and S_0 is the slope. The friction factor can be approximated by:

$$\frac{1}{\sqrt{f}} = -2\log\left(\frac{k_s}{12R} + \frac{2.5}{Re\sqrt{f}}\right)$$

Where k_s is the equivalent sand roughness factor and Re is the Reynolds number defined by:

$$Re = \frac{4RV}{v}$$

Where V is the average velocity, R is the hydraulic radius, and v is the kinematic viscosity of water.

$$k_s\sqrt{RS_0} = 2.2x10^{-5}$$

Note that sanitary sewers are designed to flow half full, with uniform, laminar flow—not turbulent flow.

However, if the pipe is turbulent (stormwater or heavy flows), the equation is as follows:

$$Q = A\frac{R^{.167}}{n}\sqrt{RS_0}$$

where n is the Manning's number and infinity is the hydraulic radius (pipe wall that is wet). The Manning's number is defined by:

$$n = \frac{R^{.167}}{\sqrt{8g}} \frac{1}{2\log \frac{12R}{k_s}}$$

where $4 < R/k_s < 500$ (Chin 2007). For sanitary sewer pipes, Table 4.2 outlines the slopes suggested for installation by pipe diameter.

Table 4.2 Minimum slope required for sanitary sewer pipe installation

Pipe Diameter	Min. Slope (ft./100 ft.)
8 in.	0.4
10 in.	0.28
12 in.	0.22
14 in.	0.17
16 in.	0.14
18 in.	0.12
24 in.	0.1

PVC and vitrified clay are not sufficient for large diameter gravity mains. Large pipes are primarily reinforced or prestressed concrete or DI (less common). Both materials are strong and provide leak-resistant joints. Both have a significant service life although both suffer from an issue—because the pipes are not full, hydrogen sulfide can create crown damage, whereby the top of the pipe deteriorates, creating the potential for collapse. Linings are helpful, but often the integrity of the same diminishes with time. Large diameter PVC is often not available for such applications.

Force Mains

Force mains are sewer lines that are under pressure. The force main network in the United States is younger than the gravity network—about 2% is greater than 50 years old, while 68% is less than 25 years old (WERF 2008). These mains normally arise from lift stations, but may arise from pressure or vacuum sewer applications. Unlike gravity lines, the pressure lines need to resist both internal and external pressures. As a result, vitrified clay is not an appropriate material since it does not work in tension. Most of the older force mains are cast iron (CI) or DI pipe, but the previously noted hydrogen sulfide problem can cause severe corrosion and dissolution of the crown near the manhole where they discharge—the result is a sudden collapse of the material around the connection point at a manhole or lift station. Once PVC started being used as a standard material in the 1970s, it replaced most of the DI pipe. New force mains are commonly PVC C900 pipe—just like the water mains (refer to Chapter 3). All issues remain the same for pressure sewers as exist for pressure water mains except the sewers are generally under less pressure and carry a far greater corrosion potential. Large diameter lines remain DI or prestressed concrete.

Manholes

There are over 12 million manholes in the U.S. sanitary sewer systems (Sterling, Wang, and Morrison 2009). Manholes are utilized in the system as points for change of direction, changes in diameter of the pipeline (allowing for greater capacity), and as access points for maintenance workers to clean and inspect the pipeline network. Manholes have a higher cost of failure than laterals or cleanouts since they are the major junction points of the gravity-fed sewer network and are large, and often, deep objects. Most sanitary sewer manholes are located within the roadway and are subjected to daily traffic loading. Manholes are often deep to accommodate the increasing depth of the gravity pipelines as the waste progresses downstream. The depth of the structure makes it susceptible to infiltration of ground water through any cracks that form due to failure conditions or improper construction methods. While manholes are important to the system, pipelines are substantially more critical to the success of the overall system.

Manholes are one of two types—reinforced concrete or brick. The standard diameter for manholes is four feet (inside) with an eight-inch reinforced concrete wall. The top area is a cone (see Figure 4.4), with a ring and 22-inch manhole cover (see Figure 4.5) on the top (note some areas in Texas now require a 28–30 inch cover). Brick manholes were the standard practice before 1970. Afterward, almost all manholes were constructed of reinforced concrete. Brick is a material similar to vitrified clay and as a result, it provides good resistance to hydrogen sulfide and other corrosion issues associated with sanitary sewer systems. Masons would lay the manholes in a near perfect circle to within a few feet of the surface, then create a cone at the top to the point where the manhole ring and cover were installed. The manholes are sturdy and last a long time with limited deterioration; but they suffer from one major concern—brick manholes have hundreds of joints. So despite the fact that brick manholes were normally coated with grout or cement on the outside, water finds its way into the manhole through any/many of these brick joints because while the brick is impervious, the grout is not. The grout softens and deteriorates in the presence of water and hydrogen sulfide.

Figure 4.4   Reinforced concrete manholes prior to installation—the exterior coating of coal tars prevents water from seeping into the manholes.

Figure 4.5   Example of typical manhole ring and cover.

The reinforced concrete manholes addressed part of this water-seepage issue. Reinforced concrete manholes are delivered in sections, normally four-feet high, plus cones of varying heights. The delivery matches the exact manhole depth. The reinforced concrete manholes only have joints where two sections come together. A variety of methods have been used to create a watertight gasket between manhole sections—most to a fairly successful degree. However, the manhole concrete is also not impervious to water, so an external epoxy or coal tar must be applied to prevent water from penetrating the wall (see Figure 4.4). In addition, the concrete is less resistant to the hydrogen sulfide than the brick, so periodic repairs are required to address deteriorating interior concrete. Manholes may leak, allowing soil fines from the surrounding ground to enter the manholes causing soil voids and surface settlement adjacent to the manholes.

Older manholes often have CI ladder rungs inside (newer ones, if present, are normally polyethylene). The CI is impacted by hydrogen sulfide just as the concrete and many ladder rungs deteriorated quickly. Some utilities attempt to address this issue by putting in plastic stairs and pick holes to vent the gas, but the pick holes permit water to enter the manhole. In addition, vibration and differential expansion and contraction mean that the cement bond between the concrete cone and CI ring is lost—a break exists almost immediately after installation, creating a leak. To resolve the ladder deterioration issue, many manholes today come without ladder rungs.

The connection between pipes and manholes is another area of potential concern. Original joints were simply grouted, which has the same concerns as noted previously—leakage through the grout. A series of *boots* were introduced in the 1980s. The boots were cast into the reinforced concrete manhole. The pipe was inserted and a stainless steel band held the pipe in place. The boot was flexible. Most of the pipe settled once the backfill was applied so the pipe may not match the invert of the manhole. The stainless steel was not robust, so it commonly deteriorated after a few years. A heavy O-ring, in use since the 1980s (see Figure 4.6), is the preferred solution.

Figure 4.6   The heavy O-ring has been in use since the 1980s for connecting pipes to manholes.

Laterals and Cleanouts

Finally, service laterals, often with cleanouts, are the final piece of the sewer piping system. Laterals tend to be small in both length and in diameter, but can have a steep slope to aid in the movement of the waste away from the customer. Service laterals have commonly been constructed with CI, vitrified clay, Orangeburg, ABS, SDR 35 PVC, and more recently Schedule 40 or Schedule 80 PVC. They are normally glued together as opposed to using gaskets. Repairs are often needed where the service connects to the gravity main.

A portion of the service lateral is normally owned by the property owner, the rest by the utility. It creates a challenge when a problem occurs in between the building and the sewer line. Often, problems cross the line of responsibility. Cleanouts are installed by some utilities to delineate where the public responsibility ends. Others have no cleanouts. The cleanouts are installed according to local plumbing code. Cleanouts provide utility workers and plumbers access to laterals for maintenance purposes. Cleanouts help with clearing the line for blockages so their absence may create a greater challenge for utilities when the blockage is in the line close to the gravity system, but still in the service lateral.

The biggest threat from cleanouts to the sanitary sewer network is the possibility of inflow should the cap of the cleanout be removed and not reinstalled properly. Cleanout caps are often damaged by cars or lawn mowers or removed to help residents drain their yards (please note that this practice is illegal in virtually all jurisdictions—but rarely enforced). Cleanouts are a significant source of inflow into the sewer system during rainstorms. These assets are numerous and cost very little, but may create significant cumulative stress to the sewer system.

LIFT STATIONS AND TELEMETRY

The goal of lift stations is to pump wastewater to higher ground. This is necessary because sewer lines deepen with distance. Eventually the lines will be too deep to effectively be repaired, so a pumping station, called a lift station, is installed. Sewers that follow the topography can avoid lift stations to a degree, but eventually a lift station will be required. The lift station *lifts* the wastewater via a force main to another gravity system. There are always two pumps (or more) in a lift station, each with the capacity to pump the peak hourly flow. This way if a pump fails, there is a backup to prevent sanitary sewer overflows. The wet well is where the gravity lines discharge. The wet well is commonly eight feet or more in diameter, depending on the size of the lift station. Construction of the wet well is reinforced concrete, just like manholes.

Lift stations come in two forms. The more common form is a wet well with submersible pumps. In such stations the pumps are in the wet well, submerged at all times in the sewage (see Figure 4.7). Rails and a quick connect fitting are used to pull the pumps if they need maintenance. No one goes into the wet well, although a vacuum truck may periodically be needed to remove grease. A separate pit that is located adjacent to the wet well contains plug or gate valves

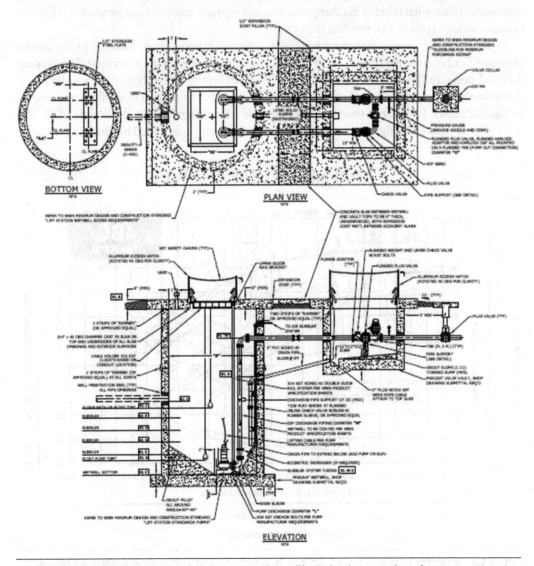

Figure 4.7 Standard lift station drawing—note that a lift station has a series of component parts; this is an example of a wet well with submersible pumps. *Source:* Broward County

and check valves. The plug or gate valves are for isolation of the pumps during maintenance; while the check valves prevent the sewage from recirculating.

The second common type of lift station is a dry well station where the pumps are in a dry pit and operate via suction pipes into the wet well. Older stations are more often of this type. These stations can lose prime, whereas the wet wells don't—which is a benefit that the submersible stations have over the dry pits. However, the pumps are not in the sewage, thereby making them somewhat easier to work on. Valves remain in the dry pit. Local preference is normally exercised with respect to which type of lift station to use.

The pumps are commonly CI and must pass a 2.5-inch ball. Wastewater pumps are different than water pumps because the wastewater has more solids, which means more wear, so the pumps must be designed and constructed to handle this load. Wastewater creates a corrosive environment, meaning that equipment wears faster as well. As a result, the materials matter. For residential sewage, there are many options—for industrial or commercial situations, the utility must understand what is being discharged so that appropriate materials can be used. Bronze, for example, deteriorates at a slower rate than CI.

To select proper pumps, one must be able to interpret pump curves. Reading pump curves identifies limitations and efficiency with any given pump. Figure 4.8 is an example. The axes are flow and pressure. There are efficiency curves based on impeller size. The efficiency should be maximized. A pump selection chart can be used to identify the desired head, flow, and impeller combinations. Pressure is lowest at the center of a centrifugal pump (Michaud 2013). The outer edge is where the higher pressure is. Pump curves can be modified by changing impeller

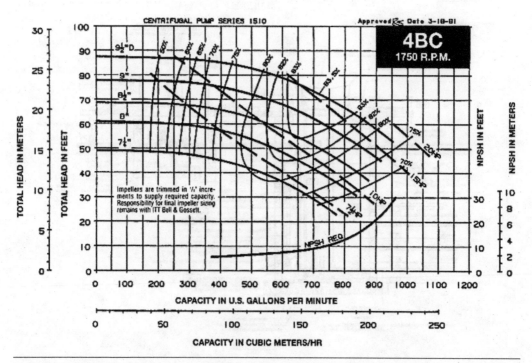

Figure 4.8   Example of pump curves. *Source*: http://jmpcoblog.com/hvac-blog/how-to-read-a-pump-curve-part-1

diameters. This can be done easily by changing the impeller, trimming the outer edge of the impeller, or trimming the vane thickness. Variable speed drives can be useful to improve efficiency by ramping pump speeds to match flow conditions. This avoids the full on/off mode. However, variable frequency drives (VFDs) can push pumps off curves—so attention is warranted.

Head loss due to valves is a consideration of pressure piping systems. Ball and gate valves have the lowest head loss; globe valves, especially small ones, have the highest. They all have significant head loss if only partially open; even at 70% open a globe valve could have 15 feet of head loss. That is a significant energy waste and indicates the need for better pump selection. Using valves to control flows should be practiced only in emergencies, not as standard practice. Valving also reduces energy efficiency. Cutting energy use on a utility system involves specifying the right pumps with the best efficiency and lowest maintenance needs possible. Pump and motor efficiency were discussed in Chapter 3. Benchmarking a plan to compare a lift station with others will help determine if energy optimization has been achieved or if further efforts may be required.

Klimes (2016) notes that pumps that fail should be analyzed to determine exact causes so that preventive measures can be taken to extend pump life. Good record keeping, work orders, and maintenance management are needed to track pump (and other equipment) performance. Common pump problems include voltage issues. If the proper voltage is not available or if the phasing is incorrect, the pump will fail prematurely. Hot pumps are an indication of a bearing issue. Bearing failures normally relate to greasing, and noise is a predecessor to bearing failure. Preventive maintenance is often the easiest solution to prolong pump life. Infrared detection equipment can be used to find hot motors and bearings, so they can be changed during off-peak hours. Flow meters are important to operations and wear detection, and they are needed to identify these issues. Table 4.3 outlines progressive pump issues (by degree of failure: 1 to 8).

Table 4.3   Progressive pump failure issues

Pump Failure Modes—Numbers Indicate Degree/Severity of Failure							
1	2	3	4	5	6	7	8
seal leaks	rotation speed	impeller incorrect or installed incorrect	impeller debris	air entrainment	bearing failure	misalignment motor-pump	improper installation
poor seal fit	impeller incorrect size	misaligned shaft	pipe defects	suction pressure (too high/ low)	packing too tight	bent shaft	motor fails
packing too loose		impeller wear	valves in wrong position/ closed/ open	cavitation from air entrainment		imbalanced shaft	control failure
			check valve failure	clearance issues		grease	
			wear	air release failure		metal shavings	coupling failure
				air binding			

Rogers and Fleming (2016) note that pumps and pumping stations should be prioritized for rehabilitation based not just on physical condition, but also on performance. High usage stations should be prioritized ahead of those with less flow. Old does not necessarily indicate condition either. Supervisory control and data acquisition (SCADA) systems and work orders are necessary to track useful data for scheduling rehab and repairs. Life-cycle analysis indicates that an asset should be replaced when the cost to maintain it exceeds the cost to own a new one. Life-cycle costs of pumps include the initial cost, energy costs, maintenance costs, potential losses due to downtime, leakage or breaks, and decommission/disposal costs.

Inside the lift station are tools to determine the level of the wastewater. The level alarms are either used to turn the pumps off, turn one pump on, turn more pumps on, or to signal that the wet well is filling faster than pumping (high level alarm), which usually indicates that pumps are not working or a check valve is stuck and the sewage is just recirculating. Having a high-level alarm trip requires a maintenance visit. Figure 4.9 shows four mercury switch alarms. The alarms trip to turn pump 1 on, turn pump 1 off (low level), turn pump 2 on, or activate a high-level alarm. Figure 4.10 shows a sonic level indicator. A sonic level indicator sends a soundwave downward. Sound bounces off the surface, and the travel time back to the sensor indicates the wastewater level in the pump stations. Sonic level indicators eliminate the issue with grease on the old mercury floats that would cause pumps to run incorrectly. Pumps can also get bound up with debris—paper towels, feminine hygiene products, cloth, diapers, and other materials—none of which should be flushed down the toilet. These are common in the collection system.

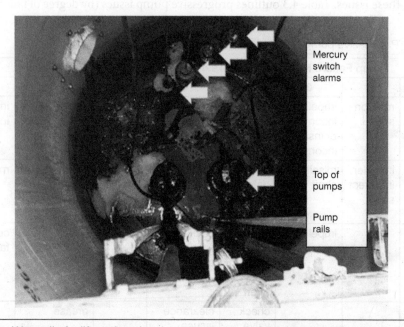

Figure 4.9 Wet well of a lift station showing two pumps and the four mercury switches that turn pumps on and off (see arrows) at the top of the pump. The pump rails are how the pumps are lifted from the bottom of the wet well to the surface so the pumps can be maintained without requiring entry into the wet well.

Figure 4.10   Wet well with sonic level indicator (see arrow) that can be used in place of mercury switches to operate pumps.

For example, a two-night investigation, opening 350 manholes of a south Florida utility in late 2017, yielded the following, with commentary:

1. Four rats—one ran across the pavement and jumped into the manhole when opened.
2. Lots of cockroaches—but not as many as expected (and people wonder how they get into the house).
3. Fatbergs (large masses of solid waste)—caused by way too much grease going down the drain, mostly from cooking. Customers need constant reminders that they must *not* put grease down the sink! Fatbergs are colored based on the grease used; in this case black, white, and tan grease were found.
4. Paper towels—also, should *not* go down the toilet!
5. More grease—see #3, except this was different cooking grease (yes, experienced personnel can tell who lives where by the grease in the sewers).
6. Handi-wipes and diapers.
7. Clothing—this was near a commercial laundry.
8. A gold watch or bracelet—the staff didn't try to figure out which or to try to remove it.
9. Condoms—please do not flush; they can easily block a drain pipe.
10. More grease—only this time it was more like auto grease mixed with paint.
11. The usual feminine hygiene products—none of them should go down the toilet regardless of what the box says, especially applicators which are plastic. Plastic does not degrade in a sewer plant; in fact, it floats. Somebody has to scoop them all off the top of the clarifier at the wastewater plant. It's the same for the *blue bags* that are used for disposal and anything with strings that could get wound around the pumps.
12. Adult diapers—near elderly care facilities and elderly apartments. Diapers go in the trash, not the toilet.

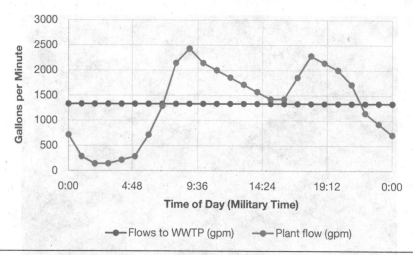

Figure 4.11   Example of peaking on the sewer—the peaks start in the early morning and after dinner; night flows are minimal. Note, it is similar to peaking for the water plant but somewhat attenuated (muted) and delayed.

For busy lift stations, controls are helpful. The controls can alert operators to problems immediately, saving on manpower required to check stations frequently. Ongoing maintenance and rehabilitation are important, as is telemetry for tracking data. It is particularly a concern if the utility purchases bulk wastewater treatment services (which means higher costs to the utility). Monitoring flows will indicate if there is a need to review reducing infiltration, preventing excess inflow, and identifying locations for the same. Like water treatment systems, there are usually peaking factors that should be understood by operations staff so they can identify changes that should be investigated (see Figure 4.11).

SANITARY SEWER PIPING SYSTEM CHALLENGES AND SOLUTIONS

Like any infrastructure system, there are a variety of failure modes for the sewer system. As pipes increase in diameter, the expected capacity of the asset increases and the associated cost of failure also increases. This is also true for longer pipes, or pipes that are deeper. Typically, pipes further downstream will have the highest cost of failure, as downtime for these assets will cause failure in other assets as well.

Frequently observed damage in gravity sewer system pipes include (WERF 2004):

- Cracks due to vibration, poor construction techniques, traffic loads, or fatigue
- Joint misalignment/separation/leakage
- Lateral connection/leakage
- Incorrect grade and alignment
- Excessive pipe deflection
- Root intrusion into the pipe

Cracks in pipes come from many issues since the pipes are subjected to a number of potential failure conditions that relate to their physical properties. Long, narrow pipes act as beams and experience loading differently than pipes of larger diameters with shorter runs and as a result, may be subject to loadings like a beam (see Chapter 3 for discussion of failure modes for pipe). Construction issues like not removing rocks and debris, laying pipe without accounting for the bell, and poor backfill can lead to breaks and cracks. Additional factors that effect pipe cracking are the depth below the surface, exposure to traffic loading, age, and material. Misalignment may be a bedding issue or associated with construction (the lack of inspection). Tree roots seek out gravity pipes because they sweat—a good source for water. The roots wrap around the pipe and can enter through breaks or joints. Leakage will result from all of these deficiencies.

Manholes and laterals may have similar issues. Infiltration and inflow are ongoing challenges. However, repairs are a challenge due to the longevity of the basic elements of many old piping materials, finding replacement parts, deep burial, the cost of repairs, and heavy traffic in the location of the repairs. As a result, much of the focus for sanitary sewer repairs are focused on utilization of the existing piping systems to extend their life (slip lining, resin repairs, pipe bursting, etc.—to be discussed in future paragraphs). The annual market for rehabilitation of wastewater infrastructure in the U.S. was reported to be approximately $4.5 billion in 2003 (WERF 2004). The sewer rehabilitation market in the rest of the world is approximately as large as that in the United States.

The repairs for sanitary sewer systems was focused on entry into the sewer system until the 1970s, and those repairs were mostly focused primarily on large diameter pipes that people could get into (>36 inches). Most of this work was *in situ* repair of damaged sections, such as mortar in brick sewers or interceptor lines. However, there is significant risk to manned entry into the system—residual hydrogen sulfide kills over 50 sewer workers every year. With lift stations, particularly, care should be taken when entering. People die in these confined spaces because they are overcome with gasses they cannot smell and did not take the time to check (Prentiss 2010). The occurrence is similar each time. The first guy goes down into the confined space. He collapses. Then the *buddy* goes to get him out. He collapses. By the time people figure out there is a problem, they are both dead. In 2017, three people, including a firefighter, died in such a situation in Key Largo, Florida. Avoidance of entry is the best option.

Lift station wet wells, manholes, and catch basins are *confined spaces*. OSHA defines a confined space in OSHA 1926.21(b)(6)(ii) as: "any enclosed or confined space having a limited means of egress and which is subject to the accumulation of toxic or flammable contaminants, or an oxygen deficiency" (Griffin 2010). When it is impossible to avoid entering a confined space, proper equipment and training must be provided. Confined space entry rules require a permit, a two-man crew with a radio, a harness, and detection of gases before entry. Many utilities issue their own permits to the crew so other crews know what is happening. Thus, permits must also be secured along with the training and equipment. A crew needs to be on hand in case there is an issue. The point is that confined space deaths are often preventable by following the rules.

Smaller sewers or sewers with more significant problems are repaired via open trench operations. Due to the difficulty in doing this, rehabilitation work is unlikely to take place unless a full collapse occurs. With the sewer system aging to 50+ years by the 1960s, the number of collapses has increased, especially in the U.K. and other Western European countries. The need to address such collapses and the associated disruption and loss of service created the idea of using cured-in-place pipe (CIPP)—invented in the early 1970s. The process of installing a lining in sewers to address breaks and cracks was developed by the enterprise that is known as Insituform. The process, and its variants, remains the market leader today (WERF 2004).

MAINTENANCE OF SANITARY SEWER SYSTEMS

Infiltration/Inflow Reduction

Getting funding for sewer collection maintenance is often difficult and may result from a failure to understand the risks related to system failures, the fact that buried pipes are out of the public view and a lack of understanding of the components of infiltration and inflow (I/I). I/I is a term that describes how water enters the sewer system—please note that while the industry commonly uses these terms together, they are actually very different concepts with very different solutions. Water entering sanitary sewers from the surface (i.e., storm drains, broken or missing cleanout caps, cross-connected sanitary and storm pipes, manhole lids or rings) is called inflow. Groundwater that enters the sanitary sewer system through defective pipe joints, broken pipes, leaks in manhole walls, tree roots, cracked or broken connections between the main line and laterals, and broken service laterals is called infiltration (Bloetscher et al. 2015). The EPA has established non-excessive/excessive infiltration criteria depending on the footage of collection sewer in the area (see Table 4.4). The criteria in this table are used as a primary indicator for the assessment and classification of collection system infiltration. To generate the infiltration data, each section of the sewer system should be flow monitored. Areas should be ranked in descending order by the measured infiltration in gallons per day per inch-mile of collection sewers, and the results compared against the EPA criteria to classify those areas having *excessive* infiltration. An easier and faster route that does not require entering the sewer is to monitor flow after midnight. Manholes in residential areas should have minimal flow at this time of day. Experienced personnel can differentiate showers, toilets, and sinks from infiltration. At the same time, to protect the current investments, the utility should monitor, televise, and line areas of the system that develop leaks. Table 4.5 shows the types of work conducted and their typical costs.

The rehabilitation costs can be significantly different in each pump-station collection area, depending upon its conditions. Experience based on past projects indicates that repair costs of

Table 4.4   EPA non-excessive/excessive infiltration criteria

Allowance Range (gpd/in-mile)	Sewage Footage (ft.)
2,000–3,000	>100,000
3,000–5,000	50,000–100,000
5,000–8,000	1,000–50,000

Table 4.5   Sanitary sewer evaluation study (SSES) procedures and associated typical costs

Procedures	Standard SSES	Modified SSES
1. Manhole inspection	$500.00 each ($0.22/LF)	—
2. Smoke testing [b]	$1.00/LF	$0.30/LF
3. TV inspection [a]	$1.25/LF	$1.25/LF
4. Evaluation [a]	$1.65/LF	$1.60/LF

[a] Assume average line segment is 250 feet
[b] Assume test performed on every 7,000 feet

Table 4.6 Preliminary selection guide for repairs of various defects

Defect	Repair Options	Defect	Repair Options
Inflow	Manhole dishes Smoke test LDL plug Cleanout caps Lateral repairs	Pipe break	Point repair Pipe bursting
		Odors	Ozone Carbon Peroxide Biological compounds
Infiltration	Televising Slip-lining LMK-style lateral liner Pipe bursting/liners		
		Blockages	Education Degreasers Grit chambers Grinder pumps Pressure jet
Grease	Degreasers Pressure jet		

construction and program management vary from $12 to $50 per linear foot (LF) and could be over $50/LF if no previous repair work has been done in the system. For those collection systems that have been through the initial rehabilitation programs, a repair cost of $12 per foot of sewer is typical. Laterals can be a significant source of I/I. Table 4.6 was developed based on field findings as a preliminary selection tool for repairs. An ongoing appropriation for I/I repair should be included in each annual budget. Borrowing should be avoided unless a huge expenditure is anticipated or the utility is trying to catch up on deferred repairs. Televising without prior review of the conditions after midnight may waste a lot of time and resources.

Inflow

For the purposes of this section, the discussion will avoid combined sewer systems and focus on designated sanitary sewers only. In the United States, 4% of the total power produced is used by water and wastewater systems (Lisk, Greenberg, and Bloetscher 2012). In both cases, as much as 90% of that power is used for pumping. In wastewater collection systems, much of that cost is lift stations, especially in flat areas like Florida where utilities routinely maintain lots of lift stations. As a result, dealing with the inflow portion of I/I is needed. In the lexicon of sewer collections, it must be clear that *infiltration* and *inflow* are not the same, and storm events do not highlight infiltration nearly as well as they do inflow because inflow is a direct connection to the surface where the rain is.

Utilities have long dealt with the I/I issues in their system by televising their pipes and identifying leak points, but this primarily addresses only the infiltration part of I/I. Inflow, which creates hydraulic issues during rain events, leads to sanitary sewer overflows, subjects the utility to fines from regulatory agencies, and can create negative publicity and legal issues from damage caused by sewage backing up into homes and businesses. Internal pipe inspection does nothing to address inflow—and searching for inflow can be costly. Where there are peaks in wastewater flows that match rainfall, inflow is the more likely candidate for the cause of the peaks than infiltration from pipes that are constantly under the water table.

Wingard (2015) reported that inflow can contribute 90% of flows under storm events, and needlessly tie up capacity of pipes and treatment facilities. As an example, the city of Austin experienced a 354% increase in flows with a 10 inch rainstorm in 2014 (Rizk 2015). The result was SSOs, plant overflows, treatment plant bypasses, and flooded homes. The city of Baltimore reported 450 basement floods a month on its 1,300 mile gravity system (Santino, Driscoll, and Hussam 2015). Neither situation is acceptable—and both subject the utility to financial risk from fines or lawsuits.

The peaking that is shown in Figure 4.12 correlates with the rainfall and thus is *inflow*, not infiltration, since infiltration is part of the base flow that creeps upward with time. Infiltration looks much like the base flow. For the utility in Figure 4.12, the average daily flow from the water distribution system is just under 6 MGD (millions of gallons per day), but the wastewater flow is over 7 MGD, indicating infiltration and inflow were present (sewer flows are typically about 70% of water pumped to the distribution system).

Even manhole tops can contribute to inflow. The National Association of Sewer Service Companies' (NASSCO's) manhole committee noted that two one-inch pick holes in a manhole cover can leak 45 gpm (gallons per minute) into the sewer system (Farruggia 2017). Manholes without pick holes can leak 12 to 27 gpm with less than two inches of water on top. Miami-Dade County found even pick hole-less manholes leaked over 10 gpm. Removing inflow reduces the risk of damages and fines to the utility, damage to property owners, and the potential for sewage in people's yards. Consequently, when plant operators see peaks in flows after rain events, it is indicative of active connections from the surface to the piping system which is *inflow*. The good news is that simple, low-tech methods can be used to detect and remove inflow, which should be the precursor to any infiltration investigation. Removing inflow to start the process often leads to a more focused plan for an infiltration correction program.

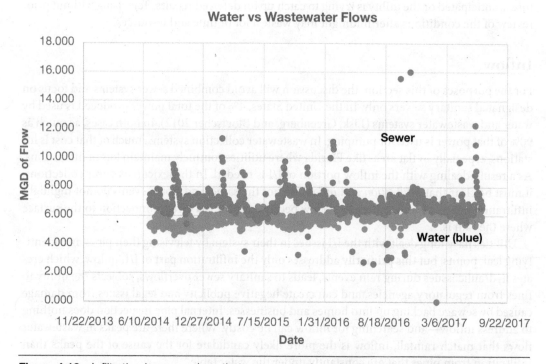

Figure 4.12 Infiltration in sewer pipes that are constantly under the water table—the utility average daily water is just under 6 MGD, but the average daily wastewater flows were over 7 MGD.

To begin inflow removal on a sanitary sewer system, the first step is to inspect all sanitary sewer manholes for damage, leakage, or other problems, which while seeming obvious, is often not performed by field crews. Manhole inspections can also document useful information like excessive grease buildup, improper disposal of feminine hygiene products, baby wipes, diapers, clothing, and other trash (see Figure 4.13a–d). It should be noted that 17% of people regularly flush baby wipes down the toilet because manufacturers say they are flushable (Stratton-Childers 2015), when in reality, they take longer to deteriorate and often cause clogging. The manhole inspection should document the condition and GPS location of the manhole and tie it to photographic data and some form of numbering. Most manholes have limited condition issues, but where the bench or walls are in poor condition, they should be repaired with an impregnating resin such as calcium aluminate. Deterioration may be an indication of wastewater

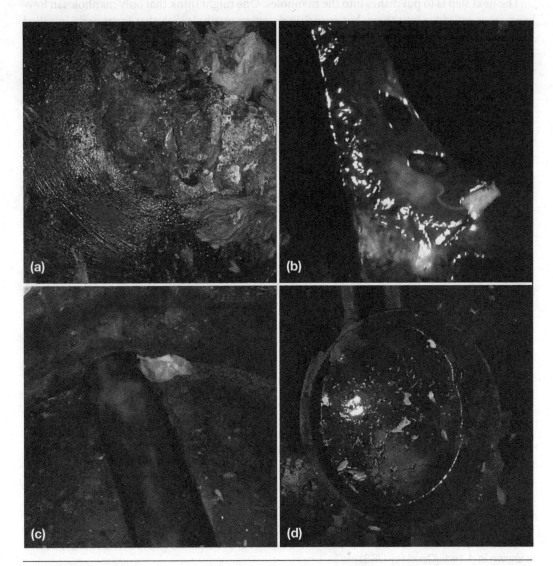

Figures 4.13   (a–d) Example of items that get into manholes that should not be there—clockwise from top left: (a) grease, (b) feminine hygiene products, (c) plastic dishes, and (d) baby wipes or hand wipes

quality concerns requiring the addition of chemicals to reduce the impact of hydrogen sulfide. White deposits are minerals that indicate leakage. Manholes should have cockroaches in them—if they do not, that should be noted for further investigation into the potential of toxins in the sewer system. Rats will also often be found in the sewer system.

Next is the repair/sealing of chimneys in all manholes to reduce inflow from the street during flooding events. The chimney includes the ring, cement extensions, lift rings, and brick or cement used to raise the manhole ring. Manhole covers are often disturbed during paving or as a result of traffic. Temperature, vibration, and traffic will break the seal between the steel ring and concrete. The crack between the ring and cover can leak a lot of water. The intent of the chimney seal is to prevent inflow from the area beneath the rim of the manhole, but above the cone (see Figures 4.14a–d).

The next step is to put dishes into the manholes. One might think that only manholes in low-lying areas get water into them, but surprisingly every manhole dish that is properly installed has water in it. As a result, utilities should assume that all manholes leak water between the rim and cover. Most collection system workers are familiar with dishes at the bottom of the manhole where they are of limited use. This is because those dishes deform when filled with water or are

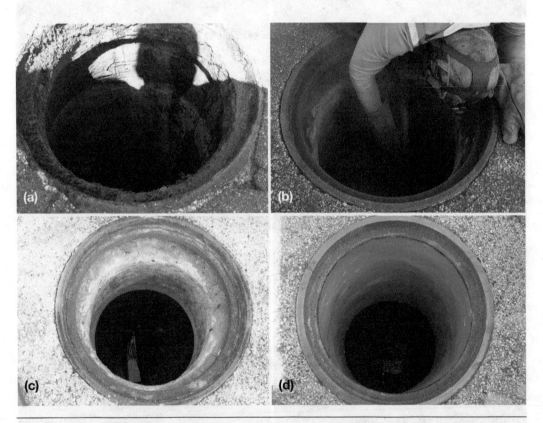

Figures 4.14 (a–d) Four-step chimney seal installation procedure—the first step is to locate the manhole and inspect it. (4.14a) The manhole must be sandblasted (4.14b) to remove debris and loose material (4.14c). Then a primer coat and a final coat of sealer are applied. The final product is shown in 4.14d. (Courtesy USSI Inc.)

constructed in such a manner that allows them to be knocked in when the cover is flipped. The solution is a deeper dish with reinforcing ribs and a gasket (Figure 4.15). The polymer-based dishes eliminate the dissimilar metals issues with stainless steel dishes and are available at a lower cost. The key is the appropriate reinforcement to prevent dishes from dropping into the manhole. The gasket seal should be made of a closed cell neoprene material with pressure sensitive adhesive on one side or a gasket set in a groove, such as an O-ring.

Once the manholes are sealed, smoke testing can identify obvious surface connections (see Figure 4.16a). During smoke testing, an odorless, non-toxic, vegetable-based smoke is blown into the sanitary sewer piping system. The smoke travels up and down the piping and should rise out of the vent stack at each house. However, if there is a connection to the storm drains, broken cleanouts, broken pipes near the surface of other entry points, smoke will rise from those sites as well (see Figures 4.16b–d). The normal protocol for smoke testing will identify broken or missing cleanout caps, surface breaks on public and private property, connection of gutters to the sewer system, and stormwater connections. All should be documented via photograph, by associated public or private address, and by GPS location. The removal or accidental breaking of cleanouts, unsealed manhole covers, laterals on private property, connected gutters or storm ponds, damaged chimneys from paving roads, or cracked pipe may cause a significant source of inflow to the system that can be identified easily with smoke testing. The public openings at cleanouts can be corrected immediately using utility funds. If the cleanout is broken, it may indicate mower or vehicle damage that can occur again. If missing, the resident may be using the cleanout to drain the yard (more common than utility personnel realize). In either case the collection system needs to be protected with a device like an LDL® plug that will be installed in the cleanout pipe below the cleanout cap (see Figure 4.17). The LDL plug will seal the cleanout even if the cap is removed.

Figure 4.15 Inflow defender manhole rain dish—note it is the ribs and depth of the dish that improves long-term strength.

Figures 4.16 (a–d) Example of smoke testing. Figure 4.16a is the smoker used to deliver smoke into the sewer line. The smoker pumps smoke into the sewer system. Figure 4.16b shows a typical street with numerous surface connections; leaky systems will have extensive smoke in the neighborhood. Figure 4.16c shows a front yard with smoke coming up, which indicates a broken line. Figure 4.16d shows a person documenting the leak for later correction. (Courtesy USSI Inc.)

At this point, the inflow problem should be addressed to the extent possible. The utility can graph rainfall with flows to verify that the flow does not change appreciably immediately after rain.

The next step should be to inspect the system to identify high priority areas for further review. The best time to do this is between midnight and 5:00 a.m. since this is the time when flows are normally lowest. Residential developments with flow at night are likely leaks. A trained engineer or inspector can differentiate a toilet flush or shower from groundwater or inflow. Where there is flow, these pipes should be noted. Figure 4.18 is an example of the results of a midnight investigation. Here the red highlighted pipes showed leakage. Leakage can be from the main line or services (see further discussion later in this chapter).

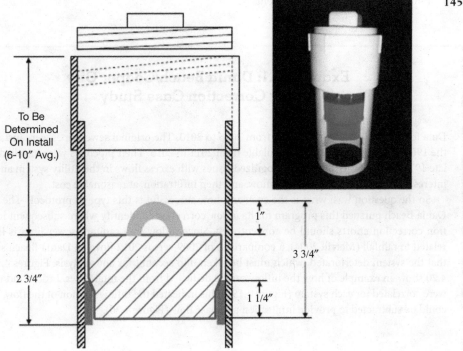

Figure 4.17 The LDL plug is placed into the broken or missing cleanouts during the smoke test. *Source*: USSI Inc.

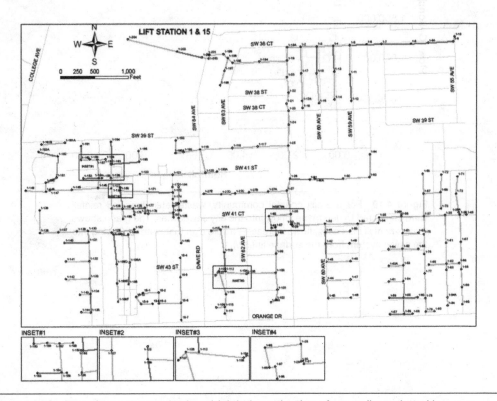

Figure 4.18 This shows the result of a midnight investigation of sewer lines plotted in a geographic information system (GIS) program for the Town of Davie, Florida. The red pipes are ones that were found to have groundwater infiltration, while the green were not. The next phase would have the red pipe televised to find the breaks.

Example 4.1: Dania Beach/Cooper City
Inflow Correction Case Study

Data for Dania Beach was gathered from 2005 to 2013. The original sewer pipes were installed in the 1960s. Cooper City data was available from 2011 to 2013. Their pipelines were installed in the late 1970s. Both utility directors recognized issues with excess flows in the utility system and were interested in resolving issues with inflow, and then infiltration, at reasonable cost.

So the question was: what is the cost, and how successful is this type of protocol? The city of Dania Beach pursued this program for its *inflow* correction to identify where subsequent infiltration correction efforts should be concentrated. Since inflow into sanitary sewer lines is linearly related to rainfall (Merrill 2004), a comparison of 2009 versus 2012 flows in Dania Beach showed that the system deteriorated, which must be taken into account for any analysis. Figures 4.19 and 4.20 show an example of how the inflow can be deducted from the total flows. Inflow and rainfall were correlated for each system (Figure 4.19). This indicated the inflow portion of the flow. Inflow could be subtracted to provide infiltration and base flow (Figure 4.20).

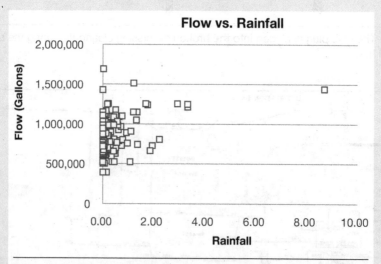

Figure 4.19 For a small coastal community, wastewater flow and rainfall were compared to determine inflow into the city sewer lines. The graph shows that as rainfall increases, flows increase—a sure sign that water is flowing into the sewer system from the surface (inflow).

continued

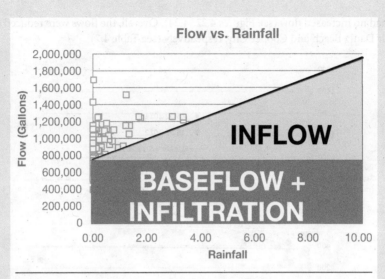

Figure 4.20   Using that same community, inflow can be separated from the base flow by determining where the typical daily flow intersects the axis at zero rainfall. When there is no rainfall, the flows are just under 800,000 GPD. Consequently, any flow above 800,000 represents inflow from the surface.

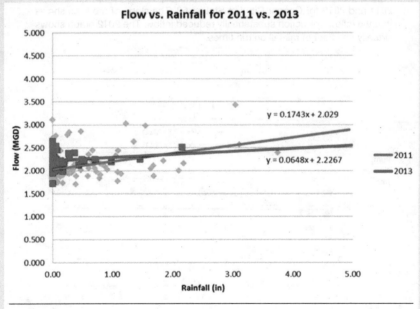

Figure 4.21   Inflow amount diminished in the city of Cooper City, Florida—note that the 2013 relationship between rainfall and flow diminished (lower slope) after the inflow correction measures were taken in 2011.

The nearby city of Cooper City conducted a similar study and once the inflow correction efforts were all completed, the inflow amount diminished (see Figure 4.21) to a point where rainfall makes limited impact on system flows. Breaking down a series of lift stations in Cooper City indicated

continued

that most had no increased flow (see Figures 4.22–4.24). Overall, the flows were reduced over 30% and 60% for Dania Beach and Cooper City, respectively (see Table 4.7).

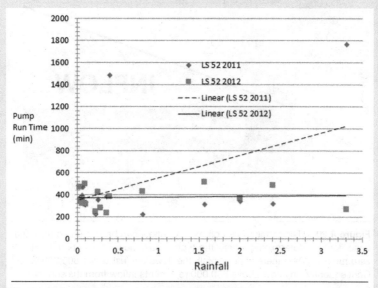

Figure 4.22 Comparison of rain events (inches) versus pump run times in 2011 and 2012 for Cooper City Lift Station 52. The slope of the lines shows that the inflow correction substantially reduced inflow. The 2012 graph shows virtually no effect of rainfall on run times.

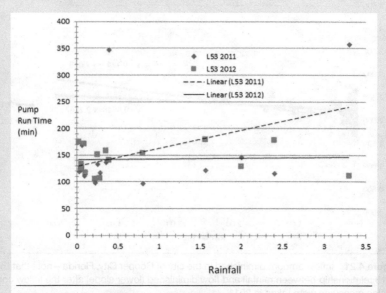

Figure 4.23 Comparison of rain events (inches) versus pump run times in 2011 and 2012 for Cooper City Lift Station 53. The slope of the lines shows that the inflow correction substantially reduced inflow. The 2012 graph shows virtually no effect of rainfall on run times.

continued

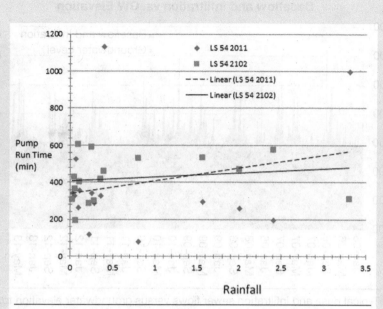

Figure 4.24 Comparison of rain events (inches) versus pump run times in 2011 and 2012 for Cooper City Lift Station 54. The slope of the lines shows that the inflow correction reduced inflow.

Table 4.7   Flow reductions for the cities of Dania Beach and Cooper City

Community	% Change (Pre vs. Post)
Dania Beach (2006 vs. 2009)	−31.90%
Cooper City (2011 vs. 2013)	−62.80%

The results of the two case studies shows that inflow is separate from infiltration, that the peaks in flows are inflow and can be removed relatively easily, that the costs are reasonable, and that the solutions are relatively simple. Getting the right technology and specifications is important.

Infiltration

Infiltration is the water entering a sewer system and service connections from groundwater leaking through defective pipes, pipe joints, damaged house lateral connections, or manhole walls (Feeney et al. 2009). Infiltration most often is related to a high groundwater table that is observed during a wet season or in response to a severe storm. Damaged lateral connections are thought to be a major contributing factor to infiltration in sanitary sewers (Swarner and Thompson 1994). Figure 4.25 shows the typical base + infiltration flows versus groundwater elevation for Dania Beach, Florida. Figure 4.25 shows that as groundwater levels rise, the flow increases (spikes). Figure 4.26 shows that deterioration continues with time and must be accounted for in any analysis. In this graph, 2013 has higher flows than 2012 (Cooper City, Florida). Figure 4.27 shows that the correlation between groundwater and base flow of the sewer

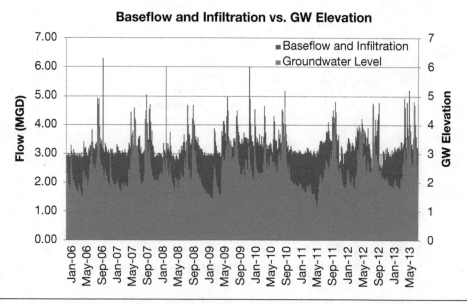

Figure 4.25   Typical base and infiltration sewer flows versus groundwater elevation for the city of Dania Beach from January 2006 to May 2013—note that as groundwater increased, so did flows. This is infiltration into the sewer pipes from groundwater. The solution is very different than when the inflow of water is from the surface.

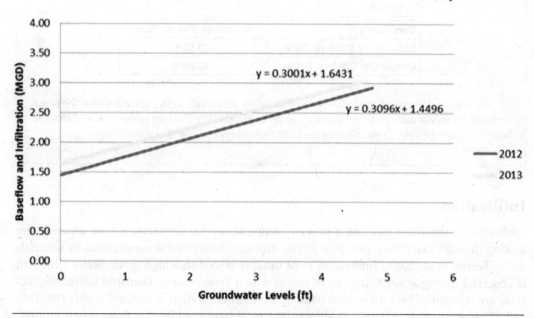

Figure 4.26   This graph provides further evidence that sewer pipe deterioration continues with time and must be accounted for in any analysis (line moved upward). This data was obtained from Dania Beach.

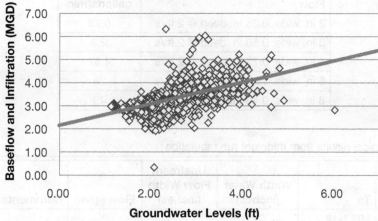

Figure 4.27   The correlation of groundwater levels (ft) versus flow is evident for Dania Beach, Florida—note the intent is to determine how much of the base plus infiltration flows are actually infiltration only. Higher groundwater levels will increase infiltration.

is not time dependent. There is a relationship between groundwater level and sewer flows that indicates that the pipe system leaks.

A low-flow investigation, which is intended to target the location of infiltration in the piping system, should be the first step to define the degree and location of infiltration issues (after the inflow has been addressed). Typically, the best time for such an analysis occurs from 1:00 a.m. to 5:00 a.m. when people are sleeping and not using water. Ground water infiltration flows do not fluctuate greatly in sanitary sewer lines during wet weather conditions. However, there are seasonal variations due to changes in the water table. For example, in south Florida, the highest levels of ground water occur in October when the water table is at its highest (E Sciences, Inc. 2013); this is different from most other parts of the country when the water table is at its highest in late winter and spring. The result will be a relationship between groundwater elevation values and flows (adjusted for days of the week in many cases). A midnight investigation event will take several days and must be planned to determine a priority manhole to start with and then sequencing. Based on a projected plan, the following is the protocol based on identifying where there is and is not flow:

- Open the manholes
- Inspect them for flow
- Determine if flow is significant, if investigation of a basin will end and a new basin will be started.
- If flow exists, open consecutive manholes upstream to determine where flow is derived from. Generally, a two-inch-wide bead of water indicates *significant* infiltration (see Table 4.8).

The results of the midnight investigation can be tabulated (see Table 4.9 for an example) and plotted in a GIS (as was shown in Figure 4.18). The red lines are the ones where leaks were found. The green lines did not show any leaks. After correcting inflow, a midnight or low-flow

Table 4.8 Estimated flow volumes for eight-inch pipe given width of flow in the invert

Flow	gallons/min
2 in. wide, 0.25 in. deep @ 2 ft./s	0.78
3 in. wide, 0.50 in. deep @ 2 ft./s	2.3
4 in. wide, 0.75 in. deep @ 2 ft./s	4.67
5 in. wide, 1.25 in. deep @ 2 ft./s	28
8 in. wide, 0.40 in. deep @ 2 ft./s	49

Table 4.9 Typical results from midnight run tabulation

From	To	Width Water (inches)	Upstream Flow Width (inches)	Flow (gpm)	Comments
MH 1-11	MH 1-12	3		2	
MH 1-15	MH 1-3	4		4	
MH 1-30	MH 1-40	3	3	0.5	
MH 1-30	MH 1-41	3		2.5	
MH 1-31	MH 1-32	2		0.35	
MH 1-4	MH 1-13	5		30	
MH 1-40	upstream	8		150	Flow coming from lift station force main
MH 1-41	MH 1-42	4	3	35	
MH 1-42	MH 1-43	3		3	
MH 1-5	MH 1-6	6	2	40	
MH 1-6	MH 1-7	5	2	30	
MH 161	lateral	2		0.35	

investigation can help utilities target the specific lines where infiltration correction is needed—negating the televising and cleaning of miles of pipe where no damage is found. This saves the utility money and brings a greater return on invested dollars in the form of reduced flows. Table 4.10 outlines the relative cost for the Dania Beach system. Note the midnight investigation saved the city $1.2 million on the infiltration phase by eliminating unneeded televising of sewer lines. However, eliminating pipe that does not need to be televised can only be done efficiently once the inflow is removed.

Once the midnight run indicates where the leaks might be, the associated rehabilitation and replacement work depends on some form of visual inspection and assessment of the sewer line. However, unless a midnight investigation is conducted first, many miles of pipe assessments will show nothing. A risk to systems that do not perform a midnight run first is that the lack of *results* from televising may cause local officials to perceive that the sewer infrastructure is *okay* as it is (Bloetscher 2011) or that staff is wasting money and, therefore, fail to provide funds for correction efforts. As a result, these piping systems may be neglected over time increasing financial risks. The midnight inspection as a part of the inflow removal process helps identify where

Table 4.10 Costs of inflow, inflow correction (infiltration), and savings for Dania Beach

Assume 800 MH	Inflow	Infiltration
Sewage reduced (gallons)	18,348,011	72,706,566
Amount saved	($55,044)	($216,665)
Cost	$472,000	$772,000
Cost/1,000 gallon	$22.72	$7.64
BUT . . .		
Avoided cost	($1,200,000)	$0
Cost/1,000 gallon	($42.68)	$7.64

these leaks might be present while eliminating the miles of pipe that may have no defects. The result is that more funds can be spent on productive activities to remove infiltration.

The concept of televising and lining gravity pipes appeals to the technological view of local officials, but this does nothing to address inflow, only infiltration. A closed-circuit television (CCTV) camera survey is the most common tool used to gauge the current condition of the sewer collection system, the presence of roots, condition of joints, and debris in the line, (EPA 2005). A 2004 research project indicated that most utilities relied on CCTV as the primary means to inspect pipes (Thomson 2004), and little has changed in the ensuing years. CCTV methods and robotic cameras have improved the quality of the video substantially. CCTV requires cleaning the sewer line to remove debris. Clogged or dirty sewers are hard to assess. Cleaning is usually accomplished by jetting the system with water under pressure, using the jet on a vacuum truck and a tank of water. The procedure is similar to that pursued to clear clogged sewer lines. Once cleaned, the camera is inserted into a manhole and it travels through the line providing live video (that is also recorded). Improvements in current CCTV systems not only provide higher resolution visuals using better lighting, but they also rotate to inspect the pipe and lateral connections. Cameras developed for lateral inspections operate by either crawling into the lateral or in some cases with side cameras that extend.

A combination of CCTV and sonar can be employed to inspect partially full or surcharged sewers. For CI and DI, there are tools that will permit the measurement of the thickness of a corroded pipe. Similar tools are available to measure the loss of concrete in the crown of concrete pipe. A means to determine the remaining thickness of a corroded pipe or to determine other defects such as cracks or pitting in the pipe wall is useful in predicting future areas that are likely to leak; then proactive activity can be taken to avoid said leaks or breaks. Other tools can measure the width and depth of a crack or hole in the pipe wall or find laterals that are not connected properly. All of these items, and more, can also be uploaded directly to GIS and mapped.

Once the lines are televised, specific breaks can be noted and service line issues identified. The CCTV exercise permits the second phase of infiltration removal—lining of the pipe. Liners are used for lateral and main lines. There are several types:

- Casting a new internal lining using permanent or removable forms
- Sprayed or centrifugally cast concrete linings
- Sprayed polymer linings (coatings or high-build structural linings)
- Cured-in-place lining systems
- Grouting and sealing approaches (including the flood grouting approach used for simultaneous manhole, mainline, and lateral sealing)

The first three are primarily used for water mains, although some spray epoxies are used for sewer systems. The liner is thin, like paint, so it does not contribute to the structural integrity of the pipe—the utility must still rely on the pipe to retain its own structural integrity. The pipe cannot be deteriorated too much or it will fatigue under pressure. The biggest issue is that in order to get the epoxy to adhere, the pipe must be clean *and dry*—and sandblasting the surface beforehand is preferred.

The fourth option listed—cured-in-place liners—are the bulk of the industry and constitute 84% of all sewer trenchless repairs. There are several reasons for this. First, the pipe must be clean, but not dry, for the liner to adhere to the pipe. Hence the wet wall is used as a part of the repair or lining strategy. Cured-in-place liners require a resin that must be active and set (or cured) inside the pipe. There are options to provide the curing of the resin: steam to heat the resin to create adherence, plain tap water, water under pressure to push the liner against the pipe wall, air pressure (same issue with pressure and the pipe wall), and the newest option, ultraviolet light that warms the resin causing it to cure. The liners can provide some structural integrity once installed correctly. Sealing the liner in place and at the manhole is a critical issue. Older liners can be found that did not cure properly and the leak now goes between the pipe wall and the liner.

There are two types of liners with ASTM standards:

- ASTM F1216—Rehabilitation of pipelines by the inversion and curing of a resin impregnated-tube
- ASTM F1743—Rehabilitation of pipelines by pulled-in-place installation of a cured-in-place, thermosetting resin pipe)

The liners are folded and inserted between two manholes (see Figure 4.28). The liner is then unfurled and the appropriate curing system is employed (see Figure 4.29). The material must

Figure 4.28   Example of liner for sealing out infiltration. The saddle (white) is on the pipe (clay). A hole is cut. However, sometimes the seal is poor and the connection starts to leak or the pipe breaks. The liner is pushed into the pipe and seals to the pipe walls, over the cracks. This stops infiltration.

Figure 4.29 The installation of the liner as seen from above ground. The liner is normally installed manhole to manhole. This crew is stringing the liner into a manhole (dark blue).

be compliant with ASTM F 2550. Third-party inspection is suggested. The third party is not employed by the contractor; instead, the engineer or city employs the third party to check the contractor's work. Any repair should have a subsequent inspection and any good contractor will be supportive of a third-party inspector. It should be noted that adherence of the liner to the pipe wall is an issue with older systems, but there have been major improvements with remote operations that appear to have lessened this concern. Curing is the key, along with a clean pipe. Curing must comply with ASTM D790 test methods for flexural properties of unreinforced plastics.

Service lateral connections are a common place for breaks to occur. There are several methods to address lateral repairs, but those that encircle the entire pipe are most likely to provide a permanent seal. One example is the LMK T-liner® with gaskets, which complies with ASTM F2561 (see Figure 4.30). Within this lateral repair process, the lateral pipe is accessed remotely from both within the main pipe and from a lateral cleanout at the surface. A resin-impregnated, one-piece main and lateral cured-in-place lining (MLCIPL) is run through the lateral and inflated with an air or water bladder or inversion process. The MLCIPL is pressed against the host pipe by pressurizing a bladder and is held in place until the thermoset resins have cured. When cured, the MLCIPL is a continuous, one-piece, tight fitting, corrosion-resistant lining extending over a predetermined length of the lateral pipe and the adjacent section of the main pipe providing a verifiable nonleaking structural connection and seal (see Figure 4.31).

For concrete repairs in manholes, it is recommended that calcium aluminate products that comply with ASTM C 109, 293, 457, 496, 596, 642, 666, and 882 be used. Calcium aluminate is much less susceptible to deterioration from hydrogen sulfide than concrete, grout, or other cement products, and provides an impregnating coating to the remaining concrete.

To take full advantage of the design life of trenchless rehabilitation technologies, it is important that the installer uses proper installation controls and that the finished quality is confirmed by good assurance protocols and/or testing. Because of the cost, depth to pipe, and potential disruption, trenchless rehabilitation of the sewer lines is normally preferred to open trench repairs

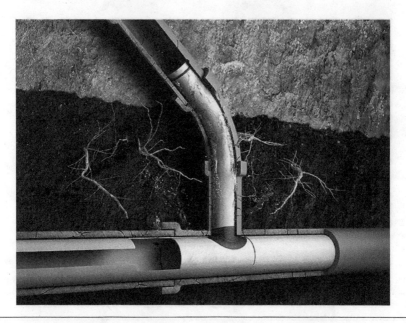

Figure 4.30   Example of T-liner in sewer lines—the liner in the main pipe is cut and a seal (black) plus a line is installed up the lateral. (Courtesy LMK Inc.)

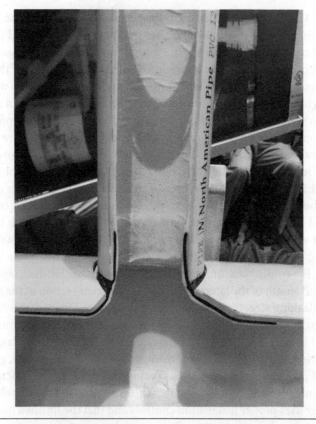

Figure 4.31   An example of a sealed service lateral connected to a sewer line (actual installation of Figure 4.30).

due to cost and disruption. There is one caveat—hot water is deleterious to thermoplastics, so certain thermoplastic liners should be avoided for the sewers with commercial discharges (laundries as an example).

Point repairs cannot be avoided in most systems. Collapsed pipes, pipe joints that are very misaligned, or where settling disrupts flow may only be fixed by open trenches. However, lining can avoid many lesser impacts and restore the sewer system to its prior condition, lessening infiltration and lowering flows. Lining all the pipe in a system is not recommended though, because the liner reduces the overall hydraulic capacity of the pipe. When all pipes are lined, it will reduce the sewer flow capacity of the system, which may be problematic in the future or may increase the potential for sanitary sewer overflows.

At times the efforts to reduce inflow and infiltration into the sewer system will be less successful than anticipated. This can be due to private lateral sewers or *laterals* (WERF 2006). The smoke test should identify these issues and the utility may need to send residents a letter instructing them to fix the problem. Most sewer ordinances require property owners to eliminate excess flows into the sewer system. Political will is required to enforce that provision. Dealing with sewer laterals in a comprehensive manner can be a significant and time-consuming undertaking for the utility. Cooperation of private property owners is required.

Example 4.2: Dania Beach Infiltration and Point Repairs Cut Sewer Flows by 32%

Figure 4.32 shows pipes identified during the midnight investigation in Dania Beach, Florida. The red pipes were the ones that had leaks of some duration that created infiltration. These lines were televised under a separate contract using a CCTV camera (after pipe cleaning). Analysis of the video indicated that half of the pipes televised had breaks and therefore needed liners, while the rest had service lateral breaks. Efforts were undertaken to repair the laterals and line the pipes as indicated by the CCTV results. Once these activities were completed to reduce infiltration into the system, the flow was analyzed over the ensuing year. The base flow plus infiltration calculations generated previously, based on the groundwater elevation and flow minus inflow, were used to calculate the cost savings from the infiltration construction activities. A comparison of the annual increase in infiltration calculated for the community based on the 2013 groundwater elevation values from Figure 4.26 shows that the amount of infiltration would have increased 8.5% if no infiltration removal activities had been undertaken. Once the inflow correction efforts outlined herein were all completed, the inflow amount diminished to a point where rainfall had limited impact on system flows. Five-point repairs were noted during the midnight investigation and repaired by city staff (one in a manhole that was noted during the midnight run). Overall, the flows decreased from 3.4 MGD in 2005 to 2.4 MGD in 2014, despite a 0.2 MGD addition of concentrate from a new membrane facility.

continued

Figure 4.32   Pipes identified as having leaks during the midnight investigation in the city of Dania Beach are in red.

Other Types of Maintenance

Maintenance staff should be aware of the issues associated with wastewater system assets during standard maintenance work. Most of these issues would be characterized as issues that arise after the inflow has been corrected because it is difficult to see any of these problems without televising the pipe. For example, unknown percentages of people flush diapers, towelettes, hand and baby wipes, paper towels, and other products down the toilet, as noted previously. Education and availability are means to address the problems. Education is the best and cheapest solution because the wipes do not deteriorate in the sewers. Instead, they clog pumps, stick to grease, and clog pipes. These problems can cause either partial blockages leading to backups and overflows upstream, which leads to surcharge and overflows downstream. Stores and restaurants can be encouraged to provide more convenient trash cans in restrooms, and to eliminate access to towels and wipes (Rabines 2015).

Grease and debris are ongoing issues in sanitary sewer gravity pipes. Grease will trap paper, napkins, feminine hygiene products, and rags—further enlarging the fatberg. A 15-ton, 2.4 meter diameter brick of grease and fat was found in London's sewer system in 2013. It was 250 meters long (Andrews 2013). Grease will cause backups and damage treatment plant operations. Grease and the bacteria that digest it can both cause foul odors. Periodic cleaning of sewer lines by jetting and the addition of chemicals that address the grease issue are helpful. BioKat® was a product that was used successfully for a number of years in Hollywood, Florida, to control grease and odors (from sulfides that contribute to pipe corrosion).

Corrosion is a problem that is encountered in every collection system and elimination of corrosion is difficult to accomplish. Concrete and iron pipe manifest similarly. In both cases, the problem involves anaerobic sulfur reducing bacteria that plays a part in creating sulfuric acid, which directly attacks the pipe, creating significant deterioration in short periods of time. This bacteria produces hydrogen sulfide gas ($H_2S_{(\uparrow)}$) from sulfate anion ($SO_4^{(2)} \cdot H_2S_{(\uparrow)}$) is dissolved by crown moisture droplets to form sulfuric acid ($H_2SO_{4(aq)}$). Hydrogen sulfide gas is:

- Toxic: hazardous to sewer workers
- Odorous: customer complaints
- Corrosive: degrades sewer pipes

$H_2SO_{4(aq)}$ eats away concrete and steel. The corrosion rate of hydrogen sulfide exposed to concrete pipe can be as high as 4.3 to 4.7 mm/yr at the sewage water level, while the deterioration at the crown can be over 1.4 mm/yr (Mori et al. 1992). Mori et al. (1992) also reported that concrete coupons (pieces cut out of a pipe or inserted specifically to measure corrosion) were reduced by 90% within one year of exposure to the sewage environment (where the concentrations exceeded 600 ppm of hydrogen sulfide). This means the pipe walls deteriorate faster, risking the integrity of the pipe. Failure along the top of the pipe, where corrosion is most likely to accumulate, is common. Solutions to hydrogen sulfide deterioration include:

- pH adjustments
- Polyphosphates and other corrosion inhibitors
- Calcium carbonate coating
- Other coatings
- Use calcium aluminate coatings that penetrate into concrete
- Use of biological additives like BioKat to control hydrogen sulfide

Steel needs constant maintenance; and while it may perform better, stainless steel will also need protection. Aluminum is more resistant to corrosion. The use of PVC and vitrified clay pipe may minimize the corrosion impact on smaller piping. In large pipe and where force mains enter manholes, corrosion from hydrogen sulfide is a significant problem. Coatings can help, but need constant maintenance because they can delaminate or detach.

Sewer Lift Stations

Many utilities face daily issues with pumping stations. All of the issues that were discussed with pumps in Chapter 3 apply to wastewater (and stormwater), so will not be reiterated. However, there are some different problems on the wastewater side that do not always occur on the water side. First, sewer lift stations are far more numerous than water pumping stations and up to 90% of power use on the sewer collection and treatment system is involved with pumping wastewater. Lift stations are designed to move flows from low points in the collection system, so the pumps are sized for large flows and low pressures, as opposed to high flows and pressure for water high-service pumps. Sewer lines that are under pressure, or force mains, are usually designed for pressures under 50 psi, but must accommodate the cycling on and off of pumps that can create a water hammer in the pipe. A water hammer can occur if the pumps start up too fast or if the check valves slam shut; thus, slow start motors, weighted check valves, pressure release valves, and air release valves are critical components of the sewer system that is under pressure. No force main should be less than four inches in diameter to prevent clogging.

Maintenance and repair of lift stations includes repairs to both pumps and motors; removal of grease in the wet wells; removal of debris from the pumps; and the cleaning, adjustment, and repair of the air and pressure release valves. Clogging is a common problem and requires physical removal of the solids. In addition to ongoing maintenance and meter reading to understand how much flow has gone through the lift station, inspecting lift stations to ensure that pumps and alarms are operating properly should be a priority. Wet wells are subject to erosion and corrosion concerns, depending on the structural materials used and the level of hydrogen sulfide produced by turbulence in the wastewater flow (Sterling, Wang, and Morrison 2009). Because lift stations are primarily concrete, steel, and CI, they have many of the same issues as water plants and manholes that must be addressed. However, lift stations serve a larger number of people than a manhole and, thus, should be monitored and prioritized accordingly.

Sizing the pumps usually means identifying both flow and head loss. For example, a lift station may need a pump to move 50 gpm of wastewater some distance, and some vertical height. The friction loss in the pipe, vertical distance, and friction losses from appurtenances all create head loss that must be overcome for the pump to operate efficiently. To size pumps, each pump will have a series of curves (see Figure 4.8), based on the impeller size, pump size, motor speed, and head. The use of these head curves will help engineers identify the proper pump for the application. In addition, certain pumps are geared for pumping clear water, as opposed to wastewater pumps that are designed to pass solids.

Lift stations always have at least two pumps, each rated at peak capacity. A series of check valves are designed to prevent backflow into the wet well. With lift stations, potential issues arise with check valves that do not seal, allowing some of the flow back into the wet well. Obstructions in the line, voltage or amperage issues, partially closed or closed plug valves on the force main, and material in the pump can all impact performance. Pump checks should include bearing lubrication, vibration testing, infrared scanning of motors, and checking the check valve operation. Pump issues are outlined in Table 4.3. Installing oversized pumps does not improve the efficiency of the system; in fact, it's just the opposite—the pumps operate off of their efficiency curves. In addition, the start and stop quickly increases wear and the force of the main flow is limited. That means added costs for maintenance.

CMOM Program

The regulatory focus under the Clean Water Act today is SSO. This SSO concern led the EPA to establish the CMOM program, which is intended to ensure that sewer collection systems, pumps, and wet wells are properly maintained in an effort to eliminate SSOs from plugged pipes or lack of pumping capacity in lift stations. The EPA estimates that $88 billion will be needed over the next 20 years to control SSOs and that this figure may be even higher if the agency takes the unlikely stance of zero tolerance on SSOs. As a part of the CMOM program, pipe cleaning and repair work is supposed to be tracked and maintenance logs are required for lift stations. Since reducing excess flows benefits the utility financially, the EPA prioritizes inflow correction programs. Most I/I programs include inspection data that can be tracked and material in the manholes that can be observed.

WASTEWATER TREATMENT PLANT FACILITIES

As noted in the beginning of this chapter, unlike water treatment, the basis for wastewater treatment is mostly a biological phenomenon, and therefore has a different set of challenges and a limited capacity for disruption (Bloetscher 2011), hence, the reason to *un*-combine sewers.

Primary Treatment

During primary treatment, primary clarifiers remove 50 to 70% of the suspended solids and 25 to 40% of the carbonaceous biochemical oxygen demand (CBOD). The primary clarifiers typically precede biological processes and can be used for flow equalization in secondary treatment facilities. These clarifiers have a detention time of only 10 to 30 minutes, resulting in the low removal rates compared to secondary clarifiers. In addition, large chunks and grit need to be removed. A bar screen and grit chamber are often employed ahead of the secondary treatment system. The bar screen should be automatic, so operators do not have to rake the material up.

Secondary Treatment

The secondary wastewater treatment concept is simple: organically-based wastes come into an aeration basin from primary treatment with trace amounts of minerals, metals, and other contaminants. The aeration basin contains bacteria that have been *trained* to use the incoming organic wastes as food. These biological processes include several types of treatment systems designated as *activated sludge* (Figure 4.33) and extended aeration systems (Figure 4.34). Activated sludge has an aeration time of 6 to 8 hours, and extended aeration systems aerate for 20 to 24 hours.

Figure 4.33 The aeration basin of an activated sludge wastewater treatment plant mixes sewage with bacteria and air allowing for the decomposition of the sewage by the bacteria. Here the detention time is 6–8 hours and the wastewater should be brown.

Figure 4.34   An extended aeration system is an activated sludge process with a detention time of 20–24 hours.

The amount of *food* for the bacteria is measured as the CBOD and is generally measured over a five-day period (CBOD$_5$). The bacteria desired to facilitate treatment require oxygen, so air is pumped into the aeration basin. Air is introduced into the wastewater in order to increase the food:air ratio to a point where the optimum number of bacteria will consume the incoming organics and use up the air. Pumping air is a significant cost to the treatment plant. Research indicates that fine-bubble aeration systems are more efficient to use, and that many treatment plants employ too much air (Stanley 2012). Dissolved oxygen control is suggested. Insuring proper operation of the aeration system will save the utility money.

The food:microorganism ratio is important in maintaining healthy bacterial populations; the goal is to keep the bacteria close to starvation, so they consume the waste and reproduce efficiently. Too much or too little food will create significant changes in the bacterial population and destabilize treatment. In the aeration basin, there will be active areas of bacteria growth (in the beginning and middle) and areas where the organics have been removed from the water thereby increasing the bacterial populations (end). The bacteria are very efficient at converting the organic waste to cells, so the carryover of waste is usually minimal.

Treatment system operators do not want to *wash out* the bacteria, so a means to capture those that exit hydraulically was developed—these are the secondary clarifiers. Secondary clarifiers (Figure 4.35) are employed after the biological processes have taken place in the aeration basin. Secondary clarifiers have a detention time that is approximately three hours. The clarifiers function to allow the bacteria—now called suspended solids—to settle to the bottom. Some are returned to the aeration basin, while others are removed as *sludge*. Secondary clarifiers often operate for years without issues, but the arms are delicate and the sludge pumps can fail, which puts the entire clarifier at risk. To address a clarifier issue, the clarifier basin must be emptied.

Some small concentration of bacteria may flow through secondary clarifiers. Secondary clarifiers are designed to achieve an effluent prior to discharge containing not more than 30 mg/L CBOD$_5$ and 30 mg/L total suspended solids (TSS), or 85% removal of these pollutants from the wastewater influent, whichever is more stringent. The TSS is a measure of the carryover solids,

Figure 4.35 Example of secondary clarifier after the biological/aeration process; bacteria and any remaining sewage settle to the bottom as sludge.

which are the bacteria that escape the clarifier. The requirement is 20 mg/L CBOD$_5$ for injection wells. Furthermore, secondary treatment standards in the United States require that these bacteria be eliminated though disinfection—chlorine being the most common choice, but ultraviolet light is another option.

Controlling the average age of the bacteria during secondary treatment is an important parameter for stable treatment outcomes. The amount of sludge (bacteria) removed from the plant controls the average age of the bacteria. The removed sludge is often further treated so when sludge is removed from the facility, it is comprised of dead bacteria and minerals—not the organic material that comes into the wastewater treatment plant.

Advanced Treatment

Beyond secondary treatment, additional chemical and physical processes may be used to enhance the treatment, like filters, membranes, and peroxide. The additional equipment needed for treatment is dictated by the effluent quality required for disposal. Advanced secondary treatment is also termed irrigation or *reuse quality water* (Figure 4.36). Reuse quality water is commonly used for irrigation and cooling towers, but other options are available. Reuse quality water requires the employment of filtration and a chlorine residual over 1.0 mg/L after a given period of time (typically 15–30 minutes). Traveling bridge filters and gravity sand filters (Figure 4.37) are common. Both are filters that allow the water to migrate vertically through the sand, removing debris. They are backwashed in the same manner as water filters. Traveling bridge filters have the benefit of cleaning a small portion of the filter constantly. Another option is cloth filters (Figure 4.38). Cloth filters allow the water to flow though the cloth and are then flushed through the cloth backward. They are slightly less efficient than sand filters depending on the filter media being compared. Regardless, all filters have mechanical parts that need ongoing maintenance to operate optimally.

The EPA has made nutrient removal of wastewater effluent a priority because effluent is released into fresh waters and estuaries, which significantly increases the amount of treatment

Figure 4.36   Advanced secondary treatment is also termed *reuse quality water*. The facility above has purple pumps that pump reclaimed water to an irrigation site or tank.

Figure 4.37   The traveling bridge sand filter shown here is a common method for filtering wastewater during advanced secondary wastewater treatment—gravity sand filters are another method, as are cloth filters.

Figure 4.38 Cloth filters at an advanced secondary wastewater treatment plant—these filters can replace sand filters (see Figure 4.37). They are nearly as efficient as the sand filters and are easily cleaned. They are sealed inside the covers for backwashing. They are common at plants that produce reclaimed wastewater.

required for those systems. The term *advanced wastewater treatment* (AWT) is generally used, and includes secondary treatment plus removal of nitrogen and phosphorous. An anoxic zone, where the amount of dissolved oxygen is less than 0.5 mg/L, is one means to address nutrient removal (see Figure 4.39), but biofilters and fixed film/rotating biological contactors may be beneficial to nutrient removal. AWT wastewater limitations indicate that the wastewater discharges may not contain more, on an annual average basis, than the following concentrations:

- CBOD$_5$ 5 mg/L
- Total suspended solids   5 mg/L
- Total nitrogen (as N)   3 mg/L
- Total phosphorus (as P)   1 mg/L

Bio-reactors or rotating biological disks need to have both motors and disks inspected and maintained on a regular basis. Dechlorination is often required where chlorine is used for disinfection. Ultraviolet light resolves the chlorination/dechlorination issue but requires much more power.

Direct and indirect potable reuse systems include all the treatment steps contemplated under AWT, plus generally reverse osmosis and/or activated carbon for removal of the remaining organics, ultraviolet light for disinfection and peroxide addition. Since these facilities are used for potable and indirect potable water, they must work properly to protect the public health of residents. The greater the treatment need, the greater the maintenance and monitoring responsibility. A number of communities have studied this option in California, Florida, and Texas. Two direct potable systems in Texas (Big Spring and Wichita Falls) were operational during a recent drought. California has been recharging aquifers for years with indirect potable reuse programs. Florida has conducted over a dozen studies, primarily in southeast Florida, but the systems most likely to come on line first may be in central Florida due to water supply limitations.

Figure 4.39   Example of anoxic zone in an advanced wastewater treatment plant—air is not applied in order to encourage anaerobic bacteria to consume nutrients (nitrogen and phosphorous).

Maintenance

One means to reduce maintenance is the use of highly durable components that do not wear and reduce efficiency as quickly as conventional components. A second energy efficiency measure is to line the interior of the pump with a material that reduces the friction. Pipe materials for aeration process piping are often stainless steel, fiberglass, or plastics suitable for high temperatures. Mild steel or CI with exterior and interior coatings can also be used (Tchobanoglous 2003). Type 304 and 316 stainless steel are most commonly used in wastewater treatment plants. Type 304L or 316L should be used when field welding is required due to the low carbon content. Type 316 stainless steel is approximately 40% more expensive than 304 stainless steel (MEPS International LTD 2012). Schedule 10S is a common thickness of stainless steel piping in aeration applications.

Wastewater treatment plant maintenance is similar to water treatment plant maintenance, except that the environment is far more corrosive due to bacteria and hydrogen sulfide. Both types of facilities are primarily steel, concrete, and perhaps stainless steel. There are numerous moving parts. Preventive maintenance is important. The operations and maintenance reports that are delivered with the completion or installation of equipment provide guidelines for preventive maintenance. Every person involved with plant maintenance should be familiar with the manufacturer's recommendations for preventive maintenance on any piece of equipment. Failure to perform preventive maintenance may void warranties and will certainly reduce the asset life. Of importance in all of this, administrators, operators, and engineers must keep in mind that wastewater plants must dispose of wastewater constantly that meets the regulatory criteria.

The utility must invest in personnel for all of these activities. All the concrete and steel discussions from Chapter 3 and earlier in Chapter 4 apply. Maintenance crews must address all equipment, structures, and grounds maintenance needs of the wastewater treatment plant on a routine basis, and as necessary—including mechanical repair and installation, plumbing, carpentry, painting, masonry work, machining of parts, electric and electronic installation, and repairs. Steel surfaces need to be repaired periodically, preferably before the damage becomes significant enough that the plant must be taken off-line to make repairs. Steel should be sandblasted and painted regularly to limit corrosion. Concrete should be rehabilitated; coatings are often a pre-ferred means to protect concrete surfaces. Note that neither steel nor concrete in a wastewater treatment plant will last as long as steel or concrete in other facilities, thus a recognition of the accelerated deterioration is required by managers and local officials. Extensive discussion could be had about plant maintenance, but plans vary so much that the operations and maintenance (O&M) manuals are the best source of information for any utility. A calendar should be created based on the recommendations of the O&M manuals. A comprehensive stock of spare parts and equipment should be maintained in order to respond promptly to emergencies.

WASTEWATER TREATMENT PLANT ENERGY PROCESSES

One area where maintenance and refurbishment can help in a wastewater plant is with energy costs. Energy costs can account for 30% of the total O&M costs of a wastewater plant (Carns 2005). Water and wastewater treatment operations are often the largest consumer of energy in a community as a result of their size, and the major demands are for the aeration and pumping processes (Stillwell et al. 2010; Bloetscher and Muniz 2012). Additional energy is used to heat, cool, or pressurize air (Barker 2010); for lighting; for heating, ventilating, and air conditioning (HVAC); and for telecommunications, as well as for transportation of employees, equipment, etc. (Mead et al. 2009).

In an active sludge facility (secondary treatment), 45 to 75% of electricity use is consumed in the aeration process (Rosso and Stenstrom 2006; see Table 4.11). Because the aeration treatment process consumes the majority of energy in wastewater treatment plants that utilize secondary treatment, improving the efficiency of aeration can result in the largest cost and energy savings to utilities. Table 4.12 demonstrates the typical energy usage at wastewater treatment facilities in the United States by the activated sludge treatment process. Most wastewater treatment plants use blowers for aeration. With blowers, the concept is to pump air beneath the surface for the wastewater in the aeration basin. Wastewater treatment plants typically use four types of blow-ers for aeration of the activated sludge process:

- Positive displacement blower
- Single-stage dual guide vane blower
- Multi-stage centrifugal blower
- Single-stage turbo blower

Positive displacement blowers have a peak efficiency of approximately 65% (Turblex 2008; Kes-kar and Liptak 2006), although they have a greater ability to be slowed with variable frequency drives (VFDs) (Stanley 2012). Dual guide vane control blowers have the ability to turn down blower speed and air flow while maintaining a relatively constant efficiency compared to multi-stage centrifugal and positive displacement blowers (Schmidt et al. 2008). The dual vane con-trol technology allows for variable flow capacity while maintaining constant high efficiency (Stanley 2012).

Single-stage centrifugal blowers range from 78 to 85% efficiency (Turblex 2008; Keskar and Liptak 2006). Schmidt et al. (2008) reports that multi-stage centrifugal blowers are the most common type of blower used in the activated sludge process with a peak efficiency between 68 to 76% (Turblex 2008; Keskar and Liptak 2006). However, multi-stage centrifugal blowers are limited in their ability to turn down flow (Stanley 2012). Turbo blowers have reported ef-ficiencies as high as 92% (Atlas Copco 2010). Rohrbacher et al. (2010) documented three life-cycle analysis case studies where turbo blowers were found to result in 10 to 15% cost savings

Table 4.11 Typical power use for wastewater treatment processes

Treatment Option	Power Use kWh/d/MGD
Trickling filter 1 MGD	1814
Tricking filter 5 MGD	983
Tricking filter 10 MGD	869
Tricking filter 25 MGD	756
Tricking filter 50 MGD	680
Tricking filter 100 MGD	680
Secondary WWTP activated sludge (very small)	2200–23600
Secondary WWTP activated sludge (small)	1100–4600
Secondary activated sludge 1 MGD	2230
Secondary activated sludge 5 MGD	1660–4200
Secondary activated sludge 10 MGD	1210–1690
Secondary activated sludge 25 MGD	1096
Secondary activated sludge 50 MGD	1058
Secondary activated sludge 100 MGD	1021
Secondary activated sludge plus filtration/pumping 1 MGD	2608
Secondary activated sludge plus filtration/pumping 5 MGD	1625
Secondary activated sludge plus filtration/pumping 10 MGD	1399
Secondary activated sludge plus filtration/pumping 25 MGD	1285
Secondary activated sludge plus filtration/pumping 50 MGD	1210
Secondary activated sludge plus filtration/pumping 100 MGD	1172
AWT 1 MGD	2948
AWT 5 MGD	1928
AWT 10 MGD	1777
AWT 25 MGD	1660–4630
AWT 50 MGD	1588
AWT 100 MGD	1550
Ponds 1 MGD	921
High purity oxygen activated sludge	1410–4100
Reuse systems	3400
Chemical advanced secondary	12230
RO low recovery	6400–8300
RO high recovery	18900–32000
Facultative pond with RIB	411
Slow rate pond	496
Pond with overland flow	619
Pond with filter	660
Aerated pond	1386

Continued

Treatment Option	Power Use kWh/d/MGD
Extended aeration	1871
Extended aeration + filter	1940
Trickling filter with digester	2145
RBC with digester	2175
Trickling filter with gravity filter	2205
Trickling filter with N removal	2296
Activated sludge with digester	2436
Activated sludge with digester + filter	2496
Activated sludge with nitrification + filter	2879

Sources: Lisk, Greenberg, and Bloetscher 2012; Bloetscher and Muniz 2012

Table 4.12   Comparison of blower motors in a wastewater treatment plant

Parameter	Positive Displacement	Multi-Stage Centrifugal	Single-Stage Dual Guide Vane	Turbo
Wire to air efficiency	45–65	50–70	70–80	70–80
Capital cost factor	1	1.5	2.5	2.4
VFD Capability	Yes	Limited	Yes but not necessary to achieve high efficiencies	Required

Source: Stanley 2012

compared to multi-stage centrifugal blowers and a 10 to 20% greater efficiency than the most commonly utilized multi-stage centrifugal blowers (Stanley 2012). Local conditions may change these findings, but they are worth a look.

The ratio of minimum to maximum oxygen demand within a typical activated sludge process varies from approximately 3:1 to 5:1 between the peak and off-peak hours, and as much as 16:1 (Tchobanoglous 2003). As wastewater flows and strengths fluctuate, the aeration volume must change to maintain a dissolved oxygen (DO) level of 1 to 3 mg/L in aeration basins to ensure adequate oxygen is supplied to sustain the microorganisms in the wastewater. There are two main alternatives for controlling the DO level in aeration basins: manual DO control or automatic DO control (Stanley 2012). When the DO gets too high, the aerators turn off, and vice versa (see Figure 4.40). Supplying excessive DO can inhibit nutrient removal of phosphorous. Conversely, supplying too little DO can also cause problems with effluent quality like TSS, BOD, and ammonia, as well as bacterial die-off. Reduced settleability and breakthrough of nitrite into the effluent causing disinfection problems are also concerns related to low DO (Ekster 2007; Stanley 2012). In comparison, automatic DO control utilizes DO sensors to continuously take DO readings and *feedback* signals to a controller that automatically adjusts airflow to maintain a predetermined DO set point, (typically 1 to 3 mg/L), by continuously adjusting the blowers and/ or air distribution control valves to each basin. Implementing automated DO control can greatly reduce electricity costs, operator workload, and help to maintain consistent effluent quality by consistently matching the amount of air supplied to the amount of oxygen required to maintain a DO set point (Stanley 2012).

Figure 4.40 When the DO gets too low in the aeration tanks, rotators will automatically start up—note the wave.

In the past, one limiting factor in implementing a well-functioning DO control system has been the unreliability of DO probes. Older galvanic and polarographic membrane-type DO probe technology using anodes, cathodes, membranes, membrane-cleaning devices, and electrolyte solutions are relatively unreliable (Keskar and Liptak, 2006). Membrane DO probes are fragile and utilize an electrochemical process that fouls the sensor, requiring frequent cleaning, maintenance, and recalibration—a huge problem in wastewater treatment plants (Hope 2005; Stanley 2012). Optical DO sensors use a light-quenching process, as opposed to membrane-type probes that utilize an electrochemical process that consumes oxygen. Optical DO sensors do not require flow across the probes and do not intrinsically foul with by-products from the oxygen-consuming electrochemical measurement process. The optical DO probes are more accurate than membrane type.

VFDs have been used by many wastewater utilities to conserve energy and reduce costs. One can find numerous success stories with energy savings ranging from 70,000 kWh/yr for smaller WWTPs (i.e., average daily flow of 7–10 MGD) to 2,800,000 kWh/yr for larger WWTPs (i.e., average daily flow of 80 MGD) (EPRI 1998; Efficiency Partnership 2009; DOE 2005). VFDs are now more available and affordable, and paybacks for VFDs range from six months to five years depending on the existing level of control and annual hours of operation (Reindl 2008). Energy is a major place to reduce operating costs in wastewater systems. Pumps and aeration are the primary areas to review when upgrading systems.

REFERENCES

Andrews, T. M. 2013. "This 'fatberg' clogging a London sewer is longer than two football fields, weighs more than 11 buses." *Washington Post*, September 13, 2013. https://www.washingtonpost.com/news/morning-mix/wp/2017/09/13/its-longer-than-two-football-fields-weighs-more-than-11-buses-and-is-clogging-the-london-sewer/?utm_term=.7eaa91210e0e.

Atlas Copco. 2010. *Variable Speed Direct Drive Centrifugal Air Compressors*. Product literature. Rocky Hill, SC.

Barker, B. 2010. *Energy Consumption Associated with Water, Office of Environmental Policy*. San Antonio, TX. http://www.sanantonio.gov/oep/Newsletter/pdf/Energy%20Consumption%20Associated%20with %20Water.pdf.

Bloetscher, F. 2011. "Utility management for water and wastewater operators," AWWA, Denver CO.

Bloetscher, F. and Muniz, A. 2012. "Where is the power to treat all the water? Potential utility-driven solutions to the coming power-water conflict." *Florida Water Resources Journal*, v. 63.

Bloetscher, F. et al. 2010. "Resolving biofilms in buildings and compounds." *Journal of Environmental Engineering Science* 27(9), pp. 767–776. doi:10.1089/ees.2009.0422.

Bloetscher, F. et al. 2015. "Defining the cost benefit of inflow removal before infiltration exploration." *Florida Section AWWA Annual Conference, Renaissance*. Seaworld, Orlando, FL Dec. 10–13, 2015. FSAWWA, St. Cloud, FL.

Carns, Keith. 2005. "Bringing energy efficiency to the water and wastewater industry: How do we get there?" *Proceedings of the Water Environment Federation 2005*, 7(2005), 7650–7659.

Chin, D. A. 2007. "Hydraulics of sewer flows." *FWRJ*, Vol. 58, No. 12, p. 55.

Dance, S. 2017. "Sewage soiling thousands of city basements, but another decade of repairs looms." *Baltimore Sun*. 4/8/17. http://www.baltimoresun.com/news/maryland/investigations/bs-md-sewer -backups-20160514-story.html.

Efficiency Partnership. 2009. Water/Wastewater Case Study: South Tahoe Public Utility District, Flex Your Power. http://www.fypower.org/pdf/CS_Water_South_Tahoe.pdf.

Ekster, A. 2007. "Forecasting energy savings achieved by automation of dissolved oxygen and sludge age controls and optimization of setpoints." *WEFTEC 2007 Conference Proceedings*. WEF Press, Alexandria, VA.

Electric Power Research Institute (EPRI). 1998. Quality Energy Efficiency Retrofits for Wastewater Systems. EPRI, Palo Alto, CA. CR-109081.

E Sciences, Inc. 2013. "Miami Beach groundwater elevation mapping and modeling."

Evans, J. and Spence, M. N. 1985. "The evolution of jointing vitrified clay pipe." *Advances in Underground Pipeline Engineering, Pipeline Division, ASCE/Madison, WI/August 27–29, 1985*. ASCE, Alexandria, VA.

Evans, M. A. 2015. "Flushing the toilet has never been riskier." *The Atlantic*. https://www.theatlantic .com/technology/archive/2015/09/americas-sewage-crisis-public-health/405541/.

Farrugia, M. 2017. "Manhole insert case study: How a small U.S. town reduced rain surges into its wastewater treatment plant by over 20 percent." https://www.linkedin.com/pulse/manhole-insert-case -study-how-small-us-town-reduced-rain-farruggia/.

Feeney, Christopher S. et al. 2009. *Condition Assessment of Wastewater Collection Systems*. U.S. Environmental Protection Agency, Washington, D.C.

Ferry, S. 2002. "Why pipes fail, plumbing engineering." http://www.ncpi.org/assets/DiscoveringWhyPipes-Fail.pdf. Accessed 2/15/17.

Griffin, J. 2010. "Confined space: Deadly if not prepared." *Underground Construction*, 65:4. p. 20.

Hope, James. 2005. "DO'ing more with less." *Water Environment Technology*. Vol. 17, No. 7. WEF Press, Alexandria, VA.

Keskar, P. and Liptak, B. 2006. "Water supply plant controls." Chapter 8 in Liptak, B. *Process Control Optimization Volume 2*. CRC Press, Boca Raton, FL. http://twanclik.free.fr/electricity/IEPOPDF/1081fm .pdf. Accessed 2/13/12/.

Klimes, M. 2016. "Pump failure analysis." *Flow Control*, Nov. 2016.

Lisk, B., Greenberg, E., and Bloetscher, F. 2012. "Implementing renewable energy at water utilities." Web Report #4424, Water Research Foundation, Denver, CO.

Liptak, B. G. 2006. *Process Control and Optimization*. CRC Press, Boca Raton, FL.

Ludwigson, M., Rago, L., and Greefield, J. 2016. "Horizontal of vertical high service pump selection." *FWRJ*. Vol. 67, No. 12, p. 54.

Mead, S. P. et al. 2009. "A water/energy best practices guide for rural arizona's water & wastewater systems." Northern Arizona University, Flagstaff, AZ. http://www.waterenergy.nau.edu/docs/Best _Practices_Guide_2009.pdf. Accessed 3/31/2012.

MEPS International LTD website. 2012. "North American stainless steel prices." http://www.meps.co.uk/ Stainless%20Price-N.Amer.htm. Accessed January 29, 2012.

Merrill, S. et al. 2004. "Reducing peak rainfall-derived infiltration/inflow rates-case studies and protocol." Water Environment Research Foundation, Alexandria, VA.

Michaud, C. F. 2013. "Hydrodynamic Design Part 10, Sizing Pumps." *Water Conditioning and Purification*, 11: 2013, p. 42.

Mori, T. et al. 1992. "Interactions of nutrients, moisture and pH on microbial corrosion of concrete sewer pipes." *Water Res.* 26, 29–37.

NASSCO Manhole Committee. 2015. Greenville, Ohio. "A gallon saved is a gallon earned." *Trenchless Technology*, Vol. 25, No. 5, p. 72.

Prentiss, D. 2010. "Lift station safety: It's a minefield down there." *FWRJ*. Vol. 61, No. 4, p. 24.

Rabines, A. 2015. "Father-son wipes creators cleanup." *Sun Sentinel* Palm Beach County Edition, 7/15/2015. http://www.pressreader.com/usa/sun-sentinel-palm-beach-edition/20150711/282097750380216/ TextView. Accessed 7/15/15.

Reindl, D. 2008. Applying Variable Speed Drives In Industrial Refrigeration Systems, Focus on Energy, https://focusonenergy.com/sites/default/files/Doug_R._Refrigeration_3_Handout_Final.pdf.

Rizk, T. 2015. "When it Rains . . ." *WE&T*, Vol. 25, No. 6, p. 43.

Rogers, K. S. and Fleming, J. C. 2016. "Hillsborough county pump station rehabilitation program." *FWRJ*. Vol. 67, No. 10, p. 18.

Rohrbacher, J. et al. 2010. "Beyond improvement or hot air? Comparison of high-speed single-stage blowers to conventional technologies." *Proceedings of the Water Environment Federation, WEFTEC 2010*: Session 91 through Session 100, pp. 6868–6882(15), Alexandria, VA.

Rosso, D. and Stenstrom, M. K. 2006. "Economic implications of fine-pore diffuser aging." *Water Environment Research, Vol. 78, No. 8* (August): 810–815.

Santino, A., Driscoll, M., Hussam, T. 2015. "The Forgotten Asset—Addressing Problematic Wastewater Lateral Connections in Baltimore, Maryland." *Proceedings of the Water Environment Federation, WEFTEC 2014: Session 210 through Session 218*, pp. 6915–6926(12), WEF, Alexandria, VA.

Schmidt, H. Jr. et al. 2008. "Balancing treatment process requirements and energy management." *WEF Sustainability 2008 Conference Proceedings*. WEF Press, Alexandria, VA.

sewerhistory.org. 2004. http://www.sewerhistory.org/chronos/convey.htm.

Stanley, E. 2012. "Identifying cost savings through energy conservation measures in mechanically aerated activated sludge treatment processes in Southeast Florida." A Thesis Submitted to the Faculty of The College of Engineering and Computer Science in Partial Fulfillment of the Requirements for the Degree of Master of Science. Florida Atlantic University, Boca Raton, FL.

Stanley, Eric. 2009. "Preserving the environment by increasing reuse supplies: blending nanofiltration concentrate and treated effluent." *Florida Water Resources Journal*. Vol. 60, No. 8, pp. 52–59.

Sterling, R., Wang, L., and Morrison, R. 2009. "Rehabilitation of wastewater collection and water distribution systems." EPA/600/R-09/048 EPA, Cincinnati, OH.

Stillwell, A. S. et al. 2010. "The energy water nexus in texas, synthesis." Part of a special feature on the energy-water-nexus, *Ecology and Society*, 16(1), 2. www.ecologyandsociety.org/vol16/iss1/art2/.

Stratton-Childers, L. 2015. "Flushing away the discord." *WE&T*, Vol. 26, No. 6, p. 24–26.

Swarner, R. and Thompson, M. 1994. "Modeling inflow and infiltration in separated sewer systems." *Proc 67th Annual Conf. Water Environment Federation, October 1994, Chicago, IL*. WEF Press, Alexandria, VA.

Tchobanoglous, G. 2003. *Wastewater Engineering: Treatment and Reuse. (Metcalf & Eddy, 2003) Fourth Edition*. McGraw-Hill, New York, NY.

Thomson, J. C. 2004. "An examination of innovative methods used in the inspection of wastewater systems." Water Environmental Research Foundation, Alexandria, VA.

Turblex, 2008. "Power-Gen 2008 Conference Coverage." Turblex, Inc. http://www.mcilvainecompany.com/ PowerGen2008/Audio/PowerGen2008-%20(25).mp3. Accessed 2/12/12.

U.S. Department of Energy (DOE). 2005. Onondaga County Department of Water Environment Protection: Process Optimization Saves Energy at Metropolitan Syracuse Wastewater Treatment Plant. U.S. Department of Energy, Energy Efficiency and Renewable Energy, Washington, D.C. http://www1.eere .energy.gov/industry/bestpractices/pdfs/onodaga_county.pdf.

U.S. Environmental Protection Agency (EPA). 2002. *Fact Sheet Asset Management for Sewer Collection Systems*, 833-F-02-001. Office of Wastewater Management, EPA, Washington, D.C.

————. 2005. *Guide for Evaluating Capacity Management, Operation, and Maintenance (CMOM) Programs at Sanitary Sewer Collection Systems.* EPA, Washington, D.C.

————. 2005. *Guide for Evaluating Capacity, Management, Operation, and Maintenance (CMOM) Programs at Sanitary Sewer Collection Systems.* EPA-305-B-05-002. EPA, Washington, D.C.

Water Environment Research Foundation (WERF). 2004. *An Examination of Innovative Methods Used in the Inspection of Wastewater Systems.* WERF 01-CTS-7. Water Environment Research Foundation, Alexandria, VA.

————. 2006. *Methods for Cost-Effective Rehabilitation of Private Lateral Sewers.* WERF 02-CTS-5. Water Environment Research Foundation, Alexandria, VA.

————. 2008. "Guidelines for the inspection of force mains" (Draft). 04-CTS-6UR. Water Environment Research Foundation, Alexandria, VA.

Wingard, R. 2015. "Infiltration/Inflow: Turning a problem into an opportunity to regain capacity and lower operating costs." *Trenchless Technology*, Vol. 25, No. 5, p. 68.

———. 2003. Guidance for Auditing Capacity, Management, Operation, and Maintenance (CMOM) Programs at Sanitary Sewer Collection Systems. EPA, Washington, D.C.

———. 2005. Guide for Evaluating Capacity, Management, Operation, and Maintenance (CMOM) Programs at Sanitary Sewer Collection Systems. EPA 305-B-05-002. EPA, Washington D.C.

Water Environment Research Foundation (WERF). 2004. An Examination of Innovative Methods for the Inspection of Wastewater Systems. WERF 01-CTS-7. Water Environment Research Foundation, Alexandria, VA.

———. 2006. Methods to Cost-Effectively Rehabilitate or Renew Lateral Sewers. WERF 02-CTS-6. Water Environment Research Foundation, Alexandria, VA.

———. 2008. Guidelines for the Inspection of Force Mains. WERF 04-CTS-6UR. Water Environment Research Foundation, Alexandria, VA.

Winkler, R. 2013. Intermittent Inflow—finding a program that is opportunistic to repair capacity and lower operating cost. Treatment Technology, Vol. 25, No. 5, 16-22.

5

STORMWATER MANAGEMENT SYSTEMS

OPERATIONAL GOALS OF STORMWATER MANAGEMENT

Stormwater, or stormwater runoff, is defined as the "precipitation from rain or snowmelt that flows over the ground." The United States Environmental Protection Agency (EPA) has defined stormwater as the main contributor to nonpoint source water pollution in the United States (EPA 2003). Runoff flows from roofs and yards, along streets, and across other impervious surfaces to reach storm drains, canals, gutters, or other stormwater infrastructure—where it is eventually discharged, often untreated, into receiving water bodies. All contaminants and pollutants, such as pet wastes and other bacteria sources, dirt, discarded trash, nutrient rich fertilizers, grass clippings, pesticides, motor oils, brake dust, tire fragments, and toxic chemicals from the road that accumulate along the way are eventually swept up by stormwater and deposited into receiving water bodies. Consequently, proper stormwater management in urban areas is very important.

But urban systems, while the focus of this book, are not the only contributor to stormwater runoff. All property creates runoff and the quality and quantity of runoff can be impacted by changes to the land surface. For example, mining and logging practices change the landscape and vegetation patterns and can increase the loss of soils, all causing an acceleration of water off the land that would have otherwise held it. This can be seen in agricultural practices as well, where plowed fields and the application of herbicides, pesticides, fertilizers, and other materials can impact downstream water quality. Therefore, stormwater management, or the lack thereof, upstream of a community can have significant effects on the community itself.

The goal of a stormwater management system is to protect public health, welfare, and safety by reducing flood impacts on a community, the potential for waterborne disease from flooding, and to lessen the potential for property damage if flooding occurs. This is how stormwater differs from water and sewer utilities where the only goal is public health protection. Stormwater systems also protect property—both public and private. Public and private property may include homes, businesses, roadways, railroads, bridges, utilities, etc., so the initial goal of a stormwater management system is always to remove excess water in a timely manner, to a place where it will not adversely impact the public. To determine how much water this is, stormwater systems are designed to remove runoff from a given storm event that has a specific amount of rainfall occurring for a specific location, such as the one-in-10-year (1:10) storm event or the one-in-100-year (1:100) event. It is noted that these frequency definitions do not mean that the rainfall amount for a specific storm event will occur only once in that time period. Using the 100 year event as an example, it is more accurate to define the event as having a 1% chance of that amount of rainfall occurring in any given year at a specified location that has a return frequency of 100 years. Additionally, storms are rated for rainfall distribution and duration for the time period, such as, a one-day or a three-day event for the specified frequency storm event. Thus, the storm would be designated as a 100-year/three-day event for a specific location, where a table or graph will provide the amount of rainfall for that event at that location. The rainfall distribution is used to develop rainfall intensity for the design of stormwater facilities. Different communities have different regulations, and those located in flood prone areas generally have more stringent requirements for protection of buildings and roads, although this is not always the case. Please note, there is recent data that suggests rainfall patterns and rainfall intensity appear to be increasing in many locations. As an example, the 1:10 year storm event of 20 years ago may in fact be the 1:5 year storm event today. For this reason, the rainfall distribution and frequency is periodically reviewed for locations by government agencies. Each community can define their level of service differently, but the expectations of the public must be kept in mind when defining the stormwater level of service.

At the same time, no community can plan for all events. The recent 50-inch rainstorm in Houston, Texas, was a wake-up call to many residents who felt safe from flood events; but Houston did not have a stormwater management system that could handle such a major rainfall event, nor does any municipality—this is far beyond most planning horizons. The cost of stormwater management infrastructure can be very high relative to the number of major rainfall events. Communities must weigh the cost versus the risk in making decisions about what level of service to provide to their residents. It has been the decision in years past for economic reasons by many local agencies to use the so called 1:100 year event to be the storm event that is used to establish flood protection for homes and businesses. Herein exists another way stormwater systems are different from water and sewer systems—a level of service is set to protect public health safety and property, but that level of service is not protection under all circumstances. Much discussion can be had about whether a community plans for the right frequency of events, and those communities that have ongoing, repetitive losses (floods), should evaluate their level of service or where construction should take place. Only a few local governments actually prohibit construction in flood prone areas due to private property rights concerns, even if those properties have been subject to repetitive losses. The tendency in these high-risk areas is to depend on the federal government to address the loss. The Federal Emergency Management Agency (FEMA) is the primary federal agency for responding to emergency situations and in addressing losses. But FEMA doesn't prohibit building in flood plains and flood prone areas either. Building in flood prone areas is one of the difficulties that both local officials and federal officials face—trying to protect those who knowingly develop or live in flood prone places where losses are likely and who expect that *the government* will protect them or make them whole.

To prevent flooding and the potential for health risks associated with stagnant water, stormwater runoff must be managed in an organized and systematic manner if property owners are to enjoy the full use of their property and roadways are to be clear. As a result, stormwater facilities must be constructed and maintained to reduce the negative impacts of runoff. The burden of managing this stormwater typically falls to a stormwater organization—typically a special district, stormwater utility, or a division of a local government. More recently, newer subdivisions are required by many governmental agencies to take the responsibility for maintenance of their own stormwater management system (defined later in this chapter). A community's stormwater system consists of pipes, catch basins, curb inlets, culverts, canals, swales, pump stations, ditches and manhole inlets, and other structures that help channel the stormwater to rivers, lakes, retention basins, or canals, but may also direct it into basins that help resupply groundwater. As a result, the majority of the stormwater system costs have traditionally been associated with maintenance activities (70% of total cost is often estimated to be operational cost) to ensure the existing facilities store or channel water as anticipated. However, little information exists about the value of the buried infrastructure associated with stormwater systems, although federal regulations are aimed at trying to capture some estimates. For many years, little information existed about the quality of the water being discharged, which conflicts with the intent of the Clean Water Act.

The regulatory efforts associated with the first 20 years of the Clean Water Act focused on point sources—typically wastewater treatment plants. Point sources are believed to account for about 50% of flows to rivers and streams and about half the pollution (EPA 2007). The rest of discharges are nonpoint sources—i.e., stormwater runoff from the land (often through pipes). Nonpoint source water pollution has been a growing concern in the United States since the 1990s as water quality improvement in rivers appeared to reach a plateau. A study of water quality indicated that nonpoint runoff and stormwater were major sources of water pollution facing all U.S. waterways. According to the EPA, 40% of surveyed impaired water bodies are impaired due to polluted stormwater runoff (EPA 2007; Mulcahy 2006), which can be from urban cities or agricultural lands. This polluted stormwater is split about 50:50 between urban and agricultural land, so efforts have been undertaken under the auspices of a federal permitting process called Municipal Separate Storm Sewer System (MS4) to address just the urban component. So far, no programs address agricultural runoff.

The MS4 program was developed by the federal government as a means to require stormwater maintenance entities to implement procedures and tracking mechanisms to help reduce the potential for water pollution from stormwater runoff and reduce flood risks caused by the lack of maintenance of stormwater systems. Since the intent of the MS4 program is to reduce pollution to waterways that impact aquatic life, the focus has increasingly turned toward managing (measuring and characterizing) stormwater discharges.

Many communities attempt to protect the health of local water bodies; thus, keeping fertilizers, pesticides, and herbicides out of water bodies is seen as a positive because they can have deleterious effects on waterways and the organisms living in those waters. Protecting the water quality while trying to discharge the same water as quickly as possible to protect property is a challenge that stormwater managers must deal with since goals are often mutually exclusive under current conditions. Outdated or nonexistent infrastructure may lead to flooding, soil erosion, and changes in natural hydrology.

To address the challenges with managing stormwater and the associated assets, over 400 municipalities in the United States have stormwater utilities in place. The revenues are typically raised from utility user fees that are designated as a dedicated funding source for stormwater management in these municipalities, or from municipal property taxes. Creating a stormwater utility is viewed as a possible solution to the common problem of underfunding stormwater management. The stormwater utilities allow communities to create a mechanism to raise funds to operate, maintain, and upgrade the stormwater system.

STORMWATER SYSTEM COMPONENTS

Archaeological studies have indicated that stormwater piping has been practiced for thousands of years. Much like sanitary sewer systems, stormwater systems work from small areas to larger ones like sewer systems (see Figure 5.1). However, unlike wastewater systems, there are rarely significant stormwater treatment works prior to discharge into receiving water bodies, therefore

Figure 5.1 Schematic of storm sewer/drainage system that shows how water drains from a property to the storm sewer and is eventually discharged into a water body. *Source*: Warren County Soil and Water Conservation District: https://www.warrenswcd.com/conservation -connection-a-blog/category/storm-drain

pollution will be discharged from the stormwater system into receiving water bodies unless efforts are undertaken to avoid it. That creates added tools in the stormwater system.

Most stormwater management systems are made of a gravity network consisting of lakes or ponds; dry detention or retention areas; conveyance solutions like swales, gravity drains, pipes, yard drains, and connection to eaves and gutters; structures like manholes; and a variety of means to limit entry of surface debris. As noted in Chapter 4, combined sewers were installed, but many have been separated or constructed separately with time.

Pipes

For the most part, stormwater sewers have historically been constructed with reinforced concrete pipe and corrugated metal pipe as these were inexpensive and available materials (brick might be found in large tunnels). However, prior to World War II, concrete pipe had no gaskets and the pipe lengths were short—only two feet long to permit hand installation. Like vitrified clay sanitary sewers, early means to limit entry points in joints involved cement mortar and burlap, commonly referred to as burlap diapers or oakum (Evans and Spence 1985). Once machines became more common, the machines allowed longer joints to be installed, but gaskets are relatively recent, which means the pipes were neither waterproof nor capable of dealing with differential settling—thus, most leaked quickly.

Today, large pipes are commonly reinforced concrete (see Figure 5.2), brick, or metal material, depending on the age of the pipe. Today, reinforced concrete pipes have gaskets to address

Figure 5.2   Reinforced concrete pipes for stormwater management—this type of pipe can also be used for sanitary sewers and/or raw water lines not under pressure (but not preferred for the latter due to potential leakage).

the potential for leaks. Reinforced concrete pipe today meets ASTM C14, C14M, C76, C76M, C443 for joints; C506 and C506M for arches; C507 and C507M for elliptical pipe; AASHTO M170 and 170M for culverts; and AASHTO M175 and M176 for perforated and porous pipe and a seal that meets ASTM C12. Typical sizes for reinforced concrete pipe are 15, 18, 24, 30, 36, and 48 inches. Reinforced concrete pipe is the most common pipe used under roadways. It performs well in load conditions, but is heavy and expensive to install.

Corrugated metal pipe comes in both steel and aluminum. Corrugated metal pipe is still used today because it is lightweight, easy to install, and is relatively inexpensive compared to concrete. For many applications it works well. The corrugations enhance its strength, making it a choice for driveway culverts in rural roads. However, the life expectancy of the corrugated steel pipe is much less than concrete. Many examples of old, rusted-out steel storm drains can be found and many failures in piping can be traced to old steel corrugated pipe (see Figure 5.3). When corrugated metal pipe deteriorates, it creates cave-ins as the soil above the pipe falls into the hole in the pipe created by the rusting out from corrosion. The Missouri Department of Transportation (DOT) (Wenzlick and Albarran-Garcia 2008) reported that concrete pipe was rated better, and could easily make the 100 years of life that was originally planned, while the galvanized corrugated steel pipe had a life of about 40 years. Improvements were made many years ago as the steel pipe was galvanized to protect it. This extended the pipe life, but soil conditions can create accelerated deterioration, much like it does with galvanized water pipe.

Aluminum corrugated pipe was developed to overcome the corrosion issues in steel pipe. The use of aluminum is more recent but the life appears to extend beyond that of steel pipe since the corrosion is not an issue. However, the strength of aluminum can be compromised because it has neither the compression load resistance of concrete nor the tension load capacity of steel. Deformation can be a concern unless construction methods are carefully implemented in accordance with the manufacturer's recommendations. Aluminum arches and other options are available for use in bridge spans as a part of an integrated design that includes reinforced concrete.

Ductile iron and asbestos cement are options that were installed after 1950 but neither were significant options in most jurisdictions. These are basically the same types of pipes used for

Figure 5.3 A rusted-out steel pipe that has been removed from service but not disposed of properly.

water lines (see Chapter 3 for more discussion). Both had gasketed joints that solved the leaky pipe issues that challenged many stormwater systems; and both were rigid materials, which was preferred under roadways. PVC pipe was introduced for storm sewer applications in the 1970s (again with gaskets like water pipe). A series of ribbed versions were, and continue to be, produced to improve resistance to loads and prevent deformation. However, PVC deteriorates in sunlight and can flex, which are issues that should be avoided in many stormwater systems.

In the 1980s, high density polyethylene (HDPE) came into vogue as a stormwater material, particularly in areas not under pavement. HDPE is light, easy to install, and does not corrode. It replaced the corrugated metal pipes in many places. Corrugated HDPE is common (see Figure 5.4), especially in non-traffic areas. While the ribs make it more rigid, the potential for loading in non-traffic applications that would flex the pipe are minimal. HDPE is preferred by many for stormwater pipe since it does not deteriorate like PVC will in sunlight. However, none of these options is sufficient for large-diameter gravity mains despite their strength and leak-resistant joints so reinforced concrete remains the choice for large-diameter pipe. For the most part, storm sewers have minimum sizes (15 inches is common), and because the flows can be much larger, will typically be larger than sanitary sewer pipes.

Catch Basins

Aside from pipe, the most common feature in stormwater management systems is the catch basin, which comes in a variety of sizes and depths (see Figure 5.5). Figure 5.6 is a typical catch basin with an open grate. They are needed to connect multiple pipes and to allow for pipe size changes, direction changes, and slope changes—as well as supplying an opening for flow and maintenance. The catch basin is where the trash, sand, silt, grass clippings, and other debris can be caught and removed. The MS4 program actually requires stormwater entities to track these activities, but actual records may be a challenge if a formal work order system is not in place. Most are made of reinforced concrete, although older ones may be concrete block or brick, just like manholes on the sanitary sewer system. Brick and concrete block catch basins will suffer

Figure 5.4   Ribbed HDPE pipe with smooth interior prior to installation.

Figure 5.5   Pre-installation of three reinforced concrete catch basins (rectangular shape) and one stormwater manhole (circular shape). An open grate will then be installed in the cut-out section of each lid.

the same issues as brick sanitary sewer manholes (see Chapter 4). Many catch basins are located in the roadway, although on the edge. That makes them subject to loads from buses and other heavy vehicles. State transportation departments normally establish standards for catch basin boxes and steel grates that other agencies follow.

Figure 5.7 shows the curb cuts for a storm grate installation that are common for roads with curbs. Local standards will define the number of catch basins—wet areas need more than dry areas. Roadways and parking areas must be designed to ensure the catch basins are located at the low points to drain optimally. The depth of the structure can make it susceptible to infiltration of groundwater through any cracks or poorly connected joints if located below the water table. The lack of flow at sufficient velocities will also cause the accumulation of silt, sand, and debris.

Figure 5.6 Typical open storm grate that is installed on top of the catch basin. The grate helps to keep debris such as leaves and trash out of the catch basin, a situation that is an ongoing maintenance issue.

Figure 5.7 This is a standard curb cut for catch basin grates that are common for DOT roadways. The grate and curb cut allow for very large amounts of stormwater to be removed quickly from roads.

Sand, surface debris, and trash can be transported to the catch basins and create significant issues with capacity. Sand and silt will often partially fill pipes that must be cleaned regularly.

Manholes

Less common are stormwater manholes (see Figure 5.8). Manholes are the major junction points of the gravity network. Like sanitary sewer systems, manholes are utilized in the system as points for change of direction, changes in diameter of the pipeline (allowing for greater capacity), and as access points for maintenance workers to clean and inspect the pipeline network. The standard diameter for manholes is four feet (inside) with pipes less than 24 inches, just as it is with sanitary sewers. Some manholes may also act as catch basins in the curb inlets or parking areas (see Figure 5.9). Like sanitary sewers, stormwater manholes will have a 22-inch manhole cover and ring on the top, but should say "storm sewer" on the cover. Larger pipes will require larger manholes, so six- and eight-foot diameter manholes are not uncommon, given that stormwater pipes larger than 24 inches in diameter are not uncommon.

Most drainage manholes are located within the roadway and are subjected to daily traffic loading, so traffic bearing lids are required. Manholes may be deeper than sanitary sewers to accommodate the increasing depth of the gravity pipelines as the stormwater progresses downstream. The depth of the structures makes the pipes susceptible to infiltration of groundwater, just like sanitary sewers. Many of the same solutions apply. Since both sanitary and storm sewers flow by gravity, there are occasions where the pipes may conflict. The solution is to create a conflict box or conflict manhole. Figure 5.10 is an example where a sanitary sewer and storm sewer crossed. As both are gravity pipe that have slopes that cannot be altered, the

Figure 5.8 Stormwater manhole top set in a roadway. Like sanitary sewers, stormwater manholes will have a 22-inch manhole cover and ring on the top, but should say "storm sewer" on the cover. This type of structure is intended only for access, not to drain stormwater.

Figure 5.9   Example of stormwater manhole installation, with inlet on top—the grate will go on top of the square box.

Figure 5.10   Example of conflict manhole for PVC (white) sanitary sewer and large gray stormwater pipes—both require gravity flow. The sanitary sewer line crossed the stormwater path, running perpendicular, and above the invert of the stormwater line.

solution was to permit the smaller gravity sewer to go through the conflict manhole for the bigger storm drain.

Brick and block catch basins and manholes were the standard practice before 1970. Afterward almost all structures were constructed of reinforced concrete. One major concern was that brick manholes have hundreds of joints, just as sewer manholes do. Stormwater structures have little risk of damage from hydrogen sulfide, but structures may leak allowing soil fines from the surrounding ground to enter the manholes, causing soil voids and surface settlement adjacent to the manholes. Those open to the street will have issues with debris. Because of low flow conditions, most storm sewer pipes have accumulations of sand and silt in the bottom of the pipe.

Heavy rains will cause a flushing of debris into the water, which can overwhelm the receiving waters. This is why retention and detention areas, bioswales, and spreader swales are preferred as treatment mechanisms prior to discharge into surface waters.

Concrete is not impervious to water, so an external epoxy or coal tar should be applied to prevent water from penetrating the wall. The ring and cover (or grate) are cast iron so vibration and differential expansion and contraction mean that the cement bond between the concrete and cast iron ring is lost. Unlike the sanitary sewer system, catch basins and storm manholes do not include *boots*, although gaskets may be available. Since concrete is connected to concrete, normally only grout and perhaps a gasket may be used. Mastic and other materials can be used as well. The connection between pipes and manholes is a potential area of concern.

Ditches and Swales

At this point, the commonality between sanitary and storm systems diverge. In developed countries, storm systems may include substantial open structures, which sanitary sewers do not. Ditches and swales are the most common stormwater conveyance systems (see Figure 5.11). Ditches are deeper than swales and tend to convey water in a given direction more than swales. Most ditches are located adjacent to roads but in rural areas they can be anywhere, and if not along a road, unlikely to be maintained by public works officials. Ditches are simply cuts in the dirt that are theoretically cut on a grade that will move water in a specific direction. In rural areas, drainage is provided most economically by open ditches that allow soil moisture to drain laterally. As a rule of thumb, the bottom of the ditch ought to be at least one foot below the base course of the pavement in order to drain the soils and keep the base course dry. This means that minimum ditch depth should be about two feet below the center of the pavement. Deeper ditches, of course, are required to accommodate roadway culverts and maintain the flow line to adjacent drainage channels or streams. Silt, sand, and debris will compromise the slope and plug swales.

In urban and suburban areas, streets without curbs will often have swales next to the road or between the edge of the pavement and the sidewalk. These swale areas are designed to facilitate the infiltration of runoff into the soil and/or move water to a receiving inlet. Typically, the swales

Figure 5.11 A swale adjacent to a road (and parking lot from where photo was taken). The swale will collect stormwater running off the parking lot and roadway to reduce flooding.

are simply shallow ditches along the road that are grassed (which must be mowed), although less urban areas may have natural vegetation. Parking on swale areas is a problem in urban areas without curbing. Parking on a swale compacts the dirt which makes the swale incapable of absorbing water. Some residents also build up their swales to keep water in the street, compromising the intent of the swale design. Such alterations need to be addressed administratively.

If water moves in a swale or ditch, some degree of vegetation may be needed to prevent wear caused by moving water. In addition, the water must be moved past the area where roads or driveways enter the roadway—this is where a roadside culvert (structure that allows water to flow under the road or driveway) is normally installed (see discussion earlier in this chapter). Culverts should be properly terminated (see Figure 5.12) to reduce risk to motorists who may accidently swerve off the pavement or need to pull over for an emergency. Many transportation departments will have some standard means to finish the end of a culvert to prevent damage to the pipe during accidents and lessen the damage to vehicles that hit the culverts.

Many ditches and swales will fill with debris, plant materials, soil, and other materials so they must be maintained (see Figure 5.13). Mowing swales and cleaning ditches on a regular basis is necessary to prevent excessive plant growth. Periodically, a Gradall excavator or other piece of equipment should be used to reestablish the ditch profile and remove the excess debris. One caution is that ditches that are close to roads may create underlying pavement damage (see Figure 5.14). Inspectors of roadways should work carefully with stormwater staff to determine where the road damage may be caused by sloughing due to stormwater damage. As will be discussed in Chapter 6, the pavement may indicate failure of the base due to saturation or ditch failure.

Retention and Detention Basins, Bioswales

Depending on local conditions, there are a number of solutions for dealing with excessive run-off aside from putting it into waterways. The first are *green* infrastructure solutions. All are commonly used and easy to construct. Retention basins are used to divert water that will not be drained into a waterway. The water is permanently *retained*, meaning it is not discharged

Figure 5.12   This stormwater culvert is properly terminated, reducing risk of car damage to a motorist who may hit it or injury to a pedestrian walking on it.

Figure 5.13 A swale filled with debris and plant material—this material will impact the ability of the swale to drain water vertically into the ground; it needs to be cleaned out.

Figure 5.14 This ditch that is very close to the road may have caused this underlying pavement damage.

off-site, but remains on-site and filters into the ground. Retention of stormwater in basins works well where there is land and an appropriate soil profile underneath. Ongoing maintenance is required to insure that vegetation is limited in the basin and that the soil surface remains permeable. Fertilizers can compromise the soil surface by accelerating microbes in the soil that can diminish percolation ability. Wetlands can be constructed within retention basins to absorb fertilizers as well.

Detention basins are much like retention basins, except that detention basins assume the water will be discharged slowly from the basin in order to have the water level return to its designed control elevation. An overflow box (Figure 5.15), a bleeder slot, or a small pipe will slowly drain down the detained water in the detention basin to an appropriate location (usually

Figure 5.15 Example of an overflow in a detention basin—these are normally concrete structures like a catch basin. The detention pond must completely fill before water can flow over the top of the overflow box.

another water body; however, flow is usually limited by the regulatory agencies to prevent downstream flooding). The detention basin requires maintenance to limit debris and keep the pipe or overflow clean. The dry detention basin should be kept mowed and grass clippings removed to prevent discharge into the receiving water. The overflow box is normally a concrete structure like a catch basin where the water must fill a detention basin before it can overflow into the top of the overflow box. However, many older overflow structures—particularly those found on agricultural sites—are constructed from corrugated metal.

Where land is limited, particularly in urban settings, but there is still a need to remove particulates such as hydrocarbons and fertilizers, a bioswale may be constructed. These occur most commonly near large parking areas or adjacent to some roadways. The bioswale vegetation is usually not controlled in an effort to maximize nutrient uptake. However, some bioswales look like a patch of unsightly weeds, so consideration of the plant palette is necessary.

Pervious Pavements/Pavers

Pervious pavements/pavers are a solution for large paved areas. These may be parking areas in commercial, institutional, or office complexes, in addition to areas adjacent to roadways where people park cars in the swale. Pavers can be pervious brick or blocks that have spaces between them to allow water to enter the underlying soil (see Figure 5.16). An example of a design for a paver driveway is shown in Figure 5.17. Pavers can also be used for residential driveways and swales for parking. However, the inability of water to infiltrate the soil because of poor percolation may make pavers an inappropriate solution in some areas. Pavers, however, can be a solution to restore the ability of soil to absorb some of the stormwater instead of becoming runoff. Pervious pavement and pavers must be periodically maintained because the pervious paver pores will fill with dirt, requiring the pavers or pervious pavement to be vacuumed out in order for infiltration into the soil to continue. Grass can grow in them as well so periodically the pavers need to be vacuumed out and re-sanded to keep inflow going. There are many

Figure 5.16 Example of a pervious paver swale that is used for parking—spaces between the pavers allow for percolation of stormwater.

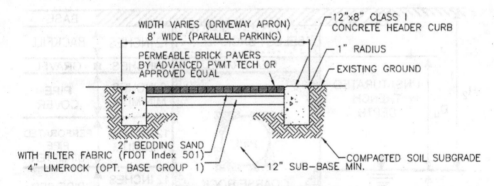

WIDTH VARIES (DRIVEWAY APRON)
8' WIDE (PARALLEL PARKING)

PERMEABLE BRICK PAVERS
BY ADVANCED PVMT TECH OR
APPROVED EQUAL

12"x8" CLASS I
CONCRETE HEADER CURB

1" RADIUS

EXISTING GROUND

2" BEDDING SAND
WITH FILTER FABRIC (FDOT Index 501)
4" LIMEROCK (OPT. BASE GROUP 1)

COMPACTED SOIL SUBGRADE
12" SUB–BASE MIN.

PERMEABLE BRICK PAVER SECTION

1. BEDDING SAND TO CONFORM TO ASTM C 33 COMMONLY KNOWN AS MANUFACTURED CONCRETE SAND. DO NOT USE MASONRY SAND.

2. PROVIDE 1" WIDE x 4" DEEP SAWED CONTRACTION JOINTS IN CONCRETE AT 10' CENTERS (MAX.) NO LATER THAN 12 HOURS AFTER INITIAL SET. COMPARABLE TOOLED JOINTS MAY BE PROVIDED IN LIEU OF SAWED JOINTS.

3. PARALLEL PARKING MUST SLOPE 2% TOWARDS THE EDGE OF PAVEMENT.

4. GRADING IS REQUIRED TO BLEND WITH EXISTING TOPOGRAPHY AND GUARANTEE POSITIVE DRAINAGE.

Figure 5.17 Example of an engineering design for a paver driveway.

options for pavers, pervious concrete pavement, pervious asphalt pavement, and the like to enhance their effectiveness, depending on the storm event that is being addressed. A concern with pervious pavements is that use in places with heavy traffic may not be appropriate and actually damage the pervious pavement.

French Drains/Exfiltration Trenches

French drains/exfiltration trenches are an option to manage stormwater where surface storage is limited or where there is a desire to recharge groundwater. The concept is to drain water into a perforated pipe below the surface, and allow the water to flow into the soil. These systems work best when the water table is not near the surface. The concept is illustrated in Figure 5.18 (which comes from the South Florida Water Management District's website for design of exfiltration trenches). Exfiltration trenches use perforated pipes to leach water into the ground. Extensive use of exfiltration is pursued in some areas like south Florida, where soil conditions permit.

Where the exfiltration trench will not work, gravity wells may be an option. The concept is to drain or pump the stormwater into the ground to a formation where the water will flow easily. Typically, the receiving rock formation should contain brackish water for the wells to work efficiently. Standards for wells are often location specific. In either case, pre-treatment to remove trash, silt, sand, and vegetative debris is required for either the exfiltration trenches or gravity wells to work. Debris must be removed from the structure on a regular basis. A two chambered box with a baffle and screen is required (see Figure 5.19).

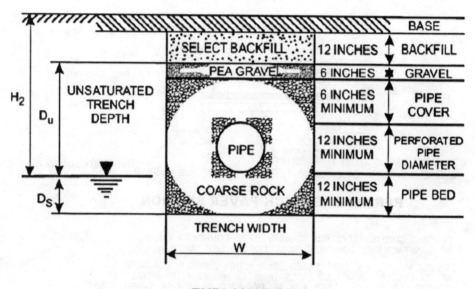

EXPLANATION

H_2 Height of ground surface above the design water table

D_u Volume of runoff that can be stored

D_s Depth of trench below the water table

If $D_u > D_s$: $L = V/(K(H_2W + 2H_2D_u - D_u^2 + 2H_2D_s) + (1.39e\text{-}4WD_u))$

Figure 5.18 Drawing of typical exfiltration trench. *Source:* The South Florida Water Management District

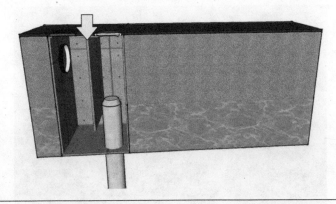

Figure 5.19   Typical configuration for gravity wells—gravity wells only work where there is salt-water beneath the surface or where the groundwater is deep. Gravity wells are located in over-sized catch basins. There is a baffle to catch debris (arrow). The water level must be below the top of the well in the adjacent groundwater. Note these require permitting because there is potential to contaminate the groundwater from gravity wells.

Canals

Canals (see Figure 5.20) are used to drain large areas and are another solution for stormwater management, albeit usually a regional option as opposed to a local one. In most cases, canals are dug in low-lying areas so that surrounding groundwater will drain off into the canal, thereby lowering the surrounding groundwater. The concept is that draining groundwater creates some soil storage capacity that will allow water to percolate faster into the soil. Examples of this type of canal can be found all over South Florida. Construction of canals in South Florida began in the early 1900s, but was interrupted by the hurricanes in 1926 and 1928 and the Great Depression. After the devastation and major loss of life around Lake Okeechobee, the Army Corps of Engineers began the construction of the Hoover Dike (see Figure 5.21). After much of South Florida was under water from extensive rainfall caused by the 1947 hurricane and due to a lack of a regional canal network, Congress authorized the Central and South Florida Flood Control Project to construct many miles of canals. This network of canals helped create the Everglades

Figure 5.20   A typical canal created for stormwater management in South Florida.

Figure 5.21   The Hoover Dike surrounding Lake Okeechobee in Florida, as seen from U.S. Route 27 in Palm Beach County.

Agricultural Area (EAA) on the south side of the lake after World War II. To function well, canals require bank and bottom maintenance or they will fill in or collapse, thereby reducing their usefulness.

An issue in waterways and canals, especially coastal ones, is the tidal action that can cause tidal water to flow upstream in the system. Some canals in South Florida were constructed without the benefit of a coastal control structure that would prevent saltwater intrusion to occur by moving upstream. In such cases, there is a need to prevent the water from moving upstream through piping systems that use the canal as their outfall. Figure 5.22 shows a Tideflex Valve, which can provide a solution to this problem. The valve only opens one way so saltwater cannot

Figure 5.22   Example of Tideflex Valves used to prevent discharged water from backflowing into the stormwater system. These are typically used in areas where stormwater is released into receiving waters subject to tidal action.

Figure 5.23 Example of a large regional stormwater pumping station in Davie, FL (operated by the South Florida Water Management District).

backflow into streets. Where the backflow may be very large, a flood gate or large pump station might be installed to ensure that no backflow occurs, along with the added ability to pump stormwater forward (see Figure 5.23). However, there is a major difference—stormwater pump stations are designed for very high flow volumes with very low head pumps that normally go directly to a waterway (sanitary sewers are usually smaller flows and higher heads). Pump stations for stormwater are similar to sanitary sewer pump stations, but at a much larger scale. Normally, some form of baffling, velocity reduction, silt retention, trash collection, or other treatment tools may be required for water quality purposes.

MAINTENANCE

All of these stormwater solutions require consistent, ongoing maintenance in order to work properly, along with some form of work-order tracking system that will allow operators and managers to identify problem areas and track the work performed. Like water and sewer, the failure to maintain these structures creates the potential for flooding, which may put the organization at risk for responsibility of damage on public or private property. Lawsuits can be lost due to negligence of ongoing maintenance of a stormwater system.

Because stormwater protection is often more regional than local, most communities participate in programs under permits secured by a regional agency (county level is common) to address the interconnectedness of water bodies through neighboring jurisdictions. The permits are issued for any system with conveyance piping discharging stormwaters into local water bodies that are:

- Owned by a state, city, town, village, or other public entity that discharges to waters of the United States
- Designed or used to collect or convey stormwater (e.g., storm drains, pipes, ditches)

- Not a combined sewer
- Not part of a sewage treatment plant or publicly owned treatment works (POTW)

To prevent harmful pollutants from being washed or dumped into specific types of waterways, regulatory agencies will require a Federal National Pollution Discharge Elimination System (NPDES) permit for a single pipe discharge, MS4 permit for maintenance, and/or stormwater management programs (SWMPs). SWMPs require communities to define their system and develop goals and objectives for both stormwater systems and the waterways. A typical community will include many of the following stormwater goals and priorities:

- Address detention pond maintenance
- Address water quality concerns in waterways in the community
- Mitigate flooding to the extent practical
- Maintain compliance with current and future regulatory requirements and permits
- Compliance with external regulations
- Maintain a functional stormwater drainage system
- Maintain the health and quality of life for residents

Maintenance of the drainage system is easier with logs, work orders, and geographic information system (GIS) integration and repair activities. GIS is a means to support staff in addressing issues associated with residents' property. Many concerns could be eliminated through improved record keeping and public education. Both will require the investment of time and resources. An appropriate amount of staff time must be available to solve some of the problems; however, access to accurate site-specific stormwater records can eliminate the need for excessive staff time. The ongoing enhancement of the stormwater program may include the following:

- A work order system
- Enhancement of staff resources to address stormwater problems
- Development of public information and education/outreach programs
- Adding equipment for maintenance activities
- Adding public involvement opportunities
- Enhancing the reporting and tracking tools—i.e., a work order tracking system
- Continuing GIS database development and management (digitizing of under-reported impervious areas, mapping of stormwater infrastructure, mapping of drainage basins and other stormwater features). Light detecting and ranging (LiDAR) data is particularly helpful in this regard as it can provide the mapper with relatively accurate ground elevations, building elevations and locations, and even vegetation (depending on the detail available from the LiDAR that is being used). LiDAR is collected by airplane and involves shooting pulsed laser light at the ground and detecting how fast it returns to the plane's sensors. The accuracy of the LiDAR results can be very good (four to six inches vertically). Figure 5.24 is an example of two-inch vertical resolution LiDAR compared to an aerial photo of the same site. LiDAR is available for many communities (see Figure 5.25).
- Updating imperviousness of the site based on acquisition of improved impervious area measurements
- Implementing a capital improvement program

Required field operations activities include the activities that must be reported upon—work orders are the easiest to track and input into a GIS. A series of ongoing inspections is

Figure 5.24 Comparison of aerial photograph (a) with LiDAR map of the same site (b). In the LiDAR map, the building, road, parking lot, curbs, vegetation, etc., can be seen. The color coding shows the elevation. Dark blue is low. The building is high (it is a 3-story building). At the top of the photo are some one-story houses (green). LiDAR allows engineers and planners to *see* elevation on the map easily.

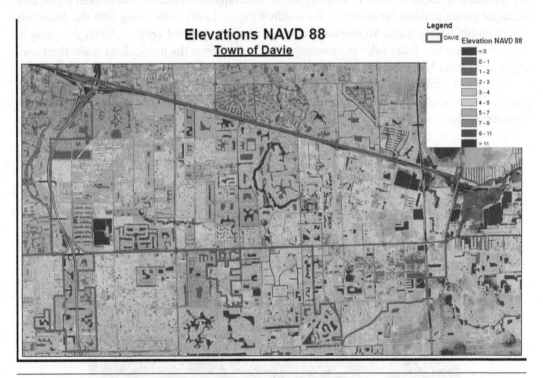

Figure 5.25 LiDAR map of a community within the town of Davie, Florida. Blue areas are low lying and will likely flood more quickly than the brown areas.

needed—inspections include annual visits and evaluations of wet retention areas and swales, and the disking of the bottom of retention areas and swales that have become compacted. Swales, dry retention areas, and exfiltration trenches should be inspected more often for sediment. Responsibilities include ensuring that new development flood control meets performance standards, reviewing proposed construction for integration and compliance with local needs and regulations, and developing mapping data related to water quality information. A GIS is useful in this regard.

Appropriate removal of material should be scheduled. Work orders should be used to track activities so that those that require more frequent maintenance can be identified. Likewise, catch basins should be inspected multiple times per year, and sediments and debris should be removed using a vactor truck (see Figure 5.26). Again, documentation will help staff identify those areas with more frequent cleaning needs and those that require fewer visits.

Maintenance activities include disking dry retention area bottoms and swales that have been compacted. Since these areas may be extensive, having the proper equipment and staff to do the work is necessary. In addition, inspections may indicate areas where actions are needed to: (1) correct areas of erosion, (2) undercut dead grass in wet and dry retention areas and swales, (3) document where swales and ditches may need to be cleaned, (4) have invasive plants removed (not by herbicide), and (5) clean sediment from exfiltration trenches. Vactor trucks are also useful for cleaning exfiltration trenches.

Other ongoing, repetitive actions include mundane tasks such as mowing wet and dry retention areas and swales where residents are not required to do so; stabilizing the banks of wet and dry retention areas; removal of trash and plant debris; repairs to catch basins, broken pipes, and damaged grates; and similar activities. Re-sodding the banks of canals, along with the detention and retention basins is also an ongoing activity in order to prevent erosion. Street sweeping is another activity that falls under stormwater maintenance since the particulates make their way to the local water bodies. Litter control, and measurements of same are often reportable.

Stormwater facilities must be constructed and maintained to reduce the negative impacts of runoff. These systems and their maintenance are expensive. Efforts to repair or rehabilitate exfiltration trenches should occur about every 10 years, while maintenance of catch basins, grates,

Figure 5.26 Example vactor truck used to suck debris, sand, leaves, etc. out of catch basins. The truck also has a tank and the ability to high pressure clean the pipes that are used for storm and sanitary sewer systems.

storm drains, structures, swales, gutters, and other features occur as needed or when damaged. Keep in mind that all infrastructure degrades with time. It does not have infinite life! Data should be kept on all maintenance and repair activities.

Compliance requirements for an MS4 permit for stormwater management programs include the submission of annual inspection and maintenance reports. This submittal includes the reporting of required inspections, the reporting of maintenance on stormwater facilities, as well as the developing and maintaining of a record-keeping system. Reviewing construction sites for compliance with runoff is important to protecting water bodies.

Case Study: Davie, Florida

The following is a series of tasks that the town of Davie, Florida, undertook to comply with MS4 permit responsibilities:

Inspections

Annual

- Wet retention area
- Swale bottoms
- Disk bottom

Semi-Annual

- Dry retention areas
- Exfiltration trenches
- Swales
- Sediment in wet retention, dry retention, and swale areas

Quarterly

- Catch basins

Maintenance

Annually

- Disk dry retention area bottoms
- Disk swale bottoms
- Correct stormwater wet retention area

Semi-Annually

- Correct areas of erosion, undercutting, or dead grass in wet and dry retention areas and swales
- Take appropriate action on petroleum or other pollution spills noted
- Swale cleaning
- Remove invasive plants
- Remove sediment from exfiltration trenches
- Clean exfiltration trench

continued

As Needed

- Mow wet and dry retention areas and swales
- Stabilize banks of wet and dry retention areas
- Rehabilitate exfiltration trenches every 10 years
- Correct wet and dry retention area equipment
- Correct dry retention area bottoms
- Stabilize banks on wet and dry retention areas
- Nutrient/pesticide management
- Clean bottom debris

Every Five Years

- Scrape bottom of retention areas and swales
- Re-sod banks of wet and dry retention areas as needed
- Inspect all retention ponds

Compliance Requirements for MS4 Permit

Stormwater Management Program

- Submit annual inspection and maintenance report
- Conduct required inspections and maintenance
- Develop and maintain record-keeping system

New Development

- Implement policies with regard to stormwater and drainage management controls
- Review land development regulations to determine where changes must be made, especially to swales, low impact development, stormwater reuse, and landscaping
- Enforce requirements of Pollution Control Code

Roads

- Litter control
- Implement best management practices (BMPs), also called best stormwater practices
- Perform maintenance of catch basins, grates, storm drains, structures, swales gutters, and other features

Flood Control

- Ensure new development flood control meets performance standards in accordance with regulatory standards
- Strengthen local comprehensive plans and submit same to county
- Maintain a GIS layer with water quality information
- Ensure flood control meets with water management district rules

Pesticides and Herbicides

- Provide certification and licensing of applicators

continued

Illicit Discharges

- Conduct assessment of non-storm discharges
- Provide copies of newly adopted ordinances prohibiting illicit discharges and dumping
- Continue BCDPEP random inspection program
- Define allotment of state and resource to stormwater program
- Report all violators to the local regulatory agency (Broward County)
- Conduct periodic training to staff on identification and reporting of illicit discharges
- Terminate illicit discharges and document same
- Develop municipal procedures for handling and disposing of chemicals and spills, including training of staff on emergency response
- Distribute brochure to public on appropriate disposal of hazardous materials
- Develop public outreach effort for oil, toxic, and hazardous waste for public
- Promote Amnesty Day
- Develop voluntary storm-drain marking program
- Continue infiltration and inflow program on sanitary sewer system
- Investigate septic tank discharges to stormwater system

Industrial Runoff

- Maintain inventory of high-risk discharges, including outfall and surface waters where discharge occurs
- Provide ongoing inspections of high-risk facilities
- Provide annual report to county
- Monitor high-risk facility discharge water quality

Construction Sites

- Ensure stormwater system meets treatment performance standards according to regulatory standards
- Continue construction site inspection program to ensure reduction of off-site pollutants
- Implement standard, formalized checklist of stormwater management and water quality inspection items
- Maintain log of stormwater management activities at construction sites
- Provide detailed description of inspection program and forms
- Provide summary of activities
- Continue inspection certification program to stormwater management, erosion, and sediment control for operators, developers, and engineers
- Develop outreach program for local professional organizations

Develop, Maintain, and Conform to Standard Operating Procedures

- Develop and maintain accurate mapping of the drainage system
- Track areas with ongoing stormwater issues and develop programs to alleviate same

The MS4 permit requires ongoing demonstration of maintenance through the use of logs, work orders, photographic documentation, and—in the best of worlds—GIS support to ensure all of these facilities not only operate properly, but also reduce pollutants. These requirements mean that the community needs funds to ensure that monies are available to properly execute the MS4 program to ensure compliance. Significant effort is required to maintain functioning of stormwater systems, many of which have been neglected with time. Extra effort may be recommended prior to rainy seasons to limit flooding potential from unmaintained facilities.

EFFECT OF DEVELOPMENT

The amount of stormwater runoff is determined by soil conditions and the amount of manmade impervious area. Stormwater runoff must be managed in an organized and systematic manner if property owners (commercial, industrial, and residential) are to enjoy the full use of their property. Initial designs must meet the level of service defined by the community through a regulatory process. Problems occur later, when property owners want to redevelop and increase impervious area, which changes the stormwater system characteristics. Because the burden of managing stormwater typically falls to the local government, local officials must be judicious in protecting the veracity of the stormwater system. Development patterns immediately affect the demands for stormwater in a way that is dissimilar to sanitary sewer or water systems.

The amount of impervious surfaces and other runoff characteristics are generally accepted as the rational nexus for the *runoff* or *burden* placed on the system by each eligible parcel. As a result, a stormwater system has two elements that need to be considered regarding runoff: the impact due to adjacent properties (how much water runs off) and the intensity of that impact (i.e., how fast water runs off due to a difference from the pre-development condition). The relative *use* or benefit that the property owner derives through the indirect or direct connection of his property to the community's stormwater management system is defined as the demand a property places on the system, services, and facilities that have been established and are maintained to protect both properties and the receiving waters. As the impacts of water and wastewater utilities have historically been based on water use volumes, stormwater impacts can use measurable runoff characteristics, such as impervious surfaces, as the *rational nexus* for the volume of runoff expected to be discharged by each property. The square footage of impervious area is used as a proxy for the amount of runoff that is collected, transmitted, or treated by the stormwater management system during a specific rain event. The greater the flow, the greater the use (benefit to the property) and the greater the impact on the system. The understanding of the imperviousness parameter addresses the fact that those parcels with a lot of impervious area will cause runoff to occur faster (depending on the geology), thereby increasing their impact on the public system.

Figure 5.27 is an example of the considerations which relate to pipe sizing. In this figure a single family house is defined by the impervious area it creates on the site (2,500 sf) and the percent of imperviousness it creates (31%). Redevelopment often ignores the parameters upon which the original piping and conveyance systems were designed—bigger houses and paved driveways and swales may dramatically increase the amount of runoff contributed at a given time to the storm sewer system. In evaluating development and redevelopment, care must be undertaken to ensure maintenance actually occurs and that the system derives the benefit of a reduction in the property's stormwater runoff.

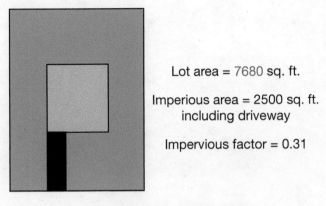

Lot area = 7680 sq. ft.

Imperious area = 2500 sq. ft.
including driveway

Impervious factor = 0.31

Runoff Generated by Statistical Median
Single Family Residential Lot

Figure 5.27   Example of calculation of impervious area on a single family lot.

REFERENCES

Evans, J. and Spence, M. N. 1985. "The evolution of jointing vitrified clay pipe." *Advances in Underground Pipeline Engineering, Pipeline Division, ASCE/Madison, WI/August 27–29, 1985*. ASCE, Alexandria, VA.

Mulcahy, L. A. 2006. "The effectiveness of stormwater utilities in mitigating stormwater runoff in the United States." Submitted in partial fulfillment of the requirements for the Degree of Master of Arts in the Program in Environmental Studies at Brown University, Providence, RI.

United States Environmental Protection Agency (EPA). "National summary of water quality conditions." EPA, Washington, D.C.

———. 2003. "Water: After the storm—weather." EPA-833-B-03-002, Washington, D.C.

———. 2007. "Total maximum daily loads and national pollutant discharge elimination system stormwater permits for impaired waterbodies: A summary of state practices." region V, EPA, Cincinnati, OH.

Wenzlick, J. D. and Albarran-Garcia, J. 2008. "Effectiveness of metal and concrete pipe currently installed in Missouri (Phase 2)." Report No. RI07-058 9. Missouri Department of Transportation P. O. Box 270, Jefferson City, MO.

Lot area = 7500 sq. ft.

Impervious area = 2500 sq. ft. including driveway

Impervious factor = 0.31

Runoff Generated by Gravitational Medium
Single Family Residential Lot

Figure 5.27. Example of calculation of impervious area on a single lot, for

REFERENCES

Evans, T. and S. Jones, M. V., 1988. "The evaluation of routing with different pipe" Albania interlaboratory Pipeline Engineering, Baptist Division, ASCE/Albania, *J Weapon* 21-22, 1988, ASCE, Albania, data, VA.

Anthony, E. A., 2010. "The effectiveness of stormwater utilities in offsetting stormwater runoff in the United States. Submitted in partial fulfilment of the requirements for the Degree of Master of Arts in the Program in Environmental studies at Brown University, Providence, RI.

United States Environmental Protection Agency (EPA). National stormwater water quality conditions. EPA, Washington, D.C.

———. 2002. "Water After the storm—weather." *EPA-833-F-03-001*, Washington, D.C.

———. 2007. "Total maximum daily loads and national pollutant discharge elimination system storm water permits for impaired watersheds: A summary of state practices" region VI EPA, Cincinnati, OH.

Wanielista, D. and Albright-Garcia, J. 2008. "Effectiveness of metal and concrete pipe currently installed in Missouri Phase 2). Report No. RI02-054." Missouri Department of Transportation, P.O. Box 270, Jefferson City, MO.

6

TRANSPORTATION SYSTEMS (ROADS AND RAILROADS)

TRANSPORTATION SYSTEM GOALS

Transportation systems differ from the water, sewer, and stormwater infrastructure discussed in the previous three chapters because they are not primarily designed to protect human health or property. Instead, transportation systems are designed to promote economic activity by facilitating the movement of goods and services from one location to another. In addition, transportation is not limited to any one service—transportation includes pedestrians, motor vehicles, railroads, air, and waterways. However, while the multi-modal aspect of transportation systems is distinctly different from other infrastructure systems, the transportation corridors are what bring the other systems together. Water, sewer, power, stormwater, and other systems are typically located in roadway corridors. In addition, there are other layers of transportation services and the locales they can reach. For example, water transportation is limited to navigable waterways, air travel is limited to where there are airports, and goods and people must have a means to move from airports to other destinations. Railroads resolve the limitation of ports and airports by extending service to communities that are not located on waterways or air terminals. But, even the railroad must remain on tracks that have limitations as to where they are located—motor vehicles resolve this limitation. Roads were developed to facilitate this multi-modal transport of goods and services, thereby connecting air, water, and rail outlets. Bicycles and motorcycles can travel places where cars cannot. Pedestrians, of course, can walk just about anywhere the climate and terrain permits. Intra-modal needs of transportation indicate the least expensive way to move goods is ship-rail-truck-bicycle-pedestrian—in that order. Each layer carries less goods than the former, and at a higher cost, but can reach more destinations and more customers.

Because the span of transportation services is vast, and the options are usually limited within most communities, this book will not delve in to airport or waterway aspects, which are often limited or lacking locally and are often under federal or state oversight. The more extensive the transportation network, the more complicated it can be, although the physical tenets have similarities because the materials are common—steel, asphalt, and concrete. For example, ship traffic requires a port, and maybe locks, dams, or other structures to facilitate movement between water bodies. These are concrete and steel, and generally were constructed by the federal government for the benefit of the nation. Likewise, airports may be community based, but major hubs are regional and constructed with concrete and steel, and generally with federal monies to benefit the region and/or nation. Hence, local infrastructure managers need limited information about them unless the responsibility has been delegated to them, as is the case now for the city of New Orleans. Then, the responsibility ramps up significantly. Usually when delegated, some federal oversight is still maintained—if not, the catastrophic events such as the destruction of New Orleans by Hurricane Katrina can happen (see Chapter 14).

Instead, this chapter will focus on local roads and, to an extent, bridges—which may be under the purview of local public works departments, with a brief discussion on rail since light rail may be associated with some local agencies.

ROADWAY COMPONENTS

The original roads were paths created by animals to traverse forests and fields. People followed these paths, and in doing so, the paths became more worn, encouraging more traffic. As commerce developed, roadways were constructed because they were the preferred means to move freight between places that were not coastal. Many were over the old paths and trails. Given development patterns of a region, at any given time, the number of roads in a country or community varied. More roads generally meant more commerce. A study by Bogetic and Fedderke (2006) indicated that paved road density was highest in areas with higher commercial opportunities. Population density was only a part of the equation.

Most goods are moved by the private sector and it is the private sector that benefits from good transportation systems, whether it is moving goods and resources, or attracting labor. The private sector wants well-built roads in order to move goods and services efficiently. Today, most *roads* in North America involve:

- *Trails*—roads that are generally not maintained and rely upon repeat traffic to ensure the path remains visible with time and are lightly traveled (Figure 6.1). These roads have no drainage systems or directional signs.
- *Unpaved, but maintained roads*—common but less-utilized public roads in rural areas that are rare in cities (Figure 6.2). These roads may or may not have a speed limit but driving conditions usually require slower speeds. Maintenance is needed to ensure usability after rain, snow, or other weather events that can wash away the surface. These roads have no drainage systems and few directional signs.
- *Rural paved two-lane roadways*—among the most commonly used type of road between communities in any region (Figure 6.3). These roads normally have high speed limits, no curbs, but extensive drainage improvements (normally ditches) to keep the road base dry. Signs are limited and stop signs are rare except on side streets. They may have water distribution systems buried in the corridor, but rarely sewer utilities. Power will often be run along the edge of the road right-of-way.
- *Local community streets*—these generally serve houses within a subdivision and may have curbs and sidewalks with water distribution, sewer collection, and stormwater systems underground, and will generally have speed limits of 25–30 mph (Figure 6.4). Stop signs are the normal means for intersection control. Speed bumps can be used to keep speeds down. Rarely are these roads more than two lanes, although often the lanes are wide enough to support curbside parking. Crosswalks may be present, although crosswalks should never be installed where there is not a stop sign.
- *Collector streets*—urban roads that are the ending point for local community streets. These will generally have stormwater systems, utilities, curbs, sidewalks, and may have more than two lanes with higher speeds than local roads (Figure 6.5). Collector streets may

include traffic devices like traffic lights, pedestrian signals at crosswalks, bus lanes, and even bike lanes may appear in some communities.

- *Arterial roads*—roads that are multi-lane each direction and intersect with collector streets. Traffic lights are normally employed for traffic control. Speeds are often 45 mph or more and traffic lights tend to be timed to show preference for arterial drivers (see Figure 6.6). These roads will have stormwater systems, curbs, utilities, sidewalks, and more.
- *Highways/interstate highways*—these are limited access roads with high speeds (55–75 mph, depending on location) and prioritize long distance driving. Many of these roads were originally built for military use in the 1950s and 1960s and continue to be used and expanded today (see Figure 6.7). Utilities are not traditionally associated with the interstate system.

Figure 6.1   A well-worn trail that relies upon repeat traffic to ensure it remains visible.

Figure 6.2   An unpaved—yet maintained—road with a limited number of signs.

Figure 6.3   Rural paved two-lane roadways with signs are the most commonly used between communities.

Figure 6.4   Streets that are paved and have signs, sidewalks, speed bumps, and so forth are found within local communities.

Note that with each expansion of use, the number of amenities increases and the character of the road changes (see Figure 6.8). Sidewalks make their first appearance with local streets—the first roads to clearly delineate multi-modal use—and signage increases. Interstate roadways break this trend. Usually water, sewer, power, and communications are not located in interstate rights-of-way because interstates are not intended to serve local communities—they serve region and state, and ultimately the nation. Hence, local infrastructure serves a limited purpose for interstates. No one lives on an interstate. However, drainage is a major component of interstates because the commercial purpose of interstates would be compromised if they flooded. Sound walls may be an urban priority along with landscaping due to the noise and visuals in local

Figure 6.5   A typical collector street complete with street lights.

Figure 6.6   Arterial roads are paved and have multiple lanes in each direction, along with medians, signs, pavement markings, sidewalks, and traffic lights.

neighborhoods, but neither is relevant to the mission of the interstate. As a result, interstates are maintained by state or federal entities, as opposed to local entities.

While the breadth of roadway options is large, the construction materials are fairly limited. Road surfaces can be dirt, rock, asphalt, or concrete. New bridges for roadways are concrete and/or steel. There are a few old bridges constructed of wood timbers that are still in use, but newly constructed bridges are rarely made of wood. Within the roadways themselves, the road has a base, sub-base, and a surface. The base and sub-base materials are compacted rock—usually 95–98% compaction. The base and sub-base provide the stability to the road. Keeping them dry keeps the base from swelling. A wet base will move and flex as there is nothing to hold

Figure 6.7 Example of highways/interstate bridges.

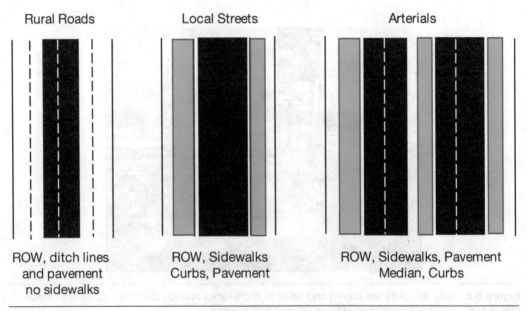

Rural Roads

Local Streets

Arterials

ROW, ditch lines
and pavement
no sidewalks

ROW, Sidewalks
Curbs, Pavement

ROW, Sidewalks, Pavement
Median, Curbs

Figure 6.8 Example of comparison of rural road rights-of-way, with rights-of-way for local streets and arterials.

it together. Hence, the road surface coat is designed to keep the base dry (see Figure 6.9). Base failure is one of the most common causes of roadway damage. Improperly compacted bases or bases that are saturated with water will have long-term maintenance issues unless corrected. Base failure will translate to the cracking of surface pavement—more quickly in asphalt than concrete. As a result, it is important to understand when roadway failure is simply due to the pavement and when it is caused by other factors that require more extensive repairs. Note that how the road is initially constructed is the best predicator of future problems. Most repairs must follow the design created by the engineer. The following sections will outline how pavements

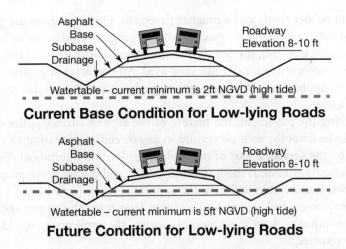

Current Base Condition for Low-lying Roads

Future Condition for Low-lying Roads

Figure 6.9 Stormwater drainage systems are designed to keep the road base dry. This figure shows that the ditches on the side of the road are designed to be below the base and sub-base rocks. In areas with sea level rise, one major concern is that the water table will rise into the base. Wet base will cause the roadbed to fail.

should be installed, and what failure modes are indicated by surface cracking so that correct solutions can be applied to ensure long-term pavement performance.

Asphalt Pavements

The most common pavement types found in North America are asphalt concrete pavements (also known as asphalt cement, blacktop, or just asphalt pavements). The term *asphalt* is used by the public, but the roadway surfaces are technically "asphalt concrete" surfaces. Asphalt concrete has two components: asphalt and the rocks in the asphalt—called aggregate. Asphalt is bitumen, a volatile hydrocarbon compound of tar that creates a sticky connection between the aggregates that exist in a solid state at normal air temperatures. It is one of the mixtures that comes off the refining process for oil at 900°F by applying a vacuum. By heating the bitumen, it coats the aggregate and then sets up as it cools, creating asphalt concrete. Ultimately, asphalt concrete pavement is a mixture of rock, pebble, or sand aggregates and 3–12% bitumen, depending on the aggregate size. Asphalt concrete pavements are generally installed at temperatures above 185°F.

Asphalt is used because it is readily available, relatively inexpensive, and wears well. The surface of an asphalt concrete road may last an average 10–15 years, depending on local weather and other conditions. It is easy to work with, and can be delivered in varying quantities, making it useful for large and small jobs. Aggregates used in the asphalt are generally local materials, further improving cost efficiency. Temperature and percent of bitumen are key variables. Weather plays a part as well—wet surfaces will retard bonding and cool the asphalt pavement too quickly. Bitumen shrinks as the temperature decreases and becomes less workable, which is a critical concern. Asphalt mixes that are either too hot or too cold create workability conditions during the rolling/compaction process that damage the ensuing pavement, leading to early failure. Recycling asphalt is common.

The aggregates hold the load in asphalt pavements. Both natural and processed aggregates are used, and there are various mixes utilizing different sizes of aggregate depending on the design need. Grain sizes can be large (0.75 of an inch) or small (through the 200 sieve). Larger aggregate leads to noisier roads and a rougher pavement. Fine aggregates are generally only a small portion of the total aggregate mix. Smaller aggregate requires more bitumen, which does not wear as well; however, it resolves the ride issues. Thin coats and touch-ups may use finer material, but rarely is a grain size like sand used. An inspector or engineer should review each load prior to application. In some cases, asphalt pavements fail due to the wrong size aggregate being used.

When inspecting the pavement, the texture of the asphalt surface is the easiest thing to look at. It should be smooth, with no texture in any direction. If tearing or ripples occur, it is likely caused by the screed, a part of the asphalt application equipment. The temperature of the mix may also be at fault. If the mix is too cold, it will not lie down properly. If there is too much bitumen (wet mix), it will have the appearance of cold mix or of a wet, black goo. This situation may create bubbles and blisters that later tear. The wrong aggregate mix can also affect the smoothness. Anytime roughness occurs, the application should cease until the situation can be resolved.

Asphalt machines are designed to spread the asphalt in a manner that creates a smooth surface (see Figure 6.10). By focusing on the surface, the imperfections and damage that has occurred to the current surface can be resolved. Temperature is critical. Asphalt cement delivered to a site is normally at 250–300°F. Mixes above 325°F are too hot to apply. Asphalt that is too hot will be difficult to compact and lack cohesion. It should be allowed to cool. Likewise, no asphalt (except cold mix or cold patch) should be applied if the temperature of the mix is below 185°F. The pavement mix becomes less fluid and more solid, making it difficult to work with as the viscosity decreases. Compaction is compromised below 185°F. Unfortunately, contractors will try to apply that last bit of asphalt and too often it fails to meet the temperature requirement, creating a maintenance issue later on.

Compaction of the asphalt is accomplished with rollers (see Figure 6.11). Rolling asphalt is designed to improve stability and to compact the mix in order to reduce void space, which reduces the space where water can enter and damage the pavement. Rollers may be heavy,

Figure 6.10 This figure shows a Caterpillar brand asphalt machine used to lay new asphalt on a roadway. A dump truck filled with asphalt mix will connect to the right side of the equipment and the asphalt machine will push it along as it spreads the new pavement.

Figure 6.11   Steel drum rollers used to compact hot asphalt after it is applied by an asphalt machine (Figure 6.10). The goal is to compact the asphalt mix enough to create a relatively impervious surface, without over-compacting and thereby damaging the mix.

vibrating, or tire mounted, depending on the job. Heavy rollers are used when the surface is rough. Light rollers are suggested where there is low viscosity and lots of fine particles. Rubber-tired rollers are suggested if the asphalt breaks under steel rollers (although this likely suggests other design concerns). Most asphalt applications do not require vibration and only use rubber-tired rollers for the last pass.

The goal for rolling is to reduce voids (air space in the asphalt surface) to around 8% (i.e., 92% is solid material). If there are more voids, disintegration and raveling are likely to occur. It may also make the pavement more susceptible to the freeze/thaw cycle damage due to the expansion and contraction of air space in the voids. Denser mixes (fewer voids) will have higher cohesion but may become unstable with time or leach asphalt in warm seasons (called bleeding—discussion shortly). Damage to the pavement can occur during rolling. Rolling compromises pavement integrity at high temperatures. Heat checking—hairline cracks in the pavement that occur during rolling—tend to occur where there are temperature differentials that change viscosity in parts of the pavement (see Figure 6.12). Heat checking occurs when the asphalt is too thin or the temperature within the mix is too great. It indicates that the asphalt is moving under the roller. The result will be cracks that curve from the high viscosity (warm) area toward the cooler edges. This should be avoided and any asphalt with heat checking should be picked up and removed from the site.

Outside temperatures, winds, humidity, and mix temperature affect how much time there is for rolling. No rolling should occur on asphalt with a temperature below 185°F as noted, so the window of opportunity can be tight. Rolling should *not* occur below an outside temperature of 40°F. Thick courses will hold heat longer and may contain more temperature differentiation. Thin courses must be rolled immediately before the temperature drops. Courses less than 1.5 inches are likely to have temperature issues.

Rolling should be limited to three to five passes. More passes tend to damage the pavement. Each pass is designed to increase density before the asphalt temperature drops below 185°F.

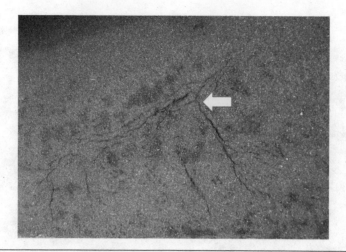

Figure 6.12   Heat checking occurs when small cracks in pavement are observed.

Weight of the roller is an important consideration since heavy rollers can break the less stable mixes, thereby negating the benefit of the repair. The final pass is designed to improve the surface and may be done with a different roller than the initial passes.

With asphalt, limiting water penetration is key. Water is an ongoing issue with pavements because water does not compress, whereas hot asphalt will expand. In cold climates, the water expands when it freezes, damaging the pavement. Hence, keeping water out of the pavement and out of the base will protect road stability. Water can percolate through worn pavement, under the edges of the pavement, in dips or corrugations in the road, and via capillary rise from the water table. All of these issues will manifest themselves as pavement damage, so understanding pavements and surface damage will help with long-term maintenance and repair planning.

Water makes maintenance needs for asphalt pavements difficult to define. Some agencies call repaving *maintenance*, while others will include only patching and sealing as part of maintenance. Either way, both activities need to occur on a regular basis. The most important aspect of asphalt maintenance is to repair the damage to the pavement promptly since the deterioration will accelerate once the surface is damaged. Overlays should not be assumed to be the solution to all underlying pavement damage because the pavement damage may indicate more significant problems. Stresses and minor defects are present in every pavement. The initial failures manifest as cracks and deformations. Water, dirt, plant seeds, etc., can get into the cracks, and accelerate damage to the pavement and underlying base.

Inspection is the best means to find problems before they become significant—especially in northern areas where frost heave can exacerbate existing defects. Many defects can be identified by tracking headlight or taillight patterns at night. In a long-exposure photograph, a good layer of pavement should create steady beams of light (see Figure 6.13). A wavy line of light indicates some form of pavement discontinuity (see Figure 6.14). Unfortunately, the solutions to pavements with wavy light patterns are expensive. The pavement must be removed through planing or grinding, application of a tack coat, and application of a new layer of asphalt.

Making repairs quickly may prevent further recurrence, if properly done. Therefore, knowing what defect the pavement damage indicates is relevant. One needs to understand the cracking to make correct repairs—the cracks indicate water, base failure, and wear patterns. Ultimately, these guidelines will permit asphalt cement pavements to be repaired or restored to extend their useful life.

Figure 6.13 Traffic taillights on a smooth asphalt pavement are not particularly wavy because the pavement is relatively smooth (time-delayed photograph). Pavement should be smooth and therefore the light patterns should be smooth as well. *Source*: Maine Department of Transportation (DOT)

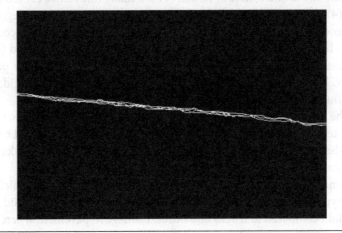

Figure 6.14 Traffic headlight patterns on a rougher asphalt pavement appear wavy (time-delayed photograph)

Properties of Asphalt Concrete

The goal of asphalt concrete pavement is to provide a hard surface that protects the roadway base as well as providing a good driving surface. Asphalt concrete pavement is also designed to be relatively impervious so that water does not get into the base material causing it to swell or heave. This is particularly critical in areas where freeze/thaw cycles may occur. As a result, there are a number of properties of asphalt concrete mixes that should be considered. Field observation can relate information about the mix, particularly if the mix is not correct. Stability is the ability of the pavement to resist rutting and shoving from traffic loads. Indicators of unstable pavements include ripples, ruts, washboarding, and other surface defects. Stability concerns may be indicated if these conditions are present as they may relate to issues with the concrete mix itself—too much bitumen, too much sand, or too much rounded aggregate.

Durability is the ability of the asphalt to resist disintegration due to traffic and weather. Poor durability can be caused by low bitumen content, insufficient compaction, or susceptibility to shedding of the asphalt from the pavement. When these conditions are present, the pavement hardens too early, creating brittleness, void space, or delamination. A cause may be that the asphalt was too cold when applied or that it was applied on a wet surface (dew can create this problem).

There are a number of reasons why imperviousness can be lost in the construction process. The first is that there is insufficient bitumen in the mix. When this occurs, the grains do not get coated and the voids are not filled. The voids create the pathway for water to get into the pavement. Voids can also be present if the compaction was improper. Too little compaction will not eliminate voids. Compaction issues may relate to the asphalt mix being too cold when rolled.

Workability of the mixture relates to the ability to place the mixture easily to create the pavement surface. Like concrete, mixes with large volumes of large aggregates tend to separate and become hard to compact. Too much bitumen will also make the mix hard to compact; instead, the mix will be gummy and, thus, stick to equipment. A central point for workability allows workers to hand place the mix around objects like manhole covers and valve boxes, while also being placed on longer roadway stretches. Cold asphalt loses the quality of workability.

A tender asphalt mix is one with four issues: (1) too many fine sands, (2) too wet, (3) smooth aggregates, and (4) too much filler. All four issues are present to varying degrees with tender asphalt mixes. Tender mixes can be manipulated too easily, so they will not compact properly. A mix that is too hot may have similar characteristics. A mix that is too cold (<185°F) will be unworkable because the bitumen is setting/hardening. The flexibility of the pavement is also related to the mix. Densely graded, low asphalt content mixes will be more rigid, while the open graded, high bitumen content mix will be flexible.

The base of the road material is critical. The base should not be flexible or move; the base is the roadway integrity, so if it flexes, the pavement will also. Expansion and contraction from freezing is a further consideration. Rigid pavement will tend to show shrinkage cracks more easily in cold climates.

Fatigue resistance is related to the ability of the pavement to absorb vehicle loading. Like parking lots, the type of traffic and repetitiveness of same is an important consideration. Passenger vehicles are light compared to trucks and buses. Damage to pavement where there are buses and trucks is common. Pavement damage needs to be dealt with in the design of both the base and the pavement mix. Voids from lack of compaction will accelerate fatigue; so will subgrade compaction problems. Thick pavements tend to bend less than thinner pavements, but the road base may be a larger factor. More asphalt concrete does not guarantee a better paved surface.

In any case, testing should be conducted of any hot mix delivered to a site. The testing should be conducted by a knowledgeable asphalt inspector or engineer to ensure that the application and mix are correct. As with most infrastructure systems, quality control in the construction process is the best indicator of long-term maintenance costs. The better the initial product, the less maintenance is needed and the longer life the asset may have.

Determining asphalt needs is straightforward if one understands what one is looking at. To start, the easiest means to find problems is a windshield survey—often from a vehicle through visual inspection. *Feeling* the pavement while driving (through the steering wheel) is a useful and inexpensive option. Repairs should be noted and corrected as quickly as they are noted so that they do not become larger maintenance problems. Drainage issues will create long-term failures so areas with water damage should be identified for corrective measures. Storm drains should be cleaned and inspected regularly to ensure they work so as to reduce impact on the

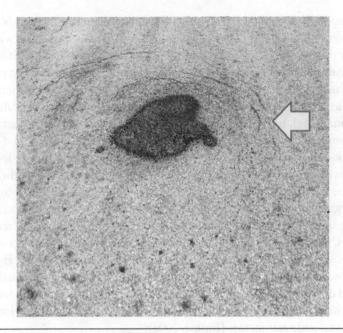

Figure 6.15   Example of a pavement where asphalt has been used to fill the hole for years. It is not the pavement that has failed—there is a pipe under the road that has failed. Note the scalloping around the patch (arrow).

roadway. However, it is important to understand what causes the defect in the pavement, as opposed to just repairing it—a common failure for maintenance staffs. Figure 6.15 shows a pavement where asphalt has been used to fill the hole for years. Just before this photo was taken, it was *patched*. The problem is that the defect indicates a leaking pipe beneath the surface—in this case a sanitary sewer line or service appears to be the problem. This location had been patched at least twice before, and will continue to be until the underlying problem—the broken pipe—is fixed. Divots and scalloping are indicators of subsurface pipe breaks.

Asphalt is classified as a flexible pavement since it will move with loads and base conditions. As the asphalt flexes, it wears. As a result, the appearance of the asphalt is a good indication of the cause of failure—freeze/thaw cycles, base condition/construction, water, surface loading, roadway construction practices, and so forth. Each failure mode can create a slightly different failure indicator, depending on related conditions; and the failure mode may suggest slightly different fixes. Therefore, understanding what one is looking at is important.

Assessing Asphalt Concrete Pavement

Assessing asphalt cracking can provide useful information into what is really happening to a roadway surface, which may affect pavement performance and determine the severity of the damage and needed improvements. It also can point out structural or material defects that could lead to reduced pavement performance and eventual failure of the pavement and base. Base failure is expensive. As a result, it is well worth it to an organization that must maintain roads to have a person who can look at failures and assess the cause. Several organizations provide certifications to people who take classes on asphalt inspection. These certifications are worth the investment to those organizations in order to have an in-house person who can diagnose and correct the defects as opposed to making fixes that will not last. Visual inspection is the best means to assess roadways, albeit it is not always necessary for the inspector to be walking the road. Groeger, Stephanos, Dorsey (2002) note that the American Association of State Highway and Transportation Administrators (AASHTO) has developed tools to assess pavement condition using remote vehicles with mounted cameras for network surveys. Two cameras are normally included—both looking downward. The cameras allow the data to be correlated by pavement type (flexible, composite), traffic volume, estimated condition, and environmental region.

Ongoing surveys of the roadway are critical because as Uzarowski et al. (2008) note, the roadway network deteriorates as the condition worsens and traffic volumes increase. Because most agencies have financial constraints placed on them, getting to every roadway segment that needs help can be a challenge. Uzarowski et al. (2008) note that in an effort to stretch limited dollars, many organizations will look at pavement recycling as a means to cut costs; recycling also reduces the impact of pavement construction on the environment by reusing depleting natural resources, reducing energy consumption, and reducing greenhouse gases.

In assessing pavements there are numerous methods to assess condition. However, the primary one used by transportation departments is the Pavement Surface Evaluation and Rating (PASER) Manual developed at the University of Wisconsin—Madison. They identify four major categories of common asphalt pavement surface distress (as discussed previously) that result from environmental and wear factors (Entine, 2002):

- Surface defects—raveling, flushing, polishing
- Surface deformation rutting, distortion—rippling and shoving, settling, frost heave
- Cracks—transverse, reflection, slippage, longitudinal, block, and alligator cracks
- Patches and potholes

Work orders should note areas repaired so a tracking tool will help identify problem areas. A comprehensive pavement management system involves collecting data and assessing several road characteristics: roughness (ride), surface distress (condition), surface skid characteristics, and structure (pavement strength and deflection). However, first there is a need to inventory all roads, evaluate the pavement surface condition, and identify corrective actions. It should be noted that the rate at which pavement deteriorates depends on its environment and traffic. The condition index may represent a single distress such as fatigue cracking or a combination of many pavement distresses, which is then compiled into a composite index (FHWA 2010). The following ratings are related to needed maintenance or repair based on data by Entine (2002):

- Rating 9 and 10: no maintenance required—yet
- Rating 8: little or no maintenance
- Rating 7: routine maintenance, crack sealing, and minor patching
- Rating 5 and 6: preservative treatments (seal coating)
- Rating 3 and 4: structural improvement and leveling (overlay or recycling)
- Rating 1 and 2: reconstruction

In any case, agencies should plan on asphalt repaving occurring every 7–15 years depending on conditions, weather, soils, and traffic. The following paragraphs identify the more common failure issues and the strategies to address them.

Mix Issues

There are a number of potential issues with asphalt mixes that shorten the life of the asphalt. These include too much or too little asphalt mix, inappropriate aggregate size or shape, too much moisture, and damage from oil on the surface. The results are washboarding, corrugations in the pavement, and shoving (pavement slippage). All of these defects—except the oil spillage—are construction issues that are caused by the acceptance and use of asphalt mixes that do not meet specifications. Oil spillage is caused by vehicular traffic or accidents. Oil issues are normally limited to parking lots as opposed to roadways.

Bleeding

Bleeding is the presence of excess bitumen on the surface of the pavement (see Figure 6.16). Most roadways will turn gray with time as the volatiles in the asphalt evaporate and the remaining bitumen retracts into the mix. With bleeding, this is not the case because there is too much bitumen in the asphalt mix, which creates a low void amount. As temperatures rise, the bitumen expands and has nowhere to go but to the surface. The bitumen is sticky in hot weather, but can become slick in cooler weather and after rain because of the oils that are present. It is not reversible. Once bleeding starts, the severity of damage is defined as medium (HPS 2001). With high-severity conditions, the aggregate is not visible.

Fixing bleeding requires effort. The pavement can be milled and reapplied, but in such cases, no new bitumen should be added. The heating will volatize some of the bitumen. Sand can be used to absorb excess bleeding in small areas without milling.

Bleeding is different than polished asphalt. Polishing is a condition where repeated traffic has made the roadway surface smooth by wearing down the aggregate. It suggests that the aggregate is too soft and friable, is too fine (sand), too smooth, or the mix is not designed for the traffic load. In any case there is a need to remove and replace the damaged asphalt.

In either case, bleeding or polished asphalt, the result is a slick surface that impacts the ability of vehicles to stop—a major concern for traffic engineers. While pavements that are smooth are desirable, if the pavement is too smooth, hydroplaning will occur when water is present, which impacts vehicle stability. Worn asphalt and rounded aggregate can also reduce skid resistance.

Obvious Base Defects

Settlement is a dip or depression in the longitudinal profile of the pavement surface. When there are significant dips in the roadway, the likelihood of base failure is high. Sometimes this

Figure 6.16   Bleeding in asphalt indicates the presence of excess bitumen.

is related to other factors (broken pipes, leaking water mains, broken sewer lines), and such defects will create a noticeable impact on ride quality. Settlement is different than pavement corrugation. In evaluating pavements, the University of Wisconsin–Madison (1992), Walker (2002), and ASTM D6433-07 provide guidance to evaluate settlement based on frequency of dips in the roadway. Minimal effect on drivers and frequency of only a couple dips per mile would be classified as low severity and create limited review (HPS 2001). Once passenger discomfort increases the severity to medium, some effort should be undertaken to understand why the base is failing. Noticeable dips (greater than 0.5 feet) of more than four per mile create a poor ride and immediate attention is needed (HPS 2001). Reconstruction of the base is the appropriate solution in such cases. Simply repaving the road will not accomplish a long-term goal of roadway comfort and protection of vehicles.

Raveling

Raveling is a condition where the top coat of asphalt is worn away with time (see Figure 6.17). In such cases the cause may be old asphalt where the bitumen has evaporated or where there was insufficient bitumen in the top coat in the first place. The aggregate separates from the road surface and the road will be strewn with stones. There is no solution to raveling except milling the road surface and applying new asphalt. All the old topcoat must be removed in these conditions or the problem will perpetuate itself.

Alligator Cracks

Alligator cracks are a series of interconnected cracks in the pavement (see Figure 6.18) that are caused by fatigue failure of the asphalt concrete surface under repeated traffic loading (HPS 2001). This occurs when the pavement has an unstable or poor subcourse. Because the base is unstable, cracking begins at the bottom of the asphalt surface where tensile stress and strain are

Figure 6.17   Raveling occurs when the top coat of the pavement wears down due to poor mix design and poor bonding.

Figure 6.18 Alligator cracking occurs when pavement has an unstable or poor subcourse.

highest under traffic loads. The cracks propagate to the surface as a series of parallel longitudinal cracks. After repeated traffic loading, the cracks connect (HPS 2001). Alligator cracking may be considered a combination of fatigue and block cracking (FHWA 2010). It can occur anywhere in the road lane. The pieces are generally small and easily displaced.

The FHWA (2010) manual defines low-severity alligator cracking as having few interconnected cracks in a limited area (one square foot) and the cracks are less than 0.25 of an inch in width. These cracks can be sealed temporarily until more permanent solutions can be funded. Medium severity occurs when the cracking pattern becomes obvious—the cracks are close together and begin to widen. High severity occurs when the complete alligator pattern emerges—the cracks spall and are wider than 0.75 of an inch (FHWA 2010). In such situations, covering alligator cracks with new pavement does nothing to prevent their recurrence. Wear and loads accelerate the damage. The combination of defects causes the pavement surface to wear and disintegrate. Even if the pavement generally stays together, the integrity of the surface is lost and gains roughness with time. Base stabilization is required.

Edge Cracks

Edge cracks occur as a result of a lack of shoulder support caused by vegetation, drying of soil underneath the shoulder, dissimilar soils, and settlement of shoulder soils. The vegetation damage is a failure-to-maintain issue. Grass and weeds will grow under and then through the damaged asphalt, further damaging it (see Figure 6.19). The other issues involve the initial installation of the shoulder, often separately from the rest of the roadway. Dissimilar soils, especially on embanked edges will tend to slough if the angle of repose is not correctly defined during construction.

Figure 6.19   Edge cracks occur as a result of poor shoulder support and vegetation is typically seen growing in the cracks.

The solution to this problem requires the removal of the vegetation from around the pavement and application of an herbicide. Next, the damaged pavement—plus some additional area around it—needs to be removed. The base should be dug out and replaced with new, clean base and sub-base materials that are appropriate to the locale. Since edge cracking is normally manifested as narrow strips along the edge of the road, the asphalt mix width is often only a few feet. The tack coat is important to ensure a better seal than the prior edge had. Compaction should be carefully undertaken to make sure that the roller does not create a separation between the new and old asphalt. Edge crack repairs are difficult. Asphalt should be added in one to two inch layers, leveled and compacted with three to five passes of a roller. Cold patch is not appropriate for long stretches.

Traverse Cracks

Traverse cracks are defects that cross the traffic lane. They are often regularly spaced along the road. They can indicate the covering of concrete pavement; but that will be covered later. The cause is movement of the asphalt due to temperature changes and hardening of the asphalt with aging. Severity of traverse cracks is related to the width and displacement of the cracks (displacement is normally low). Low-severity cracks are less than 0.25 of an inch wide. Moderate-severity cracks are less than 0.75 of an inch in width—unless there is also severe adjacent cracking, in which case the crack level is severe. Because temperature is the typical culprit with transverse cracks, the solution is in the base pavement design.

The repair of joint cracks involves cleaning the joint of the unwanted debris, applying an asphalt tack coat and a sand aggregate in low-severity situations. Applying only a tack coat will not resolve the problem since the tack coat will volatize and shrink relatively quickly, leaving the pavement with the same condition as prior to the fix. A mix of sand and bitumen is one option. If the cracks continue to open, this may indicate a base flexure failure issue.

Joint Cracks

Edge cracks and traverse cracks are both a form of joint crack, but joint cracks extend well beyond the edge of the pavement. Joint cracks can occur because of several issues: (1) when the asphalt is overlain onto concrete slabs, (2) when there is differential settlement of the base, (3) when there is a lack of appropriate initial tack coating on the vertical face of a repair that connects asphalt lanes, or (4) during the expansion and contraction or freeze/thaw cycles. Figures 6.20 and 6.21 are typical of longitudinal and traverse joint cracks respectively. Joint cracks worsen with time as debris and weeds start to grow in the cracks. Once started, removing vegetation is difficult. Persistence leads to more extensive asphalt damage.

Figure 6.20   A longitudinal joint crack within the road pavement—this is caused by the failure to seal adjacent courses of asphalt mix.

Figure 6.21   A traverse joint crack within the road pavement—this is caused by temperature changes and hardening of the asphalt with aging. Severity of traverse cracks is related to the width and displacement of the cracks.

Figure 6.22　Here only liquid bitumen was used to seal the cracks in the road. This will seal the crack temporarily and keep out plant seeds and debris, but is not a long-term fix.

The solution to the joint crack problem is the same as for transverse cracks. Too often repair crews use liquid bitumen to fix the joints (see Figure 6.22). One solution is to heat and recompact the asphalt. In such cases, the tack coat is not required.

Shrinkage Cracks

Shrinkage cracks are interconnected cracks on the surface that occur in large blocks, indicating potential cooling that occurred too quickly. They normally suggest a mix issue—a change in asphalt mix aggregate or bitumen concentration. The bond with lower layers is poor because the lower coat may be too cool. They can also be initiated between passes—starting from many layers below. The shrinkage may also be related to a lack of sufficient tack coat. The solution to the shrinkage crack problem involves either blowing out the crack with high pressure air and filling it with an injected sand/asphalt slurry or removing the damaged asphalt area, then applying an asphalt tack coat with squeegees before repaving. No traffic should be allowed to come into contact with the patch until it has cooled and cured.

Slippage Cracks and Delamination

Slippage cracks are suggested when the asphalt appears to move on the surface (see Figure 6.23). They tend to occur at intersections where traffic starts and stops (FHWA 2010). Slippage cracks indicate a poor seal with lower pavement layers and may present as delamination. The cause of slippage cracks may involve one or more of several factors. The bond may be affected by moisture, oil, or failure to clean the surface. In humid climates, water is an issue. Heavy loads, where the base-to-asphalt seal is absent or limited, is another challenge. Wheels for vehicles may cause soft asphalt to move in such circumstances. A change between the base and surface coat can create a discontinuity that can be exploited as shrinkage cracks. A high concentration of low-aggregate asphalt mixes is another offender. It also may be caused by a coat of asphalt that is too thin. The failure mode is distinctly different than where the surface coat just wears off.

Figure 6.23   Slippage cracks indicate a poor seal with lower pavement layers and may present as delamination as seen in this photo.

An issue related to slippage cracks is *delamination* (where the slippage is between asphalt layers). With delamination, the upper layer of asphalt peels away from the layer below (see Figures 6.24 and 6.25). An insufficient connection between layers is suggested. A tack coat of asphalt is routinely applied to the existing asphalt surface. Delamination suggests that the tack coat was incorrectly applied. Another likely cause of delamination is the presence of water on top of the tack coat when originally applied. Water occurs if there is rain or if the paving is being done at

Figure 6.24   Example of delamination where the upper level of the asphalt has come loose. This is generally caused by moisture and/or insufficient tack coat. It can be caused by too thick of a layer of asphalt.

Figure 6.25 Pavement delamination at an intersection—the delamination (and slippage) occurs in front of the stop bar, meaning people are pulling up closer to the intersection then starting. This indicates multiple issues—pavement, stop bar location, and visibility.

night and the temperature crosses the dew point. Both events will cause water to bead on the tack coat and not seal properly. The asphalt is not hot enough to boil away the dew, so it gets trapped between layers. The result is that there is no bond between the existing and new asphalt in such situations, thus facilitating the delamination.

Over rolling the asphalt can also create delamination by causing the cooling asphalt to stick to the roller drum and detach from the layer below. In all cases delamination is a construction issue that can be prevented with appropriate care and attention during construction.

The solution to delamination is removal of the layer of asphalt that is not sealed beyond the area of failure to improve integrity with the pavement and base. This requires an asphalt grinder (Figure 6.26) to be employed to remove the existing layer of asphalt. This material is then reheated with an appropriate amount of asphalt added, and the mixture is reapplied. A new surface, hopefully without the prior defects, is provided. It should be noted that after planing/grinding the asphalt should be allowed to dry to remove any remaining moisture. A vibratory roller should be used for compaction of the pavement.

Reflection Cracks

Reflection cracks are cracks that reflect underlying pavement damage (see Figure 6.27). In other words, someone just covered over the older pavement cracks without correcting the underlying problem. They are difficult to prevent and correct. In cases where the cracks are small, after a high pressure air hose cleans the joints of debris, they can be filled with bitumen and sand. An asphalt slurry can also be injected in the base to fill the crack from the bottom as a possible solution for small cracks. However, for anything more than very small cracks, periodic minor repairs are not only a waste of time and money, but are likely to increase the future costs and frequency of repair. For reflection cracks, the solutions used for edge and joint cracks are not applicable. In these cases, the cracked area should be removed—including the base—and patched. Related to reflection cracks are widening cracks that show the longitudinal crack from the prior asphalt layer.

Figure 6.26 Example of the type of machine used in road repair to grind off existing asphalt, which is then placed in a truck so it can be recycled. Adding more layers to a roadway will not solve base failure issues, but will improve the smoothness of the ride. Grinding off the top layers and recycling them prevents excessive pavement buildup, which impacts drainage and adjacent properties.

Figure 6.27 Reflection cracks in asphalt indicate underlying pavement damage.

Ruts, Potholes, Utility Cuts, and Other Damage

The remaining means of asphalt failure can be grouped together. Rutting is displacement of material in the travel lanes, parallel to traffic where the tires normally are. Ruts are usually manifested as waves in the pavement, centered along travel lanes (see Figure 6.28 where the pavement looks wavy). Rutting indicates sub-base or base consolidation. Typical causes are insufficient design for the applied vehicle weight on the pavement; commonly buses and heavy trucks. They may occur in conjunction with shrinkage cracking. In severe cases pavement uplift may occur along the sides of the rut—looking like a fold in the pavement (visible in Figure 6.28). The severity of rutting depends on the depth of the rut. If the rut is less than 0.4 of an inch, the severity is low; between 0.4 of an inch and 1 inch in depth is moderate. Repair of the subbase and base is the only solution for this problem as just reapplying pavement will not end the rutting issue. Rutting may also accelerate with time.

Longitudinal cracking is a companion to rutting (see Figure 6.29). If the cracks are in the wheel path it provides more evidence of base failure—fatigue failure from heavy vehicle loads (HPS 2001). However, longitudinal cracks may be edge cracks (see prior discussion) or lane cracks. The latter means there was an insufficient bond between passes of the paving machine and that the two edges did not bond properly (see Figure 6.30). This may result from temperature issues or poor compaction by the rollers. When it comes to the severity of longitudinal cracks related to crack width and depth, less than one inch wide is low severity, while anything between one inch and two inches in width is moderate. Severe longitudinal cracks require pavement removal and reapplication if related to the seal.

Depressions in the pavement (as was shown in Figure 6.15) suggest leaks in stormwater or sanitary sewer pipelines. These are common around manholes that are often built up (Figure 6.31). A common defect, as noted in Chapter 4, is the gap between the rim and structure caused by vibration and differential expansion and contraction as a result of temperature which can allow leakage and inflow of particles from the base into the sewer line.

Longitudinal streaking/corrugations (Figure 6.32) suggest a bad mix application (too cold, poorly sorted, etc.). Traverse corrugations, shoving, or rippling refers to pavement material that

Figure 6.28   Ruts in the pavement from heavy vehicles are particularly noticeable in the white stop bars.

Figure 6.29   Longitudinal cracking in the travel lane is evidence of base failure.

Figure 6.30   Lane cracks indicate that subsequent pavement layers did not bond correctly.

is displaced perpendicular to the direction of traffic. It indicates that the mix discharge was not smooth. Failure to provide the appropriate compaction is another potential cause. Rippling can develop into washboarding when the asphalt mixture is unstable because of poor quality aggregate or improper mix design. Severity of damage is deemed to be low if there is a noticeable effect on the ride, but not the comfort of passengers. Moderate severity occurs if there is

Figure 6.31 The built-up area around the manhole indicates there was settling, or a water or sewer leak, causing a depression.

Figure 6.32 Longitudinal streaking, also called rutting or corrugations, are seen here along the travel lane.

an impact on passenger comfort. High severity occurs when speeds must be slowed because the corrugations impact the ability of drivers to control their vehicles. Severe conditions need an immediate fix.

Utility cuts that are made in order to install or repair utility lines are common in roadways. However, all too often the proper means to restore the pavement is inadequate. Figure 6.33 compares two cuts—(a) one that was done well and (b) one that has failed. Compare the width of these two repairs to the one in Figure 6.34 that is a better utility cut repair. Compaction issues are common with utility lines. When a pipeline is placed under pavement or a repair is

Figure 6.33 Example of a pair of utility cuts—both photos show repaired pavement after utility cuts have been made. Photo (a) shows a well-done repair, while photo (b) shows a repair that has failed.

Figure 6.34 Example of well-done stormwater drain utility cut in an asphalt parking area.

made, the backfill is not compacted to the same degree as the initial roadway base. As a result, a depression develops.

In all of these cases, the damaged pavement needs to be removed. The edges should be saw cut and the base elements addressed. The solution to this problem requires the removal of the damaged pavement, plus some additional area around it. The pavement should be saw cut for a smooth edge. The base must also be removed, reconstituted, and compacted. Once compacted, the edge should have a tack coat applied to ensure that the new pavement and old pavement will bond. Then one to two inches of asphalt will be added, leveled, and compacted with three to five passes of a roller. Excessive rolling will damage the pavement, create sticking on the droller drum, and cause damage to the integration with the base. Depending on local codes, more layers may be required, but the same basic practice should be followed. If hot-mix asphalt is used, temperature is critical to a good bond and long life. The pavement should be allowed to cure prior to allowing traffic back on it. Where there is a leak in a structure, the structure should be repaired and the base compacted to 95–98%. Then the tack coat and asphalt should be applied and rolled.

Finally, potholes are damaged pavement from a variety of causes, but where nothing specific is indicated. The pavement gets damaged and starts to chip away. Causes of potholes are many—traffic loading, fatigue, poor drainage, inadequate strength, etc. The freeze/thaw cycle exacerbates pothole development. Figure 6.35 a–d shows the process to repair potholes. Figure 6.36 shows a pothole developing. Figure 6.37 shows an example of a pothole where the underlying cause of the failure was not investigated and, as a result, repairs have been made many times in the same vicinity. The divot in the patch near the manhole suggested a break in the sewer line

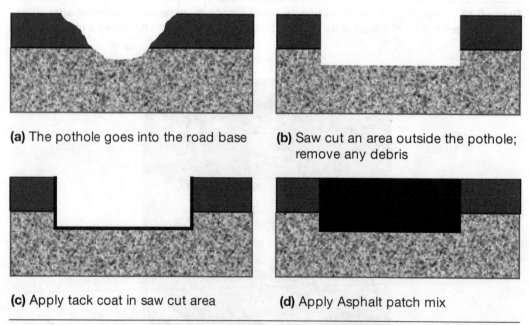

(a) The pothole goes into the road base

(b) Saw cut an area outside the pothole; remove any debris

(c) Apply tack coat in saw cut area

(d) Apply Asphalt patch mix

Figure 6.35 Example of the proper way to repair a pothole and other asphalt damage. Once the pothole is found (a), the pavement should be saw cut for a smooth edge (b). The base must also be removed, reconstituted, and compacted. Once compacted, the edge should have a tack coat applied to ensure that the new pavement and old pavement bond (c). Then one to two inches of asphalt should be added, leveled, and compacted with three to five passes of a roller (d).

Figure 6.36   A typical pothole that has not been repaired.

Figure 6.37   This pothole that has been repaired once is developing again because the issue that originally caused it was not corrected. This re-emerging pothole located in the patch is called a divot.

or manhole. Unfortunately, most people who fix potholes just put cold patch into the hole and seal it.

To correctly patch asphalt, the area must be saw cut out until the base is in the appropriate condition. Next, the tack coat should be put down, then asphalt should be applied, and the mix placed and compacted in accordance with standard practice for the locale. Base failure should be investigated since the pothole may be an indicator of a large problem. A proper patch is shown in Figure 6.38.

Table 6.1 outlines the basic failure modes for asphalt and the effect on pavement. Understanding the nature of cracks is important so that the correct fix can be applied. The potential failure modes and the options for correction indicate there is a need for roadway organizations to have professional asphalt inspectors on their staff who can assess damage and determine these fixes.

Figure 6.38   Example of a proper asphalt patch.

Table 6.1   Common asphalt concrete defects and causes

Concern	Causes	Effect on Pavement
Low stability	Excess bitumen	Washboarding
		Rutting
		Bleeding
	Excess medium-size sand	Roller tenderness
		Difficulty compacting
	Excess rounded aggregate	Rutting
		Channeling
Poor durability	Low asphalt content	Raveling/dryness
	High void content	Hardens too quickly
	Water under pavement	Stripping
Too permeable	Low asphalt content	Raveling
		Thin asphalt film
	High void content	Water gets in pavement
	Inadequate compaction	High voids

Continued

Concern	Causes	Effect on Pavement
Poor workability	Aggregate too large	Rough surface, hard to place
	Poor gradation of aggregate—too many large particles	Will not compact properly
	Low temperature	Hard to compact Loose aggregate Early deterioration Uncoated aggregate
	Too much sand	Moves under roller
	Low mineral filler content	Void spaces
	High mineral filler	Tends to stick to roller Not durable
Poor fatigue resistance	Low bitumen content	Fatigue cracking
	High voids	Pavement ages too fast Fatigue cracks
	Lack of compaction	Pavement ages too fast Fatigue cracks
	Inadequate thickness	Excessive bending Fatigue cracks
Poor skid resistance	Excessive amounts of bitumen	Asphalt bleeding Cars slide when stopping
	Poorly graded aggregate	Hydroplaning
	Smooth aggregate	No skid resistance
Cracking	Heat checking	Uneven ride, damage to vehicles
	Holes in asphalt	Uneven ride, damage to vehicles
	Alligator cracking	Uneven ride, damage to vehicles
	Bleeding	Hydroplaning
	Edge cracks occur as a result of poor shoulder support and vegetation	Uneven ride, damage to vehicles
	Longitudinal joint cracks caused by the failure to seal adjacent courses of asphalt mix	Uneven ride, damage to vehicles
	Traverse crack due to temperature issues or aging	Uneven ride, damage to vehicles
	Slippage cracks—poor seal with lower pavement layers	Uneven ride, damage to vehicles
	Delamination—occurs where the upper level of the asphalt has come loose	Uneven ride, damage to vehicles

Continued

Concern	Causes	Effect on Pavement
Cracking (*cont.*)	Cracks that reflect underlying pavement damage	Uneven ride, damage to vehicles
	Longitudinal cracking	Uneven ride, damage to vehicles
	Lane cracks	Uneven ride, damage to vehicles
	Longitudinal streaking, also called rutting or corrugations	Uneven ride, damage to vehicles
	Utility cuts	Uneven ride, damage to vehicles
	Potholes	Uneven ride, damage to vehicles
	Unstable or poor subcourse	Uneven ride, damage to vehicles

PROPERTIES OF CONCRETE ROADWAYS

Many older federal interstates and certain major roadways were constructed with concrete for durability and other reasons. The first concrete road was Woodward Avenue in Detroit in 1909 (Michigan DOT). While asphalt is a non-structural, flexible pavement, concrete is just the opposite—structural, rigid pavement. And, unlike asphalt, UWM (1992) notes that reinforced concrete is primarily a structural component designed to act like a beam that uses the rigidity of the slabs to carry the traffic load. However, cost is often a consideration; asphalt is less costly to install, but often requires more maintenance than concrete.

Concrete pavements are reinforced to control expansion and flexure via a steel mesh or reinforcing rods—the reinforcing is usually in the lower half of the slab since concrete is poor in tension but excellent in compression. Steel is excellent in tension, so reinforced concrete resolves both loading issues. The concrete can hold heavy loads and bridge gaps in the base layer as a result. The reinforcement is also designed to limit crack openings. Pavements are constructed with contraction joints to control cracking because concrete warms and expands after pouring. Cuts also are placed at regular intervals in the pavement to act as expansion and contraction joints to help with weather-related temperature changes. This gives the slab a place to crack in a straight line on a well-formed groove, which is easy to seal. Cracks that are full depth are filled with a packing material to prevent growth of plants and inclusion of dirt. The placement of expansion joints is a critical design consideration but can create a tiresome noise pattern on roadways. Asphalt is smoother because it lacks joints and therefore is less noisy. However, asphalt placed over a concrete road tends to have reflection cracks that mimic the expansion joints.

The construction and design differences used in concrete pavements causes them to support traffic loadings differently than asphalt. These principles are used to design bridge decks, which are normally concrete. Parking lots also can be concrete, and most parking garages are concrete or have concrete decks at minimum.

An engineer needs to derive the appropriate concrete mix based on the weather, temperature, and load conditions. Mix specifications include items such as bearing capacity (3,000 psi), mix content (just like asphalt the ratio of aggregate sizes needs to be established), aggregate size, and source of aggregate all need to be specified. Inspectors perform slump tests and pour cylinders to verify the mix. The concrete requires working to make it smooth and to protect the edges. Placement of the steel is done manually. Concrete cure times are usually 28 days, which is difficult to deal with when making repairs. Concrete, despite its structural differences from asphalt, has many of the same failure modes.

Epoxy-coated dowels are now commonly used to protect the steel, but epoxy coatings have less bonding capacity than uncoated rebar. Kehr and Barouky (2005) report that the bonding capacity of epoxy coated rebar is between 65 and 90%, compared to black bar. Studies vary, however, so the recommendation is to increase the development length required for splice and anchorage lengths, as opposed to an increased cross-sectional area of the steel (Clifton, Beeghly, and Mathey 1974; Brown, Darwin, and McCabe 1993; Hadje-Ghaffari et al. 1992).

Applying coatings on roadway surfaces to protect and seal the surface is difficult and will wear quickly due to traffic. Admixtures like fly ash can help harden the mixture. Steel, fiber, and other types of reinforcement are used with concrete roadways where the potential for corrosion, spalling, and edge cracking may be present. Surface damage can be accelerated with rain and ice if the mix was not correct when placed. As a result, there can be a variety of defects in concrete pavements.

Assessing Concrete Pavements

Concrete and asphalt roadways require inspection, ride resting, and periodic evaluation to observe problems prior to them becoming bigger. Once concrete is damaged, the damage accelerates, potentially creating risks to traffic. Those risks translate to losses to the owner of the road as a result of claims from people in accidents or who have damaged cars. Driving/windshield inspections are important. Concrete often wears more quickly in the winter, so a spring inspection is of benefit.

Concrete pavements are designed to last far longer than asphalt surfaces. However, like asphalt roadways, paving over base failure or poor base conditions is counterproductive. Repairs will be dictated by the base condition, surface wear, and weather conditions. Some concrete lasts for years with little maintenance; in more corrosive environments, less so.

Because concrete is structural, base conditions may initially appear less important. However, base failure beneath concrete can be a problem and a means to monitor it is needed. On roadways with significant risk, ground penetrating radar may help identify where voids may occur. Ongoing efforts to track surface and base condition should be performed and budgets adjusted accordingly to prevent failure. Drainage system maintenance, as discussed in Chapter 5, is a critical component of long-term base protection.

Like asphalt, the cracking may provide a clue as to the failure mode. Because concrete is rigid, it may mislead the investigator as well. Voids, poorly compacted bases, insufficient expansion joints, poorly constructed slabs, and inadequate or misplaced rebar are all potential issues to address. Unlike asphalt, the correction is more challenging because the damage cannot be covered over and the real culprit (issue) may be hidden.

Like asphalt, a program of evaluating pavements has been developed by Walker (2002). Walker (2002) developed the PASER manual for concrete pavements so that local road agencies can work with a comprehensive condition-rating method. An inventory of roadways is developed and information is collected on construction history, roadway width, etc. This evaluation is noted in Table 6.2. The time it takes to go from an excellent condition (10) to a very poor condition (1) depends largely on the quality of the original construction and the amount of heavy traffic loading.

Table 6.2 Concrete pavement evaluation scale

Rating	Description	Signs to Look for
10	New	No maintenance needed
9	Excellent	No maintenance needed
8	Very Good	Very good—no maintenance needed
7	Good	First signs of transverse cracking, patching, or repair; more extensive pop-outs or scaling; some manhole displacement, isolated heave, or settlement. May need some sealing or routine maintenance.
6	Good	First signs of corner cracking or shallow reinforcement. More frequent transverse cracks. Open (1/4") joints and cracks. Moderate scaling. Needs joint and crack sealing.
5	Fair	First signs of joint or crack spalling or faulting. Multiple cracking at corners with broken pieces. Patching in fair condition. Surface texturing repairs may be necessary. Some partial depth patching and joint repairs may be needed.
4	Fair	Severe surface distress requires asphalt overlay or extensive surface texturing. Multiple transverse cracks with spalling and broken pieces. Corner cracking with potholes or patches. Blowups. Some full depth joint or crack repair required.
3	Poor	Most joints and cracks are open (1"), spalled, or patched. D-cracking is evident. Severe (1") faulting. Extensive full-depth patching required plus some full slab replacement.
2	Very Poor	Pavement recycling and reconstruction necessary
1	Failure	Complete failure

Edge Cracking

As discussed previously, concrete is not an impervious material. It is full of micro-cracks and will expand and contract based on temperature throughout the year. As a result, the expansion joint cracks will change width throughout the year, which creates challenges. Transferring loads across joints is important, which is why expansion joints rarely cut through the reinforcing steel. Wide cracks required by widely spaced joints will open but cannot transfer loads. The result is the edges have higher applied loads and will flex, which makes them subject to premature wear. These higher edge loads can cause further cracking and deterioration along the joint or crack edges (see Figure 6.39). As a result, water gets into the concrete through the micro-cracks that permeate concrete. The freeze/thaw cycle creates expansion and contraction that grinds the concrete and causes the deterioration to accelerate. If salt is used, it can be drawn into the concrete as well, causing deterioration of the concrete and corrosion of the steel. The solution to fix this is to saw cut the edge and repour the slab. Dowels will need to be cut into the remaining slab to maintain the integrity of the reinforcing steel.

Pumping

Pumping is a phenomenon that indicates that the slab moves when traffic goes across it. What happens is that when a vehicle crosses the edge, the edge deflects. If there is water under the slab, it will be squirted about along with some soil, creating a *pumping* scenario. With time, soil will build along the edge, and wash away, leaving a void under the edge. At some point the edge of

Figure 6.39 Concrete pavement that has edge cracking at the joints.

the slab will fail, leading to a far worse condition that must be repaired by cutting out a portion of the slab, replacing the base, and repouring a new slab. In heavy traffic areas, with sandy soils and a high water table, this may be common. Unstable or poorly drained subgrade soils may cause pavements to settle after construction as well. Pumping can be detected by the soil stains around pavement joints or cracks. Maintaining a tight joint seal can prevent intrusion of water and reduce freeze-thaw damage and pumping. Figure 6.40 is an example of a concrete slab joint (roadway center) covered with asphalt where the pumping crack was not fixed. The concrete pumped water through the overlying asphalt (wet cracks).

One means of repair is to fill the voids with pressure grout. Slabs can also be leveled by slab jacking or mud jacking them back into place and pumping base under them. In some cases, smooth steel bars are installed with epoxy across the joint to transfer traffic loads between adjacent concrete slabs while allowing opening and closing of the joint. These bars can rust and sometimes cause problems, which is basically going back to the pumping problem. Being watchful for signs of pumping and sealing joints or cracks is a means to reduce the long-term risk.

Cracking

There are many forms of cracking with concrete pavements. One is map cracking. Map cracking is characterized by a series of fine cracks that are interconnected like a road map. Often, they are in recognizable patterns like a checkerboard (akin to alligator cracking in asphalt). They are caused by an improper cure of the concrete or overworking the surface (Walker 2002).

Traverse cracks and spalling are the most common concrete damages. Traverse cracks occur where there is a repeated bending or flexure of the slab due to loads, an uncompacted or soft base, or a failure to include shrinkage or expansion joints. Traverse cracks can be repaired by

Figure 6.40 Example of pumping through asphalt-covered concrete pavement—the way to tell that this is pumping is by noticing that the asphalt is darker at the cracks, indicating that water and other materials are being pushed through the asphalt to the surface.

sandblasting the cracks, cleaning out the joints (compressed air) to remove foreign matter, and sealing with a rubber/asphalt sealer if the cracks are small. Otherwise, the solution to this problem is cutting out the offending slab pieces, restoring the base, and repouring the slab. Dowels are usually needed to connect adjacent slabs.

D-cracking develops when the aggregate is able to absorb moisture. This is a concrete mix and poor aggregate selection issue. In such cases, the aggregate will break apart during the freeze-thaw cycle. D-cracking starts at the bottom of the slab and moves upward with the movement of moisture. Fine cracking and a dark discoloration adjacent to a joint often indicates a D-cracking problem. The only solution to D-cracking is replacement of the slab.

Cracks may widen between slabs at the expansion joints, or slabs may break at places that are not at the expansion or contraction joints (see Figure 6.41). If the slabs break at places that are not the expansion or contraction joints, it is called faulting (like an earthquake). These cracks may be transverse or longitudinal. In either case, it usually means differential settling between adjacent slabs or failed expansion joints that create spalling and may permit plants to seed within them. In such cases the joints need to be cleaned out (dug out with a joint cleaning machine or compressed air) and a new seal provided. Sand blasting is an option as well. Filling the void that created the flexure is important to avoid a repeated failure.

Faulting that is allowed to progress will create a poor ride and accelerated slab deterioration in several ways—the edges are broken, or worn, or the slab simply cracks into pieces. Minor faulting can be corrected by surface grinding to improve the ride, but this does nothing to resolve the underlying problem. Voids can be sealed or slabs mudjacked back to level position. Severe cases may need joint replacement.

Corners and edges are common places for cracking. In most cases, the location near the edge of the pavement creates the potential for the base to be washed out or damaged with time as a result of weather. Traffic loads at the edges encourage small pieces to break off, much like what happens with sidewalks.

Figure 6.41   If the slabs break at places that are not the expansion or contraction joints, it is called faulting. The arrow indicates where concrete faulting has been repaired.

Diagonal cracks are simply corner cracks farther from the corner that fail for similar reasons (see Figure 6.42). Partial or full-depth concrete patching or full-depth joint replacement is suggested. The repair is more difficult—the corner will detach, so the broken corner will need to be removed and replaced. Often it is easier to cut a swath across the entire slab to address both diagonal and joint cracks.

Figure 6.42   Diagonal cracks are simply corner cracks farther from the corner. Both diagonal and corner cracks result from problems in the base.

Meander cracks may be caused by settlement due to unstable subsoil or drainage problems (Figure 6.43). The problem is encouraged by frost heave and spring thawing. Meander cracks are often localized and may not indicate any specific pavement problems. Such cracks are normally minor and can benefit with epoxy sealing to minimize water intrusion. Extensive or severe meander cracking may require all or a combination of the following: replacing the slab, stabilizing the subsurface, or improving drainage.

Raveling or Scaling

Scaling is surface deterioration caused by the displacement of the aggregate. Usually, fines and the cement mortar are the first to go, creating a rough surface. The result is an indication of an improper concrete mix issue that causes loss of fine aggregate and mortar. Figure 6.44 is an example where this has occurred on a utility patch (larger areas are usually fixed quickly). More extensive scaling can result in loss of large aggregate. The cause is concrete that has been air-entrained, which exacerbated the impact of the freeze-thaw cycle damage. Scaling is also aggravated by the use of de-icing chemicals, heavy vehicle pathways, poor quality concrete, or improper finishing techniques. Deterioration can extend deep into the concrete. This is common on longitudinal joints.

A similar defect is excessive wear due to a poor concrete mix. The mix of the concrete in combination with salt and ice can damage the surface as well. Fines may be removed first, which makes the surface rough, or the opposite can occur—the aggregate gets polished by vehicle traffic. If the wrong aggregate mix was used, then the problem is compounded. Field observation is required to ensure the proper mix is used. The only solution to either of these issues is replacing the concrete slab with better quality concrete.

Exploding Concrete

The phenomenon called exploding pavement occurs periodically as a result of hot sun on pavement that expands but lacks the expansion capability provided by expansion joints. The

Figure 6.43   Concrete pavement cracking—meandering cracks like this may be caused by settlement due to unstable subsoil or drainage problems.

Figure 6.44   Example of raveling or scaling on concrete utility patches.

lack of joints facilitates exploding pavement and edge spalling due to grinding of the slabs. What happens is one slab pops up above another (see Figure 6.45). This is a dangerous driving issue. U.S. Highway 23 in the lower peninsula of Michigan had issues with exploding pavement in the late 1960s and early 1970s. Exploding concrete can be the result of one of these issues: (1) the lack of expansion joints, (2) improper load transfers to the subsurface, (3) steel corrosion, and (4) freezing water. The lack of joints can be either a construction or a design issue. Drawings should be reviewed to determine the assumptions used for joint installation. Both are easily resolved by adding expansion capability (larger or more frequent expansion joints).

Potholes

Like potholes in asphalt, potholes in concrete pavements occur where there is settling, water damage, unchecked damage from cracking or spalling, and the freeze/thaw cycle. Local

Figure 6.45   Concept of exploding concrete where one slab pops on top of another. This can happen in hot temperatures when the pavement heats up and expands without the benefit of proper expansion joints. *Source:* Roland White, Champaign city engineer, published by: http://www.news -gazette.com/news/local/2012-07-04/olympian-drive-reopening-after-pavement-repaired.html

subgrade issues and water can encourage pothole development. Likewise, features like curbs, storm drains, manholes, and valve boxes create a differential expansion situation where the concrete can become cracked and potholes develop. While early detection can permit agencies to repair these easily (often with cold patch asphalt), letting them fester will require a full-depth cutout, base correction, and slab replacement, just as is necessary with asphalt potholes.

Preemptive detection and repair of concrete pavement damage will permit low-cost solutions to be implemented to extend the life of concrete pavements. However, this rarely happens in practice (not that it should not happen), so the solution to most concrete pavement problems involves cutting the damaged section out with a saw, cleaning the joints and then sealing the joints, and repouring the slab—usually with doweling to the remaining slab. In some cases, surface wear and budgets may suggest that the concrete surface be overlain with asphalt to improve the ride (and noise). This can be done if a tack coat is applied first to ensure a bond between materials. While this may protect the surface, water (and salt) may still be able to penetrate the concrete, creating spalling under the asphalt.

BRIDGES

Bridges cross rivers, streams, roads, railroads, ravines, etc.—virtually anything where there is a grade differential. However, most smaller communities use these features as boundaries—they do not cross them. As a result, bridges are usually not an issue for small cities. Instead, bridges are usually the purview of large municipalities or the county or state DOT that do not have the luxury of using terrestrial features as a boundary—commerce must get across the river to get to the next community. Despite the fact that most communities do not maintain bridges, it is worth spending a few pages to discuss them. Most rural and interstate roadways include bridges. Many arterial streets may include them as well. Bridges can certainly vary, from suspension bridges to simple concrete ones that bridge over short dips in topography. New bridges are generally constructed with steel, concrete, or prestressed concrete. The decision of how to construct the bridge and what material to use depends on where the bridge is located. In older cities and areas with earthquakes, bridges are typically made of steel. Bridges may be elaborate like the Brooklyn, Mackinaw, or Golden Gate bridges (see Figure 6.46) or may be a simple structure common to most interstate systems (see Figure 6.47). Older stone and wood bridges exist, but they are rare and do not deserve more than a mention here since very few of these will ever be encountered.

At present, most departments of transportation have rigid standards for constructing bridges. Most are designed with reinforced and prestressed concrete components that link together. This creates a degree of consistency among bridges and design. Weight holds the bridges in place, which makes them susceptible to horizontal load failure. The result of bridge failure along the Gulf Coast on Interstate 10 after Hurricane Katrina shows bridge decks falling in the water like dominos and called this construction method into question (Figure 6.48). This situation was caused because the decking and span components sit on top of the pilings. They are not welded or bolted in place. Most are short spans.

Concrete bridges have many of the same issues as concrete roadways—the potential for spalling due to salt-induced corrosion. The problem is caused primarily by salt-induced corrosion of steel in concrete (Kehr and Barouky 2005), from the use of salt-contaminated aggregate, road de-icing materials or water, brackish water in the mix, or chloride-based admixtures. The chloride ion initiates and catalyzes the corrosion reaction with the iron. The iron expands and creates

Figure 6.46   The Golden Gate Bridge.

Figure 6.47   Modern concrete bridges are typically made of all precast concrete members—the weight of the concrete holds the bridge in place.

significant pressure on the concrete, which puts the surface in tension (Kehr and Barouky 2005). The pressure causes the concrete to crack and spall, increasing the ease of penetration of the salts to the iron, and accelerating the corrosion. Epoxy and a variety of other product coatings can be used on concrete bridges to reduce the potential for corrosion (Kar 2004). Painting issues include overspray damage and paint chip removal, but note that many concrete bridges are not painted at all because they do not require the same protection of the surface.

Figure 6.48 This photo shows the results of bridge failure; this bridge was along Interstate Highway 10 in Louisiana after Hurricane Katrina hit in 2005. The bridge decks fell into the water like dominos. *Source*: FHWA https://www.fhwa.dot.gov/publications/publicroads/11mayjun/02.cfm

Many older bridges are built of steel that were bolted together (Figure 6.49). Even today many large spans on the interstate system are steel with concrete decks (see Figure 6.50). The steel in bridges will encounter many of the same issues with corrosion as do water plants and stormwater systems, therefore coatings are required. The environment (especially coastal areas over saltwater) can be very corrosive; and wear, dissimilar metals, connection to concrete (which holds moisture), temperature, and other impacts can shorten bridge life. Unlike water plants, steel bridges have long made use of cathodic protection systems that reduce corrosion potential. Also, steel bridges are routinely painted with coatings designed to limit exposure to the elements and thereby reduce corrosion. Painting of a steel bridge requires sandblasting the metal to remove rust, repair the metal, prime it, and repaint—usually with both a primer and one or more topcoats. Various paints can be used—an engineer should be consulted for appropriate paints for a given locale.

Figure 6.49 Draw bridges are typically made of steel that is bolted together—these in Chicago are over 80 years old.

Figure 6.50   Composite interstate system bridges are made of steel with concrete decks.

All steel bridges are painted. Issues with painting of steel bridges involves the removal of paint chips (no one wants to contaminate the water below a bridge), safety (potential for paint, workers, or equipment to fall on traffic below the bridge), and the potential for lead paint (common pre-1980) that requires significant protection to workers and enclosure of the structure. While lead in paints was banned nearly 40 years ago, many steel bridges pre-date the ban. As a result, lead paint can create a challenge; all steel bridges that were constructed before 1985 should be assumed to have the potential for lead paint. Lead paint cannot be discharged to the environment; it must be contained and properly disposed of in a hazardous material site. Testing is required. The result is that old paint removal is expensive. Encapsulation is an option pursued by many bridge owners, but the sand blasting required to remove rust and old paint may cause the lead to be removed or uncovered. Care should be taken in this regard.

Painting steel bridges also creates the potential for accidents, which normally involves getting paint on cars or property not owned by the bridge owner. Many painting companies will wrap the structure in question to keep paint in, protect workers, and provide a safer environment. However, this is costly and is one reason that new steel bridges are less common than concrete bridges.

Inspection

Bridges have a little better grade from the American Society of Civil Engineers (2009, 2013, 2017) than most other infrastructure because bridge safety emerged as a high-priority issue, following the collapse of the Silver Bridge between Ohio and West Virginia, which killed 46 people. That collapse prompted national concerns about bridge condition, and safety, and highlighted the need for timely repair and replacement of bridges. The Minneapolis bridge failure in 2008 further indicated a significant issue might exist with the inspection and maintenance of bridges. Bridges play a huge role in the transport of goods across the country—a failure of one of these could cause a highway to be impassable at any time (ODOT 2011). Failures are usually spectacular and generally avoidable, so effort was put into the inspection of bridges. GAO (2013) noted that there has been some improvement in bridge conditions in the past decade, but a substantial number of bridges remain in poor condition. GAO (2016) further noted that:

"Of the 607,380 bridges on the nation's roadways in 2012, one in four was classified as deficient. Some are structurally deficient and have one or more components in poor condition and others are functionally obsolete and may no longer be adequate for the traffic they serve. Data indicate that the number of deficient bridges has decreased since 2002, even as the number of bridges has increased." This is a slight improvement from Kehr and Barouky (2005) who indicated that approximately half of the nearly six hundred thousand bridges in the U.S. Federal Highway system have structural deficiencies or are functionally outmoded, but there is still a long way to go.

All bridges require ongoing inspection and maintenance. Routine inspection of bridges should include gusset plates, structural components, bolts, flanges, a check for spalling and corrosion, paint evaluation, and the like. The underside of bridges should be cleaned periodically to remove detrital matter that will accumulate with time as a result of dew and particulates in the air. Bird feces is also an issue.

Steel bridge inspection can include the means to evaluate the deck and steel. A variety of tools are available including:

- Ultrasonic methods that involve the passing of sound waves through concrete slabs to assess condition
- Strain gages to measure stress
- Point-by-point echo scans to reflect sound waves through the slab to find voids
- Scanning bridge decks to find debonding of overlays
- Seismic tests to identify discrepancies in the foundation of the bridge

During an inspection, bridge inspectors rate bridge components using a numerical system to describe the condition of the bridge and its components. Oregon's Department of Transportation (ODOT 2011) developed an integrated approach to decision making for bridge improvements, after noting that when they updated the state's Seismic Lifeline Routes, bridges were often part of major highway corridors. Bridges are now a critical component. FHWA (2005) developed a recording and coding guide for the nation's bridges.

The major issue of concern is finding deficient structural members (this was a finding in Minneapolis). This means sending crews into the field to inspect and use tools to analyze what is happening. The load-rating capacity of post-tensioned concrete segmental bridges and tensioned concrete segmental bridges can be checked in the longitudinal and transverse direction with strain gages. It is also possible for transverse effects in a typical segmental bridge box section to govern the load rating for a bridge. Professional testing labs perform these analyses and should be employed on a routine basis to evaluate the condition of the bridge deck and supports. CSHRP (1995) noted that a life-cycle analysis of viable alternatives for repairs is needed to develop useful strategies for repair, rehabilitation, and replacement of concrete bridge components. Concrete repair is one component. They also recommended testing alkali-silica reactivity on concrete structures.

TRAFFIC SIGNS AND SIGNALS

The first traffic signals were put into place over 80 years ago, but street signs were in place for many years before that. Signs and signals are designed to convey information to the person using the transportation system and for safety purposes. Information such as who must stop at a four way intersection, the speed limit, locations of school crossings, and so forth must clearly be displayed. Failure to provide clear signage—and maintain it—can subject a local government to civil litigation losses.

Most roadways, as noted at the beginning of the chapter, started with only a few rudimentary signs. Most trails need few signs because there are few travelers and the probability that they arrive at the same place at the same time is very low. However, as traffic increases, stop signs are often the first signs that appear. Street signs may come just before or concurrent with the stop signs. As traffic increases, a blinking light may be required, or a standard traffic light. Busier roads will have dedicated turn lanes and four-way traffic stops. Intelligent systems have been designed to help traffic engineers determine how to time these major intersections to increase traffic flow efficiency. Of course, sometimes this does not work and traffic is worse; or the timing is off and people stop at every light. Good traffic control takes significant care and can change throughout the day and the year. So, not only do the light bulbs in the lights need to be replaced regularly, but the program logic controllers (PLCs) that regulate the lights need regular tuning and maintenance.

Street signs must be 7 feet from the bottom of the sign to the bottom of the sign post, and must be maintained in easily visible and easily readable condition (see Figure 6.51). Common issues with signs include fading (see Figure 6.52), damage from accidents, overgrowth of vegetation, and loss of site visibility due to vegetation and/or structures like fences and walls in the right-of-way (see Figure 6.53). It is incumbent on the organization that is responsible for signs to erect new or replacement signs when damaged. When driving around many communities, one will note numerous battered, broken, or unreadable signs. Signs that are not readable can become a legal liability should there be accidents caused by the condition of the sign. Lawyers will point to negligence on the part of the local government in such cases.

Most communities have many, many signs—and keeping track of all of them is difficult. A geographic information system (GIS) can help. Work orders are the best way to track damaged

Figure 6.51   This is a proper street sign—it is the correct height, clearly visible, and readable.

Figure 6.52   This sign is in very poor condition and needs to be replaced.

Figure 6.53   Sight visibility can be disrupted by structures and vegetation at intersections—it was not possible to see this car before it started turning because of the wall.

signs; and the police departments are often the best source of finding damaged signs. Police departments routinely travel streets at all hours. It is easy to find signs that do not serve their intended purposes. Calling in these sign issues and submitting a work order for repairs is easily accomplished. It makes enforcement easier, as well. Unfortunately, among the many priorities that a public works department has, it is easy to forget about signs. As a result, sign replacement tends to get put off. But there are standards for signs and markings and these must be maintained to limit legal liabilities. Striping, crosswalks, stop bars, center lanes, edges, etc., are all pavement markings that also must be maintained for the same reason. Note that the signs and markings are the major tools that will be used by future vehicles to avoid accidents, so ongoing maintenance inventories and efforts should be undertaken to identify, replace, and update signs and markings on a regular basis.

SIDEWALKS

Sidewalks are found along most streets in an urban (city) setting. Rarely are they found in rural settings. The goal is to get people to walk on the sidewalk, not in the street, and to minimize the potential for vehicle/pedestrian accidents. With respect to maintenance, sidewalks are similar to concrete roadway pavements, only with lighter loads. Many are only four to six inches thick, without reinforcement. Yet, they have all the same issues as concrete pavement damage, plus tree roots that dislodge the pavement and vehicles driving over them (such as in the case of driveways or parking in swales). Cracked, damaged, and dislodged sidewalks are a trip hazard for pedestrians and as a result, a legal liability. Many cities may not know where sidewalks are an issue, but once told, they have an immediate obligation to correct the problem or face legal liabilities. However, the failure to know about a damaged sidewalk because they do not inspect them does not exempt communities from this legal liability. Cities and counties that have responsibility for sidewalks should have an ongoing program to inspect sidewalks and get corrections made. Some local governments put this responsibility on homeowners, but sidewalks are normally owned by the local government, and as a result, they are responsible for their maintenance. If the homeowner does not correct the problem and someone falls, the community will be liable.

Note that sidewalks that do not have reinforcement will be easily damaged by vehicular traffic. The scalloped pattern seen in Figure 6.54 is typical of four-inch, unreinforced sidewalks that break easily when cars drive on them (parking or driveways). Like all other assets, they need to be inspected and maintained. Figure 6.55 is another example of a sidewalk that needs attention. The sidewalk in Figure 6.56 is on a bridge and the entire edge is spalled off. Ice and snow are

Figure 6.54   Example of failed concrete tile—this is a pedestrian trip hazard.

Figure 6.55 Example of a failed concrete sidewalk at a poorly located catch basin—a leaking catch basin will cause base failure of a sidewalk.

Figure 6.56 Spalled concrete sidewalk on bridge—the cause of this spalling is winter weather that freezes moisture in the concrete and salt.

suggested as the reason, given that the location of this sidewalk is in Colorado. Deterioration is significant. Figure 6.57 is a spalled sidewalk—the cracks are a pedestrian hazard, as is the rebar sticking out. The surface creates a trip hazard as well. A trip hazard may be defined by as little as 0.5 of an inch of vertical difference between adjacent sidewalk tiles, which is not a lot. Grinding edges is a solution to address this issue.

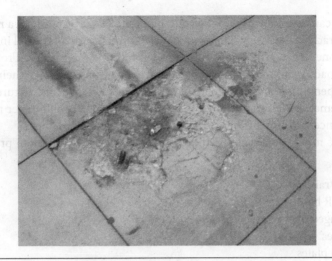

Figure 6.57   Spalling concrete sidewalk—the likely cause here is the use of salt to protect the concrete surface from ice formation during the winter.

RAILROADS

This chapter will briefly mention railroads because while most railroads in the United State are private, there has been growth in publicly owned light rail systems and some conversion of private track into public hands. The issue with many rail lines is that their construction occurred well in the past; when rail traffic, speed, and weight were significantly different. The vibration of the locomotives and cars are a confounding factor today (compare Figures 1.3, 1.4, and 6.64). The speeds are also very different (40 versus 80 mph) and the haul is higher (averaging 150 cars versus 20 in the past). Freight rail is far more damaging to rail systems than light rail, but constitutes well over 90% of all rail traffic. Very few new railroads have been constructed in the past 50 years and techniques have changed (see Figure 6.58). No longer is it creosote soaked ties, rails, and spikes driven by crews of laborers; much of it now is automated.

Figure 6.58   The new way to construct railroads—the concrete ties are prefabricated and already connected together—once in position, the rails are simply tied down.

Railroad track can be frequently used or not. Figures 6.59 and 6.60 show a rarely used and a commonly used track, respectively. The amount of *silver* on the top of the rail indicates use—the commonly used one has polished rails. Either way it must be maintained. From 1980 through 2003, Class I railroads spent more than $320 billion, approximately 44% of their operating revenue, on capital expenditures and maintenance expenses related to infrastructure and equipment (Cambridge Systematics 2007). This amounts to an average of $1 million per mile of track per year on maintenance. Most of it is performed *in situ* (see Figure 6.61).

Railroad traffic means accelerated damage to the rail base. In addition, problems that can occur include:

- Worn track/rails
- Damaged RR base
- Worn/damaged timbers/cement
- Loose RR ties
- Loose/worn plates

Note that having rails move under the weight and force of moving trains is not unusual. Having rails split with time can cause train accidents. Base failure will cause the tracks to move, bend, or dip. None of these structural issues is safe, so efforts to make repairs must be undertaken constantly. The initial effort includes inspection of all previously mentioned components on the track and an assessment of repair needs. Railroad repair equipment—like the track stabilizer shown in Figure 6.61—is used to fix the track in place. For repairs, equipment is used to pick

Figure 6.59 Example of a rarely used railroad—the top of tracks become rusty from lack of use.

Figure 6.60   Example of a frequently used railroad—here the top of tracks are shiny due to frequent train traffic.

Figure 6.61   Example of track stabilizer used to repair rails.

up the track and ties, compact the base via vibratory means (and add base as needed), then set the track back down. Ties, plates, and rails may be replaced during this time. The work must be done in low traffic times, often at night. Newer ties may be concrete, but many railroads simply replace old wood ties with new wood ties. The locomotive and rail cars must also be maintained—a 2013 North Dakota derailment was caused by a broken car axle.

Local agencies are more likely to deal with light rail for passenger use. Light rail tends to *share* existing corridors with private rail. Light rail is intended to attract commuters who work in the cities. To keep them riding, care must be maintained in this way. New York learned in the early 1990s that people will not ride cars covered with graffiti—the same with transit buses. So, cleanup and maintenance of rail cars are needed during the night when people are not riding. San Francisco has a system that they have been operating in this way for over 100 years (see Figure 6.62). With transit of any type, service must be convenient or people will not ride. At the initial stages, virtually all rail and transit is heavily subsidized by the government.

Aside from the operation of the trains and maintenance of the track, another major liability is safety. Trains create significant damage to vehicles that are crossing the tracks and can kill or injure pedestrians who are on the tracks. Positive train control systems are designed to help the trains navigate the tracks, but sensor systems are needed to alert trains of obstructions. The American Railway Engineering and Maintenance-of-Way Association (AREMA) creates standards for railway engineering. Already, most trains that are given positive train control can operate as robots, so the change is coming (the engineers are there for union reasons and to address the areas without positive train control). They note that positive train control systems are designed to (FRA, 2012):

- Create train separation
- Increase collision avoidance
- Enforce speeds
- Increase rail worker wayside safety

However, positive train controls are not available everywhere. With respect to the train wreck in Philadelphia in 2015, there was no positive train control. As a result, no warnings were received. The same was true in the derailment outside Seattle in December 2017. In both cases, positive train control was absent; reinforcing the idea that train engineers need to be very careful since a fully loaded train may take over a half mile to stop. Hence, there is a need to install and maintain train safety devices—especially at intersections with roadways and pedestrians.

Figure 6.62   San Francisco street cars are cleaned and maintained at night.

Figure 6.63 Typical railroad crossing gate—crossings are a major safety concern for railroads.

Conflicts with cars and street cars or rail cars must be addressed because many people die each year due to collisions with trains. The grade crossing must be constantly maintained as there is significant pressure on the approaches. This mean lights, noise, and gates (see Figure 6.63). Flashing red lights and lowered gates are common in cities; however, a potential problem associated with them is that people drive around them, which increases the potential to be hit by a train. Fencing is common so that people do not cross tracks. Climbing chain-link fencing is common, so solid fences may be preferred. Most residents dislike train whistles, so lighting can be useful, especially flashing lights. This also helps hearing-impaired people. Florida DOT has installed yellow bars in the train movement zone at intersections to denote where the train will be. Cameras and sensors can be used to let train engineers know about obstructions ahead of time. Using center islands discourages people from going around lowered gates. There are many options.

DRIVERLESS VEHICLES (CARS, TRUCKS, AND TRAINS)

Warren Buffett has noted that the landscape of insurance is going to change in the near future. This is because of the advent of driverless vehicles. The concept is being explored by all the major automakers and major technology firms like Ford and Google. People will want to be part of the changing landscape of transportation. Tesla, Ford, and others want to be on the cutting edge of the technology because there is money to be made. The trucking industry is looking to put 1.4 million truck drivers out of their vehicles within 10 years using driverless long-haul trucks. The same goal exists for taxis in major metropolitan areas. This not only creates a major economic challenge for the truckers and taxi drivers, but the changes to the safety, reliability, and cost to customers are anticipated to be significant. Within 30 years, the expectation is that no one will drive cars except in very rural areas. The main obstacle is the condition of streets and maintaining sensors and striping in streets. Wet pavement, snow, and flooding disrupt the current sensor systems. That will likely be overcome with added maintenance and more diligence applied to maintenance efforts. Like sidewalks, there will be a serious liability issue for failure to maintain the striping and sensors.

The change will mean that less hard infrastructure may be built, but the importance of maintaining and refurbishing the existing infrastructure, especially signs, sensors, and pavement markings, will increase dramatically. The need to understand what these assets are, where they are located, their condition, and their maintenance needs will be of paramount importance. That means a much larger focus on asset management, which is the subject of the next chapter.

REFERENCES

ASCE. 2009. 2009 Report Card for America's Infrastructure, ASCE, Alexandria, VA, http://www.infra structurereportcard.org/2009/, accessed 4/3/16.

——. 2013. 2013 Report Card for America's Infrastructure, ASCE, Alexandria, VA, http://www.infra structurereportcard.org/tag/2013-report-card/, accessed 4/3/16.

——. 2017. 2017 Report Card for America's Infrastructure, ASCE, Alexandria, VA, http://www.infra structurereportcard.org/, accessed 3/3/17.

Bakke, G. 2016. *The Grid: The Fraying Wires Between Americans and Our Energy Future*, Bloomsbury Press, New York, NY.

Bala, S. 2008. *The New AASHTO MANUAL for Bridge Evaluation*. "2008 FHWA load & resistance factor rating." FHWA, Washington, D.C.

Bertram, M. 2008. "Long-term transportation infrastructure need." *North American Construction Outlook Conference, Chicago Marriott Downtown Magnificent Mile, Chicago, IL, May 2008*.

Bogetic, Z. and Fedderke, J. W. 2006. "International benchmarking of infrastructure performance in the Southern African customs union counties." World Bank Policy Research Paper 3987, World Bank, Washington, D.C.

Brown, C., Darwin, D., and McCabe, S. L. 1993. "Finite element fracture analysis of steel concrete bond." The University of Kansas Center for Research, Inc., Lawrence, KS. November 1993.

Cambridge Systematics, Inc. 2007. "National rail freight infrastructure capacity and investment study." Cambridge Systematics, Inc., Cambridge, MA.

CBO. 2015. "Public spending on transportation and water infrastructure, 1956 to 2014." March 2015, Congress of the United States Congressional Budget Office (CBO), Washington, D.C.

Clifton, J. R., Beeghly, H. F., and Mathey, R. G. 1974. "Nonmetallic coatings for concrete reinforcing bars." U.S. Department of Transportation, February 1974.

CSHRP. 1995. "Concrete bridge component evaluation manual, Technical Brief #7." Canadian Strategic Highway Research Program (CSHRP), CSHRP, Washington, D.C.

Entine, L. 2002. PASER *Asphalt Road Manual: Pavement Surface Evaluation and Rating Manual*, Donald Walker, T.I.C. Director, author, Entine & Associates, editor.

Federal Highway Administration (FHWA). 1995. *The Recording and Coding Guide for the Structure Inventory and Appraisal of the Nation's Bridges*. Report No. FHWA-PD-96-001, Federal Highway Administration, Washington, D.C.

——. 2003. *Distress Identification Manual for the Long-Term Pavement Performance Program*. Publication No. FHWA-RD 03-031, June 2003. Federal Highway Administration, Washington, D.C.

——. 2005. *Quality Assurance Manual, FHWA-NPS Road Inventory Program, 2005*. Federal Highway Administration, Washington, D.C.

——. 2010. *Pavement Distress Identification Manual for the NPS Road Inventory Program Cycle 4, 2006–2009*. Federal Highway Administration, Washington, D.C.

Follett, A. 2016. *Feds: Cyber-Attack on Ukraine Proves U.S. Power Grid Is Vulnerable*. Read more: http://dailycaller.com/2016/01/07/feds-cyber-attack-on-ukraine-proves-us-power-grid-is-vulnerable/#ixzz4dgxM4qMX. Accessed 2/2/17.

FRA. 2012. "Positive Train Control Systems (RRR); Notice of proposed rulemaking." *Fed. Regist.* 77 FR 73589.

Government Accountability Office (GAO). 2001. "Highway infrastructure: FHWA's model of estimating highway needs has been modified for state level planning." GAO-01-299. GAO, Washington, D.C.

———. 2013. "Limited improvement in bridge conditions over the past decade, but financial challenges remain statement of Phillip R. Herr." GAO-13-713T. GAO, Washington, D.C.

———. 2016 letter from Matt Goldstein, GAO. "Transportation infrastructure: Information on bridge conditions." GAO-16-72R Highway Bridge Conditions, GAO, Washington, D.C.

Groeger, S., Stephanos, P., and Dorsey, P. 2002. "Evaluation of AASHTO cracking protocol: Quantifying distress in asphalt pavement surfaces." Pavement Evaluation Conference, 2002, Roanoke, VA.

Hadje-Ghaffari, H., Choi, O. C., Darwin, D., and McCabe, S. L. 1992. "Bond of Epoxy-coated reinforcement to concrete cover: Casting position, slump, and consolidation." The University of Kansas for Research, Inc., Lawrence, KS. June 1992.

HPS. 2001. "Pavement condition evaluation manual, pavement evaluations for use with the DRM System." Highway Preservation Systems, Ltd., Hamilton, OH.

Kar, A. K. 2004. "FBEC rebars must not be used." *The Indian Concrete Journal*, January 2004. P 56–58. http://icjonline.com/forum/forum_jan2004.pdf. Accessed 3/19/2017.

Kehr, J. A. and Barouky, F. F. 2005. "Fusion-bonded epoxy coatings: A technology for rebar corrosion prevention." *NACE Corcon 2005*, Chennai, India.

Lisk, B., Greenberg, E., and Bloetscher, F. 2012. "Implementing renewable energy at water utilities web report #4424." Water Research Foundation, Denver, CO.

MACTEC, 2006. "Pavement condition evaluation and maintenance needs assessment 2005, submitted to Nantucket Department of Public Works." MACTEC Engineering and Consulting, Beltsville, MD.

Michigan DOT. 2015. http://www.michigan.gov/mdot/0,4616,7-151-9623_11154-129682—,00.html. Accessed 3/7/17.

ODOT. 2011. "Asset management strategic plan." Oregon Dept. of Transportation, Salem, OR.

Roustan, W. K. 2017. "Tri-Rail frustrated by maintenance issues, duct tape repair." Sun-Sentinel. http://www.sun-sentinel.com/news/transportation/fl-reg-tri-rail-on-time-performance-20170224-story.html. Accessed 2/14/17.

University of Wisconsin–Madison (UWM). 1992. "Concrete PASER Manual pavement surface evaluation and rating." Department of Engineering, Madison, WI.

Uzarowski, L. et al. 2008. "Initial evaluation of foamed asphalt stabilization using modified asphalt cement." *2008 Annual Conference of the Transportation Association of Canada Toronto, Ontario*.

Walker, D. 2002. *Pavement Surface Evaluation and Rating Concrete PASER Manual*. Wisconsin Transportation Information Center, Madison, WI.

SECTION II

Assessing Assets

7

ASSET MANAGEMENT

How would modern society function without the infrastructure that surrounds it? The simple answer is that society would not function. So, as noted in the last six chapters, the reality is that society has built infrastructure for economic and public health reasons, and must rely on that infrastructure to maintain economic and public health viability. As a result, wherever we as a society have created a reliance on infrastructure to deliver the services we need, we have also created a need for the effective management of these physical assets. The public has made significant investments in the construction of infrastructure assets that they rely upon daily, and expects that agencies that own and operate them will be responsible stewards of those investments, especially where the investments were made by governments using public dollars (taxes or fees). What is more difficult to discern is the amount of effort and resources needed for the proper maintenance and operation of these infrastructure systems. This problem is not new, although the most obsolete assets may have been phased out. For example, for most of the 1800s, boot scrapers were an absolute necessity. Why? Because the streets were disgusting. If you had taken a walk through most parts of town, your shoes would have ended up caked with rancid muck that had horse manure as its primary constituent. Until the 1880s, New York City didn't have a sanitation department to collect garbage or shovel snow. So it stayed in the streets. Most neighborhoods still relied on communal privies that often overflowed into the streets. Then, there were the horses that pulled all those carts and carriages. In 1881, horses left 2.5 million pounds of manure on the streets of New York every single day that needed to be removed by the 15,000 workers employed to do so. Fortunately, we are past that.

As noted, the primary owners and operators of the infrastructure (assets) discussed in this book—roads, ditches, bridges, pipes, water and wastewater treatment facilities, tanks, pumps, buildings, and a variety of other pieces of equipment—are local governments. Government agencies build and acquire long-lived, fixed assets to provide services to their customers. These fixed assets are termed *capital assets* because they are fixed, have long lives, and have significant value. Long life means that the asset has a useful life greater than one year (Smith, Lovett, and Caldwell 2004). So, in constructing for the long term, public agencies construct assets of high quality that are designed to incorporate materials and techniques that extend their life and limit maintenance needs. As a result, many assets require minimal maintenance in their early life. Figure 7.1 shows an example of maintenance needs for a pipeline, indicating that the early years require few resources. Since many of these assets are designed to last for 50–100 years and the maintenance needs in the early years of the asset's life is minimal, the public consciousness assumes that the low maintenance condition is the norm. Many assets are old enough that few recall their construction.

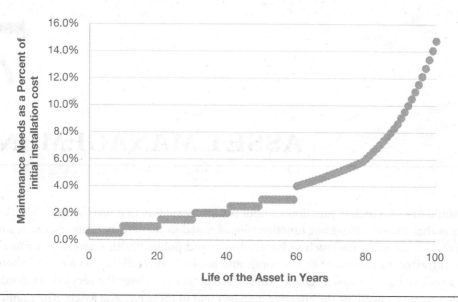

Figure 7.1   Maintenance needs for a pipeline asset (proportion of initial value).

Most agencies recognize that the public will hold them accountable if assets fail (FHWA 2016; http://www.fhwa.dot.gov/infrastructure/asstmgmt/assetman.cfm), but most systems appear to operate with little or no deterioration or maintenance needs for many years. As a result, local officials are lulled into a false sense of security with regard to the condition of the assets and maintenance needed. Their infrastructure "cannot be part of the problem" is the typical view. Yet, the American Society of Civil Engineers (ASCE) has issued a report card every four years for nearly 20 years on the major infrastructure systems in the United States. The ASCE evaluated infrastructure by function and averaged the grade across the entire country using a variety of factors, experience, and engineering judgment. The results are not good (see Table 7.1). The ASCE evaluated roads, bridges, water systems, stormwater systems, wastewater systems, solid waste facilities, airports, and other vital public infrastructure. Over five different report cards, the ASCE has added categories and drilled into more depth. Each time the results are similar—for the most part, U.S. infrastructure is graded overall as a D. The poor grade reflects that the United States has not reinvested and upgraded the infrastructure expansion from the 1930s–1960s and that this failure to reinvest creates a greater potential for failure, thereby subjecting the country to greater economic and social risk. This also suggests that the lack of investment may make the United States less competitive economically (something discussed in Chapter 2). At present, state and local governments spend about 1.8% of the GNP on infrastructure, as compared to 3.1% in 1970 (Prall 2014; Carew and Mandel 2014). A large portion of these expenditures are for growth as opposed to repair and replacement. The result is that their embedded, existing infrastructure that has been relied upon for years, tends to deteriorate further each year as local officials opt to limit budgets in the absence of good data on exactly what their infrastructure maintenance, repair, and replacement needs may be. The latter is part of the problem—where should expenditures be made?

Entities can collect data on their assets, and many times, correctly assess their condition. But often the issue that is left unanswered is what to do with the information and how to develop a needs assessment. In other words, deciding when maintenance is necessary and how much deterioration is viewed as acceptable before repairing. The implementation of repairs is often a challenge—meaning many maintenance needs are often deferred, which increases the potential

Table 7.1 ASCE infrastructure grades—average of grades for major infrastructure systems in the United States (2001–2013)

Infrastructure Category	2001 Grade	2005 Grade	2009 Grade	2013 Grade	2017 Grade
Aviation	D	D+	D	D	D
Bridges	C	C	C	C+	C+
Dams	D	D	D	D	D
Drinking water	D	D–	D–	D	D
Energy (national power grid)	D+	D	D+	D+	D+
Hazardous waste	D+	D	D	D	D
Inland navigable waterways	D+	D–	D–	D–	D–
Levees	–	–	D–	D–	D–
Ports	–	–	–	C	C
Public parks and recreation	–	C–	C–	C–	C–
Rail	–	C–	C–	C+	C+
Roads	D+	D	D–	D	D
Schools	D–	D	D	D	D
Solid waste	C+	C+	C+	B–	B–
Transit	C–	D+	D	D	D
Wastewater	D	D–	D–	D	D
Overall	D+	D	D	D+	D+

Source: Martin 2015

for premature failure. This is not to say that the means to maintain assets are not generally understood, instead it is the timing of the maintenance that is at issue. An assessment of the condition of the infrastructure (discussed further in Chapter 10) can be conducted, but there are assumptions and uncertainties built into any analysis. One concern is that in the absence of certainty about the condition of an asset or its likelihood of imminent failure, money might not be not appropriated, which increases risk of failure of the asset(s). The result is that there is often a disconnect between perceived and actual needs of the infrastructure systems by local officials and residents.

To complicate issues further, much of the discussion about asset management has been developed at the federal level, while most of the assets are local. The federal government uses the Government Accountability Office (GAO) to evaluate federal assets, but many local jurisdictions have no such agencies and lack data. As a result, many jurisdictions have limited information about their systems and little data to use to justify spending. There is a need for better tools for asset management.

Examples of statements from the ASCE and the GAO on the condition of local infrastructure is illuminating:

- The United States has 610,749 bridges. Nearly 25% of all bridges are deficient, with 10% categorized as structurally deficient and 14% categorized as functionally obsolete (GAO 2016a). The cost to repair these deficient bridges is estimated to be $76 billion (Martin 2015).

- The average age of pipe in some jurisdictions is approaching 80 years. There are an estimated 240,000 water main breaks per year in the United States. The cost to replace all of the outdated pipe is estimated to be $1 trillion (Martin 2015). The Environmental Protection Agency (EPA) estimates that nearly $300 billion will be needed over the next 20 years to repair, replace, and upgrade the nation's 50,000 community water systems (GAO 2002, 2015).
- $298 billion in capital investment is needed for the nation's wastewater and stormwater systems over the next 20 years (Martin 2015).
- Some of the nation's aging electrical grid and pipeline distribution systems originated in the 1880s.
- Many railway beds originated in the 19th century. The costs for rail maintenance are $1 million/mile/year.
- The average age of the 84,000 federal dams in the United States is over 50 years old.

This does not factor in the 100,000 miles of levees and dikes, local stormwater systems, the $46 billion in planned port improvements, or the $11 billion backlog of maintenance that the National Park Service has. Of interest, is a Center on Budget and Policy Priorities paper that notes that the total of state and local government investments in infrastructure dwarfs the federal investment in areas like drinking water and schools (see Table 7.2; *source*: McNichol 2016), but overall the total is still only 1.94% of asset value in 2014, down from over 2.5% just five years earlier. That seems counterintuitive to the reality seen every day and the needs identified by other agencies.

In theory, maintenance is carried out to preserve functionality and structural integrity of assets by reducing the rate of deterioration of those assets. Visible assets are easier to assess and will normally claim the most attention. However, over half of all infrastructure is buried, and as a result, gets little attention until it fails. Hence, it is important that local jurisdictions understand their infrastructure systems, which is why the lengthy discussions in Chapters 3 through 6 were undertaken. Many just do not comprehend the physical magnitude of these assets in a community. To begin any asset management system, the first thing needed is a means to identify all the assets that define the system. However, the level at which the systems are defined is often a barrier to successful implementation of ongoing maintenance and an asset management program. Table 7.3 is an outline of a typical summary of assets for an actual community. This is useful for defining the system, but less useful for determining any actual needs

Table 7.2 State and local governments account for nearly 75% of public infrastructure spending (billons of 2004 dollars)

	Federal	State and Local	Private
Schools	$0.40	$75.50	$23.80
Highways	$30.20	$36.50	n/a
Drinking water	$2.60	$25.40	n/a
Mass transit	$7.60	$8.00	0
Energy	$1.70	$7.70	$69.00
Telecommunications	$3.90	n/a	$68.60
Other	$16.10	$17.20	$12.10
Total	$62.50	$170.30	$173.50

Table 7.3   Summary of asset systems for Example City *A*

Component	Amount	Unit
Stormwater (piping and catch basins)	75,268	LF
Real Property (property, minus buildings)	50	ac.
Roads	149	mi.
Sidewalks	322,317	LF
Curbs	13,200	LF
Water System		
Water Lines	360,237	LF
Treatment Plants	5	MGD
Water Services	4,664	ea.
Meters	4,664	ea.
Sewer System		
Sewer Lines	327,360	LF
Sewer Services	3,814	ea.
Treatment Plant	0	ea.
Pumping Stations	16	ea.

Key: LF = linear feet; ac. = acres; mi. = miles; MGD = millions of gallons per day; ea. = each

that the community might have with respect to repairs or replacement. There is also no useful information on asset condition, but the summary gives the viewer a means to understand the magnitude of the system.

Table 7.4 shows the replacement value of a small city's assets. This city has a population of 16,000. What was surprising to the city officials was that the assets, minus buildings, was over a quarter of a billion dollars. This was far beyond what the local officials might have otherwise estimated. The problem extends to employees and the public, as few truly comprehend the magnitude of infrastructure investments in most communities. A quarter of a billion dollars is a huge asset that requires significant dollars for maintenance, which makes it easier to explain to elected officials and the public. Missing are important variables that might be of interest, such as age, material, usage, condition, and location, particularly if they are buried (more on that in Chapter 10). The lost value of those assets cannot be captured in Table 7.4 and assessing the condition of pipes that are 40–60 years old is difficult.

Table 7.4   Valuation of infrastructure assets in Table 7.3 for Example City *A*

Component	Amount	Unit	Unit Value (2015 $)	Asset Value
Stormwater	75,268	LF	$ 150	$ 11,290,200
Real Property (excludes buildings)	50	ac.	$ 20,000	$ 1,000,000
Roads	149	mi.	$ 1,000,000	$ 149,000,000
Sidewalks	322,317	LF	$ 30	$ 9,669,510
Curbs	13,200	LF	$ 40	$ 528,000

Continued

Component	Amount	Unit	Unit Value (2015 $)	Asset Value
Water System				
Water Lines	360,237	LF	$ 80	$ 28,818,960
Treatment Plants	5	MGD	$ 2,000,000	$ 12,500,000
Water Services	4,664	ea.	$ 250	$ 1,166,000
Meters	4,664	ea.	$ 100	$ 466,400
Sewer System				
Sewer Lines	327,360	LF	$ 120	$ 39,283,200
Sewer Services	3,814	ea.	$ 450	$ 1,716,300
Treatment Plant	0	ea.	$ 2,500,000	$ —
Pumping Stations	16	ea.	$ 50,000	$ 4,000,000

TOTAL CITY PROPERTY AND ASSETS $ 259,438,570

Key: LF = linear feet; ac. = acres; mi. = miles; MGD = millions of gallons per day; ea. = each

WHAT IS ASSET MANAGEMENT?

Asset management is a term that is used to describe tools that can be used to help owners track the inventory of their assets, the life and condition of the assets, the effort put into maintenance and replacement of assets, and the need for maintenance or repairs. Asset management is a business process and a decision-making framework that covers a long-term time horizon, draws from economics as well as engineering, and considers a broad range of assets. Asset management has come of age because of:

- Deteriorating condition of the infrastructure in the United States
- Increased potential for water and sewer system disruptions
- Increasing flood event frequency
- Infrastructure failures
- Changes in public expectations
- Advances in technology

As a result, asset management is part maintenance and part management, thereby placing responsibility at two levels of the organization.

For asset management programs, assets are defined as systems of components, buildings, and equipment that have been purchased to accomplish a task. The terminology applies to public works as well as information technology and medical tracking software. However, for public works, the definition needs to be more specific. Public works infrastructure, as discussed in prior chapters, involves large capital investments that have a long life, significant cost, and for the most part are not movable (although equipment and vehicles can certainly be included in asset management systems).

To be useful to the organization, an asset management program must be a systematic, coordinated, planned program of investments/expenditures. Asset management includes early decisions such as those used in design (such as material choices), along with the methods of construction. Many maintenance issues arise as a result of design decisions and the methods and care used during construction. Operations and the ongoing evaluation of facilities must integrate issues throughout the entity including budgets as they relate to adaptation to current and future assets.

Much of the drivers for asset management have stemmed from federal requirements for tracking maintenance for municipal storm water systems (MS4 permits) and the capacity, management, operations, and maintenance (CMOM) programs for sanitary sewer systems. In both cases, the concept is to demonstrate that the owner of the assets was making a good faith effort to ensure that these assets were operating as intended, and providing the benefits predicted. From an MS4 perspective, the concept is to ensure that pump stations work, catch basins are clear, and pipes are cleaned out regularly so that during heavy rains, the stormwater system will drain properly, thereby reducing risk to adjacent properties (public and private). With the CMOM program, the concept is to minimize the potential for sanitary sewer overflows and backups into homes and businesses. Many road and bridge departments also perform inspections to evaluate the conditions of roadway surfaces and bridges. In each case, one of the requirements is that there be documentation that the work has taken place. In the case of MS4 and CMOM programs, regulatory agencies require reports to be submitted and that periodic audits be conducted to verify that the work is occurring and that the tracking information is correctly collected. Water distribution programs are being initiated in many jurisdictions for the same purpose, but since CMOM programs have been around for over 15 years, the water program will need time to become more robust.

In all cases, the documentation is designed to accomplish three things: (1) to verify that maintenance is being conducted—since maintenance is often the first item cut from public budgets because it is so difficult to measure, (2) to provide some insight on the frequency of certain maintenance activities such as pipe cleaning, blockages, or breaks, and (3) to provide the owner with a legal defense against claims from properties damaged as a result of failures of the infrastructure system. Since catastrophic impacts can occur when public infrastructure fails, the goal is to track the maintenance in order to demonstrate that activities do improve resilience and reliability of the assets. Unfortunately, the area where asset management systems fail is in the documentation phase, so there must be documentation that the work has been done, which is typically handled with work orders. Too often, field crews resist the use of work orders and work tickets because they perceive them to be too much paperwork, inconvenient to use, a poor fit in the field (especially paper, since the field may be wet, requiring office effort to translate to computerized systems, education, etc.), and the biggest issue—a means to compare their work effort to others. This resistance must be overcome because work orders are the basic data collection tool for long-term management of assets. For example, if a small community has no records as to where water main breaks have occurred over the past 10 years, how can anyone evaluate pipe replacement needs? Age is often not the primary driver for replacement.

Work Orders

Work orders are a means to track work that is performed by crews on infrastructure systems. They can be simple paper documents, an app on a phone, or something in between. In any case, the data needs to be gathered and entered into a computerized asset management system that tracks the type of work performed, type of incident that occurred (routine, preventive, or reactive); number of incidents; work effort required (hours, equipment, parts used, equipment

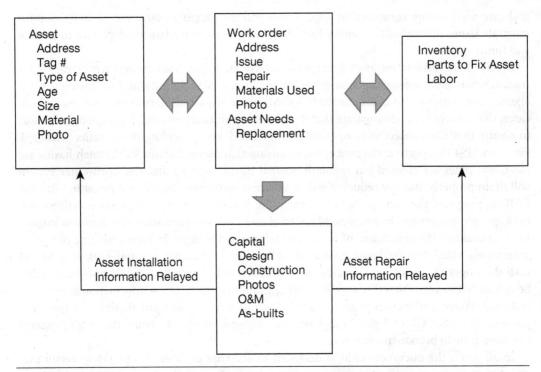

Figure 7.2   This figure shows how the data from the work orders should link the asset history (by address) with parts inventories, work schedules, and equipment usage. Capital construction creates the asset. Data is provided to the asset managers and the parts and inventory personnel. In turn, they provide data on parts used, changes to the asset, and information on conditions to all three parties. They also advise the capital group when the asset needs to be replaced. Arrows indicate information flow.

used, worker time); and preferably, photographs, commentary, and maps of the repairs that can be useful in assessing the condition of buried assets. Work orders should cover routine preventive and reactive maintenance; all work conducted should be tracked and then organized so that monthly and annual reports can be generated (like the ones required for MS4 and CMOM programs). Figure 7.2 shows how the data from the work orders should link the asset history (by address) with parts inventories, work schedules, and equipment usage. It is the work scheduling aspect that creates the resistance—workers think they are being compared to others, which if true, can create significant conflict within the organization. Supervisors should resist such comparisons because most work tasks have intrinsic differences that are unique to the task at hand. Making comparisons, even on something seemingly benign like the number of meters changed out in a day, may not capture all the issues that might arise in the field, so comparisons are often unjust. What should be explained to workers is that this data can be used to help justify budget requests, improve tools and parts supplies, and optimize crews and equipment.

Maintenance and Replacement of Aging Infrastructure

As infrastructure ages, it is normal for maintenance to increase. Figure 7.1 showed a graph of maintenance costs over time. The question is: at what point is the infrastructure system worth repairing rather than replacing? To make a call of repair versus replacement, maintenance costs must be tracked, which requires work orders. Financial data can only be accomplished by tracking work orders and the investment of time, equipment, and parts. Most small water systems do not have tools in place to track maintenance, which places them at a disadvantage with respect to information upon which to make such decisions. The risk is spending good money on an obsolete or unrepairable asset. At some point the annualized cost of replacement will be less than the annualized cost for repairs. Ongoing maintenance is needed because replacement is often not warranted. Proper maintenance and rehabilitation can increase the value and extend the life of assets—though not indefinitely, and never to *new* condition (as shown in Figure 7.3). As a result, an infrastructure system will always be constructing capital projects. Figure 7.4 shows that early in the life of a water system, the investment is for expansion. As the system ages, the costs need to shift to replacement. An example is in order.

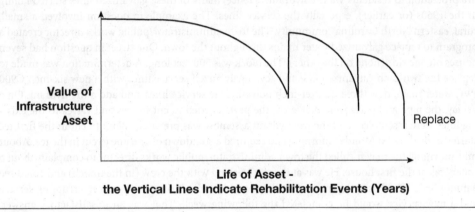

Figure 7.3   This graph shows that as the asset ages (horizontal axis), it will deteriorate. Periodically, money will need to be spent to restore the asset (vertical lines). Restoration improves the asset value but never restores it to *new* condition. Ultimately, the asset will need to be replaced.

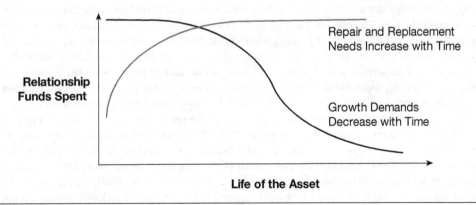

Figure 7.4   Early in the life of a water system, the investment is for expansion of the plant as the surrounding community grows. Then as the system ages, funds need to be shifted to replacement.

Example 7.1: Repair Versus Replacement—
An Example of the Costs of Excessive Repairs

Much of the infrastructure installed in rural southern communities was constructed during the Great Depression in the 1930s under the Federal Works Projects Administration (WPA) program. The WPA was a means to put millions of unemployed people to work. The south and rural areas were particularly hard hit by unemployment and generally lacked central water and sewer facilities; thus, much effort was placed into infrastructure construction in smaller rural and southern cities. The common materials for water lines were cast iron (with lead joints) and galvanized service lines (often with lead goosenecks to redirect the pipe)—these were state of the art in the 1930s. While the cast iron pipe often holds up well, the galvanized pipe does not due to the fact that it includes dissimilar metals. Small galvanized lines were also laid, but they are too small to provide fire protection to residents via fire hydrants. Hence, many of these galvanized lines started failing in the 1960s (or earlier), especially the service lines. The example in question involved a small, rural, eastern North Carolina community. The town administrator/public works director created a program to replace galvanized water mains throughout the town. One street in question had seven houses on one side (none on the other). The block was 500 feet long. A determination was made to replace the two-inch galvanized line and seven galvanized service lines with a new six-inch C900 PVC water main, new three-quarter-inch polyethylene service lines, and add a fire hydrant. On a Friday, the pipe and parts were delivered, the parts needed to cut in tees and valves to the existing lines were assembled, and the fire hydrant assembly was prepared. Work to cut in the first tee started early the next Monday morning and required a shutdown of one line to cut in the tee. About mid-morning, the mayor called the town administrator/public works director to complain about a water leak at the first house. He was at the site working with the crew (in the trench) and could see a minor leak. He advised the mayor that they would not fix it since it was not interrupting service and a new service would be completed the following week. That was an unsatisfactory answer, but work continued. On Wednesday, as the fire hydrant was being installed mid-block, the mayor called again to complain that not only had the first leak not been fixed, but now there was a second leak on the same street. Indeed there was, but like the first one, it was not interrupting service, so the mayor was again advised that the leak would not be fixed. Again, an unsatisfactory answer and the town administrator knew the commission meeting on Tuesday of the following week would be difficult. Thursday saw a third call, and a third leak, as the final tie-in was being completed. One can imagine the fuss. The line was completed and disinfected by Thursday evening. Samples were taken Saturday and on Monday, the line passed disinfection, new services were run, and the old line abandoned by 5 p.m. on Monday. The town administrator/public works director asked that a piece of the first service line, about five feet long be cut out and presented at the commission meeting on Tuesday night. On that five feet of pipe, there were 22 clamps and the leak was between two of the clamps—unfixable. That did not mollify the mayor, but the town administrator/public works director did receive a unanimous direction from the rest of the commission not to fix any more galvanized service lines; instead to replace them. It is unknown how many clamps were on these seven service lines, but extrapolating 22 in 5 feet means: 15 feet long service lines divided by 3 5-foot sections times 7 houses, there were at least 450 clamps, which meant over 3,600 man-hours (2-man crew) devoted to repairs. The new line and seven services were constructed in under 150 hours. Clearly, the old lines should have been replaced years ago, but the lack of a means to track the work created the situation faced by the town administrator/public works director. Fortunately, the commission had his back, but it was an uncomfortable position to be in and one not all that uncommon in the industry.

Any kind of *data tracking* was missing in Example 7.1. As a result, there was no idea of the condition of the service lines. The line was impossible to fix, but at some point in the past, the service line should have been replaced. This example demonstrates why a work order tracking system should be used to track all incidents on the infrastructure system under the control of local jurisdictions. In Example 7.1, there might have been a dozen or more prior work efforts that could have been accessed that would have provided the town administrator/public works director with data to indicate why the line should not be repaired, but replaced. Such arguments are often difficult to make without data. That creates stress on managers and frustration at the field level. But the field level is where the data resides and is generated. As a result, the failure at the field level to collect data means that underfunding is a likely ongoing concern, which leads to deferred maintenance obligations. The argument needs to be altered to indicate the importance of work-order tracking.

Deferred Maintenance

It is not uncommon for budget decisions to cause delays in capital expenditures or maintenance. Nearly every community has examples of the backhoe, dump truck, pipeline, repaving project, etc., that will not be funded until the following year. Yet, the next year comes along and the same thing occurs—and the item is delayed again. Years later the asset fails, becomes much more costly to maintain, or is set aside due to unreliability. Duct tape, superglue, and bailing wire only get one so far in the maintenance industry. Repair and replacement of an asset that is delayed for budget reasons is termed *deferred maintenance*.

Aging water and sewer systems have traditionally fared poorly with regard to replacement because of rate impacts to customers and the fact that half of the assets are unseen. Upgrading treatment for regulatory purposes or expansion tends to be a higher priority for local elected officials than replacement of active buried infrastructure. However, the deferral of replacement of aging infrastructure can lead to greater vulnerability to system failures, less system reliability, and ultimately, more fiscal difficulties in the future. A large number of deferred maintenance projects must be addressed at some point. Delays in addressing the problem costs the system money in the future and the growth in costs can be exponential. Large bond issues, and/or system failures, are too often the norm for addressing deferred maintenance obligations. Borrowing money drives local rates, fees, and taxes, which has long-term negative consequences for future generations of local government officials.

System reliability is directly related to maintaining and replacing components of the system. An old, inadequately funded maintenance program will increase the likelihood of failure of the system and more periodic outages, despite the best efforts of the maintenance staff. Public infrastructure systems need to make service available to all customers at all times while also protecting the public health. All the regulations are designed with this goal in mind: cost is not a factor in decision making. As a result, public sector operators orient their operations to minimize service disruptions, which have serious political consequences to local elected and appointed officials who want to reduce costs.

The questions to answer are: (1) which assets are critical and (2) which require an asset management system to provide the useful data. There must also be an appreciation of the asset's details and data on individual repair or replacement projects that have been completed at the component level from year to year. While this data is useful when it comes to maintenance planning purposes for individual facilities at the tactical level, it can be time-consuming to gather and maintain this level of information, even during a "traditional" condition assessment.

To start a traditional condition assessment, consultants who are developing asset priorities will often send detailed questionnaires to facility managers. The answers allow consultants to estimate the likely condition of the asset. While questionnaires are useful for small-to-moderately sized building portfolios, they may become cumbersome for portfolios that consist of thousands of buildings or water line segments. Furthermore, not all questionnaires may be completed in a timely manner, assets may be missed, and less may be known about certain assets than others. However, a subset of data gathered from the questionnaire can be used to develop a theoretical condition index that projects information to all assets. This can provide a simple and easy tool to predict asset condition where information is limited.

More robust systems involve visual assessments to gauge condition. Visual assessments from simple defect-finding exercises to comprehensive surveys are simple, cost-effective, data-gathering tasks that can populate models providing information that allows effective future strategic planning (Teicholz and Evans, no date). They just take time and manpower, which is expensive. A web-enabled system will allow organizations to enter their own data.

An asset management program or maintenance management system can also be populated with standard preventive maintenance tasks per component (developed from manufacturer's recommendation in the operations and maintenance manuals) and associated unit costs for maintenance. Individual preventive maintenance tasks are then triggered by preprogrammed frequencies based on manufacturer recommendations. Repair needs are identified from due dates for preprogrammed repairs or replacements of individual components. These *trigger* dates are then used to derive condition indices. There are two issues—most buried infrastructure has no preventive maintenance recommendations and a series of expensive repair or replacement projects in a single year can significantly change the condition index results from year to year.

INITIATING THE ASSET MANAGEMENT PLAN

Asset management implies change, which in an organization, can be difficult to deal with. As a result, managers must discuss strategies that can be put in place to bring about change more effectively, while improving communication among all personnel involved in implementing the asset management program. Senior leadership and management of an agency must be committed to the principles of asset management and to providing the resources to implement it.

Implementation of any asset management program should include six steps—as outlined in Figure 7.5 and discussed throughout the following pages.

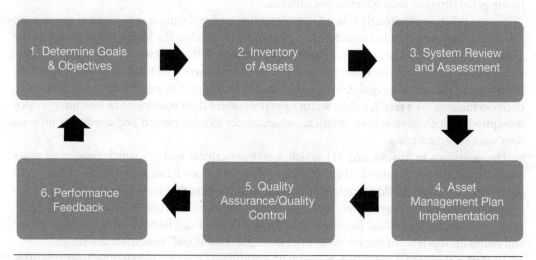

Figure 7.5 Asset management development chart

Step 1—Determine the Goals and Objectives

The first step is to define the goals and objectives of the asset management system. The major goals and objectives for an asset management program should include the following:

- *Protection of the public health, safety, and welfare*—usually a legal parameter that exists in all water-related legislation, police powers for local governments, and the based-on legal precedents for defending local entities. Without protecting the public health, safety, and welfare, few of these programs have value.
- *Reliability of the infrastructure system*—most entities strive to minimize disruptions. This means that the materials used to construct the asset is critical, and ongoing maintenance efforts should be designed to ensure the materials perform as intended.
- *Long life*—coincident with reliability; it is not intended that these assets be replaced frequently.
- *Low maintenance*—reducing costs and therefore disruption over the life of the asset.

Goal-setting for the asset management program is important. The goals, data needs, and use must be clear to the employees. Management buy-in, including appropriate allocations of resources (staff, funding, and tools) is essential. If the goals of the asset management program are unclear to the staff and the data to meet the goals is not easily obtained and/or management buy-in is unclear, the program will not be successful.

Step 2—Inventory Assets

Before any asset management plan can be created, an inventory of assets needs to be established (see Tables 7.1–7.4 and subsequent chapters herein). Depending on the accuracy wanted, the data can be gathered in many ways—ranging from onsite field investigations (which could take a lot of time); to using existing as-built maps; to using as-built maps while verifying the structures using aerial photography and video; to recollections of field staff.

Generating a summary of the assets should involve record reviews such as the following:

- Stormwater
  - Review of stormwater master plan and maps
  - Generation of system maps
  - GIS maps/databases
  - Updating existing stormwater maps and derivation of stormwater system maps for newly discovered areas from site visits and discussions with the staff
  - Determination of the approximate installation date of all stormwater infrastructure components
  - Field verification of stormwater system components and condition, including identification of storm drains, catch basins, and ponding areas
  - Compilation of prior inventory information (values, length of pipe, etc.)
- Roadways
  - Review community maps to identify community-owned streets and right-of-ways
  - Field verification of all streets to determine the lane miles and pavement condition
  - Field verification to determine the existence of sidewalks and condition
  - Field verification to determine the existence of curbing and condition
  - Field verification of street lights, signs, and other appurtenances

- Field verification to inventory landscaping and other roadway improvements
- System or geographic information system (GIS) maps/databases

- Real property
  - Determine community-owned property and existing facilities from maps
  - Verify property ownership from property appraiser's office property records
  - Property records should include:
    - Parks and recreation sites
    - Buildings
    - Beach improvements
    - City Hall
    - Parking areas
    - Miscellaneous sites
  - Visit each site for field verification to determine condition and on-site facilities
  - Verify records of insurance documents
  - GIS maps/databases

- Water and sewer
  - Update utility infrastructure (water and sewer) from master or facilities plans
  - Update system mapping records/GIS maps and databases
  - Determine approximate installation date of all infrastructure components via discussion with staff, aerial photos, and available drawings

Step 3—System Review and Assessment

Once the asset inventory has been developed, it needs to undergo a condition assessment (see Chapter 10). The system review and assessment (Step 3) is designed to ensure that all assets are included; that assumptions across the system are reasonable; and that data on maintenance needs, damage, visual assessments, etc., are developed to understand their true condition. The assets' initial level of service is recorded as well as physical conditions that are apparent. This will help assess the remaining effective life of each asset. The importance of the asset must also be noted; for example, one outfall pipe may serve a large geographic area that includes many catch basins, while another outfall pipe may only serve the road drainage basins. Through condition assessment, the probability of failure can be estimated. Assets can also fail due to a growing area that may exceed its maximum capacity. Operation and maintenance of the assets are important for reassuring a longer life span as well as getting the most out of the money to be spent.

Part of the system review and assessment is vulnerability. A goal of an asset management plan is to find the most critical items of the system. Prioritizing the assets through a defined system will allow the community to see what areas are most susceptible to vulnerability/failure, which assets need the most attention due to their condition, and where the critical assets are located in relation to major public areas (hospitals, schools, etc.) or having a high population (see Chapter 9). The presence of a GIS and *as-built* mapping is a great help in this endeavor. This portion of the work results in an inventory of all assets owned by a community. Getting an inventory can be difficult where maps and as-builts may be lacking. Even in communities with data, it is not surprising to find assets that cannot be located or were lost with time (covered over during paving, buried, broken below ground, etc.).

Step 4—Implementation

There are two different approaches for the implementation of an asset management program: policy-based or performance-based. The former tends to support a long-term, life-cycle approach to evaluating investment benefits and costs to define the directions and overall priorities for an agency's infrastructure management. It does not establish priorities. Performance-based asset management programs use identified measures and targets. This is good business, but in the infrastructure maintenance industry, the lack of effective performance measures for maintenance, operation, and engineering has resulted in performance approaches based more on budget cost management than on asset performance management (FHWA 2015). That is not the goal. As a result, most asset management systems are policy based. That means understanding more about the system than just the inventory. It means concepts like value, vulnerability, and condition—the concepts discussed in the next three chapters—are required.

While there are two means to implement asset management strategies, neither carries with it a standard approach or strategy for implementation; in part because while standards are developing, the *one-size-fits-all* model will rarely work. That is the reason for the need for communication and buy-in. The reason to develop goals for the asset management program is so the organization can determine what information they already have available and where they want to go before determining (or continuing) an approach to implement an asset management program. As a result, asset management programs generally start slowly and in stages. Effort can be initiated for one set of assets to demonstrate principles and requirements of the asset management program within the organization to establish the organizational linkages needed to make the approach work. Such a staged implementation can serve as a model for subsequent implementation and expansion of the asset management program to address cross-organizational needs and address organizational processes. For example, implementation of a work order program that addresses construction techniques used to install the asset, repairs needed, asset condition, work effort, inventory usage, and site condition information would be useful for tracking repair information as well.

For many systems, retroactive data is less necessary once data gathering has started. One issue is that the context of past issues may be misunderstood—a pipe might have been replaced, a bad section lined, or a portion of road base replaced, thereby eliminating the need to investigate further. Tracking retroactively requires knowledge of the changes in the system, something that may not be easy to acquire.

Implementing an asset management system has many challenges aside from the work orders. It is these challenges that often undermine that implementation:

- Budget limitations put in place by administrators or elected officials
- Political climate that favors lowering costs over public health and safety
- The lack of useful decision criteria
- The lack of maintenance policies and O&M manuals
- The lack of best practice standards
- The lack of data on failures
- The lack of a realization for the need for environmental stewardship (noting that the term environment does not include only ecosystems, but the environment in which residents live and work)

A part of a mature asset management program is the ability to use analysis and modeling technology to project life-cycle costs, lifetime of the asset, failure modes, and material solutions.

Step 5—Quality Assurance and Quality Control

Quality assurance and quality control (QA/QC) are designed to ensure the program works as anticipated by creating a series of checks and balances. Checks and balances may include reviews by outside auditors, consultants or internal budget, public works engineers, or other staff. The concept is to ensure that the tasks are undertaken correctly, the metrics used for evaluation are consistent and meet with industry guidelines, and that the guidelines used are appropriate to the assets being reviewed. The goal of the QA/QC process is to place responsibility on some part of the organization to check the process to make sure that the expectations for data reporting, work order implementation, communications, and fiscal information are working. This is done by having some form of oversight that checks each part of the process to make sure that the data being reported is within acceptable accuracy (it will never be perfect and two people may look at the same thing and develop slightly different opinions). Training may be needed if repetitive differences occur. Revising the process may be necessary if certain data is consistently hard to acquire or takes significant time to acquire. For example, a utility may find much information consistently missing. Someone in the organization needs to understand and communicate as to which pieces are most important to acquire. For example, the type of pipe may be more important than the date installed, so acquiring pipe type may be a higher priority (and easier to obtain) than date installed; or soil conditions may be a better indicator of condition for that certain pipe type. Data used in the analysis is important, and accuracy of this data is the most important to review for quality assurance. Bad information is actually worse than no information—it may lead the agency to the wrong conclusions, while no information may allow the agency to get lucky part of the time from sheer randomness.

Step 6—Performance Feedback

The final part of the process is performance feedback—how well this process works, were the recommended solutions meeting the goals, what adjustments must be made, and what changes to the process are needed. The results of each part of the process create a feedback loop to reset the process. The feedback portion of the six steps will indicate where changes must be made to improve or streamline the process. Questions should be asked about barriers to data, limitation in data collection, new sources of data, which data is not useful, etc. The process should be adjusted to account for these findings, which should streamline efforts for all parties, in time. Any proposed revision must avoid adverse impact to the local environment, where the "local environment" is defined as "where people live." Changes should help the agency improve decision making to protect public health. Changes to cut costs may not meet this tenet.

CHALLENGES WITH IMPLEMENTING ASSET MANAGEMENT PROGRAMS

There are a number of challenges to implementing an asset management program. The first is setting strategic goals. An asset management decision-making framework is guided by the setting of strategic objectives. Agencies must coalesce with stakeholders who have different missions, agendas, and values. These differences lead to a range of differing objectives, which are often in competition both within and outside of the organization. The strategic goals should be in place to help with such decisions as prioritizing when the water, sewer, stormwater, and paving are installed in a given street. To do otherwise creates the all-to-common concern where the paving goes first, followed by the pipelines that go under the recently placed pavement. Competing ideas must be resolved through alignment with policy goals, an understanding of the tradeoffs, and consensus-building. Focusing too much on annual budgets often makes it difficult to properly plan the long-term capital investment needs for a proper asset management plan. The mantra "to do more with less" creates challenges as infrastructure deteriorates and the need for maintenance, repair, and ultimately, replacement increases, not decreases.

As noted previously, probably the most challenging aspect of implementation is addressing the needs of people within the organization. Because asset management needs to be holistic throughout the organization in order to be useful, there is a need for communication and coordination among organizational structures that are often functionally segregated into silos within the organization. This is where the good management of the organization becomes paramount. Silos are often embedded within an organization; it is management's responsibility to overcome these silos to ensure that different groups communicate and exchange needed information. That means engineers, operators, field personnel, and finance staffs must be involved and communicate. Too often this does not occur naturally. The challenge for management is to help the staff, especially field staff, to understand the benefits of the process and build an organization-wide commitment to change. Creating buy-in at both the executive and operations levels of the organization is critical to success.

The latter challenge comes from the expectation that management will create a set of metrics that will be used to determine whether certain goals are reached. Most management schools are private sector oriented, where profits matter. They teach that metrics must be quantifiable, which can constitute a problem when trying to apply business metrics to the public sector because public health should be the priority, not profits. To compound the challenge, half of the assets are not visible and performance is difficult to measure. Concurrently, while there is a tendency to invest heavily in data acquisition, if there is no one to analyze the data, the data is useless, and the investment is of little value—the result is the community is data rich but information poor. Defining what is designed for performance measures is useful—and how disparate systems must be consolidated. The following example helps illustrate this point.

Example 7.2: Meter Reading

A small local government had 5,000 water meters on its system. They employed one meter reader who read the meters monthly. The perception was that the meter reader was not very efficient—reading only 250 meters per day. The metric for the reader's efficiency was the number of meters read. What was not collected was data on re-reads due to incorrect readings, data on stopped meters that might need to be replaced, replacing those meters, work orders for repairs, or work orders for a number of other services provided (turn-ons, turn-offs, etc.). The latter tasks actually consumed a large portion of the meter reader's day; but work orders were not tracked—nor was the time involved. Data was lacking. The question was: what should be done about it? A discussion ensued involving

continued

switching over to automated meter reading. The concept was to secure funding to have the meters read monthly, but also to act as data logging devices so that a better understanding of the water being used by the system could be undertaken. As more discussion ensued, the more advanced the proposed system became. Data logging, auto-reads via antenna at the water department, etc., were part of the plan. The costs kept climbing, which was the first concern of management. The costs to maintain the meters were obtained from a contract vendor. It was greater than the reader's salary and benefits. Next it was noted that the ancillary duties—work orders, turn-ons, etc., were not factored into the discussion and someone else would need to perform those duties. Finally, it was noted that water department staff had never run zero read reports and over 10% of the meters in the system showed no usage. All were defective and needed to be replaced. These four factors killed the project.

This example is a classic case where the key to assessing program effectiveness is measuring the right things. In this example, a major problem included identifying appropriate and meaningful performance indicators and the lack of capacity to use the data already in hand to make decisions. A lot of time was spent on what would have been a bad decision because sufficient, useful information was not in hand. A lack of work-order tracking was a huge issue. The lack of work-order tracking, and therefore a lack of understanding of the reader's real work, was why the reader's productivity seemed low. No one factored in the other issues initially. Inability to use existing data compounded the problem. The utility wanted a lot of interesting data, but they did not use the most basic piece of information they already had—the number of meters that were reading zero each month (zero reads) because they were no longer functioning. Meters reading zero significantly reduce revenues. When the zero reads were as high as 10%, that meant that revenues should probably be 8–10% higher than they were. So, if the community did not have anyone to review the basic current data, of what value would data logging be? It would have been money wasted because no one was there to use it. So, despite having an idea that might improve meter reading, the goals and data use did not suggest this was going to be a successful program.

Many of the performance measurements are extremely difficult to analyze. For example: comparing the efficiency of crews repairing water leaks, replacing meters, reading meters, repairing storm drains, or sealing cracked sewer pipes can be difficult. The conditions for each job may be different so comparisons will spawn dissension, not cooperation. The issue is compounded when performance reviews are tied to these poor performance measures. It is no wonder that field crews resist documentation of their jobs.

In addition to performance measures, updating databases can be a challenge. Many organizations find that different departments or programs have their own databases, and those databases do not communicate. Data integration and sharing for asset management involves bringing data from various sources into a framework that incorporates all of the items needed to perform the desired asset management functions. A lack of consistent definitions across sectors complicates the issue, making coalescing and analyzing data a challenge. For an asset management program to be successful, the organization needs to navigate these issues to develop solutions to overcome these challenges (FHWA 2015a; http://www.fhwa.dot.gov/asset/if08008/amo_04.cfm).

Example 7.3: Florida Atlantic University Assets

How many organizations have a 3-D GIS map of their infrastructure? Not many—but Florida Atlantic University (FAU) does, as a result of a recently completed project with students and the Facilities Department staff at FAU (see Figure 7.6). The facilities staff needed better mapping, better asset locations, and some evaluation of conditions. The resulting map and associated GIS database created the list of assets (water, sewer, stormwater, lights, power, communications, etc.), the locations, conditions, and a photograph. The question though, is how all of this data will integrate with their work order system that is not GIS interfaced. As that process sorted itself out, there was an excellent opportunity for two groups within one organization, that otherwise seem to have little in common, to work toward a great project. It was noted that GIS and other computer programs are powerful tools to help infrastructure organizations with data compilation, and one they should embrace wholeheartedly. But there is so much more than mapping to do—and GIS platforms are fantastic tools for hosting this data. Since data gathering is critical, survey tools like GPS-based Leica and Trimble units can permit staff to gather a lot of data easily. Light detecting and ranging (LiDAR) data can be expensive, but the value is tremendous. It is easy to see that the FAU system is laid on a 3-D LiDAR topographic map (six-inch vertical accuracy). Asset condition assessments were also done concurrently, which adds a lot of information to the system (all assets were also photographed and linked). Drawing files can be downloaded and extruded from 2-D to 3-D. Engineers know GIS or can learn it, which makes a fully expanded GIS system for the utility easy to derive if the time is spent. This is a valuable tool when linked to work orders and asset management programs.

Figure 7.6   FAU 3-D utility map from beneath the ground (looking up)

REFERENCES

Carew, D. G. and Mandel, M. 2014. *Infrastructure Investment and Economic Growth: Surveying New Post-Crisis Evidence*. Progressive Policy Institute, Washington, D.C. http://www.progressivepolicy.org/wp-content/uploads/2014/03/2014.03-Carew_Mandel_Infrastructure-Investment-and-Economic-Growth_Surveying-New-Post-Crisis-Evidence.pdf. Accessed 6.15.16.

Federal Highway Administration (FHWA). 2007. Federal Highway Administration's (FHWA's) Office of Asset Management is pleased to present this *Asset Management Overview*. FHWA-IF-08-008. U.S. Department of Transportation Federal Highway Administration Office of Asset Management. http://www.fhwa.dot.gov/asset/if08008/amo_04.cfm.

———. 2015. http://www.fhwa.dot.gov/asset/if08008/amo_05.cfm. Accessed 11/14/16.

———. 2015a. http://www.fhwa.dot.gov/asset/if08008/amo_04.cfm.

———. 2016. *Asset Management*. http://www.fhwa.dot.gov/infrastructure/asstmgmt/assetman.cfm. Accessed 11/14/16.

Government Accountability Office (GAO). 2002. *Water Utility Financing and Planning*, GAO-02-764. GAO. Washington, D.C.

———. 2015. *Critical Infrastructure Protection, Sector-Specific Agencies Need to Better Measure Cybersecurity Progress*. GAO-16-79. GAO, Washington, D.C.

———. 2016a. *Water Infrastructure EPA and USDA Are Helping Small Water Utilities with Asset Management; Opportunities Exist to Better Track Results*. GAO-16-237.GAO, Washington, D.C.

Martin, James R. 2015. *ASCE Report Cards for America's Infrastructure 2001, 2005, 2009 and 2013*. ASCE. http://maaw.info/ASCE2001Infrastructure.htm.

McNichol, E. 2016. *It's Time for States to Invest in Infrastructure. Center of Budget Priorities*, Washington, D.C. http://www.cbpp.org/research/state-budget-and-tax/its-time-for-states-to-invest-in-infrastructure. Accessed 11/14/16.

Prall, D. 2014. "Government infrastructure spending spills over into local economies." *American City and County*. March, 2014. http://americancityandcounty.com/economic-development/government-infrastructure-spending-spills-over-local-economies. Accessed 6/15/16.

Smith, Michael P., Lovett, Michael W., and Caldwell, John T. 2004. "Designing an asset management system." *Florida Water Resources Journal*, Vol. 53, No. 5, pp. 35.

Teicholz, E. and Evans, G. (no date). *Theoretical Condition Indices*. http://www.graphicsystems.biz/gsi/articles/Condition%20Indices%20and%20Strategic%20Planning0707.pdf. Accessed 6/15/16.

8

VALUATION

Good valuations require an accurate inventory of all assets of the system and a good understanding of the value of the same. However, most public works and utility enterprises do not have data about the costs for infrastructure construction that occurred 50 or 60 years ago. Some may have limited data going back 20–30 years, while others have difficulty finding data on assets constructed five years ago. That creates a challenge for asset managers who attempt to develop their *condition index* by comparing the total value of these projects with the replacement value of the system. As a result, proper, systematic valuation becomes an important part of any asset management project.

The good news is that the Governmental Accounting Standards Board Guideline 34 (GASB 34) made local and state governments start collecting both cost and asset inventory data in the year 2000. The GASB is an independent private sector organization, formed in 1984 by accounting professionals, that establishes financial accounting and reporting standards for state and local governments. This chapter discusses valuation and GASB 34—and how the information developed for GASB 34 might be useful to a local infrastructure organization. The GASB 34 rules are only required to report from 1980 (<40 years) onward and many of the cost valuations developed under the initial GASB 34 efforts are hopelessly outdated; consequently, any use of these valuations to create future investment needs, will underestimate those needs. This is one possible reason why monies for investments are lacking.

One difficulty with GASB 34 was determining which inflation algorithm to use. Older costs, even when in hand, cannot be easily inflated because construction costs are often changing at a different rate than the consumer price index (CPI), are less available, and can vary considerably locally. The CPI is the easiest index to use because it is published and updated regularly by the federal government. Engineering News-Record (ENR) develops and publishes a monthly construction cost index (CCI) and a building cost index (BCI) for 20 cities across the United States. Both current and historic (back to 1908) indices can be accessed at ENR's Web site (http://www.enr.com). Places like south Florida and Las Vegas experienced major changes in construction costs annually (>20%) during the early 2000s while the CPI was under 3%. Construction costs tend to increase faster than inflation except in economic downturns. Also, in the early 1980s, the CPI was adjusted to look at prospective costs, which created a much lower projection than the years prior to 1980. Replacement costs are the easiest values to develop since bids for similar equipment or assets are often available or can be developed through engineering estimates or the use of online software to develop costs (like RSMeans®). Local labor markets and material availability alter construction costs, so recent costs are a better indicator of future costs than the past and may represent a better estimate of future inflation. As a result, managers can develop the replacement value of an asset, which allows the value of the asset to be projected backward in time. Figure 8.1 shows the cumulative CPI from 1920 to 2017.

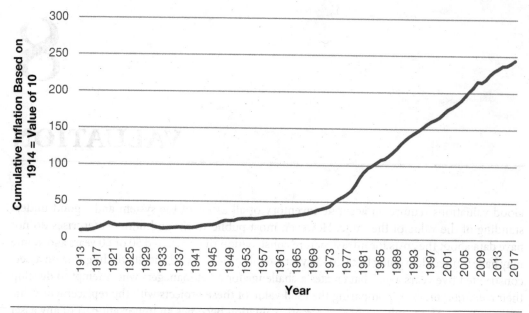

Figure 8.1   Cumulative CPI value from 1913 to 2017.

METHODS FOR ASSET VALUATION

In Chapter 7 (Table 7.4), the assets of a small city (City A) were valued. However, to evaluate assets more accurately, more detail is needed than the estimate made in Chapter 7 because different sizes of pipelines, the depth of pipelines, roads on different soils, groundwater levels, and other factors will affect the value of assets within any given system. Table 8.1 shows some construction costs for various sizes of water mains for the same city, assuming current construction dollars. This is the unit price approach, one of two appraisal methods that can be used to evaluate the value of a system. The unit cost approach is more concerned with the physical assets acquired whereas the market approaches are about economics. The unit costs use an inventory of assets developed as a part of the asset management system to develop actual costs, and then using concepts of time and depreciation to determine what the value might be today. Replacement value is used as a comparison.

The economic approach is completely different, and less useful for asset management purposes. The major use for the economic approach is for the acquisition of revenue-making systems such as water and sewer utility systems. Recent private system acquisitions have focused on market approaches as they have generally been undertaken by lawyers and finance people as opposed to engineers and managers. The market system advantage is that the uncertainties associated with the infrastructure and its condition do not require the same level of due diligence. The failure of this approach is that the condition of the infrastructure may be much poorer than assumed, meaning the acquisition costs are higher than they should be thereby increasing the rate more than expected. Deferred maintenance obligations are completely unknown.

Unit Cost Method

The most reliable methods of estimating the value of assets fall under what is termed the *unit cost method*. Unit valuation means that each asset within the larger infrastructure system is

Table 8.1 Summary of water distribution assets for City A

Asset	Number	Unit	Unit Price	Value New
2" Pipe	15,085	LF	$ 50	$ 754,250
4" Pipe	8,760	LF	$ 65	$ 569,400
6" Pipe	214,922	LF	$ 80	$ 17,193,728
8" Pipe	80,220	LF	$ 95	$ 7,620,900
10" Pipe	5,025	LF	$ 125	$ 628,125
12" Pipe	36,960	LF	$ 150	$ 5,544,000
14" Pipe	2,345	LF	$ 150	$ 351,750
16" Pipe	2,510	LF	$ 175	$ 439,250
20" Pipe	2,200	LF	$ 200	$ 440,000
Fire hydrants	403	ea.	$ 5,500	$ 2,216,500
Valves	967	ea.	$ 1,500	$ 1,450,500
Services ¾"	3,485	ea.	$ 450	$ 1,568,250
Services 1"	525	ea.	$ 550	$ 288,750
Services 1.5"	221	ea.	$ 750	$ 165,750
Services 2"	199	ea.	$ 1,050	$ 208,950
¾" meter	3,485	ea.	$ 150	$ 522,750
1" meter	525	ea.	$ 175	$ 91,875
1.5" meter	221	ea.	$ 250	$ 55,250
2" meter	199	ea.	$ 450	$ 89,550
				$ 40,199,528

There are 8,192 units/customers served $ 4,907.17

Key: LF = linear feet; ea. = each

valued in order to develop a valuation of the entire system. The question is how detailed the asset list needs to be. There is a point of diminishing returns whereby the cost to develop the asset values exceeds their actual value, so most infrastructure system valuations look at pipelines, hydrants, valves, tanks, wells, treatment plants and pump stations, road pavements and bases, property values, etc., as opposed to every nut and bolt of say a clarifier or traffic box. Therefore, the first task in using the unit cost method of estimating the value of an infrastructure system is to determine the asset to be appraised and then to extend that unit cost estimate to the value of the entire system. The unit should include all the property as well (which is why roadways have a high cost). Valuations that are useful include:

- *Original cost*—is the actual acquisition cost of an asset of property when it was first acquired or constructed. This includes the cost of items such as pipes, fittings, and so forth, and the cost of their initial installation. It is nearly impossible to determine these values for most assets.
- *Book value*—is the original historical cost of a property less the accrued depreciation. Book value is used as a rate base component for setting the maximum allowable earnings for regulated utilities. This is also known as net plant assets in local government audits

and may reflect an approximation of the true value of the assets of the system. However, because the original costs are generally not known, the book value estimates are suspect as well.

- *Reproduction cost*—is the present dollar cost to produce a replica of the infrastructure system. The reproduction cost requires the use of identical materials and embodies all the same deficiencies, super-adequacies, and obsolescence present in the system. This method may be most similar to the actual system value, but how the costs for obsolete assets are valued is a challenge. For example, asbestos concrete pipe has not been available for 35 years, so trying to determine a cost to reproduce this material can be difficult. Consequently, this method also relies on data from the past, which is often unknown.

- *Replacement cost*—is the cost in current dollars required to replace all of the assets in an infrastructure system with assets having similar or equal utility, i.e., meets current standards. Replacement cost is based upon modern design materials and technology. This method overvalues the investments made in the system, but provides much more useful information with respect to the funding needed for future investment. Most GASB 34 analyses used replacement value, deflated to the past to find the sunk costs (see Chapter 13 for a discussion on discounting). However, the inflation rate used was often the CPI, which may underinflate the asset investment.

Any valuation method should include valuing the property including vacant land, along with the value of improvements including the roadways, road bases, bridges, plants, equipment, piping systems, etc. Since no system is *new*, the next step is to determine how much of the value of the system has been lost with time. These include physical deterioration and certain types of asset losses due to obsolescence. In most instances, one or two forms of depreciation or obsolescence are present:

- *Physical deterioration*—this form of depreciation is a loss in value caused by wear and tear due to the normal aging process. Inadequate maintenance over the physical life of an item can have a direct bearing on the amount of deterioration present.

- *Functional obsolescence*—this is loss in value caused by factors outside the asset itself. This loss in value due to functional obsolescence is in addition to physical deterioration. Rapid technological changes within the past few years have accelerated functional obsolescence of many mechanical and electrical components of infrastructure systems.

- *Economic obsolescence*—this loss in value is caused by factors outside the system itself. This loss in value is in addition to physical deterioration and functional obsolescence.

All forms of cost are poor evidence of value unless adjusted to eliminate all forms of accrued depreciation.

In many instances, loss in value caused by functional obsolescence and economic obsolescence is greater than the loss in value caused by physical deterioration. The person performing the valuation must be familiar with the various forms of depreciation and the means of measuring it to arrive at a meaningful estimate of the remaining value of the asset. Example 8.1 illustrates this concept.

Example 8.1: Small Utility

Something many local officials fail to realize is the value of the assets they must manage. Once a valuation has been conducted and the value of the assets documented, the numbers can be overwhelming. Table 8.1 shows the same utility as Tables 7.3 and 7.4, only here the water distribution system is detailed. The data shows that different assets have different values, so a system with all small or large pipes may have very different costs. Likewise, different-sized meters will adjust the cost. In reviewing Table 8.1, a series of assets is listed. The number of linear feet (LF) of pipe, number of fire hydrants, etc., is included. The value of this system is about $40 million. Given that there are 8,900 customers connected to the utility, the value of the water distribution utility system per customer is about $4,900 for the water distribution system. The water plant and wells (Table 8.2) was estimated to be $31 million, bringing the per person replacement cost to nearly $8,800 per unit connected. Expand this to a small system—say Clewiston, Florida, with 3,500 water customers—the water system replacement value might be just over $31 million. Doubling this value for the sewer system results in a number of around $60 million, approximating the city's estimates of its system value. In a system like Miami-Dade County, with nearly two million people served, and 1.2 million customers, the system value would be over $11 billion. That is a big asset that would be expected to require a lot of maintenance, something that is often difficult to grasp. State governments and the federal government have assets in the trillions of dollars, so hundreds of billions should be anticipated to maintain these assets. The maintenance numbers sound big, but the asset values are huge.

Table 8.2   Water treatment plant assets for City *A*

Asset	Number	Unit	Unit Price	Value New
Water Plant - Lime softening	3	MGD	$ 2,500,000	$ 7,500,000
Nanofiltration Plant	2	MGD	$ 4,000,000	$ 8,000,000
Water Plant - Land	8	ac.	$ 20,000	$ 160,000
Water Plant - Tank	2	MGD	$ 3,000,000	$ 6,000,000
Water System - Wells	3	ea.	$ 150,000	$ 450,000
High service pumps	12	MGD	$ 750,000	$ 9,000,000
				$ 31,110,000
			cost/customer	$ 3,797.61

Key: MGD = millions of gallons per day; ac. = acres; ea. = each

As noted previously, replacement value is easy to acquire; actual values are not—especially when roads, pipelines, and structures may be 50 years old. Knowing which pipelines were constructed in a given period, and where they are located in the system, may be useful in identifying where obsolescence and significant deterioration may have occurred. Therefore, to be more useful for managing assets, individual assets must be identified within an infrastructure system. From Table 8.1, one can assume that the individual components of the water distribution system are desired. Table 8.3 shows a snippet of a large table of assets of individual water lines for the same community. The pipe for each block is a separate asset. The assumption is that the

Table 8.3 Portion of a table of assets for City A

Asset Description	Street	From	To	Diameter (inches)	LF	Date Installed	Replacement Value
Water Main	Dania Beach Blvd (DBB)	SE 1st Ave	SE 5th Ave	2	2,520	1952	$ 126,000
Water Main	Alley SW 6th/7th	SW 4th Ave	US 1	2	1,715	1953	$ 85,750
Water Main	Alley SW 5th/6th St	SW 4th Ave	US 1	2	910	1953	$ 45,500
Water Main	Alley 4th/5th St	SW 4th Ave	US 1	2	840	1953	$ 42,000
Water Main	Hitching Post Alley	US 1		2	700	1955	$ 35,000
Water Main	US 1 Alley	SE 4th St	SE 7th St	2	1,260	1956	$ 63,000
Water Main	US 1 Alley	Sheridan	SE 10th Terr	2	1,750	1956	$ 87,500
Water Main	Alley SW 8th/9th	SW 4th Ave	SW 2nd Ave	2	1,190	1959	$ 59,500
Water Main	Alley SW 7th/8th	SW 4th Ave	SW 2nd Ave	2	1,085	1959	$ 54,250
Water Main	Alley 6th/7th	Alley US 1	SE 3rd Ave	2	1,190	1978	$ 59,500
Water Main	Alley 3rd/4th St	SW 4th Ave	SW 2nd Ave	2	1,470	1995	$ 73,500

Key: LF = linear feet

characteristics of the asset (pipeline) are the same by block—installation date, materials, etc. Age data can be used to support assumptions about conditions for the infrastructure, but as noted earlier, may not be the most important factor. Unfortunately, the exact installation date for much of the infrastructure cannot be determined easily due to the lack of as-built drawings and lost plans.

However, approximate dates can be determined based on the experience of existing staff and often, aerial photographs. If the infrastructure is broken into many components, it is likely that true deviations will balance out. Having estimated dates for individual assets allows assessors to estimate costs for installation.

When deflated based on the CPI, Table 8.4 shows how individual assets can be decreased to estimated initial costs. The means to create this table results from the following process:

1. Identify the block in question—assume the first one on Table 8.4—there are 2,520 LF of pipe that was installed in 1952
2. The replacement cost was estimated—in this case the replacement cost was estimated in year 2000 dollars at $126,000
3. Determine the inflation that has occurred between 1952 and 2015 (factor in this case was 8.94)
4. Divide the $126,000 by the inflation that has occurred in order to create original installation costs in 1952 of $14,088
5. Continue for all assets until complete—Microsoft® Excel is an excellent program in which to accomplish this

Depreciation

Determining how much the asset has declined in value is the next step. The lessening of the value of an asset with time is the concept of depreciation, but depreciation requires some discussion and an estimate of the asset life. Keep in mind that depreciation is an accounting tool that is used to estimate the loss of value of an asset over time for tax purposes by private entities. As a result, it has limited utility for governmental entities that are exempt from taxes. Depreciation also never actually relates to the value of the asset even in the private sector—the market does that. What depreciation is used for is to stimulate economic growth during a recession by allowing faster write-off of capital equipment; governments will typically pass legislation allowing for accelerated depreciation of capital assets like equipment, buildings, and machinery. Accelerated depreciation encourages companies to invest in big ticket items or large pieces of equipment because they can deduct the depreciation faster for tax purposes. It is an incentive to make big purchases, which are the backbone of the economy. This is an important point. For people and governments, depreciation is not tax deductible, so its real importance is in the life-cycle analysis.

So how is *depreciation* calculated? It is actually easy. Since depreciation is defined as the reduction in value of an asset, it is generally based on initial cost, also called *first* or *adjusted cost*. The remaining value is termed *book value*, a term defined earlier. At the end of the useful life of the asset, the book value is called the *salvage value*. The depreciation rate is the fraction of value removed each year. There are a variety of depreciation models of varying complexity, including:

- Straight-line
- Declining balance
- Modified accelerated cost recovery (MACR)

Table 8.4 Portion of a table of assets with initial installation costs for City A

Asset Description	Street	From	To	Diameter (inches)	LF	Date Installed	Replacement Value	Deflated Value
Water Main	Dania Beach Blvd (DBB)	SE 1st Ave	SE 5th Ave	2	2,520	1952	$ 126,000	$ 14,088
Water Main	Alley SW 6th/7th	SW 4th Ave	US 1	2	1,715	1953	$ 85,750	$ 9,660
Water Main	Alley SW 5th/6th St	SW 4th Ave	US 1	2	910	1953	$ 45,500	$ 5,126
Water Main	Alley 4th/5th St	SW 4th Ave	US 1	2	840	1953	$ 42,000	$ 4,713
Water Main	Hitching Post Alley	US 1		2	700	1955	$ 35,000	$ 3,958
Water Main	US 1 Alley	SE 4th St	SE 7th St	2	1,260	1956	$ 63,000	$ 7,230
Water Main	US 1 Alley	Sheridan	SE 10th Terr	2	1,750	1956	$ 87,500	$ 10,042
Water Main	Alley SW 8th/9th	SW 4th Ave	SW 2nd Ave	2	1,190	1959	$ 59,500	$ 7,305
Water Main	Alley SW 7th/8th	SW 4th Ave	SW 2nd Ave	2	1,085	1959	$ 54,250	$ 6,661
Water Main	Alley 6th/7th	Alley US 1	SE 3rd Ave	2	1,190	1978	$ 59,500	$ 16,368
Water Main	Alley 3rd/4th St	SW 4th Ave	SW 2nd Ave	2	1,470	1995	$ 73,500	$ 47,260

Key: LF = linear feet

Given that declining balance and MACR depreciation methods are complicated business practices that are used to accelerate depreciation for tax purposes—they are more important to business and taxes than engineering—they will not be discussed further here (see Blank and Tarquin 2014 or Bloetscher and Meeroff 2015 for more information if desired).

Straight-line depreciation is the most common and simplest method to use for depreciation. The concept is shown with the orange line in Figure 8.2. The formula for straight-line depreciation is as follows:

$$D_j = (C - S_n) \div n$$

where:

D_j = annual depreciation cost
C = initial cost or initial book value (BV)
S_n = salvage value
n = recovery period
Using Excel: SLN(BV,S_n,n)

With Excel, book value (BV) reflects the depreciated (value) at a given point in time. As noted, it is unlikely to reflect the true market value of the asset, since that depends more on what a willing buyer and a willing seller can agree to for a sale or exchange, which is more related to supply and demand or emotional attachment than book value. Recovery period is the depreciable life of the asset in years (n). Often there are different n values for book value and tax depreciation, and both values may be different from the asset's estimated productive life. Salvage value (S_n) is the estimated trade-in or market value at the end of the asset's useful life. Normally for depreciation purposes, salvage value is estimated at the time of purchase. It can be negative due to dismantling and removal/disposal costs, but one does not typically depreciate an asset below its estimated salvage value.

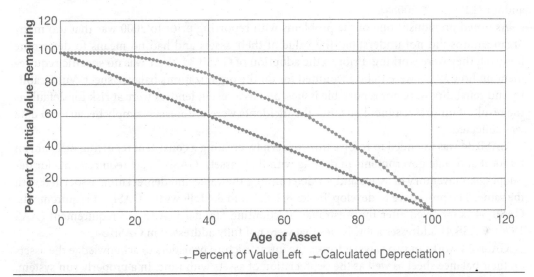

Figure 8.2 A comparison of *actual* versus *calculated* depreciation—this figure demonstrates that the actual value of the asset (percent of value left) at any point during its life may be very different than the depreciated value of the asset according to accounting methods (calculated depreciation).

Figure 8.2 shows a typical cost curve for depreciation and illustrates that while most finance experts depreciate the infrastructure using straight-line depreciation, the reality is that the in-service condition generally stays significantly above the depreciated value, as illustrated by the curve on top. This figure shows that deterioration starts slowly, and then accelerates. Moreover, as a result of the deterioration occurring more slowly than anticipated with straight-line depreciation, the ability to maintain infrastructure condition increases if improvements or rehabilitation are made at the appropriate time.

The value of the asset may deviate from depreciation records if obsolescence comes into play. In a water treatment plant for example, upgrades generally replace older, outdated mechanical equipment and controls for new equipment, which allows the systems to operate much more efficiently. Many times, these upgrades are required in order to deal with changes in technology that address concerns with age and regulatory requirements that improve treatment quality, efficiency, or reliability. They also generally improve the operation and maintenance of the system by lowering costs and simplifying operations in order to create conditions that are less likely to need repairs. With pipelines and roadways, this may not be as critical of a concern, given that the technology does not change significantly with time.

GASB 34

Asset management has had a much larger role in infrastructure systems as a result of GASB 34. Initially the goal of GASB standards was to help lenders assess the ability of governmental entities to provide services and repay debt in the form of easily comprehensible financial reporting. Auditors, finance directors, bankers, and bond and lending agencies are all familiar with the standards developed by GASB, which are recognized as the official source of generally accepted accounting principles (GAAP) for state and local governments. GASB 34 was developed to require local officials to report the value of their infrastructure assets and the cost of deferred maintenance to obtain a *clean opinion* (i.e., a good credit rating) from an auditor (EPA 2008, 2008a).

As noted previously, one of the problems with reporting prior to 2000 was that too many organizations did not understand the value of their assets and had no means to determine how well they were working. Prior to the adoption of GASB 34, there was no specific requirement on how assets were to be accounted for on the organization's balance sheet. Vulnerability and reliability were not reportable issues, and as a result, lenders were at risk for default as the result of unforeseen incidents. But the incidents were unforeseen simply because no data was collected.

GASB 34 was intended to resolve this concern by creating a consistent reporting mechanism for local and state governments in dealing with their assets. GASB 34 set requirements for developing an inventory of assets and for determining the value and depreciation associated with the same. The procedure to develop Tables 8.2, 8.3, and 8.4 follows the GASB 34 requirements. GASB 34 is modeled after the Australian Accounting Standard (AAS) 27 requirement (AAS 1996). GASB 41 addresses a few issues that were not fully addressed in GASB 34.

GASB 34 works in two ways. It forces the infrastructure providers to acknowledge the assets on their balance sheet as well as the depreciation of assets with time. In a properly run system, the total value of the assets should continue to increase with time as investments are made and older assets are retired. The annual audits of finances or comprehensive annual financial reports (CAFRs) should show a general increase in total net asset value with time. If significant borrowing goes on at the same time, the liabilities associated with the assets through borrowing are also recorded.

To meet the GASB 34 requirements, the following assumptions are generally employed:

- All assets included in the inventory are of significant value and long life
- Disposable infrastructure, such as signs, should be deemed expenses, not capital assets, regardless of cost
- The inventory captures all assets, despite age, as best as possible, with some estimate of age, assuming they were initially installed as having significant value (e.g., utility piping that may be as much as 50 years old)
- Depreciation should be straight-line for all assets
- The asset value should tie to the *net plant assets* in the annual audit
- Since most infrastructure age cannot be accurately determined, averaging should be employed
- Property records should be used to develop the value of real property
- Pipeline and utility infrastructure should be based on replacement value—deflated and depreciated—for estimated installation dates using bid information and engineering judgment for new improvements
- Roadway infrastructure should be based on condition surveys and devalued from replacement values determined from bid information for new improvements, discussion with engineers, and engineering judgment for consistency among municipalities
- Depreciated value for roadways will be based on the following formula: right-of-way value plus the remainder of pavement and base multiplied by the present cost of installation
- With regard to landscaping, only landscaping installed and maintained by the organization should be included

The valuation is based on using the depreciated value of the replacement value for infrastructure, unless otherwise noted. The results of the valuation methods use the condition assessment for roadways and the age of assets for utility infrastructure. Table 8.5 shows the depreciated values for the assets in Tables 8.3 and 8.4 (note that these are only a part of much larger tables). Of interest for this set of assets is that the actual installed value was only 18% of the replacement cost (take the deflated cost and divide it by the replacement cost, noting that different assets will have different values. For example, for the 2,520 LF pipe in the first line, the deflated or installed cost is only 11.2% of the replacement cost—$14,088/126,000). When depreciation was factored in, the costs were 8.4% of the replacement value. For the same 2,520 LF pipe asset, the depreciated cost (value remaining) was only $2,994, which is only 2.37% of the replacement value. This should be done for all assets in the system. Newer assets will have higher deflated and depreciation values. Older pipes will be much smaller like this 2,520 LF pipe. For the system represented in these tables, both the 18% and 8.4% indicate a significant deferred maintenance obligation. Based on research by Bloetscher (2017, 2018), the value comparing deflated or installed cost to replacement value should be around 43.5%.

Table 8.5 Depreciated infrastructure for City A

Asset Description	Street	From	To	Diameter (inches)	LF	Date Installed	Replacement Value	Deflated Value	Depreciated Value (from 2015)
Water Main	Dania Beach Blvd (DBB)	SE 1st Ave	SE 5th Ave	2	2,520	1952	$ 126,000	$ 14,088	$ 2,994
Water Main	Alley SW 6th/7th	SW 4th Ave	US 1	2	1,715	1953	$ 85,750	$ 9,660	$ 2,174
Water Main	Alley SW 5th/6th St	SW 4th Ave	US 1	2	910	1953	$ 45,500	$ 5,126	$ 1,153
Water Main	Alley 4th/5th St	SW 4th Ave	US 1	2	840	1953	$ 42,000	$ 4,713	$ 1,060
Water Main	Hitching Post Alley	US 1		2	700	1955	$ 35,000	$ 3,958	$ 990
Water Main	US 1 Alley	SE 4th St	SE 7th St	2	1,260	1956	$ 63,000	$ 7,230	$ 1,898
Water Main	US 1 Alley	Sheridan	SE 10th Terr	2	1,750	1956	$ 87,500	$ 10,042	$ 2,636
Water Main	Alley SW 8th/9th	SW 4th Ave	SW 2nd Ave	2	1,190	1959	$ 59,500	$ 7,305	$ 2,192
Water Main	Alley SW 7th/8th	SW 4th Ave	SW 2nd Ave	2	1,085	1959	$ 54,250	$ 6,661	$ 1,998
Water Main	Alley 6th/7th	Alley US 1	SE 3rd Ave	2	1,190	1978	$ 59,500	$ 16,368	$ 8,798
Water Main	Alley 3rd/4th St	SW 4th Ave	SW 2nd Ave	2	1,470	1995	$ 73,500	$ 47,260	$ 35,445

Key: LF = linear feet

Example 8.2: Florida Atlantic University (FAU)— Boca Raton Campus Electrical and Communication System

FAU is one of 11 universities in the Florida State University System. The school has 30,000 students. Over 25,000 of those attend classes on the 800 acre Boca Raton, Florida, campus. The Facilities Department was interested in gathering information on the buried power and communications infrastructure on campus. FAU engineering students investigated these assets for the FAU Boca Raton campus's power and communications infrastructure and each asset's useful life was estimated based on its asset type. For this assessment, the useful life used for each asset type is based on recent assessments of power distribution infrastructure conducted by utility companies. The useful life values used for this assessment are summarized in Table 8.6. These results were gleaned from the literature and adjusted based on engineering judgment for the campus for local conditions and input from the Facilities Maintenance staff. From each asset's useful life and installation date, the asset's remaining useful life can be determined. A summary of the remaining useful life of the campus's power and communications infrastructure is provided in Figure 8.3. Negative values represent assets that are past their estimated remaining useful life.

Table 8.6 Infrastructure useful life by asset type at FAU, Boca Raton, Florida Campus

Asset Type	Useful Life (Years)
Conduit	100
Copper Line	100
Electric Light	60
Electric Line	100
Emergency Hatch	20
Light Pole	50
Manhole	80
Meter	10
Optic Line	100
Siren	75
Telecommunications Box	15
Traffic Signal Light	60
Transformer Vault	50

FAU is a relatively young school (just passing its 50th anniversary). However, many of the older assets are not well understood and their conditions are unknown. The resulting analysis permitted the Facilities Department staff to better understand the assets they were tasked with maintaining. Mapping (as shown in Chapter 7) and the data from the asset analysis showed that certain assets were beyond their useful life. This analysis was the beginning of the department's efforts to argue for funding to maintain critical infrastructure on campus.

continued

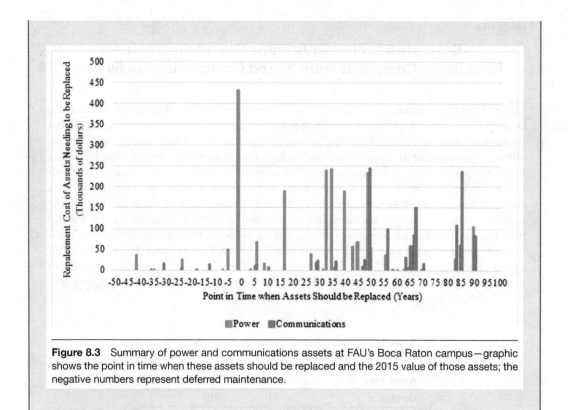

Figure 8.3   Summary of power and communications assets at FAU's Boca Raton campus—graphic shows the point in time when these assets should be replaced and the 2015 value of those assets; the negative numbers represent deferred maintenance.

PRIVATE SYSTEM VALUATION

Note that for public infrastructure systems, the reason to perform valuation methods is to determine the needed investments going forward and to determine the amount of deferred maintenance that might be needed. This is not the goal of valuation for private systems. In the case of a private system, the task of the appraiser is to estimate the most probable price at which the property will sell under a set of given conditions or to determine the rate of return on investment made by private equity providers. For example, most state public service commissions regulate private power, water, and sewer utilities. Under this regulatory framework, private utilities recoup their operating costs plus an annual percent of the depreciated value of their investments on the system. This requires a thorough analysis of the operating characteristics of the private entity and the factors peculiar to a particular industry. Under the sale scenario, once depreciation is removed, cost then becomes a meaningful indicator of market value. However, most utility sales are based on profit over a period of time using the economic valuation method. As a result, despite significant efforts required by public service commissions to generate real asset data to determine rates and provide acceptable rates of return to private investors, much of this information is ignored as a part of the ownership transfer process—the emphasis is on potential profits, not on infrastructure needs or condition. Local officials should understand this issue when dealing with such private entities or entities trying to privatize otherwise public infrastructure systems.

REFERENCES

AAS. 1996. *Financial Reporting by Local Governments.* Prepared by the Public Sector Accounting Standards Board of the Australian Accounting Research Foundation. 211 Hawthorn Road, Caulfield Victoria 3162, Australia.

Blank, L. T. and Tarquin, A. J. 2014. *Basics of Engineering Economy.* McGraw-Hill Higher-Education, New York, NY.

Bloetscher, F. 2018. Risk and economic development in the provision of public infrastructure, *Journal of Environmental Protection.* Vol. 09 No. 09, Article ID:86955, 18 pages. 10.4236/jep.2018.99061.

Bloetscher, F. 2017. "Risk and Economic Development in the Provision of Public Infrastructure." *Florida Section of the American Water Works Association Annual Conference Proceedings—November 30, 2017.* FSAWWA, St. Cloud, FL.

Bloetscher, F. and Meeroff, D. M. 2015. *Practical Design Concepts for Capstone Engineering Design.* J. Ross Publishing, Plantation, FL.

GASB 34. 1999. "Statement No. 34 of the Governmental Accounting Standards Board Basic Financial Statements—and Management's Discussion and Analysis—for State and Local Governments." Governmental Accounting Standards Board of the Financial Accounting Foundation. Norwalk, CT. http://www.gasb.org/jsp/GASB/Document_C/GASBDocumentPage?cid=1176160029121&accepted Disclaimer=true. Accessed 3/15/2002.

GASB. 2001. "Summary of Statement No. 34 Basic Financial Statements—And Management's Discussion and Analysis—for State and Local Governments." (Issued 6/99), Governmental Accounting Standards Board of the Financial Accounting Foundation, Norwalk, CT. http://www.Gasb.Org/St/Summary/Gstsm34.Html. Accessed 2/15/2016.

United States Environmental Protection Agency (EPA). 2008. "Check up program for small systems: Manage your water and wastewater assets, tasks, and finances with one easy-to-use tool: Trainer's Kit." EPA. Washington, D.C.

———. 2008a. "Asset management for local officials." EPA, Washington, D.C.

9

VULNERABILITY AND RISK MANAGEMENT

Having gathered the assets and valued them via the valuation and depreciation methods discussed in Chapter 8, the next thing to analyze is the potential *risk* that the failure of any given asset poses to the infrastructure system. Higher priority assets are just that—ones that need to be kept in service and are more important than other assets (not that any of them are unimportant). The reason some are more important is that their failure presents a greater consequence to the system. Therefore, in describing risk, what is generally meant is that higher risk assets are those that produce greater consequences if they fail and should therefore be prioritized over assets with lesser consequences. For example, a raw water line presents a higher risk from failure than a service line; a bridge on an interstate is of higher concern than an old, historical, wooden covered bridge. As a result, most entities apply a surplus of resources to the higher consequence assets, but first it should be noted that there are no zero-risk options in the world. The old axiom that "if something can go wrong, it will," is frequently supplemented by public works operations and maintenance staff to include the most likely time for the failure is *during rush hour on Friday evening, a holiday, after midnight, or all three*. Risk is defined as the probability of some adverse consequence, or impact. With all infrastructure systems, the consequences are potential public health and safety impacts, and potential risks of injury to employees working around large equipment or within road rights-of-way.

The challenge is to try to minimize the potential for impacts as a result of breaks or failures—or to be able to predict where and when failures are likely to occur so that appropriate steps can be taken to repair or replace assets before they fail. Since infrastructure staffs are aware of the potential risks associated with providing service to the public, they often include redundancy, additional personnel, extra parts, and sensors to maintain equipment and be prepared for the inevitable failures. Thus, very few public health incidents occur in the United States in any given year as a result of infrastructure system failures, despite the potential. Elected and appointed officials need to recognize that it is the lack of spectacular failures with infrastructure systems that should be recognized and appreciated, not the infrequent failures that no system can avoid. But note, the failures will be well publicized—the most recent example is the failure of the Florida International University (FIU) pedestrian walkway bridge in Miami in March 2018. The internet is full of reports pondering the cause of the failure and questioning local officials who were involved with the project. The National Transportation and Safety Board (NTSB) will provide a report, engineers and contractors will be sued, and CNN covered the issue for several days (newspapers even longer).

The events in the recent past, including the attacks of 9/11 and Hurricanes Katrina and Sandy, revealed the lack of plans to protect infrastructure systems (Solano 2010). Cutter et al.

(2008) outlines gaps in the research field of vulnerability and the means by which managers can select infrastructure systems for protection. There is still progress needed to develop useful standards and metrics for measuring disaster resilience (Cutter et al. 2008) and to decide how communities will be protected against potential threats (Solano 2010). The U.S. Environmental Protection Agency (EPA 2002) indicates in its vulnerability assessment factsheet, that single points of failure could be common to many facilities, but there are few attempts to combine all factors that contribute to vulnerability.

A number of authors have attempted to develop conceptual models for hazard vulnerability. Most methods propose mathematical modeling through network theory (Lewis 2006; Ouyang et al. 2009; Arboleda et al. 2009; Eusgeld et al. 2009). Most use linear optimization to solve for minimum-cost scenarios, not human risk. Graphs are used to represent the asset vulnerability and interdependencies. In addition to network theory, Lewis (2006) uses principles of logic, probability, and cost minimization. Some authors mention analysis of uncertain values for model parameters (Ezell 2007; Lewis 2006; Cooke and Goossens 2004). Physical and human component interdependencies in infrastructure seem to be very strong and complex, but most models do not address these issues nor the coupled human-environment system associated with the proximity to a risk factor (Solano 2010). Human factors—such as decision making related to infrastructure operations and maintenance—are considered in just some of the reviewed papers (Baker 2008; Hellstrom 2007; Egan 2007). Bea (2009) suggested that no system is either purely physical or purely technical, suggesting that the reliability of both and the interface between both must be factored into to any vulnerability analysis.

There are multiple factors that can contribute to failures. As an example, when dealing with direct potable reuse projects (wastewater treated for water supply purposes), the technology for treatment is proven but the reliability of those processes and ability to monitor them are less robust. Hence, numerous back-up systems are required to be put into place to protect public health.

CASCADING EFFECTS ON DISRUPTIONS

Cascading disasters are defined as "sequences of events governed by cause-effect relationships, as coincidence plays a significant, although rather unpredictable, role in crises" (Stallings 2006). Events such as Hurricane Katrina (2005), the 2003 power outage in the northeastern United States, the eruption of the Eyjafjallajökull volcano (2010), the Tōhoku earthquake and tsunami (2011), and Hurricane Sandy (2013) demonstrated their effects upon a society that relied on infrastructure systems that were interconnected. In strategic sectors such as energy, telecommunication, and transportation, a disruption in one sector of the infrastructure network can rapidly proliferate to much broader effects by cascading throughout the network and often disrupting other infrastructure systems (Amin 2002). Examples abound. The loss of power in the northeast in 2003 created issues with traffic signals, water supply, and wastewater treatment. The inundation from flooding due to dike failure after Hurricane Katrina created damage to roads, inundated sewers, and shut down the water system in New Orleans. After Hurricane Sandy hit New York, the saltwater shorted out the power system underground, thereby incapacitating power, water, traffic signals, and other services for months (or in some cases, years). Buried power lines are great when trying to reinforce against wind damage, but are useless against saltwater inundation that creates rapid corrosion and damage to the copper cables. Pescaroli and Alexander (2015) note that cascading disasters are an emerging area of study, but those in the infrastructure field know that cascading disasters are more common than many realize.

From a theoretical perspective, cascading disasters are those where a series of decisions or events combine to exacerbate a situation. According to Pescaroli and Alexander (2016), this kind of phenomenon is distinguished by its "high level of complexity and the presence of non-linear paths that lead towards secondary events." Van Eeten et al. (2011) found that more than 47% of all cascades originate within the energy sector. Luiijf et al. (2009) found that the energy sector accounts for 60% of all cascades. A well understood example is the 2003 power outage, which will be discussed in Chapter 14. The focus on the power industry is not surprising since the power grid across the North American continent is interconnected, just as it is in Europe. A failure of a component causes the system to redirect power to compensate for the failure (more in Chapter 14). This is why infrastructure systems build in redundancy, and why, in the power industry, power lines are constructed in a grid so power can approach a property from multiple directions. Likewise, power providers have multiple plants and built-in capacity excess to deal with routine shutdowns of plants for maintenance. However, in 2003, the power plants were not the problem—an overloaded switch failed, causing other switching systems to become over-loaded, each of which proceeded to fail, expanding the area of the blackout. Ultimately, the cascade extended as a result of the joint effects of production pressures in different areas, the lack of assessment of potential risks, and a lack of coordination among power providers, suppliers, and transmission entities (Dien and Duval 2014). This is somewhat surprising since the power industry, most notably the nuclear power industry, is the primary infrastructure system where risk assessment and management was first developed thoroughly in the form of Program Evaluation Review Technique (PERT) charts that used probability of failure and redundant systems to manage the risk of nuclear catastrophe. They included redundancy and probability assigned to the potential for failure of any component, and the risk that a series of components might fail.

The power outage of 2003 in Italy can be used as a study in a whole infrastructure system collapsing due to a simple, foreseeable incident. On the night of September 28, 2003, electricity was being imported into Italy from Switzerland via three routes. One of the transmission wires apparently touched a tree branch (not unusual, but preventable through trimming programs), creating a short circuit. Transmission automatically switched to the other two lines, increasing the amount of power being transmitted, which increased the temperature of the lines. When they reached a certain temperature, they shut down automatically to prevent overheating. A series of blackouts ensued from the Swiss–Italian border as far away as Sicily and Geneva, affecting 56 million people. However, the impact was not confined to power—it affected trains that were marooned in tunnels and people trapped in elevators of buildings. Civil aviation was briefly shut down. Roadway signals and lights did not work—shutting down the roadway transportation system. Hospitals and clinics without generators failed—shutting down portions of the health care system. The Internet failed—meaning that sensors, signals, telemetry (water and sewer), and cameras did not work. The loss of power put food at risk. The loss of electrical power to the Internet propagated failure at power stations due to an inability to transmit control data (Bacher and Näf 2003). The extent of the impact to Swiss and Italian residents was the result of cascades across infrastructure systems and borders. Flint, Michigan, and Walkerton, Ontario, are two examples of cascades in the water industry. The recent Flint water-quality debacle had its genesis in part from the inability of the city of Detroit to supply Flint with water in the aftermath of the 2003 blackout, thus, over 10 years later, the 2003 cascade effects had continued, creating a further cascade in an entirely different infrastructure sector (see Chapter 14 for more discussion). The potential for incidents that cascade is why most public infrastructure systems have built-in redundancy—even if they do not quite understand what they are planning for—to kick in if another component fails.

In line with Holling (2001), Pescaroli and Alexander (2016) have endeavored to show how cascading disasters can be seen as an *alignment of stars* in socio-ecological systems. The breadth of consequences appears to be determined by vulnerabilities that are latent within society (Helbing 2013), which may manifest in complex and secondary events (Pescaroli and Alexander 2015). Because of their high complexity and cross-scale dynamics, cascading incidents cannot be prevented unless the latent vulnerabilities can be understood and addressed before the trigger events occur (Pescaroli and Alexander 2016). Perrow (1999) noted that the numerous interconnections within infrastructure systems make cascades inevitable.

The UK Cabinet (2010, 2011) defines the term *criticality* as it relates to infrastructure assets in terms of (a) the impact of the interruption of essential services, (b) the impact of the interruption in terms of the resulting economic impact and (c) the impact of the failure on daily life. These three factors help define the severity based on (1) the degree of disruption to essential services, (2) the extent of the disruption in terms of population impacted or geographical spread, and (3) the length of time that the disruption persists. The U.S. Government (White House 2013 and FEMA) defines the term *criticality* as it relates to infrastructure assets in terms of 16 categories (see Table 9.1), as opposed to the three used in the UK. Both use a one to five scale to measure significance regarding the delivery of vital services (Pescaroli and Alexander 2016). However, different from local governments, these federal agencies are mostly focused on widespread disasters as opposed to the more localized latent impacts hidden in local networks that may trigger the larger events. The 2003 blackout in the northeast United States and in Italy were very localized issues

Table 9.1   Critical infrastructure sectors as identified by the U.S. Government

Infrastructure Sector	Federal Responsibility
Chemical	Department of Homeland Security, Office of Infrastructure Protection
Commercial Facilities	Department of Homeland Security, Office of Infrastructure Protection
Communications	Department of Homeland Security, Office of Cybersecurity and Communications
Critical Manufacturing	Department of Homeland Security, Office of Infrastructure Protection
Dams	Department of Homeland Security, Office of Infrastructure Protection
Defense Industrial Base	Department of Defense
Emergency Services	Department of Homeland Security, Office of Infrastructure Protection
Energy	Department of Energy
Financial Services	Department of the Treasury
Food and Agriculture	Department of Agriculture, Department of Health and Human Services
Government Facilities	Department of Homeland Security and General Services Administration
Healthcare and Public Health	Department of Health and Human Services
Information Technology	Department of Homeland Security, Office of Cybersecurity and Communications
Nuclear Reactors, Materials, and Waste	Department of Homeland Security, Office of Infrastructure Protection
Transportation	Department of Homeland Security and Department of Transportation
Water and Wastewater Systems	Environmental Protection Agency

that would be unlikely to be revealed through a federal-level assessment. Cutter et al. (2008) mention important gaps in the research field of vulnerability. As a result, there is still progress to be made in the identification of standards and metrics for measuring resilience.

VULNERABILITY AND RESILIENCE

Because assets fail, how the consequences of failure are managed is vital. Not every asset presents the same failure risk, or is equally critical to operations. Therefore, it is important to know which assets are required to sustain the system's performance and which are not—that is the consequence dilemma. Critical assets are those that pose the greatest risk to system failure—the highest consequences when they fail—whether as a cascade or a distinct incident. An example is the raw water transmission main for a water utility. If it fails and there is no redundant system, the entire water system may be without water unless the main is fixed quickly. The only bridge into an island community would be another example. In Alaska, the western communities are all accessible only by air or water. In the winter, water is not an option so the protection of airstrips is vital. Those become the most vulnerable parts of the infrastructure for the community. The most critical assets are those that pose a high risk to the system because they have a greater likelihood of failing (old, poor condition, etc.) and major consequences if they do fail (major expense, system failure, safety concerns, etc.). An asset team should decide how critical each asset is and rank them accordingly.

Many water utility systems may have already accomplished this type of analysis as a result of vulnerability assessments of their utility systems per the Public Health Security and Bioterrorism Preparedness and Response Act of 2002 (PL 107–188). This Act was approved by Congress in the wake of the 9/11 incidents. The intent of the Act was to force water utility systems to identify key system components that are vulnerable to threats and then to create plans to protect the assets and guard against those threats, including appropriating funds to mitigate the most serious vulnerabilities. The reports that went to the EPA were not deemed to be public records. Unfortunately, many water utility systems have not been updated or have reviewed their vulnerability assessments since that time. With the passage of time, only a few of the original authors remained in place and even fewer public officials were ever involved in the process.

The Homeland Security Act of 2002 and the National Infrastructure Protection Plan (NIPP) call for the Department of Homeland Security (DHS) to integrate condition assessments and vulnerability assessments of federal assets to identify priorities, but a Government Accountability Office (GAO) analysis of their assessment tools and methods found that they consistently included some areas, such as perimeter security, but other areas, such as cybersecurity, were not consistently included in the 10 tools and methods (Pub. L. No. 107–296, §101, 116 Stat. 2135, 2142 2002).

IDENTIFYING RISK

Risk assessments generally include two parts: (1) assessment of the risk and (2) management of the risk. Assessment methods include measuring the effects of the exposure or the activity to the ecosystem or humans, determining the level at which the impacts are negligible, and creating methods to replicate and measure the impacts. Assessing risk involves four steps:

- Hazard identification
- Response relationships

- Exposure assessments
- Risk characterization

Hazard identification involves identifying specific hazardous situations. Hazards may range from natural disasters—such as heavy snow, rain, or storm activity (depending on the location)—to the potential for disruption as a result of accidents (train, traffic), sabotage, or human error. The reason to identify the hazard is to promote analysis that determines what can be done to mitigate the impacts of the hazard. Some hazards cannot be avoided—such as traffic accidents that may damage a fire hydrant or knock down a power pole—but realizing the possibility of the hazard will encourage assessment of its consequences and perhaps aid in the development of tools or solutions to mitigate the impact or lessen its probability. For example, fire hydrants could be located away from intersections, farther behind curbs, or have bollards placed in front of them in high traffic areas or parking lots. The response relates to the consequences. The significance of the hazard is important—unlikely events tend to have very large consequences, whereas more common events may have lesser consequences. Vilfredo Pareto studied this concept 100 years ago. If the true risk that pipe damage can have on the community is underestimated, then the potential for economic disruption increases.

Exposure assessments are geared to measuring the potential impact on the population. Contamination to the water distribution system would be the appropriate arena for this type of analysis. In the Walkerton, Ontario, incident in 2001 and the Milwaukee incident in 1993, the reason a problem was detected was that numerous residents became sick, while no one in the neighboring cities (people not served by those utilities) became sick. In both cases, the source water had become contaminated and the treatment plants failed to fully remove the contaminants, which is why source-water protection is so important.

Risk characterization relates to identifying how the risk will be manifested. Characterizing the risk is often calculated using the likelihood-cost equation—where the risk, likelihood of the risk, and costs are optimized. Once the risk characterization is defined, decisions can be made with respect to preferred actions in order to manage the risks.

IDENTIFYING VULNERABLE COMPONENTS

A community has a series of critical assets, each with specific goals. For example, the goals to be used in developing a vulnerability assessment of a water system include:

- Provide safe water quality
- Provide sufficient water quantity to meet community needs
- Provide adequate pressure for fire protection

Drinking water has a direct public health link—as was seen in Milwaukee (Wisconsin), Walkerton (Ontario), Alamosa (Colorado), and most recently in Flint (Michigan). Treatment, including disinfection, is a high priority. Without adequate treatment, impacts will happen quickly in the community—people will not be allowed to drink the water and flushing bad water out of the system becomes a challenge. When contamination occurs, notifying the public is an important aspect of managing the risk—including letting people know that they have to boil any water that is used for consumption.

Potable water is an essential resource that people expect to receive continuously. As a result, the high service pumps that create the flow and pressure in the water distribution system are deemed to be critical assets and are normally perceived to be a vulnerable asset, which is why utilities try to interconnect with neighboring systems. By having connections with neighboring water systems, the vulnerability of the system is lessened because there is another source of pressure and flow. The interconnections permit water supplies to continue to flow, reducing the threat of not maintaining the appropriate water quantity and fire pressure goals. A broken water line is likely of lesser concern, unless it is a raw water supply line or the main pipe leaving the plant (in which case a total loss of water in the system is likely). Water service breaks affect only one customer and are usually of very short duration, therefore they are perceived to create far smaller vulnerabilities to the water system.

Table 9.2 outlines a series of assets for a water system, their exposure (vulnerability), and the potential for damage (risk). Since most of these facilities are located at the water plant site, the current security measures are noted. It should be noted that the water plant sites are normally gated with controlled access at all hours. However, most water treatment plants are not restricted to allowing only water plant employees on site. This has been identified as a concern. Treatment facilities located outside of buildings (most of them) are more vulnerable than those that are not.

Table 9.2 Example of assets for Water System *A* where the current level of protection of the asset is provided, with an estimate of the current site vulnerability and risk for damage

Facility	Protection Offered	Vulnerability	Risk
High Service Pump #1	Secured WTP Site	Exposed	Accessible outside to damage
High Service Pump #2	Secured WTP Site	Exposed	Accessible outside to damage
High Service Pump #3	Secured WTP Site	Exposed	Accessible outside to damage
High Service Pump #4	Secured WTP Site	Exposed	Accessible outside to damage
Reactor #1	Secured WTP Site	Exposed	Contamination threat
Reactor #2	Secured WTP Site	Exposed	Contamination threat
Well G	In fenced offsite bldg	Exposed	Contamination threat
Well H	In fenced offsite bldg	Exposed	Contamination threat
Well I	Does not exist now	Exposed	None - Offline
Bulk Water Line	Secured WTP Site	Buried	Inaccessible
Raw Water Line	Offsite1	Buried	Inaccessible
Generator	Secured WTP Site	Inside WTP building	Limited mech disruption
FPL Service	Secured WTP Site	Exposed	Accessible outside to damage
Cl Regulators	Secured WTP Site	Inside chlorine building	Limited mech disruption
Cl Tanks	Secured WTP Site	Inside chlorine building	High chemical impact
Sodium Hypo Tanks	Secured WTP Site	Exposed	No real disruption opportunity
HSiF6 Tank	Secured WTP Site	Exposed	No real disruption opportunity

Continued

Facility	Protection Offered	Vulnerability	Risk
Phosphate Tanks	Secured WTP Site	Exposed	No real disruption opportunity
Polymer Tank	Secured WTP Site	Exposed	No real disruption opportunity
Lime Silo	Secured WTP Site	Exposed	Accessible outside to damage
Acid Storage	Secured WTP Site	Exposed	None at present
Lab Chemcials	Secured WTP Site	Inside WTP building	No real disruption opportunity
170,000 Clear Well	Secured WTP Site	Buried	Contamination threat
200,000 Clear Well	Secured WTP Site	Buried	Contamination threat
250,000 EST	Secured WTP Site	Exposed	None - Offline
Water Distribution System	Secured WTP Site	Buried	Inaccessible
Backwash Pond	Secured WTP Site	Exposed	High potential for contamination
Monitoring Well 1	Secured WTP Site	Exposed	Contamination threat
Monitoring Well 2	Secured WTP Site	Exposed	Contamination threat
Monitoring Well 3	Secured WTP Site	Exposed	Contamination threat
WTP Building	Secured WTP Site	Exposed	Collateral damage
Admin. Building	Secured WTP Site	Exposed	Personal injury
Fuel Pumps	Secured WTP Site	Exposed	Collateral damage
Chlorine Building	Secured WTP Site	Exposed	Collateral damage
SCADA	Secured WTP Site	Inside WTP building	Limited mech disruption

NOTE: Water plant site has fencing and gate access and therefore has been termed *secured*.

For sanitary sewer systems that are primarily collection and treatment facilities, potential impacts to the sewer system include (1) hazardous materials entering into the system that might explode and spill sewage, (2) rainfall that overloads the system creating overflows, and (3) grease and other debris that creates blockages resulting in overflows that allow raw sewage on the ground, in yards, or in houses. Overflows are a public health risk and may be caused by excess inflow, pump failures, plugged pipes, and pumps or pipes that are too small. Small-line issues impact few people, whereas a failure on a major sewer line or at the plant might have severe consequences. Gas leaks can create a problem whereby the pipe fills with gas and then explodes, damaging houses as well as the piping system. One of the more dramatic examples of this was in San Bruno, California, in 2010, which killed eight people and destroyed about 50 homes. Pumps at the plant, major lift stations, and trunk lines/interceptors might also be high-priority assets since their failure could have widespread impacts. A gravity line or service line that provided service to one or two people might not be a high priority since the impacts are limited.

The type of failures on stormwater systems include blockages in the pipes, debris filling the pipes or catch basins, blocked catch basins, retention/detention basins that do not drain, plugged paver systems, blocked swales, and collapsed pipes. For flood protection, all of these features are expected to function properly. Therefore, with a stormwater system, the highest risks might be large drainage interceptors, tunnels, or pumps located in a downtown commercial district—because if they fail, it may create significant economic impact for the community. In addition to preventing commercial flooding, avoiding the failure of assets under major roadways or tunnels that are relied upon to function properly for other economic reasons would be a priority. Loss of an interstate, for example, could be catastrophic for a community. Smaller drainage pipes, pipes to retention ponds and detention ponds, and pipes under local roadways, would be lower priorities since they impact only localized populations with limited consequences.

Roadways are mostly correlated with the amount of traffic on the road and the impact on vehicle trips. Local roads have local impacts, so if an incident occurs, the impacts are local. For roadways, the most common issue is that pavements fail, adding to the maintenance costs for vehicles. This means a bumpy ride, alignment issues with vehicles, and the host of pavement issues that were discussed in Chapter 6. These are less costly to fix, but they tend to be neglected, while the maintenance to vehicles increases with time. Energy contingency plans from the 1980s indicated that poor pavement conditions and poorly timed traffic lights were major causes of lowered fuel mileage in developed communities.

However, the situation can expand far beyond vehicle impacts. For example, if a pipe fails under a major roadway, the economic impact may be significant to the transportation sector if there are limited other means to access the community. A recent enlightening example of a roadway that failed was east of Estes Park, Colorado, in September 2013. A 22-inch rainstorm concentrated just northwest of Estes Park released floodwaters through the community that washed out over 20 miles of Routes 34 and 36 downstream of Estes Park. This was problematic for the community because of the four routes into the city, these were the two coming from the populated Denver area. A third route was through Rocky Mountain National Park, but this route was closing for the winter due to snow along the nation's highest route (and is never open for truck traffic). That left one two-lane road through the foothills—Route 7 from Nederland—as the only open route. The prospect of Estes Park and neighboring Lyons being cut off from civilization for the winter caused Gov. Hickenlooper to immediately send Colorado Department of Transportation crews and contractors to rebuild 20 miles of roads and bridges. They successfully rebuilt these roads over the winter and ensuing spring.

Larger disruption is often caused by the loss of bridges (which are normally located on larger roadways as discussed in Chapter 6). Fortunately, the failure of bridges is rare. An example of the loss of a bridge on a major interstate was I-10 after Hurricane Katrina. The loss of the major interstate route from the east into New Orleans and the Gulf coast slowed the access for crews seeking to help, and supplies coming in to support residents. The loss of the I-10 corridor required the alteration of many planned trips for goods and services, having significant consequences. The loss of I-10 bridges after Hurricane Katrina was symptomatic of a larger impact to society along the Gulf coast since they were cut off from the rest of the world due to the hurricane aftermath and remained at risk to food, health, shelter, and other basic needs.

PROBABILITY OF THREAT

A vulnerability assessment attempts to answer what the likelihood is that the system will fail based on the vulnerabilities identified. However, to answer the question as to the risk that the vulnerability poses, the probability of the threat occurring must also be estimated. The likelihood of a threat relates to the probability that it might occur. An asset team should be employed to help with this analysis—overestimating certain threats while underestimating others is common. Insurance agencies and certain government sites may have data on the risk of events like storms, earthquakes, vehicle accidents, and the like. However, the probability of a threat like an accident, sabotage, or downed powerlines is less certain. In any vulnerability assessment there is the likelihood of an event occurring and therefore it must be measured or estimated. Large events are uncommon, but may have very serious effects. Smaller events are common and local officials must determine how to better lessen their likelihood. For example, water service line leaks are more common than main breaks on a water system. Rather than repair services, many water utilities simply replace them as a standard solution (see the previous discussions in Chapters 3 and 7) because that lessens the likelihood of future impacts. But a preemptive change out of all service lines is usually not undertaken. An older raw water transmission line however, would be viewed differently. Their failure would disrupt the entire water system. So would the loss of the only well for water supply. Such impacts have high consequences, so local officials usually preemptively address those impacts with redundancy or early replacement.

Potholes are more common in places with ice and snow. Less common—but more significant—are roadways that wash out from storms or damage from earthquakes. Road crews plan on fixing potholes that happen during normal weather—they are easy to fix, so redundant alternate roads are not needed. The rainfall that caused the damage in Colorado in September 2013 was unlikely, but the risk was not zero. The fact that there were only four roadway routes into Estes Park (including the route over the mountains from the west) was an indication of redundant routes. The rain was unlikely; the loss of one route was less likely, even with the rain; the loss of two roads in was much more unlikely; and the loss of all three roads does not appear to have been evaluated, but the results would have been catastrophic for Estes Park in the winter.

Hurricanes have a certain incidence. Florida, for example, experiences a hurricane every 10 years on average; North Carolina and South Carolina actually have higher occurrences—1:6 years for North Carolina and 1:9 years for South Carolina (Kelleher 2016). Because of the significance of the risk, utilities plan for storms by providing backup power, storage, redundant facilities, and 24-hour emergency crews on standby. Hurricanes are unlikely, but cause major disruption. Operations errors are more likely, as is vandalism, but both are normally of limited consequence. Automation tends to reduce operator error. Electronic surveillance and fencing tends to reduce the potential for vandalism.

Table 9.3 outlines the likelihood of a threat to the plant assets described in Table 9.2 for the example Water System A. Vandalism is the most likely effect. The value of the damage would be minimal. Given there are no existing reports on vandalism or other incidents at this plant in the past three years based on engineering judgment, the likelihood of a threat was assigned a value of 10%. Outsiders, hackers, and terrorists would be magnitudes smaller in size with respect to

Table 9.3 Example of threat probability (P_A) for assets for Water System A

Threat Grouping	Likelihood	Probability
Insider	Slight	0.1
Outsiders	Minimal	0.005
Cyber	Minimal	0.001
Terrorist	Minimal	0.000001

occurrence. Foreign terrorist threats were valued at 1:1 million which is similar to the federal estimates.

Table 9.4 outlines the definition of the magnitude of the consequences for the system. The magnitude of the consequences is required to define the risk. The same factors were used by other neighboring cities. Likewise, the probability of a threat entry to the site was evaluated depending on current security on hand. These were divided into fenced areas, restricted access areas, buildings, and supervisory control and data acquisition (SCADA) monitoring. These factors $(1-P_e)$ are noted in Table 9.5.

Table 9.4 Consequence identification (C) for assets for Water System A

Loss			
Economic Loss	$250,000	$100,000	$25,000
User Health	more than 500 people impacted	5	0
Derivation of Loss	48 hrs. +	24–48 hrs.	0–24 hrs.
Value Assigned to C given the Loss Category			
High	0.9 or 0.8		
Medium		0.4	
Low			0.1

Table 9.5 Protection factor (P_e) for assets for Water System A

Prevention Measure	Likelihood of Success	Likelihood of Detection	Speed of Response	1-P_e Factor
Fencing/Gate	0.5	0.1	0.2	0.5
Enclosed Building	0.3	0.1	0.2	0.7
Buried Entry	0.1	0.1	0.1	0.9
Equip. Damaged	0.2	0.5	0.1	0.7
Security Gate	0.2	1.0	0.8	0.8
SCADA	0.3	0.9	0.8	0.9

SCADA THREATS

At the present time, cybersecurity is a major concern for large infrastructure systems. Water plant operations, pumping stations, power grids, traffic light controllers, air traffic systems, trains, and changing highway signs can all be hacked. This is the one area where not having a lot of computerized data reduces risks (although information and efficiency are sacrificed). Poarch et al. (2015) note that keeping up with cybercriminals and hackers is an ongoing challenge for local infrastructure managers. In part this appears to be due to the number of cybercriminal activities, their coordination, and the volume of hacking attempts that can be perpetrated. A major challenge within the cyber world is the race to protect against the hacker—hackers become more sophisticated and create better tools each year, requiring cybersecurity entities to do the same to protect their clients—the cyber race. Trying to understand the business processes and identifying vulnerable spots in cyber infrastructure is the same as for a water or sewer system.

Electronic monitoring and data collection is a part of most infrastructure systems, meaning they all exist in the cyber world. Another example of a challenge is a traffic operations department that must prevent hackers from altering signal timing, or making all signals green at one time, creating chaos on the roadways.

Parks (2007) reports that the North American Electric Reliability Corporation (NERC) adopted Critical Infrastructure Protection (CIP) standards in 2006 (NERC 2006). The standards establish the minimum requirements needed to ensure the security of electronic information exchange to support the reliability of the bulk power system. Labeled CIP-007, the guidelines call for: (1) a vulnerability assessment of all cyber assets within the bulk power industry, (2) a document identifying the vulnerability assessment process, (3) a review of services required for operation of the cyber assets, (4) a review of controls, (5) documentation of the results, (6) a mitigation plan, and (7) a reporting plan (Parks 2007). However, cybersecurity is a poorly understood and poorly implemented concept. Reports indicate uncertainty about implementation of the standards. Even the federal bureaucracy has challenges in dealing with cybersecurity as noted in the 2016 election and reports from the GAO (2014). The GAO (2014) notes that even the Department of Homeland Security does not have a process to share data or coordinate assessment activities among the various offices and components. But, simple things like changing passwords regularly, making passwords over 16 digits long, using encryption software, and keeping malware and other anti-hacking software current are important for sophisticated systems. If SCADA is not protected, the ability to protect the operations will be compromised.

CONSEQUENCE IDENTIFICATION

Table 9.6 identifies the total risk calculation for a water system based on the prior data collected from Tables 9.2 to 9.5 (the example Water System A) with priorities. Table 9.6 also identifies the relative priority of potential countermeasures based on the facility in question. The shaded items include everything that has a likelihood of exceeding a 1% potential for incident. The high service pumps at the water plant, which pressurize and supply water for the entire water distribution system, are the highest consequence of concern. Without them, this utility will fail to meet most of its objectives for water supply, fire hydrant flow, and water quality. As a result, this area should be protected with back-up parts and equipment. Mitigation strategies are required to address higher priority items (all items actually).

Table 9.6   Total risk calculation for assets for Water System A

Facility	C Factor	P_e Factor	Insiders	Outsiders	Cyber	Terrorist	TOTAL Risk	Priority
High Service Pump #1	0.9	0.5	0.045	0.002	0.00045	0.0000005	0.048	2
High Service Pump #2	0.9	0.5	0.045	0.002	0.00045	0.0000005	0.048	2
High Service Pump #3	0.9	0.5	0.045	0.002	0.00045	0.0000005	0.048	2

Continued

Facility	C Factor	P_e Factor	Insiders	Outsiders	Cyber	Terrorist	TOTAL Risk	Priority
High Service Pump #4	0.9	0.5	0.045	0.002	0.00045	0.0000005	0.048	2
Reactor #1	0.8	0.5	0.040	0.002	0.00040	0.0000004	0.042	3B
Reactor #2	0.8	0.5	0.040	0.002	0.00040	0.0000004	0.042	3B
Well G	0.4	0.7	0.028	0.001	0.00028	0.0000003	0.030	4
Well H	0.4	0.7	0.028	0.001	0.00028	0.0000003	0.030	4
Well I	0.1	0	0.000	0.000	0.00000	0.0000000	0.000	
Bulk Water Line	0.4	0.9	0.036	0.002	0.00036	0.0000004	0.038	
Raw Water Line	0.4	0.9	0.036	0.002	0.00036	0.0000004	0.038	
Generator	0.8	0.7	0.056	0.003	0.00056	0.0000006	0.059	1A
FPL Service	0.8	0.5	0.040	0.002	0.00040	0.0000004	0.042	3
Cl Regulators	0.8	0.5	0.040	0.002	0.00040	0.0000004	0.042	3A
Cl Tanks	0.8	0.5	0.040	0.002	0.00040	0.0000004	0.042	3A
Sodium Hypo Tanks	0.1	0.5	0.005	0.000	0.00005	0.0000001	0.005	
HSiF6 Tank	0.1	0.5	0.005	0.000	0.00005	0.0000001	0.005	
Phosphate Tanks	0.1	0.5	0.005	0.000	0.00005	0.0000001	0.005	
Polymer Tank	0.1	0.5	0.005	0.000	0.00005	0.0000001	0.005	
Lime Silo	0.8	0.5	0.040	0.002	0.00040	0.0000004	0.042	3B
Acid Storage	0.4	0	0.000	0.000	0.00000	0.0000000	0.000	n/a
Lab Chemicals	0.1	0.7	0.007	0.000	0.00007	0.0000001	0.007	
170,000 clear well	0.8	0.1	0.008	0.000	0.00008	0.0000001	0.008	
200,000 Clear Well	0.8	0.1	0.008	0.000	0.00008	0.0000001	0.008	
250,000 EST	0.1	0	0.000	0.000	0.00000	0.0000000	0.000	
Water Distribution System	0.1	0.1	0.001	0.000	0.00001	0.0000000	0.001	
Backwash Pond	0.8	0.5	0.040	0.002	0.00040	0.0000004	0.042	3C
Monitoring Well 1	0.4	0.5	0.020	0.001	0.00020	0.0000002	0.021	7

Continued

Facility	C Factor	P_e Factor	Insiders	Outsiders	Cyber	Terrorist	TOTAL Risk	Priority
Monitoring Well 2	0.4	0.5	0.020	0.001	0.00020	0.0000002	0.021	7
Monitoring Well 3	0.4	0.5	0.020	0.001	0.00020	0.0000002	0.021	7
WTP Building	0.9	0.7	0.063	0.003	0.00063	0.0000006	0.067	1
Admin. Building	0.4	0.7	0.028	0.001	0.00028	0.0000003	0.030	5
Fuel Pumps	0.4	0.5	0.020	0.001	0.00020	0.0000002	0.021	6
Chlorine Building	0.8	0.5	0.040	0.002	0.00040	0.0000004	0.042	3
SCADA	0.1	0	0.000	0.000	0.00000	0.0000000	0.000	

NOTE: Highlighted areas are the highest risk assets

While water system vulnerability assessments were required in 2003 as a result of 9/11, the same analysis should be performed for sewer, stormwater, and transportation infrastructure. The other infrastructure systems would use similar tools as water systems, but are not required by federal law to be assessed (except power systems). Not performing a vulnerability assessment simply because there is no requirement to do so is not an acceptable arrangement. A community may be surprised to find out what its most vulnerable assets are. Cyber-hacking may be the biggest threat, yet receives the least amount of attention. Operations personnel and local officials need to know the answer.

RISK MANAGEMENT

Management of the risk includes taking the steps necessary to limit exposure. For local officials, limiting exposure includes the proper training of employees, maintaining appropriate records for operations, and providing facilities with the tools needed to minimize risks to the community. In looking at Table 9.7, the measures that should be undertaken to improve protection of the facilities for this water utility (the example Water System A) based on the likelihood of damage to the facility and its importance in continued operations are as follows:

- *Security cameras, motion detectors, and SCADA*—should be provided at the water plant site and remote well sites. The video feed should be directed to the police department, as well as the water plant during nonbusiness hours. These modifications should be made a part of the water plant upgrades.
- *More on-site storage*—10 States Standards recommends that storage equals average daily flows. The utility had only 380,000 gallons of storage on site and none in the distribution system. A 1.5-million-gallon ground storage tank was proposed to meet this need. Having storage will provide help during short-term disruptions and provide a place where neighboring utilities can discharge its incoming treated water, if needed, for emergencies.
- *Gas chlorine is currently used on the site*—the system is over 50 years old and given security and risk issues, the utility desired to move to a hypochlorite system to reduce off-site risk to the public.

- *Fuel tanks are in the middle of the water plant compound*—the tanks should be moved with the rest of the public works traffic and cordoned off for protection.
- *Unlimited access for all employees to the water plant site should be discouraged*—the water plant and public works campuses can easily be separated—and should be.

For this utility, most of this work was proposed to be done at relatively minor cost. Because the water system was already planning improvements at the water plant, the cost of these additions above the current planned improvements (except the tank and new high service pumps, with redundancy, which cost $3.8 million) were minimal. If all security measures were undertaken, they would add less than $250,000 to the total bill.

Table 9.7   Proposed countermeasures for assets for Water System *A*

Facility	Priority	Countermeasures	Cost
WTP Building	1	Increase SCADA and security; separate WTP area from PW compound	$25,000
Generator	1A	As part of existing WTP upgrade; spare generator parts	n/a
High Service Pump #1 High Service Pump #2 High Service Pump #3 High Service Pump #4	2	Have pumps/parts on hand to replace, provide back-up transfer pump, additional camera security wired to police department, SCADA, separate WTP compound from access of rest of PW compound	$500,000 incl w 2 MG tank
FPL Service	3	Rehab and wall-off FPL transformer, currently exposed	$150,000
Cl Regulators Cl Tanks Chlorine Building	3A	Complete hypochlorite system and remove gas cylinders	<$35,000
Reactor #1 Reactor #2 Lime Silo	3B	Additional camera security wired to police department, separate WTP compound from access of rest of PW compound	<$75,000 incl w/ WTP upgrade
Backwash Pond	3C	Need fencing to secure site from public access, cameras, control of return	$25,000
Well *G* Well *H*	4	SCADA	$25,000 incl included w/ existing WTP upgrade
Admin. Building	5	Additional camera security wired to police department, access control	included w/ existing WTP upgrade
Fuel Pumps	6	Move from WTP building proximity	included w/ existing WTP upgrade
Monitoring Well 1 Monitoring Well 2 Monitoring Well 3	7	Locking caps on wells/test frequently	$5,000

PRIORITIZING REPAIR AND REPLACEMENT

Once the vulnerability assessment and mitigation measures have been determined, the next step is to implement the plan to address these issues—in other words, it is often possible to add mitigation measures to existing capital improvement programs at minimal cost. As noted in Table 9.6, many improvements were added to an existing water plant upgrade project for limited cost. However, the tank was a very large addition that was not on their plant list.

Unfortunately, the costs to address a critical vulnerability may be delayed by local officials for cost reasons. An incident like this occurred in Burlington, North Carolina, 15 years ago. The city did not have backup power at either of its water plants because they figured the likelihood of both plants being out of service at the same time was very small, despite being advised that backup power would be a wise investment. An ice storm put both plants out of service and falling temperatures froze many appurtenances in the system. Water service was lost for days and the costs to repair the broken fixtures due to freezing were significant. "If only the generators had been bought," the Mayor said when he made a public apology (Bloetscher 2011). Fortunately, no one died, but the problem could easily have been avoided. This is why cost should not be the overriding issue with public infrastructure.

Table 9.8 outlines the capital plan that the water utility (the example Water System A depicted in Tables 9.2–9.7) had in place and how these projects were added to it. Note the tank and high service pumps were added to the plan. Generator upgrades were included as was an electrical upgrade. The cost ended up being significant for the utility, but the enhancements for the vulnerability were not compared to the planned expenses. As a follow-up note, this utility implemented all of the recommendations.

Table 9.8 Capital improvement planned projects including vulnerability projects for Water System A

Facility	Priority	Project/Countermeasures	Cost
Water Plant Upgrade		Construction of nanofiltration plant and new control room	$5.5 million
WTP Building	1	Increase SCADA and security; separate WTP area from PW compound	$25,000
Generator	1A	As part of WTP upgrade; spare generator parts	$500,000
Cl Regulators	3A	Complete hypochlorite system and remove gas cylinders	<$35,000
FPL Service	3	Rehab and wall-off FPL transformer, currently exposed	$50,000
Reactor #1	3B	Additional camera security wired to police department, separate WTP compound from access of rest of PW compound	<$75,000 incl
Reactor #2	3B		
Well G	4	SCADA	$25,000 incl
Well H	4	SCADA	
Admin. Building	5	Additional camera security wired to police department, access control	

Continued

Backwash Pond	3C	Need fencing to secure site form public access, cameras, control of return	$25,000
Fuel Pumps	6	Move from WTP building proximity	$200,000
2 MG Tank and Piping, Transfer Pumps			$3.3 million
High Service Pump #1	2	Have pumps/parts on hand to replace, provide back-up transfer pump, additional camera security wired to police department, SCADA, separate WTP compound from access of rest of PW compound	$500,000
High Service Pump #2	2		incl w 2 MG tank
High Service Pump #3	2		
High Service Pump #4	2		
Rehab Existing Plant		Repaint, metal correction, replace launders and catwalks	$1.3 million
Downtown Pipeline		Plant to downtown to increase water supply for redevelopment	$2 million
Eastern Pipeline		Pipeline east of downtown - 1 mile	$500,000
Hypochlorite System		Replace existing gas chlorine system	$210,000
Local Water Main Upgrades		Replace pipelines on 15 blocks w 8-inch mains and fire hydrants	$250,000

How do we encourage customers to invest in their/our future? That is the question as the next 20 years play out—but it should be obvious that part of the answer requires an education of both the general public and local officials concerning the consequences for the lack of planning for vulnerable assets. Risk assessments, like vulnerability assessments, involve identifying where the most likely negative impacts can occur and then taking steps to ensure that a failure cannot affect the public. Due to reduced funding and higher demands, asset management and consequence (vulnerability) play a vital role to help minimize unnecessary spending while focusing spending on the projects that are most needed in order to protect the health and environmental needs of a community. Note that emergency repairs can cost many times as much as the cost of a planned repair, so the goal is to provide strategic, but continuous maintenance before total failure occurs.

REFERENCES

Amin, M. 2002. "Toward secure and resilient interdependent infrastructures." *J Infrastruct Syst.* 8(3), 67–75.

Arboleda, C. A. et al. 2009. "Vulnerability assessment of health care facilities during disaster events." *Journal of Infrastructure Systems.* 15(3), 149–161.

Bacher, R. and Näf, U. 2003. "Report on the Blackout in Italy on 28 September, 2003." Swiss Federal Office of Energy, Berne, Switzerland.

Baker, G. H. April, 2008. "A vulnerability assessment methodology for critical infrastructure sites." Department of Homeland Security symposium: R&D partnerships in homeland security. Boston, MA. Retrieved from: http://works.bepress.com/george_h_baker/2.

Bea, R. et al. 2009. "A new approach to risk: The implications of E3." *Risk Management.* 11, 30–43. doi:10.1057/rm.2008.12.

Bloetscher, F. 2011. *Utility Management for Water and Wastewater Operators*, AWWA: Denver CO.

Cooke, R. M. and Goossens, L. H. J. 2004. "Expert judgment elicitation for risk assessments of critical infrastructures." *Journal of Risk Research.* 7(6), 643–656.

Cutter, S. L. et al. 2008. "A place-based model for understanding community resilience to natural disasters." *Global Environmental Change.* 18(4), 598–606.

Dien, Y. and Duval, C. 2014. "Near misses and influence effects: Indirect factors of cascading in critical infrastructures." In: Pescaroli, G., Alexander, D., Sammonds, P. (eds.) *Pathogenic vulnerabilities and resilient factors in systems and populations experiencing a cascading disaster*—deliverable 2.1. FORTRESS Project, p. 57–69. http://fortress-project.eu. Accessed 2/16/17.

Egan, M. J. 2007. "Anticipating future vulnerability: Defining characteristics of increasingly critical infrastructure-like systems." *Journal of Contingencies and Crisis Management,* 15(1), 4–17. doi:10.1111/j.1468-5973.2007.00500.x.

ENR. 2003. *ENR Construction Index.* http://www.enr.com.

Eusgeld, I. et al. 2009. "The role of network theory and object-oriented modeling within a framework for the vulnerability analysis of critical infrastructures." *Reliability Engineering & System Safety.* 94(5), 954–963.

Ezell, B. C. 2007. "Infrastructure Vulnerability Assessment Model (I-VAM)." *Risk Analysis.* 27(3), 571–83.

GAO. 2014. *Critical Infrastructure Protection DHS Action Needed to Enhance Integration and Coordination of Vulnerability Assessment Efforts.* United States Government Accountability Office, Washington D.C.

Helbing, D. 2013. "Globally networked risks and how to respond." *Nature.* 497, 51–59.

Hellstrom, T. 2007. "Critical infrastructure and systemic vulnerability: Towards a planning framework." *Safety Science.* 45(3), 415–430.

Holling, C. S. 2001. Understanding the complexity of economic, ecological, and social systems. Ecosystems 4(5), 390–405.

Kelleher, S. R. 2016. "How often do hurricanes hit North Carolina and South Carolina." http://travelwithkids.about.com/od/South-Carolina-Family-Vacations/fl/How-Often-Do-Hurricanes-Hit-North-Carolina-and-South-Carolina.htm. Accessed 4/1/17.

Koopowitz, D., Gutierrez, N., and McIndoe, R. 2007. "Infrastructure Vulnerability Assessment." ISACA. *SF Chapter 2007 Fall conference, California State Automobile Association—AAA.* Northern California. file:///E:/CGN%206506%20Infra%20Mgmt/4%20T2.pdf.

Lewis, T. 2006. *Critical Infrastructure Protection in Homeland Security: Defending a Networked Nation.* John Wiley and Sons, Inc. New York, NY.

Luiijf, E. et al. 2009. "Empirical findings on critical infrastructure dependencies in Europe." In: Setola, R., Geretshuber, S. (eds.) CRITIS 2008, LNCS 5508, pp. 302–310.

NATO. 2007. *162 CDS 07 E REV 1—the protection of critical infrastructures.* Parliamentary Assembly of the North Atlantic Treaty Organisation, Brussels, Belgium. http://www.nato-pa.int.

NERC. 2006. *Status Report for Critical Infrastructure Protection Advisory Group.* North American Electric Reliability Corporation, Atlanta, GA.

Ouyang, M. et al. 2009. "A methodological approach to analyze vulnerability of interdependent infrastructures." *Simulation Modelling Practice and Theory.* 17(5), 817–828.

Parks, R. C. 2007. *Guide to Critical Infrastructure Protection Cyber Vulnerability Assessment,* Sandia National Laboratories, Albuquerque, NM.

Parks, R. C. and Rogers, E. 2008. "Vulnerability assessment for critical infrastructure control systems." *IEEE Security & Privacy,* 6(6), 37–43.

Perrow, C. 1999. *Normal accidents: living with high risk technology.* Princeton University Press, Princeton, NJ.

Pescaroli, G. and Alexander, D. E. 2015. "A definition of cascading disasters and cascading effects: going beyond the "toppling dominos" metaphor." Planet@Risk Glob Forum *Davos.* 3(1), 58–67.

Pescaroli, G. and Alexander, D. 2016. "Critical infrastructure, panarchies and the vulnerability paths of cascading disasters." *Nat Hazards.* 82, 175–192. doi: 10.1007/s11069-016-2186-3, file:///E:/CGN%206506%20Infra%20Mgmt/3%20art%253A10.1007%252Fs11069-016-2186-3.pdf.

Poarch, D. et al. 2015. *8 Steps to an Effective Vulnerability Assessment Forsythe Focus.* Forsythe Solutions Group, Inc. Skokie, IL. http://focus.forsythe.com/articles/211/8-Steps-to-an-Effective-Vulnerability-Assessment. Accessed 11/24/16.

PL 107–188, *Public Health Security and Bio-terrorism Preparedness and Response Act of 2002* (PL 107–188), Pub. L. No. 107–296, §101, 116 Stat. 2135, 2142 (2002).

Solano, Eric. 2010. "Methods for assessing vulnerability of critical infrastructure, institute for homeland security solutions." file:///E:/CGN%206506%20Infra%20Mgmt/1%20IHSS_Solano.pdf.

Stallings, R. A. 2006. "Causality and 'natural' disasters." *Contemp Sociol.* 35(3), 223–227.

U.K. Cabinet Office. 2010. "Strategic framework and policy statement on improving the resilience of critical infrastructure to disruption from natural hazards." Cabinet Office, Whitehall, London, UK. https://www.gov. uk. Accessed 11/11/16.

———. 2011. "Keeping the country running: Natural hazards and infrastructure." Cabinet Office, Whitehall, London, UK. https://www.gov.uk. Accessed 11/11/16.

U.S. Environmental Protection Agency (EPA). 2002. "Vulnerability assessment factsheet." Office of Water (4601M); EPA-816-F-02-025. Retrieved from: www.epa.gov/ogwdw/security/index.html.

Van Eeten, M. et al. 2011. "The state and the threat of cascading failure across critical infrastructures: the implications of empirical evidence from media incident reports." *Public Adm.* 89(2), 381–400.

White House. 2013. "Presidential policy directive. Critical infrastructure security and resilience." Directive/PPD-21. White House, Washington D.C. http://www.whitehouse.gov. Accessed 11/11/16.

10

CREATING A CONDITION ASSESSMENT

The ultimate goal of asset management is to provide quality infrastructure by identifying the system's needs and addressing those needs appropriately. Vulnerability analyses and valuations tell managers what is at risk, the magnitude of that risk, the system vulnerability, and the associated replacement cost. But this information does not indicate anything about the potential for failure of a given asset at any given moment in time. For example, it would be useful to estimate if a pump were going to fail in January 2018—or just in 2018. But this is impossible to know without some information as to its current condition. To determine the potential for failure, the United States Environmental Protection Agency (EPA) recommends that condition assessments are performed for the following reasons: (1) to identify assets that are underperforming or at risk of failure, (2) to determine the reason for the current condition, (3) to predict when failure is likely to occur, and (4) to determine what corrective action is needed and when (EPA 2002). In conducting condition assessments, the intent is to create a hierarchy that will lend itself to prioritization of repair and replacement needs.

To do a condition assessment, some idea of the age and the physical condition needs to be generated. The types of materials used and information on those materials throughout the system are also important requirements (Sanford 2013). When the condition of an asset is understood, it can provide insight into that asset's service life, due to greater understanding of the integrity of the asset and means to identify and correct deficiencies. Risk is reduced because the potential for future failures is better understood and preemptive repairs of replacement can be undertaken (Sanford 2013). Optimization of investments can lower costs of asset renewal by targeting rehabilitation/repair of assets that are likely to be in poor condition or of high vulnerability for failure to the community (see Chapter 9).

It is clear that when there is known data or the ability to see and/or test the condition of an asset, it is much easier to conduct an accurate assessment. Roadway systems are the most advanced in dealing with condition assessments, in part because the infrastructure is visible from the surface (although as noted in Chapter 6, it is important to understand what exactly the visible signs are indicating—base failure, structural damage, underground utility damage, etc.). Table 10.1 shows the EPA guidelines for simple assessments (EPA 2017). These guidelines are a simple one to five scale that is mostly judgment based. Table 10.2 shows the EPA guidelines for more complex, intermediate assessments. The difference between the two assessments is the level or amount of information required. Table 10.2 assumes more specific information—the amount of life left, level of service, etc. Assessments of condition can be further enhanced by using predictive maintenance tools such as vibration analysis, ultrasonic monitoring, electrical power analyses that includes voltage testing and 3-phase leg tests, thermograph surveys, mechanical system oil analysis, and pump efficiency tests. Predictive tests are used on equipment (pumps and motors mostly) to judge whether the bearings are worn (vibration or thermographic tools), the power is unbalanced (3-phase leg tests), the motor efficiency is reduced (power analysis), or the pump is worn (pump efficiency). Predictive maintenance activities are most commonly used on critical or costly assets where the additional cost of performing routine monitoring is warranted. Ellison et al. (2014) suggested that for high-consequence assets, if the likelihood of failure is judged to be high, the decision to perform condition assessment is relatively easy. For noncritical or less costly items, predictive maintenance activities may not be cost effective because the monitoring costs may exceed any potential benefits (EPA 2017).

Table 10.1   Simple grade condition/description for a given asset

Scale	Condition
0	Abandoned: no longer in service
1	Very good: operable, and well maintained
2	Good: superficial signs of wear and tear
3	Fair: significant wear and tear, minor deficiencies
4	Poor: major deficiencies
5	Very poor: obsolete, not serviceable

NOTE: This table could be for any asset, but the most common application is roadways.

Table 10.2   Intermediate grade condition description that expands on the scale in Table 10.1

Scale	Condition
0	Abandoned asset: no longer in use or no longer exists.
1	Very good: sound physical condition. Meets current needs. Operable and well maintained. Asset expected to perform adequately with routine maintenance for 10 years or more. No work required.
2	Good: acceptable physical condition. Shows minor wear that has minimal impact on performance. Minimal short-term failure risk. Potential for deterioration or impaired performance over next 5–10 years. Minor work (if any) required.
3	Fair: functionally sound but showing wear and diminished performance. Moderate short-term failure risk. Potential for further deterioration and diminished performance within next 5 years. Renewal or major component replacement expected within next 5 years. Minor work required but asset is serviceable.
4	Poor: asset functions but requires a high level of maintenance to remain operational. High risk of short-term failure. Likely to have significant deterioration in performance within next 2 years. Renewal or replacement expected within next 2 years. Substantial work required, asset barely serviceable.
5	Very poor: asset failed, or failure is imminent. Excessive maintenance required. No further service life expectancy. Significant health and safety hazard. Major work or replacement is urgent.

Source: EPA 2017

When combined with vulnerability, condition assessments can help identify the priority of projects. Conducting a good condition assessment requires a linear process that is understood by assessors, repeatable, and can be updated as data is developed. A problem with most asset management programs is the difficulty in updating data in the database; too many people feel the data is static, so changes are not made—another reason for work orders. There are ways to generate data, especially with roadways. In the best case, Sanford (2013) noted that predefined *ask* and *answer* questions—lumped together by asset type and filled out or answered by experienced field maintenance or engineering staff who have training related to the asset condition—yield the best information. Field staff probably have the best idea of condition because they have seen the assets and worked with them. It would be useful if that data, along with photographs, was incorporated into the work-order system and a geographic information system (GIS) database. Unfortunately, this is rarely the case. But whatever information is available can be used in the condition assessment to identify specific deficiencies in the assets, along with their relative condition (Sanford 2013).

Shekharan et al. (2006) noted that the concepts and principles associated with the measurement of data quality have evolved significantly in the last few decades. The traditional approach to data quality could be called an *error approach*, whereas current methods could better be characterized as an *uncertainty approach* (ISO 2008). With the traditional error approach, there was an assumed single *true* or reference value, and the objective of the measurement was to get as close as possible to that true value. With the newer uncertainty approach, there is an assigned interval, or range, of reasonable values with an acknowledgment of the uncertainty and finite amount of detail that can be measured (ISO 2008). There will be more discussion on the uncertainty approach later in this chapter.

Of importance is the recognition that just because an asset, such as a water main or pump, has reached its operating life on paper, does not mean that it automatically must be scheduled to be replaced. It may require additional monitoring or assessment, some of it destructive, to determine the remaining life (Stadnyckyj 2010). For example, coupons (pieces of the pipe wall cut out for analysis), ground-penetrating radar, leak detection, televising of pipelines, and corrosion testing may be options to analyze older assets. Remote field transformer coupling (RFTC) technology has become the standard for assessing the condition of large diameter prestressed concrete cylinder pipe (PCCP). Technology provides pipeline owners with information on the location and number of wire breaks within individual lengths of pipe, which identifies structurally weak areas—a critical concern for pressure pipe that relies on prestressing wires to maintain its integrity. After the experiences of failures with prestressing wire using Interpace® PCCP in the 1980s, a means to evaluate this pipe became critical when assessing the condition. Age is not the major factor in predicting failures—an issue that is often difficult to convey to local officials and managers.

As noted, roadway pavement analysis has the most robust means for asset assessment. However, Pierce, McGovern, and Zimmerman (2013) noted that effective pavement management systems depend on reliable, accurate, and complete information. Pavement condition data is one of the key components of a pavement management system, which means some form of visual inspection must be undertaken and the results understood. Many agencies use follow-up assessments for roadways sections that have potential concerns to insure repeatability by seeing if multiple inspectors reach the same conclusions about a pavement section (Pierce, McGovern, and Zimmerman 2013). Shekharan et al. (2006) suggested that implementation of a quality management plan should provide:

- Better compliance with external data requirements
- Better credibility within the organization

- Better integration with other internal agency data
- Cost savings from more appropriate treatment recommendations
- Improved accuracy and consistency of data
- Improved decision support for managers
- Increased accuracy in reporting deficient assets
- Increased accuracy in reporting existing condition indices
- Increased accuracy of budget need determinations.

Condition assessment of pavements is an easy means with which to create quality data to support decisions that may have direct and indirect consequences on how an agency analyzes and prioritizes its investments (Shekharan et al. 2006). Through development of characterization of current pavement conditions and trends, models can be developed to predict pavement deterioration, adjustments in pavement design standards, and allocation of resources (Shekharan et al. 2006).

Roadway assessment methods are advanced enough that a series of American Society for Testing and Materials (ASTM) and American Association of State Highway and Transportation Officials (AASHTO) standards have been developed for their evaluation. ASTM D6433 describes manual surveys that are conducted by walking or traveling at a slow speed and noting the existing surface distress. The extent of these manual surveys is limited to selected segments or may span the entire roadway length. By comparison, automated surveys typically incorporate the use of vans fitted with equipment (e.g., lasers, high-speed cameras, and computers) that is specifically designed for collecting pavement and roadway features. In both cases, but especially automatic surveys, data must be reviewed by engineers who understand the significance of what the automated systems *see*.

Yet, AASHTO and ASTM are not the only agencies to develop manuals. Other manuals have been created to help assess roadways including the Minnesota Department of Transportation (DOT) Distress Identification Manual (MNDOT 2003); the Nebraska Department of Roads (DOR) Pavement Maintenance Manual (NDOR 2002); the North Carolina DOT Pavement Condition Survey Manual (NCDOT 2010); the Oregon DOT Pavement Distress Survey Manual (ODOT 2010); the Texas DOT Pavement Management Information System Rater's Manual (TXDOT 2010); the Utah DOT Pavement Preservation Manual—Part 2, Pavement Condition Data (UDOT 2009); the British Columbia Ministry of Transportation and Infrastructure (MoTI) Pavement Surface Condition Rating Manual (BCMoTI 2012) and the Metropolitan Transportation Commission (MTC 2002). And of course, the PASER manuals noted in Chapter 6 are developed with these guidelines in mind as well.

Tiecholz and Evans (no date) note that visual condition assessments of other assets can be expensive, but there are organizations that are used to performing visual on-site condition assessments with teams of specially-trained, highly skilled assessors to identify capital repair and replacement projects for years. Visual condition assessments are appropriate for large mission-critical facilities, but frequent visual assessments of smaller, geographically diverse facilities are too costly and time consuming. In all cases, the more information desired, the more it will cost. At some point, the cost will exceed the value of the information gained.

While roadway and visual methods of assessment are robust, and mechanical equipment operators have predictive tools at their disposal; with buried infrastructure, assessments are far more difficult. There is no similar reference list of standards for assessing water main conditions. In fact, there is not *one* reference or standard. The problem is access. Access is difficult with water lines without significant expense—water pipes cannot be dug up to be assessed, so other methods are needed. Coupons may be useful, but are destructive and costly. Data gathered as a part of breaks or connections may help significantly. Tracking the location of breaks using work orders, mapping, GIS, or other tools is invaluable, but rarely done. Assessing corrosion data using electrical current technology may be useful, but access is still an issue.

All sewer and drainage lines can be televised, but only inside pipe information can be gathered via closed circuit television (CCTV). A visual survey will show that most pipes have few damaged sections (see Chapter 4). Sewer services are rarely assessed and they are often major causes of inflow and/or infiltration. The question to ask is, "When is the cost exceeding the value of data gained?" Sewers are at least accessible, so those willing to pay the costs can gather significant data. Unfortunately, too often the result is a nice cabinet full of videos, but the money runs out before the repairs are made. Five-year-old videos of a sewer system are basically worthless. Breaks, pump performance, stoppages, and manhole condition are useful for analyzing both sanitary and storm sewers.

LIMITED DATA ASSESSMENTS

Condition assessments are performed to provide some degree of understanding of the potential need for maintenance or replacement of an asset. Given the list of all assets, a large portion may be fully or partially visible, but specific information on condition is often hard to judge. As a result, many utilities that are pursuing asset management will use some sort of priority scale (1–5) as suggested by the EPA (see Tables 10.1 and 10.2), or something similar that would indicate a relative condition.

Assets can also be separated into groups—fire hydrants, valve boxes, meters, guardrails, sidewalks, etc., as noted in Chapters 7–9. The reason to do this is to be able to compare conditions across a given set of similar assets. To begin, the group of people who might be involved in the condition assessment should be gathered and some degree of effort placed so that the assessors are in general agreement of what a condition might be. This is called normalizing. Normalizing may require a series of iterative field efforts for different assets. Keeping assets grouped (i.e., storm drains) helps the assessor focus on the comparative conditions. Judgment is used, so there can be disagreements, and because portions of the assets are not seen, failures can be missed. But across hundreds or thousands of assets, the overall condition will be fairly accurate. An example is worth looking at.

Example 10.1: Stormwater Assessment

The entire stormwater infrastructure in the town of Davie, Florida, was evaluated to determine the overall system condition, operation, and its level of service. The approach began with an inventory and location of each asset. The stormwater system assets were field inspected and assessed for condition. Additional findings were included. To realize the inventory, a list of all the stormwater structures and the collection of the information included, but was not limited to:

- Taking images for all inlets, catch basins, swales, canals, and any other stormwater asset
- Collecting the longitudinal, lateral, and elevation data for all inventory collected using a global positioning system (GPS) unit

continued

- Noting the condition of the structure and of each asset
- Noting any failures or conditions that require attention
- Assessing the inventory using an assessment scale of good, fair, and poor

A numbering system and photographic tools were used to document the asset condition. This was accomplished by physically locating each asset in the field and marking it with a GPS coordinate.

An evaluation of each asset was performed based on the current condition of the asset. A condition index for the stormwater system was created with three conditions; the three criteria were based on the idea of an easy system that is quantifiable, with a clear determination of the difference between the rankings. The rankings were meant for a five-year period; this short period was selected because of future changes in the climate patterns and sea-level rise, both of which could have a major effect on stormwater systems in southeast Florida. The following simple strategy was developed and adopted by all of the assessment groups to be consistent in dealing with so many unknown parameters:

- *Good*: a good condition asset does not need repair for three to five years
- *Fair*: a fair condition asset may need repair in three to five years
- *Poor*: a poor condition asset needs repair this year

A *good* ranking is an asset that is good for future use, showing no signs of wear, and in like-new condition. It is expected to meet the stormwater needs now and for the next five years. They appear to be able to withstand severe weather events. An example of a good condition catch basin is shown in Figures 10.1 and 10.2.

Figure 10.1   Example of a catch basin in good condition.

continued

Figure 10.2   Examples of infrastructure in good condition: (a) a drainage inlet in a Florida DOT right-of-way, (b) a manhole, and (c) a canal are all in good condition and function as intended.

The *fair* ranking is an asset that is still functioning for now, but shows signs of wear or deterioration. They are still sound and show no signs of failure, but rehabilitation will be needed in the next three to five years. Examples of a fair condition infrastructure are shown in Figures 10.3–10.5.

Figure 10.3   Example of catch basin in fair condition—this catch basin is full of trash and debris that will compromise its ability to work properly in the long term.

continued

Figure 10.4 This catch basin is only in fair condition since it is not draining water properly.

Figure 10.5 Examples of stormwater assets that are in fair condition: (a) leaves and debris are in the catch basin but not clogging it; (b) the swale has not been maintained properly; and (c) the culverts are mostly full of debris.

The *poor* ranking is an asset that is not functioning at the proper level or shows major signs of deterioration. They have a very high likelihood of failure during a normal or severe weather event. This asset needs immediate attention for cleaning or rehabilitation within the next year. An example of a poor condition catch basin is shown in Figure 10.6(b).

continued

Figure 10.6 Examples of stormwater assets in poor condition: (a) the culvert has a hole in it; (b) the grating on top of the catch basin is full of mud so you cannot tell it is even there; and (c) the pipe is submerged in water.

Typically, a visual inspection will indicate that the majority of assets are in fair to good condition and meet a predefined level of service. Neither condition warrants significant work. For many local governments, the poor/failed condition will likely be in the 4–6% range, meaning that there are immediate issues to address (deferred maintenance in many cases). Many are small items that take limited time to correct—broken meter boxes, missing valve box covers, etc. However, as noted, all information is not visible from the surface.

Example 10.2: Dania Beach Downtown Assessment

The city of Dania Beach is located on the coast of southeastern Broward County. Dania Beach was incorporated under the laws of the State of Florida in 1904. It is the oldest incorporated city in Broward County. Much of the major infrastructure installation began in the 1950s. The city has annexed several large areas—doubling its area to nearly six square miles and increasing its population to nearly 30,000 people, up from 13,700 persons in 1995. The community is primarily residential. Dania Beach has small concentrations of light industry, shopping, offices, and some beachfront property existing within the corporate limits. A small industrial sector was added in the most recent annexation, but none of the industries would be considered *intensive*.

Dania Beach is looking to increase its economic opportunities. The age of infrastructure and management of infrastructure assets are key to encouraging development. Reliability and sufficiency are considerations for any development proposal. The Downtown Assessment Project was to assess the assets and the condition of those assets in the downtown areas of Dania Beach. Assets the city has in their downtown area include:

- Roadways
- Curb locations

continued

- Sidewalk limits
- Street signs
- Water mains (as defined by valve box locations)
- Sewer lines (as defined by manhole locations)
- Service laterals and meters
- Stormwater catch basins
- Green infrastructure
- Location of poles, street lights, and traffic devices (owned by others)
- City buildings

The data collection included an assessment of the condition of the infrastructure, a database of the information, and a photograph of the infrastructure. All data was included on a GIS map developed for the downtown area.

The data was developed using a GIS database that was provided to the city. Within the defined downtown area, defined as the area from the railroad tracks to Northeast 1st Avenue, and from Old Griffin Road to Stirling Road (Figure 10.7), there were a total of 1,780 points gathered that involved over 1,280 distinct asset points, 235 trees, and a series of topographic points. The points were categorized by asset type and then by condition. Table 10.3 outlines the number of assets by

Table 10.3   Asset type for the downtown area for the city of Dania Beach, Florida

Asset	Number	Asset	Number
Backflow DW	35	Manhole elect	12
Bike lock	5	Manhole SS	69
Bus bench	6	Meter DW	106
Catch basin SW	57	Observation well	5
Cleanout SS	26	Pole elect	105
Control box	10	Pole w transformer	16
Crosswalk signal	20	Pull box	97
Curb	87	Reclaimed water box	10
Curb inlet	20	Sidewalk/curb	289
Do not enter	10	Sign	68
Fire zone	1	Stop light pole	12
Riser	4	Stop sign	24
One way	2	Street light	42
Bollard	1	Swale	5
Electric meter	1	Transformer	1
Fiber optic riser	8	Trash can city	9
Fire hydrant	26	Valve DW	77
Fire sprinkler	1	Valve gas	11
Irrigation control	8	Yield	1

Source: City of Dania Beach downtown asset inventory study, 2015

continued

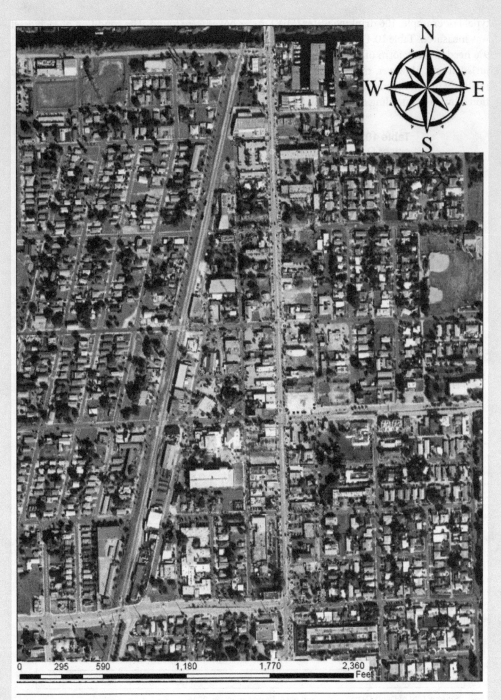

Figure 10.7   Data points gathered during the city of Dania Beach downtown assessment survey—the colored dots indicate different assets by type.

continued

type. For example, 106 drinking water (DW) meter boxes were found. The condition of each asset was measured. Table 10.4 outlines the asset condition: 6% of the assets were in poor condition, but 9% needed some form of maintenance, which is higher than the 4% reported in several adjacent jurisdictions. However, most of the repairs will take limited time to correct—they are small and relatively inconsequential. These data points were depicted in Figure 10.8, which shows the location of all assets needing repairs. The total cost of repair is just over $330,000 (see Table 10.5).

Table 10.4   Asset condition for the downtown area for the city of Dania Beach, Florida

Category	Frequency	Percent
Good	364	55%
Fair	261	39%
Poor	42	6%

Assessed points	667	

Source: City of Dania Beach downtown asset inventory study, 2015

Table 10.5   Number of defects by type and cost for the downtown area for the city of Dania Beach, Florida

Type of Asset	Number with Defects	Estimated Cost to Repair
Backflow DW	3	$750
Bus bench	2	$5,000
Catch basin SW	8	$32,000
Cleanout SS	2	$650
Electric meter	1	unknown
Local flooding	19	$200,000
Meter DW	15	$11,500
Pavement failure	2	$1,000
Pole electric	3	by others—$30,000 ea.
Pull box	2	by others—$3500
Pavement failure	3	$25,000
Sidewalk top	1	$2,500
Signs	7	$2,000
Transformer	1	$25,000
Valve boxes needing repairs	24	$ 26,000
Total with defects	93	$ 331,400

Source: City of Dania Beach downtown asset inventory study, 2015

continued

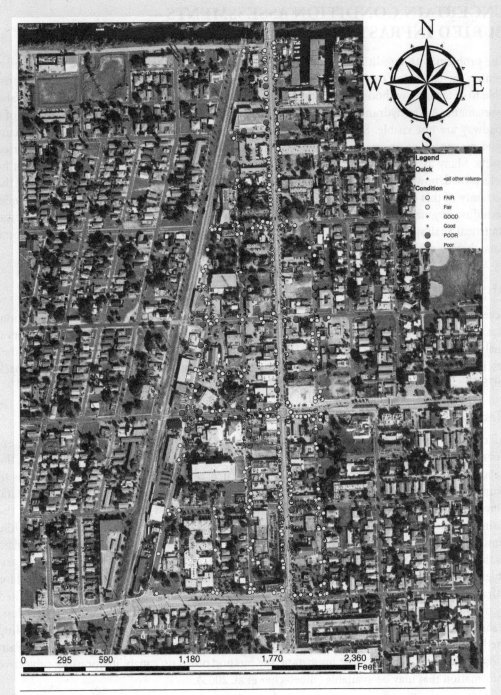

Figure 10.8 Assets gathered during the city of Dania Beach downtown assessment survey requiring maintenance—the legend shows the asset condition (red is poor).

UNCERTAIN CONDITION ASSESSMENTS—BURIED INFRASTRUCTURE

The problem with condition assessments is that for many of the infrastructure assets, determining the condition can be quite difficult. Buried infrastructure condition is nearly impossible to determine without unburying it. Even infrastructure that is visible may provide a false assessment. A fire hydrant is partially buried. The foundations for bridges and the base of a roadway are not visible. Stormwater pipes may be visible only at outlets. As a result, many asset management programs stall when there is a need to assess the condition of the buried assets. Many assume that since the buried infrastructure is unseen, the condition cannot be determined; however, this is not true. There is usually some information that is known, but the certainty of this information is the challenge. The uncertainty is a concept that many people, including engineers, operations staff, and administrators are uncomfortable with. However, statistical analysis is the mathematical means to address this uncertainty since even with some data, there is still uncertainty.

Methods to Measure Risks with Limited Data

Several statistical methods have been developed to attempt to exploit limited information: resampling (bootstrap and jack-knife methods), fuzzy set theory (Zimmerman 1985), interval analysis, information theory, and Bayesian methods. Information theory is a branch of mathematical theory within probability and statistics and can be applied to a variety of fields (Kullback 1978). Each of these methods is discussed in the following paragraphs.

Resampling

All data can be developed into a statistical probability distribution—for example: normal, exponential, and gamma distributions are common in the engineering world. In most cases, the data is limited, so an estimated distribution is created by using the likelihood as a means to estimate the underlying distribution parameters—like the mean and range of possible data. Once the estimated distribution is created, more data is created by repetitively sampling the estimated distribution (Bloetscher et al. 2005). Ultimately, the goal is to develop data while simulating the expected uncertainty and variability in the original set of data. As much data as the investigator wants can be created. The process is relatively straightforward for two variables. Unfortunately, Haas, Rose, and Gerba (1999) report that the method is tedious—and for infrastructure, more than two variables are needed. In addition, the entire analysis is limited by the quality of the original data. If the original data is flawed, the results may be meaningless. Haas, Rose, and Gerba (1999) note they experienced some severe limitations in the resampling methods with respect to risk assessment when using Monte Carlo methods, but did not explain those limitations. Perhaps most important, the method offers no capability for using subjective information that may be available (Bloetscher et al. 2005).

Fuzzy Set Theory

Fuzzy set theory/logic was formalized by Professor Lofti Zadeh at the University of California in 1965 (Bloetscher et al. 2005). The theory proposed was a paradigm shift in logic that involves a set of rules that define boundaries and solve problems within those boundaries. As the name suggests, fuzzy logic is the logic of underlying modes of reasoning that are approximate rather that exact, where everything is a matter of degree used to handle the concept of partial truths—somewhere between true and completely false (Bloetscher et al. 2005). Fuzzy set analysis is usually applied to subjective, verbal information that is divided into two sets of data. Fuzzy set theory measures the intersection of the two sets. A typical example would be determining how hot a cup of coffee is. It is not possible to define the point where a cup of coffee is hot versus warm—it is a matter of degrees. It could be somewhat warm or nearly hot, but the point where hot becomes warm is a *fuzzy* line. Fuzzy set theory defies the probability that the cup of coffee is hot and the probability that it is warm. The intersection is the fuzzy set pairing value (Bloetscher et al. 2005).

Fuzzy logic is applied as a subset of conventional Boolean algebra. Its applications may be far ranging and provide an opportunity for modeling conditions that are imprecisely defined. The most obvious limitation noted in the literature is that the data cannot be mutually exclusive from the two sets. In addition, real data are not helpful, nor is new data. Some prior information or data is necessary to create the subjective opinion that sets up the degree of membership in the two sets. In addition, there are issues concerning the stability of using fuzzy set theory in dynamic control expert systems, the most common application of fuzzy set logic, between the threshold triggers. As such, the method has limited use except in fail/not fail situations (Bloetscher et al. 2005).

Interval Analysis

Interval analysis is an approach to the analysis of systems when the value of the quantity measured is uncertain, which sounds good for buried infrastructure. Interval analysis defines the value of the quantity by specifying the interval that the value is guaranteed to fall within. Other infrastructure researchers use this method, but the *guarantee* may be problematic. The methodology provides a correct formal method for measuring the upper and lower bounds required for the worst possible case (Bloetscher et al. 2005). Interval analysis is not as powerful as other statistical methods when empirical information is available. Some prior information or data to create the subjective opinion is required (Bloetscher et al. 2005). Like the methods previously discussed, updating with new data is not feasible, and with limited data, the ability to refine the distribution with prior data is a desirable ability. As a result, this method does not work well except in limited circumstances for infrastructure assessments (Bloetscher et al. 2005).

Bayesian Theory

Entropy is another solution that stems from theory by Thomas Bayes over 200 years ago, but not resurrected until Shannon's Information Theory and a theorem that he proved in 1948: "The probability distribution having maximum entropy (uncertainty) over any finite range of real values is the uniform distribution over that range" (Shannon 1949). Information theory is a branch of mathematical theory within probability and statistics that can be applied to a variety of fields (Kullback 1978). The roots of information theory are found within the concepts of disorder or entropy in thermodynamics and statistical mechanics (Shannon and Weaver 1949). Since the turn of the century, significant literature has been devoted to studying the mathematical form of information entropy (Shannon and Weaver 1949). The literature assumes a sample space S, with a series of events i, characterized by individual probabilities p_i. Shannon (1949) developed the following measure to indicate the information content of a probability distribution, which is in the form of the negative of the information content of the data set:

$$H = -\sum_i p_i \ln p_i$$

Definitive observations do not play an important part in information entropy theory since once the definitive observation is made, the underlying uncertainty is greatly reduced or eliminated. The uncertainty defines the confidence in the observation (Englehardt 1993).

By maximizing information entropy, the most conservative or broadest distribution consistent with the available information can be derived—such as the mean, variance, and range (Englehardt and Lund 1992). A uniform distribution is the least informative. In general, if the mean or expected value of any function $F(X)$ of a quantity is known, the entropy can be maximized subject to:

$$\sum_i p_i = 1$$

and

$$\sum_i F(X)p_i = E(F(X)) \text{ for all functions of } F$$

The concepts of information entropy are a useful theoretical underpinning in the application of Bayesian methods, which are useful in many aspects of the analysis. The principles of information entropy are addressed through the use of Bayesian methods. The selection of Bayesian methods assumes that the absolute or unconditional probability density function $p(x)$ on X is the underlying distribution found through curve-fitting. Its form, as defined by Aitcheson and Dunsmore (1975) is:

$$p(x) = \int p(x|\theta)p(\theta)d\theta$$

where $p(x)$ and $p(\theta)$ are completely different, and independent functions and the function $p(\theta)$ is the prior distribution. The Bayesian approach is to assume that while the true value of θ is unknown, there are probabilities that can be assigned for a series of possible values of θ (Aitcheson and Dunsmore 1975). More precisely it is assumed that $p(\theta)$ is a density function.

The posterior distribution incorporates observations from x into the sample space S. The outcome of the observations conveys additional information about the true content of S through a series of informed assumptions (Aitcheson and Dunsmore 1975). The basis for the information obtained is the influence of the proper distribution and the attachment of same to the possible distributions for x. Updating the plausibility in light of the observations of the prior, using Bayes notation leads to the posterior probability function (Aitcheson and Dunsmore 1975):

$$p(\theta|x) = \frac{p(x|\theta)p(\theta)}{p(x)}$$

where $p(x)$ can be shown to be equal to 1 and therefore neglected.

The information derived permits a plausible assessment on the outcome of future observations. The interest in these outcomes is a result of the prior information so that as new information is gathered, the predictive function can be refined. This is the benefit of the Bayesian approach: given uncertainty about a density function $p(y|\theta)$, some data can be deduced from the assessment of $p(\theta \mid x)$ over θ when the experimental results of x are known (Aitcheson and Dunsmore 1975).

Development of a predictive Bayesian model for incident assessments was originally proposed by Englehardt (1993). Predictive Bayesian methods include the development of a predictive distribution that evolves in shape in response to data (Bloetscher et al. 2005). Predictive Bayesian methods permit the use of subjective information with limited data availability. Uncertainties in the data are incorporated into the parameters of the prior distribution(s) so that the resulting equation incorporates the uncertainty (Bloetscher et al. 2005).

The benefit of Bayesian methods is that subjective information is rigorously integrated with available numeric data in the assessment. Press (1989) notes that advantages of the Bayesian approach include:

- Practical experience and subjectivity can be accounted for explicitly
- Uncertainty is factored into the analysis and the cumulative density function (CDF)
- Confidence intervals are small
- Can be applied where objective methods are required, while using vague priors
- Hypotheses can be tested without predetermining the outcome
- Forecasts are accurate, errors are small—if reasonable subjective information can be elicited.

Previous application of Bayesian inference for risk estimation performed by Jarabek and Hasselblad (1991), Englehardt (1993), Englehardt and Lund (1992), Bloetscher et al. (2005, 2019) has focused on the development of probability distributions for incident size and dose response functions. However, the concepts of Bayesian analyses can be incorporated into a condition assessment quite easily. Kleiner and Rajani (1999) used the concept of grouping infrastructure into periods to characterize condition and propensity for breaks to gain a perspective on condition, which is acceptable when lacking more detailed information. This tool is useful where some idea of development patterns and infrastructure information exist, without the need for all the mathematics. The following example will demonstrate this.

Example 10.3: Determining the Amount of Data that Is Useful

Assume that there are five arbitrary levels of condition available to analyze the asset—excellent/new, good, fair, poor, and failed. If there is an asset and there is no information about it, the condition could be any one of those (see Figure 10.9); the probability is 20% for each (the uniform distribution). Assume that one data point is known—that would change the analysis considerably (see Figure 10.10). Or what if the data were *sort of* known—say a probability that the asset was possibly good, or fair based on some factor? Then, the probability would be altered toward the good/fair condition—but, less so to the poor, failed, or excellent. Still there is uncertainty involved. This is precisely what Bayesian statistical methods are trying to get at. The assessor has a lot more data than one thinks even though much of it may not be known with complete certainty. The uncertainty is contained in the judgment of the assessor about certain factors.

continued

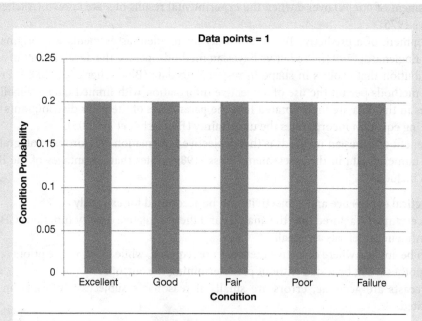

Figure 10.9   With only one data point—the answer to the question "What is the condition of the pipe?" is "It could be any condition"—all options are equally possible.

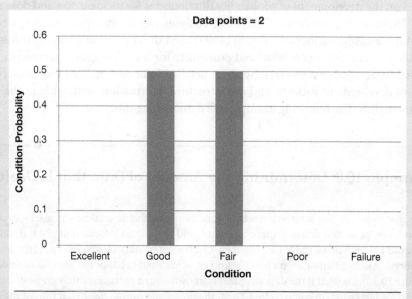

Figure 10.10   With two data points—the answer to the question "What is the condition of the pipe?" can be better known.

Continuing the example, most utilities have a pretty good idea about the pipe materials. Worker memory can be very useful, even if not completely accurate. In most cases the depth of pipe is fairly similar—the deviations may be known. Soil conditions may be useful and most soil information is

continued

readily available. Aggressive soil causes more corrosion in ductile iron (DI) pipe, so if the soil and pipe materials are used, one could identify where corrosion might shorten asset life. Bloetscher et al. (2016) worked on such an example, but suspected that it will be slightly different for each utility. Also, in smaller communities, many variables (DI pipe, PVC pipe, soil condition, etc.) may be so similar that differentiating would be unproductive.

Construction may have altered the soils; for example, muck and rock likely were replaced during construction with good fill. Tree roots will wrap around pipes, so their presence may indicate damage to the pipe. But no one can know this with certainty without digging up the pipe, something most communities would prefer to avoid. The presence of trees is easily noted from aerials. Roads with truck traffic create more vibrations, causing rocks to move toward the pipe and joints to flex. Most of these high traffic routes are well known. Another possible variable: the field perception—what do the field crews recall about breaks? Are there work orders? If so, do they contain the data needed to piece together missing variables that would be useful to add to the puzzle? So, with a little research there are many known variables.

To develop this concept further, assume there are nine variables for which data is developed. Each one has an assessment of adding to the excellent, good, fair, or poor condition of the pipe. These probabilities are added each time to build an understanding of overall condition. Figures 10.11 through 10.19 show how the graph changes as more information is accumulated. There is a big change from one to two points, but notice how from four to nine points the graph does not really change. This asset has a condition that is most likely good, maybe fair. It is probably not poor or excellent.

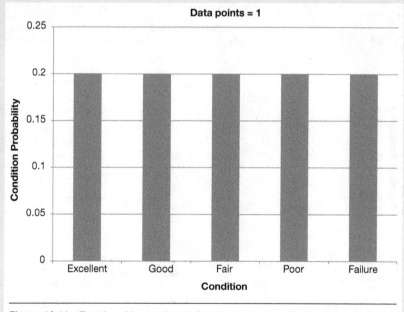

Figure 10.11   Results with one data point where data is unknown or inconclusive, all options are equally possible.

continued

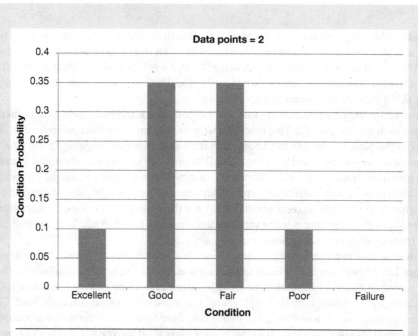

Figure 10.12 Results added together where there are two data points—where data is unknown or inconclusive for one, but better known for the second (good or fair).

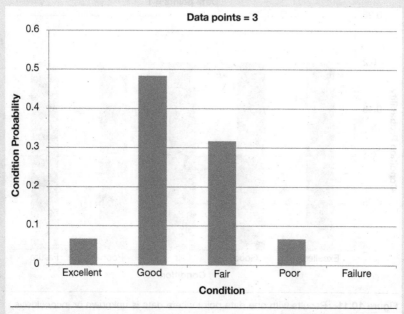

Figure 10.13 Results with three data points—by adding the third, it is suspected of being good.

continued

Figure 10.14   Results with four data points—by adding the fourth, it is suspected of being good or excellent.

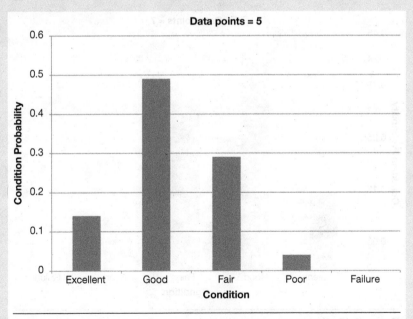

Figure 10.15   Results with five data points—by adding the fifth, it is suspected of being good or fair.

continued

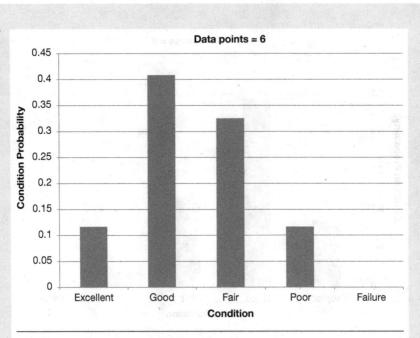

Figure 10.16 Results with six data points—by adding the sixth, it is suspected of being good or fair.

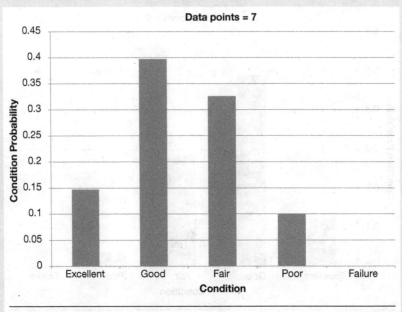

Figure 10.17 Results with seven data points—by adding the seventh, it is suspected of being good or excellent.

continued

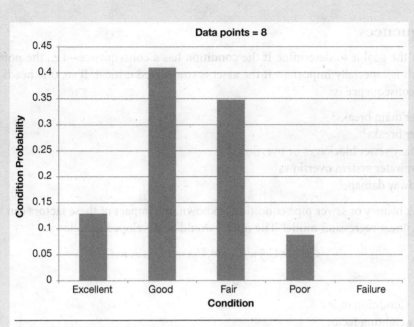

Figure 10.18 Results with eight data points—by adding the eighth, is it suspected of being good or fair.

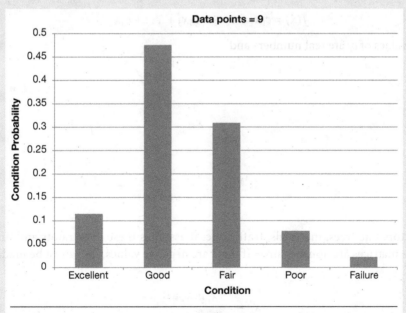

Figure 10.19 The results after adding a ninth data set still indicate that the asset is likely in good or fair condition

Consequences

Ultimately the goal is to determine if the condition has a consequence—i.e., the potential for failure. This is especially important if the asset is considered critical. If so, one needs to know what that consequence is:

- Water main breaks?
- Sewer breaks?
- Sanitary sewer blockages or overflows?
- Stormwater system overflows?
- Roadway damage?

If the break history or sewer pipe condition is known, the impact of these factors can be developed via a linear regression model. The model would be developed as follows:

$$CI = w_1 C_1 + w_2 C_2 + w_3 C_3 + w_4 C_4 + \ldots w_i C_i$$

where:

- CI = Condition index
- w = weighting factor
- C is condition factor

If one knows the consequence, the weights can be found:

$$f(x) = c_1 x_1 + c_2 x_2 + c_3 x_3 + \ldots + c_n x_n$$

where the values of c_n are real numbers and

$$x = \begin{matrix} x_1 \\ x_2 \\ x_3 \\ . \\ . \\ . \\ x_n \end{matrix}$$

are the factors line trees, materials, traffic, etc. It assumes these constraints and linear variables in the matrices are non-negative. If there are negative values, they must be made positive as follows:

$$x_i^+ = \begin{cases} x_i \ if \ x_i \geq 0 \\ 0 \ otherwise \end{cases}$$

$$x_i^- = \begin{cases} -x_i \ if \ x_i > 0 \\ 0 \ otherwise \end{cases}$$

Example 10.4: Identifying Vulnerable Pipe in a Small Water System

In this example, a water system was analyzed. The city has hundreds of separate sections of water mains from two inches to 12 inches (note this example will use 92 randomly chosen pipes from a larger distribution system). The assessment team was asked to analyze the water system. Their task was to develop a condition index for the water system; the assessors identified a number of factors where data could be estimated (Bayesian inference):

- Pipe size
- Materials
- Estimate age from aerials, as-builts, development patterns, etc.
- Condition based some on age, as-builts, field crews
- Soils—United States Department of Agriculture (USDA)—can assume better soil put back in
- Pressure (55 psi)
- Traffic (road it is on)
- Trees (using Google Earth or going out and looking)
- Rumor (what the crew members think)

Table 10.6 outlines a partial list of piping, which is an important asset for a water distribution system. An extended dataset is included, as shown in Table 10.7, which defines the following categorical variables:

Table 10.6   Partial list of assets for a water system (note: the full list may be hundreds or thousands of pipelines)

Asset	Location	Diameter
Water main	Alley SW 14th and 15th	2
Water main	Alley SW 13th and 14th	2
Water main	Alley SW 12th and 13th	2
Water main	Alley SW 10th and 11th Ct	2
Water main	SW 11th St	4
Water main	SW 8th St	6
Water main	SW 7th St	6
Water main	SW 6th St	6
Water main	SW 5th St	6
Water main	SW 4th St	6
Water main	SW 3rd St	6
Water main	SW 2nd St	6
Water main	SW 2nd Ave	8
Water main	SW 3rd Ave	8
Water main	SW 1st Ave	8
Water main	NW 1st St	8
Water main	NW 2nd St	8
Water main	NW 3rd St	8
Water main	SW 16th St	12
Water main	Main St	12
Water main	Water Plant to Main St	16
Water main	NW 5th St	8

continued

- Soil 1 = sand, 2 = muck
- Depth 1 = <6, 2 = >6
- Trees: 1 = no, 2 = yes
- Traffic: 1 = low, 2 = high
- Material: 1 = DI, 2 = galvanized iron (GI), 3 = polyvinyl chloride (PVC) C900, 4 = asbestos cement (AC), 5 = high density polyethlyene (HDPE)
- Estimated Condition: 1 = good, 2 = fair, 3 = poor

Table 10.7   Partial list of water distribution assets with attributes

Asset	Breaks in 10 Years	Diameter (in.)	Age (years)	Soil Type	Traffic Volume	Trees (2 = yes, 1 = no)	Depth (1 = under 6 ft., 2 over 6 ft.)	Pressure (psi)	Material	Est. Condition
Water main	17	2	45	1	1	2	1	55	4	3
Water main	11	2	45	2	1	2	1	55	4	3
Water main	12	2	45	1	1	2	1	55	4	3
Water main	10	2	45	1	1	2	1	55	4	3
Water main	2	4	50	1	1	2	1	55	1	2
Water main	3	6	60	2	2	2	1	55	1	2
Water main	1	6	60	2	2	2	1	55	1	2
Water main	1	6	60	2	2	2	1	55	1	2
Water main	0	6	20	1	1	2	1	55	3	1
Water main	0	6	20	1	1	2	1	55	3	1
Water main	0	6	20	1	1	2	1	55	3	1
Water main	2	6	20	1	1	2	1	55	3	1
Water main	0	8	20	1	1	2	1	55	3	1
Water main	0	8	20	1	1	2	1	55	3	1
Water main	0	8	60	1	1	2	1	55	1	1
Water main	0	8	60	1	1	2	1	55	1	1
Water main	3	8	60	1	1	1	1	55	1	2
Water main	1	8	60	1	1	1	2	55	1	1
Water main	2	12	60	1	2	1	2	60	1	2
Water main	2	12	60	1	1	1	1	65	1	2
Water main	0	16	10	1	1	1	1	65	1	1
Water main	3	8	50	1	1	1	1	55	1	2

Note: For *material* of pipe: 1 = DI; 2 = AC; 3 = PVC C900; 4 = GI; 5 = HDPE
For *soil type*: 1 = normal soil; 2 = acidic soil; 3 = muck; 4 = bedrock or caprock; 5 = sand
For *traffic volume*: 1 = residential street; 2 = heavy traffic; 3 = a railroad crossing
For *condition*: 3 = good; 2 = fail; 1 = poor

continued

Note that for each section, the number of breaks in the last 10 years was documented. Table 10.8 is a partial list of assets with the breaks noted. This is a raw data table. One significant issue is that

Table 10.8 Expansion of Table 10.7 where the numerical attributes of assets were converted to integers to permit the use of factorial and principal component analysis

Asset	Breaks in 10 Years	Dia.	Age	Sand	Clay	Low Traffic	Heavy Traffic	Trees	No trees	Shallow Under 6 ft. Bury	Deep Bury	Pressure	Ductile	AC	PVC	GI	HDPE
Water main	17	2	45	1	0	1	0	1	0	1	0	55	0	0	0	1	0
Water main	11	2	45	0	1	1	0	1	0	1	0	55	0	0	0	1	0
Water main	12	2	45	1	0	1	0	1	0	1	0	55	0	0	0	1	0
Water main	10	2	45	1	0	1	0	1	0	1	0	55	0	0	0	1	0
Water main	2	4	50	1	0	1	0	1	0	1	0	55	1	0	0	0	0
Water main	3	6	60	0	1	0	1	1	0	1	0	55	1	0	0	0	0
Water main	1	6	60	0	1	0	1	1	0	1	0	55	1	0	0	0	0
Water main	1	6	60	0	1	0	1	1	0	1	0	55	1	0	0	0	0
Water main	0	6	20	1	0	1	0	1	0	1	0	55	0	0	1	0	0
Water main	0	6	20	1	0	1	0	1	0	1	0	55	0	0	1	0	0
Water main	0	6	20	1	0	1	0	1	0	1	0	55	0	0	1	0	0

continued

the categorical variables noted cannot be mixed with descriptive variables in an analysis. So, each of these descriptive statistics must be modified to create absence or presence (a 1 or 0). Table 10.8 achieves this. The conversion to all descriptive statistics permits the creation of Table 10.9, which is a summary of standard statistics for the pipes, while Table 10.10 shows correlation between the factors. Using XLStat® software, an analysis was conducted using principle component analysis (PCA), a scree plot was developed for the factor groupings (see Figure 10.20), with factors and their components shown in Table 10.11. The first two factors have the most weight in explaining variance among variables, and each subsequent factor has less weight. Five factor groupings account for over 80 of the variances. Figure 10.21 shows a plot of the first two factors. Those variables within 45 degrees are correlated. Those at 180 degrees are inversely correlated. Little can be gleaned from this graph, which is not surprising given the nature of the data. Table 10.12 outlines the least squares approximation of the probability associated with the regression coefficient weights, which are shown in Figure 10.22.

Table 10.9 Summary statistics for the pipes

Variable	Observations (Obs.)	Obs. with Missing Data	Obs. without Missing Data	Minimum	Maximum	Mean	Std. Deviation
Diameter (in.)	93	0	93	1.000	16.000	5.011	3.255
Age (yrs.)	93	0	93	5.000	60.000	29.194	18.212
Sand	93	0	93	0.000	1.000	0.946	0.227
Clay	93	0	93	0.000	1.000	0.054	0.227
Low traffic	93	0	93	0.000	2.000	0.968	0.231
Heavy traffic	93	0	93	0.000	1.000	0.043	0.204
Trees	93	0	93	0.000	1.000	0.903	0.297
No trees	93	0	93	0.000	1.000	0.215	0.413
Shallow under 6 ft. of bury	93	0	93	0.000	1.000	0.849	0.360
Deep bury	93	0	93	0.000	1.000	0.032	0.178
Pressure (psi)	93	0	93	55.000	65.000	55.323	1.616
DI	93	0	93	0.000	1.000	0.419	0.496
GI	93	0	93	0.000	1.000	0.054	0.227
PVC	93	0	93	0.000	1.000	0.247	0.434
AC	93	0	93	0.000	1.000	0.065	0.247
HDPE	93	0	93	0.000	5.000	1.075	2.065

continued

Table 10.10 Correlation analysis for the factors developed for the water mains

Variables	Diameter (in.)	Age (yrs.)	Sand	Clay	Low Traffic	Heavy Traffic	Trees	No Trees	Shallow under 6 ft. Bury	Deep Bury	Pressure (psi)	Ductile	GI	PVC	AC	HDPE
Diameter (in.)	1	0.232	-0.043	0.043	-0.101	0.228	-0.482	0.499	-0.370	0.394	0.557	0.455	-0.295	0.298	-0.082	-0.648
Age (yrs.)	0.232	1	-0.326	0.326	-0.110	0.214	-0.376	-0.056	0.305	0.143	0.064	0.627	0.011	-0.463	0.229	-0.410
Sand	-0.043	-0.326	1	-1.000	0.589	-0.654	-0.078	0.125	-0.100	0.044	0.048	-0.087	0.057	0.137	-0.326	0.125
Clay	0.043	0.326	-1.000	1	-0.589	0.654	0.078	-0.125	0.100	-0.044	-0.048	0.087	-0.057	-0.137	0.326	-0.125
Low traffic	-0.101	-0.110	0.589	-0.589	1	-0.894	-0.204	0.188	-0.059	0.026	0.028	-0.165	0.033	0.081	0.037	0.074
Heavy traffic	0.228	0.214	-0.654	0.654	-0.894	1	0.069	-0.111	-0.059	0.261	0.122	0.249	-0.051	-0.122	-0.056	-0.111
Trees	-0.482	-0.376	-0.078	0.078	-0.204	0.069	1	-0.625	0.066	-0.352	-0.500	-0.385	0.078	0.188	0.086	0.171
No trees	0.499	-0.056	0.125	-0.125	0.188	-0.111	-0.625	1	-0.731	0.201	0.302	0.033	-0.125	0.367	-0.137	-0.274
Shallow under 6 ft. of bury	-0.370	0.305	-0.100	0.100	-0.059	-0.059	0.066	-0.731	1	-0.434	-0.103	0.175	0.100	-0.525	0.111	0.220
Deep bury	0.394	0.143	0.044	-0.044	0.026	0.261	-0.352	0.201	-0.434	1	0.342	0.215	-0.044	-0.105	-0.048	-0.096
Pressure (psi)	0.557	0.064	0.048	-0.048	0.028	0.122	-0.500	0.302	-0.103	0.342	1	0.236	-0.048	-0.115	-0.053	-0.105
Ductile	0.455	0.627	-0.087	0.087	-0.165	0.249	-0.385	0.033	0.175	0.215	0.236	1	-0.203	-0.487	-0.223	-0.445
GI	-0.295	0.011	0.057	-0.057	0.033	-0.051	0.078	-0.125	0.100	-0.044	-0.048	-0.203	1	-0.137	-0.063	-0.125
PVC	0.298	-0.463	0.137	-0.137	0.081	-0.122	0.188	0.367	-0.525	-0.105	-0.115	-0.487	-0.137	1	-0.151	-0.300
AC	-0.082	0.229	-0.326	0.326	0.037	-0.056	0.086	-0.137	0.111	-0.048	-0.053	-0.223	-0.063	-0.151	1	-0.137
HDPE	-0.648	-0.410	0.125	-0.125	0.074	-0.111	0.171	-0.274	0.220	-0.096	-0.105	-0.445	-0.125	-0.300	-0.137	1

continued

Table 10.11   PCA factors

Variable	F1	F2	F3	F4	F5
Diameter (in.)	0.553	0.699	0.117	0.077	−0.166
Age (yrs.)	0.638	−0.040	−0.537	0.324	−0.076
Sand	−0.710	0.489	−0.306	−0.159	−0.214
Clay	0.710	−0.489	0.306	0.159	0.214
Low traffic	−0.619	0.422	−0.375	0.334	0.203
Heavy traffic	0.759	−0.283	0.346	−0.394	−0.069
Trees	−0.340	−0.628	0.395	0.009	−0.296
No trees	0.106	0.789	0.274	0.109	0.190
Shallow under 6 ft. of bury	−0.002	−0.611	−0.630	0.034	−0.120
Deep bury	0.325	0.453	−0.024	−0.336	0.297
Pressure (psi)	0.328	0.502	−0.177	−0.260	0.294
Ductile	0.616	0.213	−0.540	−0.104	−0.378
GI	−0.148	−0.162	−0.114	0.064	−0.154
PVC	−0.265	0.357	0.748	0.236	−0.252
AC	0.145	−0.269	−0.003	0.659	0.516
HDPE	−0.467	−0.382	−0.072	−0.552	0.495

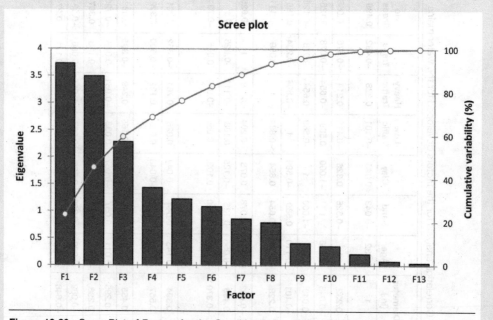

Figure 10.20   Scree Plot of Factors for the Condition Index. From this Scree plot, the first 5 factors (F1-F5) account for 80% of the variability in the data.

continued

Table 10.12   Type I—least squares analysis

Source	DF	Sum of Squares	Mean Squares	F	Pr > F
Diameter (in.)	1	18.251	18.251	13.956	0.000
Age (yrs.)	1	112.600	112.600	86.098	< 0.0001
Sand	1	11.300	11.300	8.640	0.004
Clay	0	0.000			
Low traffic	1	10.454	10.454	7.994	0.006
Heavy traffic	1	3.136	3.136	2.398	0.126
Trees	1	2.338	2.338	1.787	0.185
No trees	1	0.876	0.876	0.670	0.415
Shallow under 6 ft. of bury	1	9.483	9.483	7.251	0.009
Deep bury	0	0.000			
Pressure (psi)	1	16.176	16.176	12.369	0.001
Ductile	1	100.485	100.485	76.834	< 0.0001
GI	1	35.060	35.060	26.808	< 0.0001
PVC	1	24.698	24.698	18.885	< 0.0001
AC	1	352.104	352.104	269.231	< 0.0001
HDPE	0	0.000			

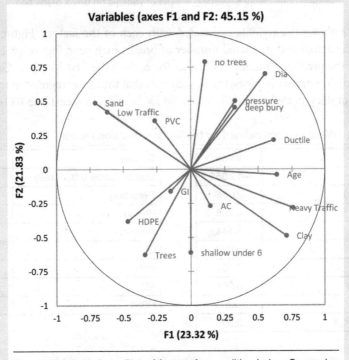

Figure 10.21   Varimax Plot of factors for condition index. Comparing the first two factors, those within 45 degrees of one another are correlated; the more variability, the longer the line.

continued

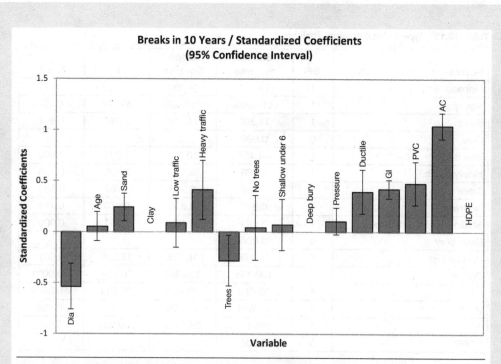

Figure 10.22   Standard coefficients explaining impact of variable on break prediction.

Table 10.13 shows the weights associated with each of the factors. Figures 10.23 and 10.24 show the actual and predicted number of breaks each year. Figure 10.25 shows the deviation for the predicted number of breaks for each of the first 93 pipes. Given limited data and a limited number of pipes, the fit indicates that the linear regression model can be used to predict the pipes where breaks are most likely to occur (see Table 10.14).

Table 10.13   Weight values defined by the linear regression model

Source	Value	Source	Value
Intercept	−12.355	Shallow under 6 ft. of bury	0.580
Diameter (in.)	−0.489	Deep bury	0.000
Age (yrs.)	0.008	Pressure (psi)	0.194
Sand	3.144	Ductile	2.342
Clay	0.000	GI	5.473
Low traffic	1.151	PVC	3.229
Heavy traffic	5.961	AC	12.428
Trees	−2.819	HDPE	0.000
No trees	0.297		

continued

Table 10.14 Predictions (partial table) on the number of likely breaks in the coming 10 years

Asset	Breaks in 10 Years	Predicted (Breaks in 10 Years)	Residual	Std. Residuals	Studentized Residuals	Std. Dev. on Pred. (Mean)	Lower Bound 95% (Mean)	Upper Bound 95% (Mean)	Std. Dev. on Pred. (Obs.)	Lower Bound 95% (Obs.)	Upper Bound 95% (Obs.)
Water main	17.000	12.193	4.807	4.204	4.913	0.592	11.014	13.371	1.288	9.629	14.756
Water main	11.000	9.049	1.951	1.706	2.331	0.779	7.498	10.600	1.384	6.295	11.803
Water main	12.000	12.193	-0.193	-0.169	-0.197	0.592	11.014	13.371	1.288	9.629	14.756
Water main	10.000	12.193	-2.193	-1.917	-2.241	0.592	11.014	13.371	1.288	9.629	14.756
Water main	2.000	1.172	0.828	0.724	0.749	0.293	0.589	1.755	1.180	-1.178	3.522
Water main	3.000	1.945	1.055	0.922	1.110	0.636	0.679	3.212	1.309	-0.659	4.550
Water main	1.000	1.945	-0.945	-0.827	-0.995	0.636	0.679	3.212	1.309	-0.659	4.550
Water main	1.000	1.945	-0.945	-0.827	-0.995	0.636	0.679	3.212	1.309	-0.659	4.550
Water main	0.000	0.826	-0.826	-0.722	-0.756	0.337	0.155	1.497	1.192	-1.547	3.199
Water main	0.000	0.826	-0.826	-0.722	-0.756	0.337	0.155	1.497	1.192	-1.547	3.199

continued

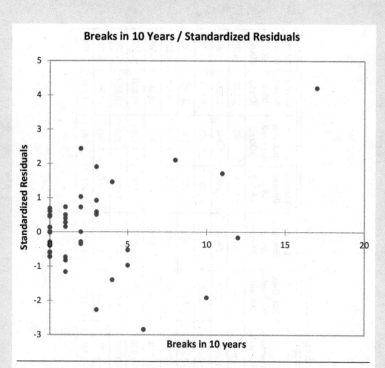

Figure 10.23   Prediction of breaks based on factors from linear regression.

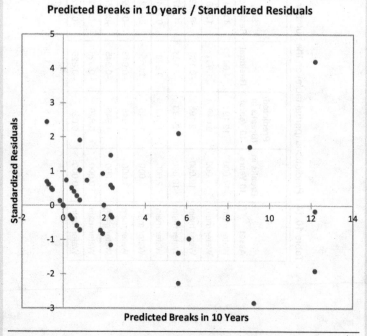

Figure 10.24   Comparison of *actual* versus *predicted* breaks over a ten year period.

continued

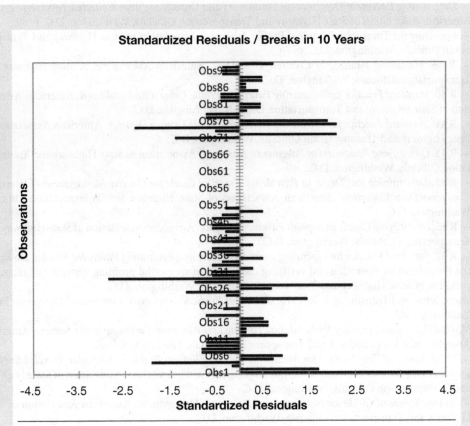

Figure 10.25 For the first 93 pipes, this graph shows the variability toward the likelihood of a break occurring.

Note that the biggest issue with Example 10.4 is not the lack of data. The Bayesian mindset can be used to create the data. The biggest barrier is useful data on consequences. This is where having a work order system to track information is vital to establishing a condition assessment. Condition information is useful, but in the absence of the consequence, it reveals little. A pipe can be old, but still in good condition and a low risk for failure. In Example 10.4, it was the AC pipes that were significant problems. Without knowing that the breaks occurred consistently on the AC pipe from work order data—and the frequency of the breakage—the condition assessment is just data that may or may not reveal anything about the true condition, and therefore increase the risk of failure of the buried asset.

REFERENCES

AASHTO. M 328. *Standard Specification for Inertial Profiler.* American Association of State Highway and Transportation Officials, Washington, D.C.

———. *Quantifying Cracks in Asphalt Pavement Surfaces from Collected Images Utilizing Automated Methods.* American Association of State Highway and Transportation Officials, Washington, D.C. p. 67.

———. *Collecting Images of Pavement Surfaces for Distress Detection.* American Association of State Highway and Transportation Officials, Washington, D.C. p. 68.

———. *Determining Pavement Deformation Parameters and Cross Slope from Collected Transverse Profile.* American Association of State Highway and Transportation Officials, Washington, D.C. p. 69.

———. *Collecting the Transverse Pavement Profile.* American Association of State Highway and Transportation Officials, Washington, D.C. p. 70.

———. R 36. *Evaluating Faulting of Concrete Pavements.* American Association of State Highway and Transportation Officials, Washington, D.C.

———. R 40. *Standard Practice for Measuring Pavement Profile Using a Rod and Level.* American Association of State Highway and Transportation Officials, Washington, D.C.

———. R 41. *Standard Practice for Measuring Pavement Profile Using a Dipstick.* American Association of State Highway and Transportation Officials, Washington, D.C.

———. R 43. *Quantifying Roughness of Pavements.* American Association of State Highway and Transportation Officials, Washington, D.C.

———. R 48. *Determining Rut Depth in Pavements. Practical Guide for Quality Management of Pavement Condition Data Collection.* American Association of State Highway and Transportation Officials, Washington, D.C.

———. R 55. *Quantifying Cracks in Asphalt Pavement Surface.* American Association of State Highway and Transportation Officials, Washington, D.C.

———. R 57. *Standard Practice for Operating Inertial Profilers and Evaluating Pavement Profiles.* Describes the procedures for operating and verifying calibration of an inertial profiling system. American Association of State Highway and Transportation Officials, Washington, D.C.

Aitchison, John, and Dunsmore, I. R. 1975. *Statistical Prediction Analysis.* Cambridge University Press, Cambridge, MA.

ASTM. D6433. *Standard Practice for Roads and Parking Lots Pavement Condition Index Surveys.* American Association of State Highway and Transportation Officials, Washington, D.C.

———. E950. *Standard Test Method for Measuring the Longitudinal Profile of Vehicular Traveled Surfaces with an Accelerometer Established Inertial Profiling Reference.* American Association of State Highway and Transportation Officials, Washington, D.C.

———. E1166. *Standard Guide for Network Level Pavement Management.* American Association of State Highway and Transportation Officials, Washington, D.C.

———. E1926. *Standard Practice for Computing International Roughness Index from Longitudinal Profile Measurements.* American Association of State Highway and Transportation Officials, Washington, D.C.

———. E1656. *Standard Guide for the Classification of Automated Pavement Condition Survey Equipment.* American Association of State Highway and Transportation Officials, Washington, D.C.

———. E1703. *Test Method for Measuring Rut-Depth of Pavement Surfaces Using a Straightedge.* American Association of State Highway and Transportation Officials, Washington, D.C.

———. D6433. *Standard Practice for Roads and Parking Lots Pavement Condition Index Surveys.* American Association of State Highway and Transportation Officials, Washington, D.C.

BCMoTI. 2012. *Pavement Surface Condition Rating Manual.* British Columbia Ministry of Transportation and Infrastructure (MoTI). Victoria, B.C.

Bloetscher, F. 2019. Using Predictive Bayesian Monte Carlo—Markov Chain Methods to Provide a Probablistic Solution for the Drake Equation. *Asta Astronautica*, Vol. 155, pp. 118–130. 10.1016/j .actaastro.2018.11.033.

Bloetscher, F. et al. 2005. "Comparative assessment municipal wastewater disposal methods in Southeast Florida." *Water Environment Research.* Vol. 77, pp. 480–490.

Bloetscher, F. et al. 2016. "Developing asset management systems—You know more than you think." *2016 Florida Section AWWA conference, Orlando, FL.* FSAWWA, St. Cloud, FL.

Ellison, D. et al. 2014. "Answers to challenging infrastructure management questions." Report #4367 WRF, Denver, CO.

Englehardt, J. and Lund, J. 1992. "Information theory in risk analysis." *Journal of Environmental Engineering.* American Society of Civil Engineers, Vol. 118, No. 6, pp. 890–904.

Englehardt, J. D. 1993. "Pollution prevention technologies: A review and classification." *Journal of Hazardous Materials.* 35, 119–150. doi:10.1016/0304-3894(93)85027-C.

Haas, C. N., Rose, J. B., and Gerba, C. P. 1999. *Quantitative Microbial Risk Assessment*. John Wiley and Sons, New York, NY.

ISO. 2008. *Principles and Terminology: The Concepts and Principles Associated with the Measure Approach* International Organization for Standardization. Geneva, Switzerland.

Jarabek, A. and Hasselblad, V. 1991. "Inhalation reference concentration methodology: Impact of dosimetric adjustments and future directions using the confidence profile method." *Presented at: 84th Annual Meeting and Exhibition of the Air and Waste Management Association, June, Vancouver, B.C., Air and Waste Management Association*, paper no. 91-173.3.

Kleiner, Y. and Rajani, B. B. 1999. "Using limited data to assess future need." *JAWWA*, 91(7), 47–62.

Kullback, S. 1978. *Information Theory and Statistics*. Peter Smith, Gloucester, MA.

MNDOT. 2003. *Minnesota DOT Distress Identification Manual*. MDOT, Minneapolis, MN.

MTC. 2002. *Pavement Certification Program*. Metropolitan Transportation Commission, San Francisco, CA.

NCDOT. 2010. *North Carolina DOT Pavement Condition Survey Manual*. NCDOT, Raleigh, NC.

NDOR. 2002. *Nebraska DOR Pavement Maintenance Manual*. NDOT, Lincoln, NE.

ODOT. 2010. *Pavement Distress Survey Manual*. ODOT, Salem, OR.

Pierce, L. M., McGovern, G., and Zimmerman, K. A. 2013. *Practical Guide for Quality Management of Pavement Condition Data Collection*, U.S. Department of Transportation Federal Highway Administration, 1200 New Jersey Avenue, SE, Washington, D.C.

Press, S. J. 1989. *Bayesian Statistics: Principles, Models and Applications*. John Wiley & Sons, Inc., New York, NY.

Sanford, J. 2013. "Asset Management and Condition Assessment Best Practices." Powerpoint Presentation, 2/28/2013. http://ch2mhillblogs.com/water/wp-content/uploads/2013/04/Presentation_condition assessment_Sanford.pdf. Accessed 11/13/16.

Shannon, C. E. 1949. "A mathematical theory of communication." *The Bell System Technical Journal*. Vol. 27, pp. 379–423, 623–656.

Shannon, C. E. and W. Weaver, eds. 1949. *The Mathematical Theory of Communication*. The University of Illinois Press, Urbana, IL.

Shekharan, R. et al. 2006. *The Effects of a Comprehensive QA/QC Plan on Pavement Management*. Transportation Research Board, Washington, D.C.

Stadnyckyj, M. 2010. "Condition assessment: Bridging the gap between pipeline investments and risk reduction." *Water Utility Infra Management*. E-Newsletter. May 3, 2010.

Teicholz, E. and Evans, G. (no date). *Theoretical Condition Indices*. http://www.graphicsystems.biz/gsi/articles/Condition%20Indices%20and%20Strategic%20Planning0707.pdf. Accessed 6/15/16.

TXDOT. 2010. *Texas DOT Pavement Management Information System Rater's Manual*. TDOT, Austin, TX.

UDOT. 2009. *Utah DOT Pavement Preservation Manual—Part 2, Pavement Condition Data*. UDOT, Salt Lake City, UT.

United States Environmental Protection Agency (EPA). 2002. *The Clean Water and Drinking Water Infrastructure Gap Analysis*. EPA, Office of Water, EPA-816-R-02-020. EPA, Washington, D.C.

———. 2017. Effective Utility Management A Primer for Water and Wastewater Utilities, EP-C-11-009 with the Office of Wastewater Management at U.S. EPA.

Zimmerman, H. J. 1985. *Fuzzy Set Theory and Its Applications*. Kluwer Academic Publishers, Boston, MA.

11

FINANCIAL RESOURCES FOR REPAIR AND REPLACEMENT

Once an understanding of the need for maintenance, repairs, and replacement has been obtained, financing these improvements is the next step. Obtaining revenues to address failing infrastructure issues starts with understanding the environment in which most infrastructure systems exist. The majority of infrastructure agencies serve small locales, although there are some very large ones serving a sizable sector of the population. The plethora of small communities providing services provides an indication of the population at risk, since smaller communities have less access to borrowing, less ability to respond quickly to expenditure needs, and less ability to raise taxes, rates, fees, or other charges to pay for those costs. To wit, there are 19,492 incorporated municipalities in the United States according to the 2010 census, but only 257 of those are larger than 100,000 people. These 257 communities serve nearly half the population of the United States. Another 1,607 municipalities have between 10,000 and 99,999 people, and 2,752 more have between 2,500 and 9,999 people. Another 3,593 municipalities serve between 1,000 and 2,499 people, and nearly half, 9,283, serve under 1,000 people (U.S. Census 2010). In comparison, there are 3,142 counties and county equivalent administrative units, including the District of Columbia, although in Alaska, most of the land area of the state has no county-level government. Half of the U.S. population lives in only 146 of these 3,142 counties, according to the 2010 Census.

Community water and sewer systems are similarly distributed, and because many counties are so large, they may have several water and sewer systems located within them. Community water systems are defined as systems that serve at least 15 service connections or serve an average of at least 25 people for at least 60 days a year. The United States Environmental Protection Agency (EPA) breaks the size of systems down as follows:

- *Very small* water systems serve 25–500 people
- *Small* water systems serve 501–3,300 people
- *Medium* water systems serve 3,301–10,000 people
- *Large* water systems serve 10,001–100,000 people
- *Very large* water systems serve 100,001+ people

In 1960, there were about 19,000 community water utilities in the United States according to a National Research Council report published in 1997 (NRC 1997). About 80% of the U.S. population was served by those 19,000 systems, but approximately 16,700 of those water systems served communities with populations of less than 10,000 (NRC 1997). Over half of the population was served by the 500 largest systems. By 1993 the number of systems had more than tripled—to 54,200 systems. Updating these numbers in 2015, there are still over 54,000 systems in the United States (three per political jurisdiction), and growth is almost exclusively in the very small sector. Today, 93% of systems serve fewer than 10,000 people; 84% serve fewer than 3,300 people and account for only 10% of the population. Most critical is the 30,000 new, very small systems that serve only 5 million people (averaging 170 per system). In contrast, the very large systems currently serve 45% of the population. Large and very large systems together account for 80% of the population, a number that mirrors 1960. Growth in large systems continues—they added 90 million people to their service areas, but the sheer number of new systems is at the micro level.

What is the point of these statistics? Whether water or wastewater systems, roadways, or stormwater assets located in cities, counties, or other districts, the point is that the infrastructure relied on daily by the citizenry is increasingly located in small systems with limited revenue sources. That portends poorly for maintaining infrastructure systems at the micro-level. While the large and very large population centers have the ability to raise funds to deal with infrastructure needs (as they have historically), arranging for funding may be a significant issue for smaller, rural systems that have grown up using federal funds over the past 50 years. This statement applies to general purpose local governments as well as water and sewer utilities. The problem will manifest itself when the infrastructure in these rural communities starts to come to the end of its useful life. Given dwindling federal funds, rural customers, who are already making 20% less income than their urban counterparts, with higher unemployment, and who are used to very low rates and taxes that generally do not account for replacement funding, will experience major sticker shock when the bill comes due. As noted, average pipeline costs $100 or more per foot to install in rural communities. In an urban area with say, 60-foot wide lots, that is $6,000 per household. In rural communities, the residents may be far more spread out. As an example, for a system in eastern North Carolina, a two-mile loop serves 50 houses—that is a $1.05 million pipeline for 50 houses or $21,000 per house. Another mile-long loop serves 30 houses located in a rural setting—that is over $16,000 per house. The cost for rural installations is significantly higher than that for urban communities. Roadways, drainage systems, and sewer systems have very similar cost differences between rural and urban settings—one of the main reasons being that rural drainage is almost exclusively related to ditches and pipes at driveways or intersecting roadways and sewer systems are primarily septic tanks.

In addition to funding limitations, expenses are curtailed so employees with greater expertise are likely to move to larger systems that offer better salaries. The result is that the large number of relatively small communities may not have the operating expertise, financial and technological capability, or economies-of-scale to provide services or raise capital to upgrade or maintain their infrastructure. Small systems generally have less in-house expertise and fewer resources to acquire needed expertise, which compounds the problem.

DEFINING THE COSTS OF DOING BUSINESS

Every infrastructure agency will spend money to operate and maintain the entity. They will also spend money on debt and capital. These factors are brought together in annual budget documents. Budgets are a necessary part of operations and are statutorily required for most jurisdictions. In most cases, all infrastructure agencies should be set up as an enterprise fund in order to allow the organization to pay its own way, which will also make it easier to evaluate the operational aspects of an infrastructure system. Water, sewer, and stormwater funds should be separated to allow officials to understand the allocation of funds. The exception is roads, which may be funded with state-shared gas taxes and other revenues that are not locally generated.

Budgeting is a process, and the input of field staff is important to consider. To properly manage the system, line managers use field staff input to create an expenditure plan. The budget is then reviewed by supervisors, department directors, a budget/finance person, and the management of the infrastructure entity. The line managers and their staff are in a position to know what the operations and maintenance manuals suggest for preventive maintenance; which pipes, pumps, or bridges need attention; what roads need repairs; and which stormwater systems are not working properly. They will also know what equipment is best put in the field to improve workforce efficiency. It is useful to have work order data to support their budget requests. As a result, line personnel usually submit a budget that exceeds the *target* set by the budget person, manager, or local officials. This is where the difficulties can start.

Numerous methods of budgeting have been developed for the public sector over the years. Incremental budgeting is used to account for each increment of growth. It adds to the prior year appropriate in the projected growth increments. A variant is *line item budgeting* which simply looks at historical costs and projects them; it is also easy to do. *Zero based budgeting* develops the budget from nothing, adding services and costs to create the budget. Each year, every activity must be justified during the budget development process. It takes time and effort to create all of the programs and is less useful for infrastructure than social services. Zero-based budgeting was popular in the 1970s, but the amount of information needed makes it difficult to use with infrastructure systems. *Performance budgeting* looks at outcomes and assumes increments of work can be developed, which is hard to do with infrastructure systems, but may be better than zero-based budgets. Performance budgeting does provide more useful information for legislative consideration and for evaluation by administrators (NCES 2003). *Outcome focused budgeting* attempts to deal with issues associated with fiscal austerity and competition for limited funds. The focus on outcomes—again difficult to apply to infrastructure—is better suited to social services.

The budget is a planning tool and an organizational tool. Although limitations on tax rates may be statutorily set in stone—water and sewer rates, stormwater, and various other fees and charges for services rarely are. To add to the difficulties, many local officials, finance personnel, and managers have trouble understanding that the budget is simply a plan, and especially with an enterprise fund, the plan needs to be adjusted to accommodate unforeseen conditions.

Table 11.1 is an example of an operating budget for a water utility (with prior years shown) that does not include capital. What's also important to note is that for many of the items, a simple increase in percentage each year is not appropriate. Past costs may be indicative to some extent of future costs, but at the same time, one can note certain line items within this budget that have varied significantly, mostly in the maintenance area. For example, in a small system, if a pump goes down, the pump needs to be repaired, but multiple pumps in a given year may not be planned for. If no pumps break, the line item will look unspent. That may mean the money is taken away in the budget process—then it breaks. The process becomes more difficult when the number of pumps, pipes, bridges, etc., is limited because the probability of any one of them failing is low, but the cost is high. The high number is rarely included in a budget.

By contrast, in larger systems, while the probability of any given item failing is small, there are so many of them that better budgeting can be employed and costs are more consistent. It is similar to a homeowner who has to budget for a new roof. Typically, a homeowner will replace a roof every 30 years. But what if another person owned 60 houses? Then, on average, the owner would replace two roofs a year. So, the single homeowner has larger expenses

Table 11.1 Operations budget for a water utility

Description	Actual 2015	Actual 2016	Estimated 2016	Budget 2017
Water Distribution				
Regular salaries and benefits	$ 435,026	$ 391,915	$ 399,274	$ 487,814
Overtime			$ —	$ —
Longevity pay			$ —	$ —
Accrued leave buyback			$ —	$ —
FICA taxes			$ —	$ —
Medicare taxes			$ —	$ —
Retirement contributions			$ —	$ —
Life and health insurance			$ —	$ —
Worker's compensation	$ —	$ 15,580	$ 25,933	$ 25,023
Professional service	$ 49,416	$ 17,649	$ 15,000	$ 15,000
Other	$ 53,642	$ —	$ 10,000	$ —
Travel and per diem			$ 2,000	$ 1,000
Telephone	$ 4,400	$ 4,511	$ 4,700	$ 1,000
Operating equipment	$ 3,961	$ 2,180	$ 8,000	$ 5,000
Utility fees/services	$ 2,508	$ 15,277	$ 1,000	$ 1,000
Miscellaneous rentals			$ —	$ —
Mtn. of equipment	$ 24,000	$ 22,300	$ 47,000	$ 44,000
Mtn. of communication/MIS equip.	$ 8,000	$ 5,600	$ 4,000	$ 6,000
Mtn. of vehicles	$ 36,800	$ 42,000	$ 28,000	$ 28,000
Mtn. of buildings			$ —	$ —
Training	$ —	$ 1,755	$ 5,500	$ 2,000
Mtn. of utilities	$ 337,346	$ 32,082	$ 71,750	$ 55,200
Gasoline	$ 8,854	$ 5,845	$ 20,100	$ 28,900
Miscellaneous supplies	$ 1,410	$ 1,870	$ 4,000	$ 4,000
Uniforms	$ 6,487	$ 8,002	$ 10,380	$ 10,500
Office supplies	$ 360	$ 690	$ 2,500	$ 2,500
Books, subscriptions, publications, membership	$ 1,876	$ —	$ 300	$ 1,000
Depreciation/R&R	$ 78,249	$ 5,970	$ 85,000	$ 20,000
Total Water Distribution	$ 1,052,335	$ 573,226	$ 744,437	$ 737,937

periodically, while the owner of multiple houses will have consistent expenses for roof replacement each year.

Budgeting to revenues works with individuals—but not infrastructure. Things may cost more because the infrastructure is deteriorating, so managers need to avoid the common practice used too often within local governments in times of stress where the instructions are that "everyone needs to curtail expenses by a certain percentage." That strategy ignores service demands. People will use however much water they need—the same with sewer. Stormwater and roads are expected to function without inconveniencing residents. One cannot turn off traffic lights one week a year to save power costs for an artificial budget goal, or cut power at water plants without some indication that water use will be substantially less. Nor can the number of water plant operators be easily cut because there are rules that require a certain number of operators at a plant. Few plants have extra operators hanging about. As a result, when imposing artificial budget constraints, maintenance and replacement are the common cuts. Across-the-board cuts to meet artificial objectives is not the way to budget.

A recent Water Resource Foundation (WRF 2014) report notes that operating expenses constitute the majority of all expenses for utilities, including the somewhat intangible cost of current asset depreciation and amortization. For most utilities, fixed costs far exceed the variable costs, particularly when debt service payments and other capital expenditures are included in the calculation. However, the median annualized increase in operation and maintenance expenses dropped from +5.8% per year between 2004 and 2010 to +1.0% per year between 2010 and 2012 (WRF 2014). Prior to 2012, expenses were generally rising faster than operating revenues—only recently did the trend reverse. It is not because the assets stopped deteriorating.

The reduction in expenditures ignores the fact that local infrastructure agencies will always be undertaking the construction of capital projects to develop and/or continue safe and reliable service and will therefore need funds for those projects. For efficient operation, these new facilities must be developed in accordance with the latest technical and professional standards to protect the health, safety, and welfare of the customers served now and in the future. Some improvements may be constructed by developers to upgrade the system to serve their projects. Other large capital projects must be constructed internally using staff, or externally using consultants and contractors to meet long-term goals.

Once the budget negotiation process is complete, the budget will be adopted by a resolution of the governing board and can only be modified through amendments approved by the governing board. However, the mechanism for approving the budget should not make each line item a not-to-exceed amount that needs to be modified through governing board action. In this manner, operations managers and personnel can move costs between line items, many of which are small, without having to go through the process of gaining approval for moving monies within a given budget. Capital should be part of this annual budgeting process—but located in a nonoperating budget.

Considering budgets—especially capital budgets—what needs to happen is that operations staff supervisors need to figure out what is needed to do their assignments, and then figure out the budget strategy to get it. Does the staff pad their budget to ensure that the budget office doesn't arbitrarily cut their request, because *that's what they do*? Do elected officials delay capital projects because it is an election year and they do not want to raise rates? Does the city manager delete the proposed new hires because he or she needs more money to be diverted to the general fund? Do these questions sound familiar? Welcome to the game that plays out every budget cycle because in most cases, there is no long-term plan or vision and fiscal policies are limited.

In comparison, when running a private business, people are expected to know what they need to do the job. They should be able to ask for what they need and get it without a lot of conflict. The comparison indicates that budget and finance directors of utility/governmental entities should be support positions, not gatekeepers. Their job is to find money to pay for operations. Operations should set the need, and budget and finance should be tasked to find the funds, but too often it doesn't work that way. The budget battle is a huge expense for every community, and one that largely provides no real benefit but instead detracts from productivity. Yet, it is funny that over time, governing bodies that are hiring city managers have moved away from people with technical backgrounds in public works and public administration and leaned toward people with business experience. The argument is that we need to *run the city more like a business*, so this should be a good fit. The conflict within this thought process is that government provides services that most private sector companies do not find to be cost effective. Businesses are designed to make a profit to reward shareholders, while governments provide social services. Those two issues are often mutually exclusive goals that cannot be resolved. Either profits suffer, or service suffers. In government, the latter cannot without major consequences.

Local Funding/Expenditures

Aside from borrowing money, it is necessary that the organization develop a means to create revenues sufficient for the operation and maintenance requirements of the system. It suggests that budgeting should start with the expenditures, and derive revenues to meet those needs, as opposed to projecting revenues and limiting the budget to the revenues. The latter is a particular favorite of local officials, elected and appointed, but often serves to delay the infrastructure needs of the system, creating deferred maintenance obligations. Part and parcel to the budgeting process, monies should be allocated to provide for capital costs associated with renewing the current infrastructure investments on an annual basis.

Taxes, User Fees, Charges, and Other Revenues

Repair and replacement funds are replenished or funded with taxes, fees, and charges. Gas taxes are typically used for roadways, but may supplement stormwater and transit. They are rarely used for water and sewer—in fact, most financial policy guidelines produced by the Governmental Accounting Standards Board (GASB) or the Government Finance Officer's Association (GFOA) discourage this practice. Repair and replacement funds are normally collected from existing customers to repair and/or replace the existing infrastructure at the requisite time. These customers create the wear, thus, they should bear the cost to repair/replace them. These collections are normally assessed as periodic service charges. Periodic charges for service are the costs collected on a regular basis from existing customers for the amount of service they receive. Each of these fees should have a basis for the charge generally consistent with the financial policy of the system. The case law defining the employment of user fees varies from state to state, but is underlain by the basic concept of fairness. Any fees imposed on constituents must not only be reasonable, they must be nondiscriminatory, although different user classes can be charged differently, provided a valid rationale exists for the difference.

Water and sewer systems are almost always set up as enterprise funds. They should bill for all costs, including operations, maintenance, administrative costs, and capital. Water and sewer systems bill based on customer water usage (monthly, bi-monthly, or quarterly—although an argument could be made that monthly billing not only collects revenues faster but also permits better tracking of unaccounted-for water). The base bill utilizes single-family users, and equivalent residential connections or units (ERCs or ERUs) to bill other customers. Note that water and sewer utilities collect money after the product has been consumed. Revenues flow 45 to 105 days after water use.

Stormwater utilities that are not funded with taxes are also commonly funded by a monthly or annual fee. User fees are becoming more the norm as municipalities move way from funding stormwater systems with property taxes. Determining which is preferable has more to do with local preferences and statutory limitations. In either case, taxes or an annual fee usually occurs on the tax bill. This charge is normally allocated based on the projected budget divided by the number of equivalent residential units served. The definition of ERUs is normally intended to represent the typical single-family house, with a given property lot area and amount of impervious area (percent of house, driveway, etc.). Some systems use an additional factor—the percent of impervious area or a lot compared to the single-family lot (imperviousness), as well as amount of impervious area. This system tends to force commercial areas that are mostly impervious and thus contribute more to the public stormwater system, to pay more for their usage (see Chapter 5 for a diagram of how this works). Operations, maintenance, administrative, and capital costs are included in many stormwater fees, although local and state statutes may impose certain limitations on the application of stormwater fees. Most stormwater fees are prospectively collected.

Toll roads and transit services collect fees prior to using the services—fares are paid before entering the bus or train (prospective). Toll roads also require the tolls to be paid prior to traveling on the road. The fees are set by local or state agencies and are usually based on a business model. Transportation services may or may not receive tax dollars to underwrite costs. If they are underwritten, the reason is normally to keep fares low to encourage more passengers. They are also underwritten to increase availability of service; a major barrier to transit ridership is that the service is too infrequent to be used regularly.

One issue that has arisen as a result of ongoing political goals to keep taxes down is the transfer of money from enterprise funds to the general tax fund in order to lower property taxes. This way enterprise fees may go up, but taxes do not, and little impact is made on deferred maintenance needs. Enterprise funding for capital infrastructure should increase as revenue increases because the increase in revenue should allow for the expansion of capital funding, but too often it does not, despite historically low borrowing rates and lowered costs of construction. Especially in light of the 2008–2011 recession, many public-sector systems have been caught in one or more of several traps: deferring capital to pay current expenses without raising rates/fees, revenue losses from defaults on housing that is vacant and not paying a bill, use of utility fees to overcome ad valorem tax losses in the general fund, or political pressure to reduce rates. While the economic recovery after 2012 has alleviated the vacant property issue to some degree in most communities, all four situations can be crippling because it not only reduces current revenues, but the loss of revenue may carry forward to the future. This is a problem, given the expanding needs for rehabilitation and replacement. Hence, most utility rate studies recommend increasing user rates. The challenge for utility management is to build and maintain public support for the utility operation and to facilitate the adoption of such increases (Ori and Mantz 2013). Keep in mind, undoing poor decisions (that transfer money out) is difficult.

Given that over 85% of the population uses public water, over 65% uses public sewer, and nearly everyone benefits from stormwater management and public roads, why is it so difficult to secure funding? One issue is that while nationally only 62% of households pay directly for water and wastewater, overall per capita water use peaked in the mid-1970s; it has reduced since that time due to increases in the efficiency of plumbing fixtures (Pacific Institute 2009). Since these entities are set up as enterprise funds, whereby revenues are gained from provision of a measurable service, the lack of increased demands can be limiting.

The enterprise designation is intended to convey the idea that these funds are designed to be operated more like a business than a government. One would think that given how many people are supported by infrastructure in their jobs and homes, support would translate to public opinion for increased revenues, especially since most systems are primarily not-for-profit entities—they don't make money for anyone. But, if the allocation of funds is altered through the political process, such as diverting one revenue stream to another cost center, or artificially keeping rates, fees, or taxes low to satisfy political objectives, this can frustrate the efforts to run an efficient and effective business-like organization. The question is then, "Where is the leadership to reverse this trend?" Unfortunately, the political leadership focus is on elections, two to four years out, not the 20- or 50-year life of the utility's assets. As a result, short-term benefits sacrifice long-term needs. This leadership void is a discussion for Chapter 16.

Repair and Replacement Funds

As discussed in Chapter 7, deferred maintenance references those known projects that should have been undertaken in the past, but still have not been funded. Deferred maintenance increases risk of failure and increases the costs for repair and replacement as projects are deferred. To avoid large, deferred maintenance costs that often can only be funded via large bond issues, provision for renewal and replacement set asides from existing revenues should be undertaken and such funds need to be grown to permit pay-as-you-go budgeting for repair and replacement to the extent possible. In this manner, existing customers who are using the infrastructure will be burdened with the cost, as opposed to new customers who normally pay for infrastructure as a part of their real estate acquisition or impact fees at the time of construction.

Annual renewal and replacement fund allocations should be calculated not only on major infrastructure, but also the cost for replacement of older subdivision infrastructure—including old water mains and sewer lines, pump stations, drainage ditches, retention basins, meters, cleanouts, storm pipes, road bases, pavement, service lines and taps, and many more. This may mean that a sizable amount of money is collected and held in the repair and replacement fund for such things as major improvements in a wastewater treatment plant, a replacement bridge or major culvert, or upgrades to a major transmission line or pump station. The accumulation of funds for large projects requiring a significant commitment of financial resources will minimize funding from external sources, thereby minimizing any effect on charges (taxes, fees, service charges, etc.) to the user in the future. Over time, all users will benefit because accumulated funds will be available for the renewal and replacement of infrastructure. In many older communities, the use of services to line sewers reduces infiltration to a point that wastewater plant expansions can be deferred. However, replacement funds should only be used to repair and replace existing infrastructure, not for growth-based facility needs. Growth-based customer needs should be met with impact fees.

For any repair and replacement fund, a financial policy for allocation and budgeting should be established. Financial policies are a set of long-range business principles, goals, and guidelines approved by the governing board to provide direction to staff on the courses of action to take with regard to the finances, establish policy guidelines for developing budgets, and provide standards upon which to measure performance. Financial policies serve a number of purposes for both management and the governing body. Among these purposes are:

- Allowing local officials to approach financial questions from a holistic, long-range vantage point to ensure that the components that may need coordination are coordinated ahead of time.
- Outlining how the capital improvement plan should be developed and updated (typically annually for a five-year period).
- Providing guidance on how financial decisions are to be made. Better direction permits more holistic financial decision making, as opposed to making financial decisions on a case-by-case basis.
- Saving time by providing direction and focus on long-range policy objectives.

- Improving financial stability of the organization by allowing it to plan and prepare for financial emergencies through the establishment of long-term reserve funds and mitigation of future increases in the expenditures or reductions of revenues.

Debt

A benefit of using repair and replacement revenues from providing service is that less money will need to be borrowed. However, a disadvantage of doing this could be that a project may wait longer to begin or move at a slower pace than it ought to, which may increase project-related costs. If repair and replacement funds are not available for the replacement or major refurbishment of an asset, debt may be an option. The good news about debt is that large expenditures that may take many years to fund with repair and replacement monies, can be funded at one time, and paid back over time. That makes the projects move along more cost effectively and faster, something that may be of significant benefit. The down side is that debt often creates immediate cost implications. That means that rates, fees, taxes, or charges must be raised in order to demonstrate that the loan can be repaid. At the same time, debt should never be used for operations, which is why it is necessary to understand the maintenance and repair and replacement needs of the system. These are actually operations issues that should be part of the current rate or tax base expenditures.

Options for debt as a means of funding infrastructure maintenance include: revenue from services/taxes in other funds (short-term only), certificates of participation, short term loans/lines of credit, revenue bonds, general obligation bonds, and state revolving funds (water, sewer, stormwater only). General obligation bonds are bonds where the taxing power of the government is used to guarantee repayment of the bonds. This may be the only option for roadway projects, but because in many jurisdictions general obligation bonds require voter approval, they are often difficult to get approved. For water, sewer, stormwater, transit, and toll roads, revenue bonds are a good way to fund projects because they provide all the needed capital upfront, but pledge only the revenues generated from the enterprise fund charges. Revenue bonds work by distributing a portion of collected revenues to the bond holders for a set period of time, typically 20–30 years. These bonds are untaxed investments that are placed with banks or other investors and generally equate to a lower interest rate loan, if insured.

Short-term loans are loans issued by banks and are typically paid back in 1–5 years. The advantage to these loans is that the money can be borrowed when other sources are impractical or unavailable. The downside to short-term loans is that they typically have higher interest rates than other bonds. State revolving funds are beneficial because they have the lowest interest rates—often well below the prime rate. The payback period is typically limited to 20 years. The downside to these funds is that they are becoming increasingly less available as federal and state governments look to cut their budgets. Since debt issuance can come in a variety of forms, depending on the size of the issue, the expediency required for the proceeds, and whether the obligation is intended to be long- or short-term, a financial analyst can make a recommendation as to the appropriate method of debt for any given issuance.

The grant option is virtually dead unless a community qualifies as one that is at risk or has significant populations of low-income people, a depressed economy, or high unemployment. Sometimes monies can be gained from state legislatures or Congress, but these funds are more difficult to obtain, as well, as line item veto power increases for governors.

All government entities should have a debt policy because it is rare that enough funds can be collected ahead of time to be able to fully pay for all services as they come due without borrowing (Miranda and Picur 2000). The GFOA recommends that a debt policy should address the following (GFOA 2016):

- Types of debt permitted to be issued—revenue bonds, line of credit, certificates of participation (COPS)
- Method of sale of debt instruments—negotiated sale, competitive bids, or private placement
- Selection procedure for consultants to help with issuance of debt
- Disclosure to investors
- Use of debt proceeds
- Debt capacity limitations
- Integration of debt and capital planning activities in the capital improvement program
- Structure of the debt issuance—terms, redemption polices, etc.
- Investment of debt proceeds
- Maintenance responsibilities
- Credit policies and compliance with existing laws
- Policy for refunding debt

Impact Fees

Only two fees have major legal constraints—impact fees and assessments. Impact fees are charges imposed against new development or connections to provide the cost of capital facilities made necessary by that growth. The driving force behind impact fees is the sentiment to have growth pay for growth. Of no surprise, litigation was the result of the imposition of impact fees, and the results of litigation have defined the limitations associated with the imposition of impact fees. The use of impact fees is based on a Florida case law derived from *City of Dunedin versus Contractors and Builders Association of Pinellas County*, where the judge ruled that a utility's "water and sewer facilities would be adequate to serve its present inhabitants were it not for drastic growth; it seems unfair to make the existing inhabitants pay for new systems when they have already been paying for the old ones." This case is the basis for much of the impact fee law that currently exists both in Florida and nationwide (Bloetscher 2009). Extensive subsequent litigation has occurred in Florida and other jurisdictions, but the Florida case law is cited in impact fee cases throughout the nation and its basic tenets have been consistently upheld.

As developed under this case, impact fees must meet the *dual rational nexus test*. To meet the dual rational nexus test, impact fees are typically based on the incremental or marginal costs of providing the service, an average cost to provide an incremental portion, or an estimate of the cost of the construction to be provided. The first prong of this test requires that there be a reasonable connection between the anticipated need for additional facilities and anticipated growth (*Hollywood, Inc. versus Broward County*). The second prong requires that there be a reasonable connection between the expenditure of impact fee revenues and the benefits derived by new connections (*Hollywood, Inc. versus Broward County*). In addition, case law requires that these fees be just and equitable (Bloetscher 2009). As a result, a profit cannot be earned on impact fees; they must be related to the actual cost of providing the service as defined in the second prong of the dual rational nexus test (Bloetscher 2009).

Other infrastructure systems are newer to the impact fee system. As constituted via litigation, impact fees act as a method of generating revenue from new customers to finance major facility construction made necessary by the addition of those new customers. All infrastructure systems must meet the dual nexus test. Because facility planning timelines may be extensive, and because of the geographical variance in growth demands, a multi-year estimate is utilized to forecast needed expenditures and proper impact fee amounts. Roads, stormwater, parks, schools, emergency medical service, and fire are among the services that have had impact fees associated with them. Local and state case law and statutes should be consulted for application and limitations on impact fee imposition.

Assessments

The cost for improvements that serve a limited geographical area can be assessed against the benefiting properties in most jurisdictions. This would include the installation of a project to address drainage issues; roadway projects to address outdated or older infrastructure like pavement, sidewalks, or curbs; street lights or beautification projects; and water or wastewater service where it is currently not available or in need of major repair or replacement. If budgets are limited, the appropriate way to fund these improvements in limited areas (neighborhoods) is through assessments against the benefiting properties, even if the funding for the project is borrowing. For most communities, home rule and statutory authority often permit residents to petition the local entity to undertake the project using assessments.

There are limitations in most states related to the imposition of assessments and a specific process through which they can be levied. As a result, there are procedures, notifications, public hearings, due process, and other requirements that must be followed in order to have a valid assessment program. Assessments have the caveat that the cost assessed to the properties must be proportional to their benefit, and that all those assessed must directly benefit from the improvements. It is normal for any bonds related to assessments to be validated by a court of law. This means that the courts agreed to the concept and methodology and that the due process requirements have been met. It is difficult to challenge assessments that have been validated by a court of law. Once put in place, the most common method of collecting assessments is on the tax bill, since both taxes and assessments are applied to the property. If they are not paid, a lien is placed against the property. This ensures that the costs can be collected to support any debt instruments.

FINANCIAL POLICIES

A formal, adopted financial policy provides a written set of guidelines upon which local officials can base decisions that avoid inconsistent financial decisions that inevitably lead to conflicts with current, past, or future policy practices. Formal adopted policies generally promote continuity regardless of changes in personnel or local officials, which help the community and financial institutions because they provide long-term fiscal procedures that can be relied upon. Formal policies can increase efficiency by standardizing fiscal procedures while informing new employees and officials of the expected courses of action. Bond rating agencies look favorably upon agencies with adopted financial policies that are in place and followed.

Governmental entities create financial policies for a number of functional areas (Kavanaugh and Williams 2004):

- Operating budget
- Revenues
- Expenditures
- Capital improvements
- Debt

- Procurement
- Investment of assets
- Risk management
- Human resources (from a compensation, pension, and classification perspective)

Contingency Funds

With any infrastructure system, things can go wrong at any time. Sometimes this is just a failure of the asset, but sometimes an accident or natural event creates the failure. In any case, funding is needed immediately to address the problem. Waiting on insurance to pay for a fix is of little consolation to those relying on infrastructure that is not available. The September 2013 floods that wiped out two major roadways in Colorado are an example.

To ensure access to cash, local officials should consider a contingency fund. Local officials and managers should develop guidelines for creating and maintaining contingency funds, including the amount of funding to be included in the fund, and when the funds can be used (i.e., in the event of emergencies, natural disasters, etc.). A catastrophic event can seriously hamper the ability of a community to serve its customers and can seriously impact the local economy along with its own financial condition. While the risk and vulnerability assessments discussed in Chapter 9 will identify those areas where emergencies might occur, funds are still required to address those problems. Equipment, inventories of parts, and contracts for services can be arranged for the most common problems, but large sums of money are needed to address the potential for major failures or natural disasters. The use of a line of credit could substitute for a contingency fund, but the amount available is critical and may be difficult to secure after an event.

An analysis of past budgets will indicate the likelihood of large, unanticipated expenses for smaller systems, but having a contingency amount in the budget (that may accumulate in a contingency fund with time) will permit the agency to fund such emergencies without having to amend the budget each time a problem occurs. If the money is not spent, it can be reappropriated in following years. However, no entity can bank enough funds for every emergency. For large cost items, a line of credit through a local bank would be prudent.

Fund Balance/Reserves

Grandma always told you to save money for a rainy day. She wasn't really talking about rainy days, but days when you had less or no income. The press talks about the huge percentage of Americans who have little or no savings, and how compared to other countries, we are at a disadvantage during tough economic times due to a lack of savings to see us through. That same argument can be translated to governments, which must provide services during economic downturns. Everyone has heard the political discussion about allowing people to keep their money versus the government, but if the government has no savings, how does it provide services? Local officials do not want to raise taxes and fees during economic downturns, but won't the loss of services just make things worse? Congress and many state legislators did not listen to grandma—and far too many states do not have sufficient funding as a result.

A recent Pew report (2015) suggests that the states *had about half the reserves necessary to address budget gaps during the first year of the Great Recession.* The 50 states had about $60 billion set aside in the summer of 2008; but in fiscal 2009, budget gaps across the country totaled $117 billion, about twice what the states had in reserve (Pew 2015). The budget gaps continued to grow in 2010 and many states struggled with shortfalls for years afterward. This is bad news, and the news really does not improve. They report that 37 states have legal caps that prevent them from saving enough to weather recessions or even enough to substantially offset revenue losses, and most of those are based on some percentage of the prior year's revenues. Minnesota and Virginia recently raised their caps to address this problem (Pew 2015). In other words, the legislatures created rules that limited themselves during hard times. That is like grandma telling you that you must save for a rainy day, but you can only save $20.

Why is this? Politics is part of it. Far too few state governments recognize the importance of saving; figuring that cutting taxes during times of plenty and giving back to taxpayers is a better use of funds, so they pass laws to reduce taxes during good times, when revenues are plentiful, as opposed to socking the excess funds away in reserves as was done prior to 1990. Too often, no reserves are funded. Too many policy-makers hope the good times continue and the reserves will not be needed; or if they are, it occurs under someone else's watch. But the good times always end. The choice then is to raise taxes or cut services. The answer is obvious—for example, Florida chopped hundreds of millions of dollars from education funding after the 2008 recession, a time when one would think that education might help get the state out of the recession faster. The Florida legislature continues to raid the affordable housing trust fund at a time when affordable housing simply is not available in many communities.

The limits to revenues apply across the board from cigarettes and alcohol taxes when people have less money to purchase cigarettes and alcohol. Additionally, as oil prices drop, so do revenues. Ask Alaska, Louisiana, Kansas, Texas, North Dakota, and others that are oil-rich states about their budgets since 2010. All have faced difficult times and no reserves (except Alaska). Legislators were begging tax-reduction advocate Grover Norquist to let them out of their "no tax increase" pledges. He said "no," of course, so those legislators were stuck in either the *do-the-right-thing* or the *get-whacked-by-Grover-in-the-next-election* conundrum. You can guess what they did—because they want to get re-elected. That doesn't help the citizens of those states. An example of the result of such actions: Standard & Poor's revised its outlook on Alaska's general obligation and appropriation-backed debt from stable to negative. That will cost them in the future.

But the issue is not just a state issue. It is a local and a utility issue as well. Local governments are closer to the ground, have less leeway in their budgets, and often have far too little funding as a result of resistance to raising property taxes, user fees, and over-dependence on state shared sales taxes, which often drop precipitously during a recession or, worst of all, limits placed on them by state legislatures. Some government leaders figure they should only keep enough cash on hand to pay bills during tax seasons. That accounts for 60–90 days of funds—far too little for dealing with economic impacts. As an example: in Florida, when property values plummeted and tourism and consumer buying diminished in 2009, the taxes related to all three went down as well. Some communities saw general government revenues fall 50% or more and face years before those values return to their pre-2008 levels. The 2008 recession caused well-funded communities to spend down what reserves they had as a means to avoid the hard and unpopular decision of raising taxes to collect the same revenues as before the mid-2000s cuts. Now the lack of reserves creates an issue going forward—as costs increase faster than revenues, there are now no reserves to tap into. It is a problem that just keeps on giving.

Despite the post-2008 recession recovery, a new Government Accountability Office (GAO) report suggests that the short- and long-term future for state and local revenues may be more difficult than currently anticipated because many public entities chose to reduce tax rates to balance the budget as opposed to restocking reserve funds. Infrastructure projects did not get funded. Another example: Moody's downgraded the City of St. Louis's credit rating one step to

A1, citing "the city's weak socio-economic profile and reliance on earnings taxes, which are due for voter reauthorization in 2016."

In 2017, over half the states had lower-than-projected revenues despite low unemployment and a record stock market (The Economist 2017). Many conservative states were having difficulty avoiding tax increases: Kansas, Oklahoma, and Texas among them (McNichol and Waxman 2017). 2018 has not improved things for many of these states, which has engendered support for other types of revenue (Setliff Law 2018). Diversity in industry and taxes is beneficial. Too often this gets lost in the desire to do more with less, but doing more means you need more funding! We need to listen to grandma and to economist Paul Krugman. Put money in savings (reserves); don't cut your income (revenues)!

Since many infrastructure systems are operated like a business enterprise, the revenues collected by the entity can remain with the entity. In many years, the full budget may not be spent, leaving surplus funds. These unexpended funds should be accumulated in reserves, a function known as a fund balance in the annual audit. Fund balances are those monies not encumbered for other purposes. Agencies maintain fund balances to cover those potentially volatile expenses like power, chemicals, unexpected overtime costs, and emergencies (so contingencies are often not funded since fund balances remain). Like your grandma said, you need funds in your savings account for that rainy day that inevitably comes along. That is what a fund balance is—a savings account. The funds are collected and invested until they are needed. When significant amounts of fund balance are collected over a period of years, they can be appropriated for capital projects, but a fund balance should not be budgeted to balance operations expenses unless some unforeseen revenue shortfall occurs or an increase in taxes, fees, or charges is anticipated in the future and some portion of the fund balance is used to ease into the increase. A fund balance should be invested to earn interest, which is added revenue. The appropriate level of a fund balance should be based on the volatility of revenues, the variability of expenses, the need to replenish debt, rate stabilization, and unreserved fund balance funds.

In addition, all organizations should have certain funds to cover immediate expenses. In most cases, revenues will generally lag expenditures simply because the service is often provided before the user pays for it (toll roads and trains being an exception), or in the case of taxes, they come in once or twice per year. As a result, the organization must have an unreserved fund balance to meet this discrepancy. The GFOA (2016) recommends that the unreserved fund balance be a minimum of 15% of the annual user fee revenues. That is typically less than 60 days, insufficient for most operations. If the unreserved fund balance falls below the prescribed policy level, a mechanism should exist to restore this amount. Unreserved fund balance monies should be highly liquid investments.

Many depression-era economists and modern-day economists note that when the economy downturns, there is a need to continue spending levels (hence, why reserves serve the purpose). The problem is that when the economy becomes difficult, local officials need to raise taxes and/or raise fees precisely when their constituency has little ability to pay for it or change levels of service. Additionally, when the economy is in a down cycle, there is a tendency for construction projects to be deferred due to the lack of revenue and reserves. Having significant retained earnings or a fund balance allows the agency to operate as it might and perhaps even act as a mechanism for increasing jobs in the construction industry. An economic downturn for those who have project funds is a means to address a backlog of projects (often at lower construction costs). As a result, the organization may find itself saving a significant amount of money by constructing needed infrastructure during poor economic conditions. Dania Beach, Florida, found they saved about 30% of the cost of their water plant construction by building in 2010 when compared to the cost estimate in 2007 (before the economic downturn).

ANNUAL REVIEW OF THE PRIOR YEAR'S FINANCIAL BOOKS—THE AUDIT

The annual audit—Comprehensive Annual Financial Report (CAFR)—comes after the budget year has been completed. The finance director normally controls the process. The CAFR is many, many pages long and includes information on revenues and expenses for the current year, prior information about the entity, as well as many other things like assets, depreciated assets, transfers to other funds, outstanding long- and short-term debt, fund balance, and reserves. CAFRs were redesigned by the GASB about 15 years ago to provide more useful information to lenders and oversight agencies. It was redesigned to help with management, discussion, and analysis of the entity's financial position. Because staff can be bogged down with day-to-day activities, the intent of the audit is to have an external accounting group evaluate whether the revenues and expenditures were appropriately categorized. Additionally, it is intended to show that money was maintained in the appropriate accounts and that revenues and expenses are fully accounted for and appropriately spent. The audit will contain useful information about the revenues and expenditures for the prior year's amount of debt, capital assets, and retained earnings. Auditors will note areas where the community can improve its accounting and finance methods. It also notes issues associated with bond revenues, expenditure tracking, and asset management, and the auditors make suggestions as to where these areas may be improved. The CAFR may also provide information on financial stability in the form of operating and revenue ratios. The auditors rarely opine on adequacy of infrastructure investments.

The CAFR should not be viewed simply as a compliance tool to submit and forget about. It is designed to be a management tool to help with tracking the performance of the entity over time. For example, assets should include the value of all installed infrastructure (fixed) and all mobile equipment (non-fixed). There should be a clear delineation of outstanding debt. The value of the entity's debt plus depreciation should be less than its asset values; otherwise the entity is underwater with respect to its assets. Transfers to other funds should be outlined with the justification for same.

But the reserves are key. Some of these reserves may be restricted, which means they are likely impact fees. Examples are reserves to cover debt coverage requirements, or covenants for repair and replacement or other purposes. Most organizations do not have a separate repair and replacement reserve, but this would be useful for planning those capital expenses. Likewise, operating reserves—for use to balance the budget in lean periods—should be identified. The reporting of reserves for rate stabilization should be separate from the operating reserves. Unfortunately, most auditors and most finance directors do not separate reserves, so tracking becomes a challenge.

Ultimately, audits provide information about whether the financial operations are in compliance with generally accepted accounting principles and adequately present the financial condition of the organization and whether appropriate laws and regulations have been complied with. They also provide some indication of whether or not expenses appear to be appropriate for labor, energy, chemicals, and other aspects of the organization. From an operating perspective, there are a series of measures that should be evaluated each year and a trend plotted. Those reviewing annual financial assessment and bond agencies look for these trends.

First, revenues should meet or exceed expenditures. The expenditures should include all relevant expenses based on the full cost accounting of system operations. Second, determine whether there is more or less cash at the end of this year than at the end of the previous year. An increase in receivables, along with an increase in payables and a decrease in cash, could be the result of normal operations (CRG 2011). A ratio of the revenues to expenses should be determined—which should be greater than one. Most lenders want to see 10 or 15% above the potential proposed debt service amount (Jarocki 2003). There should be no subsidies from other funds. There should also be a minimum of 15% of the annual operational expenses for emergencies, payment delinquencies, and cash flow purposes (Jarocki 2003). Third, the ratio of current assets to current liabilities should be well in excess of 2:1 (Jarocki 2003). Low ratios indicate that the entity has significant amounts of debt and may not have the capacity to incur more. Fourth, determine the cost recovery ratio (CRR), which compares the revenue that utilities generate from residential customer rates to the costs incurred for supplying the water. Typically, those costs fall under three categories: operating expenses, debt service (annual payments on loans), and interest expense (interest on loans) (Rahill-Marier and Lall 2015). The ratio of total sales to receivables should be in excess of 10:1 (Jarocki 2003), while the ratio of sales to working capital should be greater than one. The typical ratio is 6:1 (Jarocki 2003).

The ratio of total sales to net fixed assets should also be considered. Medium-sized utilities have a ratio of 0.3 (Jarocki 2003). If the ratio is high, it is likely that the utility is in deteriorated condition. However, a low ratio may be misleading if a series of capital improvements have just been completed. The ratio of sales to total assets should also be considered in the same light as the ratio of sales to net fixed assets and, for the same reasons, have a ratio of 0.3 (Jarocki 2003). With respect to capital assets, the ratio of net plant assets (depreciated value of original asset costs) to replacement value can be an indicator of significant deferred maintenance. Other ratios are shown in Table 11.2. Written policies and procedures are considered desirable, while the absence may lead to concerns about the preparedness of management in the event of an emergency.

A strong financial position entails efficient and effective utilization of financial resources and will increase the bonding capacity and increase bond marketability while lowering interest rates, which is of benefit to the public. The standards for evaluating financial condition are set by the three rating agencies in New York (Fitch, Moody's, and Standard and Poor's). Their evaluations of fiscal condition translate to ratings applied to the bonds, which drive the interest rate for the bonds. The objective of ratings for bond issues is to identify significant economic strengths and weaknesses. For the most part, they are looking for stress—indicators that might portend

Table 11.2   Typical finance ratios employed by enterprise funds

Financial Ratio	How to Calculate	Expectation
Current ratio	Current assets/current liabilities	Comparison of monies-in-hand and to be paid in 30 or 365 days, Ratio > 1.0 minimum, 1.5 preferred
Debt to net worth (equity)	Income to support debt vs debt obligations	1.15 minimum in most bonds, 1.5–3 preferred
Revenues to receivables	Income/monies owed	Higher value means faster payment. Useful to pay-off project capital
Revenues to working capital	(Current assets—current liabilities)/work-in-hand	Larger ratios will have a higher value

difficulty of repaying the debt in the future, but their evaluations go further than just finances. The community as a whole is part of the rating.

AFFORDABILITY OF SERVICES

Studies analyzing the effects of the Tax Reform Act of 1986 (TRA86) have emphasized that a large part of the response observable in tax returns was due to income shifting between the corporate sector and the individual sector (Slemrod 1995). However, it is how this transfer occurred that may play a part in infrastructure funding. As a result of the shift in tax burdens, one of the concerns that has arisen over the last 10 years relates to the affordability of water, sewer, wastewater, and other infrastructure services by the citizenry. The affordability concern complicates efforts to spur spending to upgrade infrastructure because there is a serious concern on the part of local governments regarding the disproportionate financial impact on households at or below the poverty level or with low, moderate, and fixed incomes at the lower end of the income distribution. This is because fees command a greater percentage of their annual incomes than they do for wealthier people (Anderson, Gatton, and Sheahan 2013). If, as Grigg (2016) suggests, water is underpriced in most markets, it raises the question, "can lower income household utility ratepayers afford to pay their utility bills?"

While one could debate economic theories for years (we have), the practical result is that when the purchasing power of the lower and middle class goes down, it is more difficult for local officials to raise taxes, fees, rates, and charges to pay for those needed infrastructure upgrades, which will put more assets at risk of failure and stress operations budgets further. As a result, increasing income disparity is a major area of concern. Economists have noted that real losses in purchasing power have been occurring for lower and middle class people since the federal tax reforms of 1986, despite the fact that jobs have increased (Piketty, Hess, and Saez 2004). From 2002–2007, the incomes of the bottom 99% grew only 6.8% (Saez 2013), then the Great Recession hit. People lost significant value in their homes, retirement accounts, and businesses. Unemployment exceeded 10% and many of those displaced workers could only find jobs at lower income levels. During the Great Recession, from 2007 to 2009, average real income per family declined by 17.4% (Saez 2013). Since the 2008 fiscal crisis, the U.S. Conference of Mayors (2012) reported that the income of the bottom 99% grew only by 0.4% from 2009 to 2012. Saez (2013) reported real income per family grew modestly by 6.0%, with most of those gains coming when income grew 4.6% from 2011–2012. The Congressional Budget Office (2013) reported that personal income grew 1.6% in 2012 and 2.8% in 2011. All reported that average incomes were below 2007 levels. In August 2012 the Pew Research Center report noted that only half of American households are middle class, down from 61% in the 1970s. More people are moving down in economic standing, not up (Pew 2015), which is why 47% of U.S. households did not earn enough to pay income tax in 2013. Only the top wage earners are seeing consistent wage increases and Bee (2013) reports that these are located primarily in the coastal areas of the Pacific, Middle Atlantic, and New England states. That does not help the rural Midwest and South.

The economy presents local governments with a variety of challenges. The first is a social issue where concern has been raised about the increase in the costs for infrastructure services exceeding the cost-of-living inflation. Very low-income users are regressively affected; they are paying much larger percentages of their income for services. Others will alter demands. This can be seen in gas tax revenues and water use. Both have declined with time, but both are used to allocate/generate funding. Financial forecasts that fail to consider this effect on demand will therefore overstate anticipated system revenues and potentially lead to realized shortfalls (Enouy et al. 2015). Therefore, infrastructure managers cannot ignore the economic realities around them.

The loss of wages is felt locally more than nationally. It means that local officials hear about costs more because water, sewer, power, etc., competes for an ever-larger portion of the shrinking paycheck. So, more attention is paid to affordability indexes. The concept of affordability is to take the cost of infrastructure (water and sewer has been at the forefront of this effort) and divide it by the average or median local income. The goal is for the cost of water plus wastewater to be under 3.5% of the median income. Keeping the percent low is great—and easy when people are making more money—but creates a lot of difficulty when incomes are static or dropping. In many communities, this means that residents have less income to pay for increasing needs for infrastructure and that infrastructure is needed to keep the economy that employs them going. Costs are rising due to the increasing need to maintain and upgrade infrastructure that has been neglected since 1980. The need is to invest above 2.4% to keep up according to the GAO (2015).

When incomes drop, local infrastructure costs (taxes and fees) are often easier targets to limit than groceries, rent, power, telephone, cable, or other services that are not subject to local voters. Local officials are sensitive to the constituency, making increases harder, but infrastructure must be maintained to preserve the quality of life. As a result, most water and sewer rate studies should include a comparison of infrastructure costs with power, cable, cellphones, etc., rates along with other basic services. Most infrastructure services come in at the bottom. But that works when everyone has access and uses those services. Several years ago, a study indicated that cable television was in 87–91% of homes. Cellphones now saturate the market, with average costs for a family upwards of $200/month. Water and sewer are somewhat under $75 a month. We need water, sewer, roads, and drainage. Not so sure about cable television or the internet. Great to have, but needed to survive? It is going to take communication and leadership to alter the current situation (see Chapter 16). Public works entities must institute programs for educating the general public about the necessity for increasing revenues and maintaining large reserves. In addition, in times of plenty, excess revenues need to be placed into reserves instead of lowering taxes. The following section explains part of the reason why.

BOOMS AND FALLS—CREATING FINANCIAL RISK

The discussion about affordability of service requires a discussion about rises and falls. The history of mining in the western United States is full of boom and bust stories. But booms and busts are not just a 19th-century phenomenon. They exist today—in the 21st century; we are just less cognizant of them for many of the same reasons our ancestors didn't pay attention to them 150 years ago. Booms are times when very fast growth occurs to address a single industry. Local officials love the booms because the growth and local revenues fill the local coffers. Communities may be flush with money. Then, the economy changes and industry abandons the community. They leave behind infrastructure that must be maintained (or abandoned), waste products that must be cleaned up (think mine tailings), equipment that must be disposed of, and impacts that will last for years (oil wells), and no money. A western mining town discussion is useful.

Colorado really got its start in Leadville. Gold miners were relatively unsuccessful finding gold, but they found lots of *black sand* that was seen as a waste product (they used it on the roads). An enterprising metallurgist figured out the sand was actually high-grade silver

(Cerussite). Mining in Leadville took off and many made millions mining silver. Tens of thousands of miners lived in and around Leadville at the time. Those who were mining owners moved to Denver for the winter and became early leaders for the state. The names like Horace Tabor, Mollie Brown, Baby Doe McCourt, August Meyer, and Doc Holliday dot the landscape. However, after 20 or 30 years, the silver ran out, the miners left, and the city shrank to under 3,000. So, who maintains the roads? Who deals with the mine tailings? Who deals with the *red* water that is caused by metals leaching from the tailings? Who deals with the old mine shafts? For over 100 years—no one. Finally, after water with high metallic concentrations threatened to contaminate the Arkansas River, the federal government declared the area a Superfund site. From a local perspective, there was no capacity to address the infrastructure after the boom.

Moving to the early 2000s, it was very clear that there was a property boom in communities like central Arizona, Las Vegas, southeast Florida, and central Florida. Property values skyrocketed, house flipping for profit became common, and developers planned thousands of new homes. And then, the 2008 financial crisis, built on the back of the real estate boom, halted those activities. Local government budgets were slashed and property values dropped by 50%; yet the amount of infrastructure to maintain had increased. Local officials were extremely reluctant to raise revenues of any type. But it was not as though no one had warned them about this. There was discussion on the news, through financial channels, in the *Wall Street Journal*, and even in columns by economists like Paul Krugman. This real estate boom resulted from the bust of the tech stock bubble (boom) in the late 1990s. People on Wall Street knew that the investments had moved to real estate and bankers where busy loaning money out with no interest for two years, no money down, with adjustable rate mortgages, and the like. A few people benefited from the boom, but many did not.

Fracking is another mid-2000 boom, just like housing and mining. The oil and gas industry boomed in North Dakota, Kansas, Oklahoma, and Texas—until it busted. The Plains states planned for the long-term boom, but the boom has settled in some places. The oil and gas industry shed 100,000 jobs (many of them high salary) between 2008 and 2015. While the need for more fracking wells decreased (at least temporarily) because the profits were not there, the loss of jobs created a decrease in local and state revenues. But that doesn't mean the infrastructure that resulted from the 2010–2014 boom no longer continues at the local level. As people in North Dakota, Oklahoma, Texas, and elsewhere have found out, boomtowns are a huge challenge to local officials who are trying to provide infrastructure to support the boom, whether it is oil, gas, mining, or housing. When jobs go, property values drop and there is no one to pay for upkeep, demonstrating that decisions that are made today absolutely affect tomorrow's operators. Such a situation is not sustainable.

A recent *Governing* magazine article notes that a dollar drop in a barrel of crude oil means a $7.5 million decrease in revenues for the State of New Mexico (Farmer 2016; Varela 2016). Since oil lost about $30 a barrel in 2015—that was a $200 million loss for New Mexico. Louisiana experienced a $12 million revenue decrease per dollar drop in the price of crude oil, so they had $360 million less to work with (Farmer 2016). Alaska, perhaps the most oil dependent budget (90%), had a $3.4 billion shortfall, but $14.7 billion in reserves (which is depleting quickly). It is unclear what the plan is after the oil reserves are used up. Texas, North Dakota, Oklahoma, and Kansas are other states that were facing losses (Farmer 2016). Fast growing states like North Dakota and Wyoming now have hard decisions to make. Growth in Texas, Oklahoma, Louisiana, and Arkansas may be cut by over 60% of prior estimates as a result—a double hit on anticipated revenues.

So, is it a surprise that some communities fight the boom times? Booms create disruption and uncertainty, a need for technology (and costs), and they rarely occur in areas prepared for such booms. Conversely, there are times a boom can help communities in financial distress. Detroit and Flint would love a boom—both have the infrastructure in place to support it, as opposed to rural communities in the Plains. There is a lot of older, underutilized infrastructure out there. Detroit, Flint, Cleveland, Akron, Toledo, Pittsburg, Cincinnati, and Philadelphia are among the older industrial cities that have stable populations—people who have lived there most of their lives, have a trained and educated workforce, and normally have lots of water and infrastructure, and lots of potential employees, all of which are underutilized and at risk due to economic losses. But that's the key—they already *have* the infrastructure in place, although much of it may need updating.

Clearly people have not learned the lessons of the many mill towns in the south or the Rust Belt cities of the Midwest that encountered difficulties when those economies collapsed. Everyone refused to believe the good times would end. Today, Detroit is half of its former self and Akron, Ohio, has the same population as it did in 1910. The moral of the story is that booms can be great, but short term. Diversity in the economy is key. Florida will continue to be subject to economic downturns more severe than other states when it relies primarily on tourism and retirees to fuel its economy. Detroit relied on automobiles, Akron rubber and chemicals, Cleveland steel, etc. Someday Silicon Valley will suffer when the next generation of technology occurs that makes current technology obsolete. This economic downturn is what happens when you believe the booms are normal and fail to financially plan by putting money aside during the boom to soften the subsequent period.

RURAL AMERICA AND INFRASTRUCTURE

The U.S. Department of Agriculture (2014) released its report of rural America. The findings are interesting and counterintuitive to the understanding of voters in many of those communities. Their findings include:

- Rural areas grew 0.5% versus 1.6% in urban areas from mid-2011–mid-2012
- Rural incomes are 17% lower than urban incomes
- The highest income rural workers (95th percentile) earn 27% less than their urban counterparts
- 17.7% of rural constituents live in poverty versus 14.5% in urban areas
- 80% of high-poverty-rate counties are rural
- All high-income counties are urban

That paints a very different picture than the political discourse at the federal level—rural areas are much more at risk than many urban areas. So, it is worth reviewing the real impact of the future in rural America.

First, let's look at where these rural counties are. Figure 11.1 is a map that shows (in gray) rural counties in the United States. The U.S. Census data indicates the 100 highest and lowest income counties in Figure 11.2. For the most part, the poorest counties (shown in blue) are rural, with the exceptions being a few areas in south Texas and west of the Albuquerque/Santa Fe area of New Mexico. Figure 11.3 expands upon Figure 11.2, however, red now indicates counties with low household median incomes, which includes much of the rural Deep South, Appalachia, more of Texas and New Mexico and part of the central valley in California.

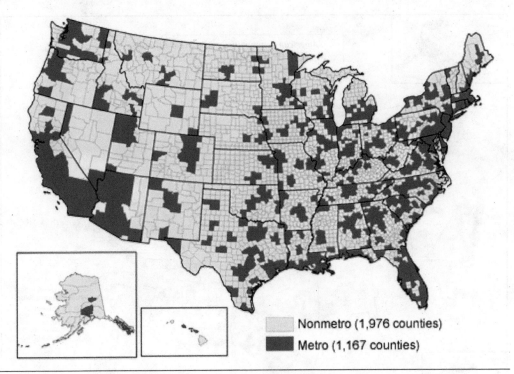

Figure 11.1   The rural counties of the United States are shown in gray in this figure. These rural counties have smaller populations and less capacity to fund public projects. *Source*: USDA, Economic Research Service using data from the U.S. Census Bureau

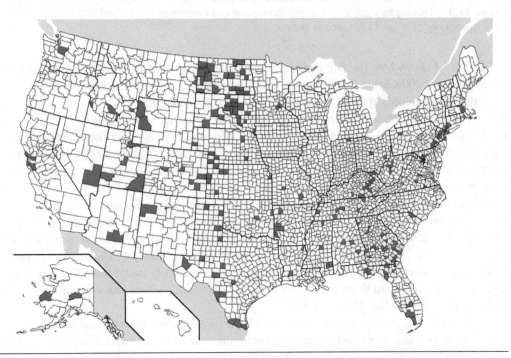

Figure 11.2   This figure indicates the one hundred poorest and the one hundred wealthiest counties in the United States. The wealthiest counties are in red, while the poorest counties are in blue. *Source*: U.S. Census Bureau

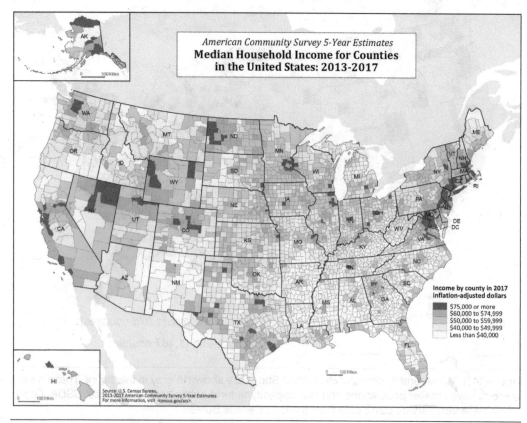

American Community Survey 5-Year Estimates
Median Household Income for Counties in the United States: 2013-2017

Income by county in 2017 inflation-adjusted dollars

$75,000 or more
$60,000 to $74,999
$50,000 to $59,999
$40,000 to $49,999
Less than $40,000

Source: U.S. Census Bureau,
2013-2017 American Community Survey 5-Year Estimates
For more information, visit <census.gov/acs>.

Figure 11.3   This figure indicates the median household income in the United States in the year 2011, by county. *Source*: U.S. Census Bureau

Figure 11.4 shows how the number of young people living in rural counties has changed between 2000 and 2009. Most rural counties have seen a decrease in the number of young people (counties shown in red) as they tend to move to more urban areas to find work (counties shown in gray) or to other rural counties that have more jobs (counties shown in green). Figure 11.5 shows that young people are moving to the Rocky Mountain states—and elsewhere—vacating high-poverty counties. Young people tend to not seek out jobs in rural America where unemployment is higher and the jobs that are there pay less. Figures 11.6 and 11.7 show unemployment by county in 2008 after the start of the Great Recession and in 2013, respectively. What these figures show is that with exception of the Plains states and the Rockies, many of the areas with high poverty also had high unemployment and that unemployment remains stubbornly high in many rural areas in the Deep South, Appalachia, and New Mexico, plus parts of the Great Lakes. Education may be a factor as to why the Plains states and the Rocky Mountain states have less unemployment—despite being rural, their students are far more likely to graduate from high school than those in the deep South or Appalachia where unemployment remains high and incomes low (see Figure 11.8).

Rural communities, as noted, have received a lot of federal grants to fund their infrastructure. The good news is that much of the rural infrastructure may be newer when compared to much of the urban infrastructure of the Midwest and Northeast. As a result, there is an argument to be made that local investment has time to develop the funds that are needed. However, the community needs to be engaged in this discussion sooner rather than later when problems occur.

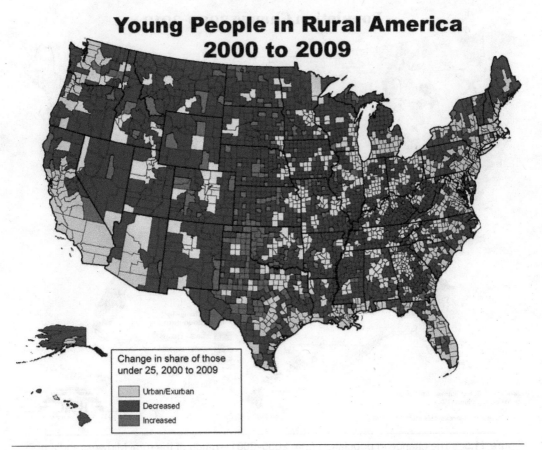

Young People in Rural America
2000 to 2009

Figure 11.4   This map shows that for most of rural America, the population of young people is decreasing (red) as they pursue careers in urban areas (gray) that have greater job potential. The green areas are the few rural counties where the population of young people has increased. *Source*: Rural Health Information Hub

Saving for infrastructure may be the best course since rural utilities will have limited access to the borrowing market because of their size. However, that means raising rates now and keeping those saved funds, as opposed to using them to defer rate increases—something that is politically difficult in these mostly conservative communities. There are larger percentages of lower income and unemployed people in rural communities. That means rate increases will be even more difficult to secure and affordability will be an issue. If ongoing efforts in the U.S. House of Representatives deplete federal funding further, the pinch will be felt sooner by rural customers who will lose the federal dollars from State Revolving Fund matching grants and Federal Housing Administration programs.

Infrastructure agencies need to understand this problem, as it demands some real, on-the-ground leadership. Again, while drainage and roadways have the same issues, water and sewer utilities have studied this further. The findings are that small and rural utilities are more costly to operate per thousand gallons than larger utilities. Bloetscher (1999) showed that economy-of-scale manifested itself to a great extent with water and wastewater operations. A similar study 20 years later (Bloetscher 2017) showed the exact same thing. The differences were not close—it is a lot less costly to operate large utilities versus small ones, per thousand gallons. This is why the EPA and professionals have long argued that centralized

Population Change 2000-2010

Percent Change by
County, 2000-2010

■ 25.0% - 110.4%
■ 10.0% - 24.9%
□ 0.0% - 9.9%
□ -46.4% - -0.1%
□ No Data

Source: U.S. Census Bureau, 2010 Census.
2010 Census Summary File 1.

Note: Alaska and Hawaii not shown to scale

rAc
Rural Assistance Center

Figure 11.5   The change in population from 2000–2001—much of the rural Midwest and a portion of the rural south decreased in population, while urban areas gained in population. *Sources*: U.S. Census Bureau and U.S. Rural Assistance Center, USDA

infrastructure for water and sewer utilities makes sense from an economy of scale perspective. Centralized drinking water supply infrastructure in the United States consists of dams, wells, treatment plants, reservoirs, tanks, pumps, and two million miles of pipe and appurtenances. In total, this infrastructure asset value is in the multi-trillion dollar range. Likewise, centralized sanitation infrastructure in the United States consists of 1.2 million miles of sewers and 12 million manholes, along with pump stations, treatment plants, and disposal solutions in 16,024 systems. It is difficult to build small reservoirs, dams, and treatment plants because they each cost far more per gallon to construct than larger systems. Operations, despite being required to have more on-site staffing, are far less per thousand gallons for large utilities when compared to small ones (Bloetscher 1999).

The issue is complicated further because not only is the number of customers limited, but the capital investment per customer is far higher than in urban areas. So, while local governments are under pressure to reduce rates, taxes, and fees, their costs are increasing and infrastructure demands are incrementally higher than their larger neighbors. That scenario cannot be sustained. As a result of receiving grants and low-interest loans, many small communities have rates that have been set artificially lower than they should. The federal and state governments are unlikely to step in to replace their initial investment, meaning that the billions of rural

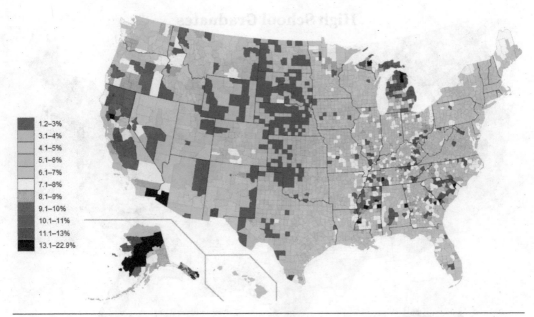

Figure 11.6   This map from 2008 shows that unemployment in rural areas was generally higher than in urban areas except in the Great Plains states, mostly due to the gas and fracking boom at the time. Areas of the rural South, Appalachian, rural Michigan, and rural Northern California and Oregon were especially hard hit. *Source*: U.S. Bureau of Labor Statistics

Figure 11.7   A map of unemployment in 2013 shows that despite much better economic conditions nationally, rural areas still had generally higher unemployment than urban areas. *Source*: U.S. Bureau of Labor Statistics

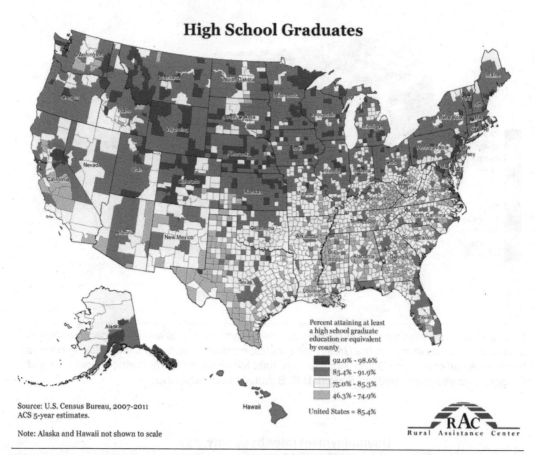

High School Graduates

Percent attaining at least
a high school graduate
education or equivalent
by county

- 92.0% - 98.6%
- 85.4% - 91.9%
- 75.0% - 85.3%
- 46.3% - 74.9%

United States = 85.4%

Source: U.S. Census Bureau, 2007-2011
ACS 5-year estimates.

Note: Alaska and Hawaii not shown to scale

Figure 11.8   The map indicates that in the United States between 2007 and 2011, rural areas and parts of the south generally had lower education achievement than urban areas (except in the upper Midwest). *Sources*: U.S. Census Bureau and U.S. Rural Assistance Center, USDA

investment dollars that will be needed in the coming years will have to be locally derived, and rate shock will become a major source of controversy in areas that are largely very conservative politically and tend to vote against projects that will increase costs to them.

So, what to do? This is the challenge. Rate hikes are a tough sell in areas where residents are generally opposed to increases in taxes, rates, and charges and who use voting to impose their desires. Consolidation is another answer, but this is in contrast to the independent nature of many rural communities. Onslow County, North Carolina, figured this out with their water system 15 years ago, but it is a tougher sell in many, more rural communities. Infrastructure banks might help, but the questions are: who will create them? And will the small organization be able to afford to access them? Commercial financing will be difficult because there is simply not enough income to offset the risk. The key is to start planning now and realize that water is more valuable than iPhones, the internet, and cable TV. Private utility industries have been very successful in convincing people to buy their products. They carefully design plans to accomplish their goals. Cable and satellite TV and phone companies compete for the same dollars as water, sewer, and stormwater utilities do. The only issues are that they spend billions on their marketing, while governments spend nothing. It is no surprise that their bills are higher and have been growing to permit them to update technology. Water utilities need to do the same.

SUSTAINABILITY EXAMPLES

The prior sections outlined risk issues, but nothing is better than some examples to show how these risk issues can manifest themselves on an infrastructure organization. For the sake of simplicity, the examples will be for water utilities. The message is that infrastructure providers may need to look at the broader picture of sustainability in their community and extend the definitions to a wider range because no one else is—and the community is looking for leadership. The agency needs to look at infrastructure and its financial outlook as a part of an overall sustainability strategy.

There are certain assumptions that most organizations make, so perhaps some of these assumptions need to be revisited in light of potential future realities. For example, what happens to communities that do not grow? Current assumptions generally assume that there will be an ongoing increase in population or use that will drive increases in revenues without specific increases on customers. However, what if you are a city where the population has dropped by 50% in the past 50 years? How do you deal with aging infrastructure and demands for increased water quality and reliability while maintaining fees at affordable levels for customers? This is a particular problem when there are economic disruptions that create a large group of disenfranchised people who become more economically disadvantaged than they might otherwise already be. The competition for sustaining water rates, infrastructure condition, and water supplies can be a difficult conundrum.

Let's take a look at some scenarios. Let's assume a water and sewer utility that serves 20,000 people (8,000 customers), with 60 miles of water pipe, 60 miles of sewer pipe, 17 lift stations, and a water and wastewater plant. Replacing this infrastructure might be valued at $90 million for pipe and $35 million for treatment plants, water supply, and pumping equipment (current day dollars). Let's also assume that their annual budget is $11 million and the typical demands are three millions of gallons per day (MGD) yielding a monthly bill of $115 per month (water and sewer).

Let's make some general assumptions, such as the pipe infrastructure might last 100 years, but clearly the treatment and mechanical parts would not. These assets would need ongoing maintenance and replacement. Fifty years is probably too long for replacement, but it is a useful assumption to use. Also, assume the overall costs increase at 3% per year and money is set aside for repair and replacement. The utility will see fairly steady rates if the customer base grows 2 to 3% per year. Ten years out, the budget will be $16 million.

If the customer base grows at 3% per year, customers will increase to almost 27,000 in ten years. This increase in demand (from 3 to over 3.4 MGD) needs to come from a new water source and requires new capacity. Many utilities will use impact fees to offset this cost to current customers so as not to adversely impact current customers too severely. That's the current assumption for most organizations—growth will occur. The result looks like Table 11.3 at 10 and 20 years.

But what if the new treatment and supply are 50% more costly and impact fee collections are less? The cost for the budget and for the infrastructure replacement increases (with the delta from debt). Costs are 50% higher as shown in Table 11.4. The utility in Table 11.4 will have to

Table 11.3   Budget for a utility that is growing (10 and 20 years)

Component		Value Today	10 Years	20 Years
Customers		20,000	26,878	36,122
Accounts		8,000	10,751	14,449
Water pipe replacement value (RV)	60 mi.	$ 45,000,000	$ 98,509,418	$ 215,646,786
Sewer pipe RV	60 mi.	$ 45,000,000	$ 98,509,418	$ 215,646,786
Treatment plants and pumping RV	3+ MGD	$ 35,000,000	$ 76,618,436	$ 167,725,278
Operations budget		$ 9,000,000	$ 12,095,247	$ 16,255,001
Capital budget		$ 1,600,000	$ 3,502,557	$ 7,667,441
Debt		$ 400,000	$ 400,000	$ 400,000
Monthly bill amount		$ 115	$ 124	$ 140
Increase to monthly bill per year			1%	1%

Assumptions: customer growth rate of 3%/yr compounded with a construction cost increase of 5%/yr; operating budget increase of 3%/yr; and capital budget increase in replacement costs of 1%/yr for pipes and 2%/yr for treatment plant and pumping.

Table 11.4   Budget for a utility with costs that are higher than expected

Component	Assets	Value Today	10 Years	20 Years
Customers		20,000	24,380	29,719
Accounts		8,000	9,752	11,888
Water pipe replacement value (RV)	60 mi.	$ 45,000,000	$ 118,965,912	$ 314,508,626
Sewer pipe RV	60 mi.	$ 45,000,000	$ 98,509,418	$ 314,508,626
Treatment plants and pumping RV	3+ MGD	$ 35,000,000	$ 92,529,043	$ 244,617,820
Operations budget		$ 9,000,000	$ 14,744,039	$ 24,154,077
Capital budget		$ 1,600,000	$ 4,229,899	$ 11,182,529
Debt		$ 400,000	$ 400,000	$ 400,000
Monthly bill amount		$ 115	$ 166	$ 251
Increase to monthly bill per year			4%	5%

Assumptions: customer growth rate of 3%/yr compounded with a construction cost increase of 7%/yr; operating budget increase of 3%/yr compounded with a customer increase of 2%/yr; and capital budget increase in replacement costs of 1%/yr for pipes and 2%/yr for treatment plant and pumping.

raise rates to meet the demands of the budget, but perhaps not too much since growth is increasing as well.

The normal assumptions are that growth will continue, but what if it does not? (See Table 11.5.) What can be gleaned as a result of a nongrowth or net decrease scenario? How is sustainability affected? Let's look at the no growth scenario. In this light, rates will need to increase at least 5% per year to ensure that the utility remains rate neutral. If there is significant deferred maintenance—which is typical of many utilities—that cost will be added to the bill. This scenario is doable, but the only real assumptions that changes can be made are related to the lack of growth. Deferring maintenance will only exacerbate the problem because there is no guarantee that growth will return. Rate neutrality becomes a public relations issue, but not insurmountable.

Table 11.5   Budget for a utility that is not growing

Component		Value Today	10 Years	20 Years
Customers		20,000	20,000	20,000
Accounts		8,000	8,000	8,000
Water pipe replacement value (RV)	60 mi.	$ 45,000,000	$ 73,300,258	$ 119,398,397
Sewer pipe RV	60 mi.	$ 45,000,000	$ 73,300,258	$ 119,398,397
Treatment plants and pumping RV	3 MGD	$ 35,000,000	$ 57,011,312	$ 92,865,420
Operations budget		$ 9,000,000	$ 12,095,247	$ 16,255,001
Capital budget		$ 1,600,000	$ 2,606,231	$ 4,245,276
Debt		$ 400,000	$ 400,000	$ 400,000
Monthly bill amount		$ 115	$ 157	$ 218
Increase to monthly bill per year			4%	4%

Assumptions: no customer growth but construction costs still increase at 5%/yr; operating budget increase of 3%/yr; and capital budget increase in replacement costs of 1%/yr for pipes and 2%/yr for treatment plant and pumping.

And what if growth stagnates or goes backwards as many did in 2009/2010? That was a severe problem for most entities, causing higher prices and lesser service, pay cuts, and deferral of needed improvements—mostly because no one had reserves since people thought the good times would roll on forever. Layoffs, price hikes, pay cuts, and deferral of needed improvements do not help society (of course, if the entity had lots of reserves, they weathered the recession without a problem, but too many did not). Keep in mind the repair, replacement, and maintenance needs, along with ongoing deterioration, do not diminish with time or lack of new customers. Infrastructure organizations have relied on new people to add money to solve old problems—as well as new problems—for many years. What is the contingency if growth stops?

A growth scenario makes the organization feel more confident when funds need to be borrowed; but it also has a downside. If growth does not stop, where is the water to come from? What resources will be used faster? Where does the power come from to treat the water or cool the houses? And where does the cooling water to cool those power plants come from? Dry areas like the American Southwest need to answer these questions—many others do as well. Even renewable resources are limited—most metals and oil have likely passed their peaks as far as production, and rainfall is not always consistent. The United States has overstressed aquifers and over-allocated surface waters throughout the country, especially in the West (Reilly et al. 2009). So, while growth makes local officials feel good financially, there is a need for answers to the growth scenario despite the fact that there may be more funding. Many resources are not limitless, but an exponential growth pattern ignores this fact.

Now, let's look at the decline issue. If the population decreases by a third over 20 years, what occurs? The costs will remain relatively constant, but the number of customers and demands for water will drive the rates up significantly (see Table 11.6). In 10 years, the rates could double in a community that is likely economically disadvantaged. The higher rates may begin to discourage economic development, while rate neutrality exacerbates the infrastructure condition problem and may increase in costs if regulatory or deferred maintenance obligations become a significant issue.

Table 11.6   Budget for a utility that is shrinking

Component		Value Today	10 Years	20 Years
Customers		20,000	16,341	13,352
Accounts		8,000	6,537	5,341
Water pipe replacement value (RV)	60 mi.	$ 45,000,000	$ 73,300,258	$ 119,398,397
Sewer pipe RV	60 mi.	$ 45,000,000	$ 73,300,258	$ 119,398,397
Treatment plants and pumping RV	3 MGD	$ 35,000,000	$ 57,011,312	$ 92,865,420
Operations budget		$ 9,000,000	$ 12,095,247	$ 16,255,001
Capital budget		$ 1,600,000	$ 2,606,231	$ 4,245,276
Debt		$ 400,000	$ 400,000	$ 400,000
Monthly bill amount		$ 115	$ 193	$ 326
Increase to monthly bill per year			7%	7%

Assumptions: a declining customer base but construction costs still increase at 5%/yr; operating budget increase of 3%/yr; and capital budget increase in replacement costs of 1%/yr for pipes and 2%/yr for treatment plant and pumping.

WHAT HAPPENS IF NO ONE WANTS TO FIX IT?

We are all cognizant of the low grades on infrastructure given annually by the American Society of Civil Engineers and periodically by the EPA. Americans spend about 1.8% of our gross national product on infrastructure. We used to spend twice that much; and it is likely that we need to spend upwards of 2.4% to stay even. Much of our infrastructure is *forgotten* because it is buried. The American Water Works Association (AWWA) published a book to highlight this problem—*Buried No Longer* (AWWA 2012). But is it helping? In a recent *Roads & Bridges* article, they noted that the bridge system continues to age faster than the repair rate. The states with more than 15% deficient bridges are mostly in the Great Plains and Northeast (Jansen 2017). The latter is no surprise because the infrastructure is generally much older in the Northeast but the Plains are largely rural. What was also interesting was that in a recent issue of *American City and County* magazine, many of the states that have bridge issues also have below average trust among the public (Barkin 2106). Most of the areas with bridge issues are rural states, such as North Dakota and West Virginia, that are often cash strapped (just like the water system with no or low growth). Those combinations do not portend well for the future and may cripple those economies further.

More interesting was the response to how some of these agencies may deal with this backlog of deferred maintenance. The Army Corps of Engineers, state transportation agencies, state land agencies, and the federal government say that they are figuring out means to prioritize the assets and dispose of those not needed. Here are some of their suggestions:

- *Abandon state roadways and let local governments deal with them*—of course, these are the roads that are challenged; such as ones that flood constantly and the cost to raise them is cost prohibitive. But, what if a city has development along the corridor?
- *The state has low value wetlands that they will donate to the underlying county*—not that the county can do anything with this land: it is not developable or provides tax revenues, but still needs to be monitored and maintained.
- *There is a waterway that has leaking dikes but serves very few people*—give it to the local community since they are the only ones who use it (but lack any resources or skills to fix it).

- *We have monitoring equipment, but it really provides more information locally than region-ally, so let's give it to them to use*—most of those agencies have little experience with data networks and limited or no access to the satellites that relay that data.

How does the *recipient* deal with these problems? The "low-value assets" are low value because they serve limited people and are deemed to have little economic or useful value or are too ex-pensive to maintain. However, they may be regionally or locally critical. If a large state or federal agency cannot afford to maintain the asset, how would a local agency? They do not have nearly the resources that larger governmental entities have. With regard to abandoning infrastructure, the answer appears to be "yes, the states might be willing to abandon it," but the problem is that the infrastructure *is* being used by people, so the reality of full abandonment is impossible. The result will be that underlying local entities will be stuck with a bill that may prove difficult to pay.

REFERENCES

Anderson, R. F., Gatton, D., and Sheahan, J. 2013. "Growth in local government spending on public water and wastewater—but how much progress can American households afford?" The United States Con-ference of Mayors. April 2013. Washington, D.C.

AWWA. 2012. "Buried no longer: Confronting America's water infrastructure challenge." AWWA, Den-ver, CO.

Barkin, R. 2016. "Repairing the bridges to nowhere." *American City and County.* http://americancityand county.com/american-city-and-county. Accessed 11/11/16.

Bee, C. A. 2013. "The geographic concentration of high-income households: 2007–2011." American Community Survey Briefs. file:///E:/CGN%206506%20Infra%20Mgmt/9968%20acsbr11-23%20 (2015_03_08%2017_53_48%20UTC).pdf.

Bloetscher, F. 1999. "Deferred maintenance obligations due to aging utility infrastructure. *WEFTEC An-nual Conference Proceedings—Orlando, FL.* October 1999.

———. 2009. "Water basics for decision makers: What local officials need to know about water and waste-water systems." America Water Works Association, Denver, CO.

———. 2017. "Risk and economic development in the provision of public infrastructure." *Florida Section of the American Water Works Association Annual Conference Proceedings—November 30, 2017, Cham-pionsgate, Orlando, FL.* FSAWWA, St. Cloud, FL.

Congressional Budget Office. 2013. *The Distribution of Household Income and Federal Taxes, 2010.* CBO, Washington, D.C.

CRG, 2011. *Small System Guide: Understanding Utility Financial Statements.* Community Resource Group, Fayetteville, NC. AR. file:///E:/CGN%206506%20Infra%20Mgmt/9%20small_system_guide _to_understanding_financial_statments.pdf.

The Economist. 2017. "Despite a strong economy, American states are desperate for revenue; Even con-servative governors and legislatures are contemplating tax increases." https://www.economist.com/ graphic-detail/2018/04/06/despite-a-strong-economy-american-states-are-desperate-for-revenue. Accessed 12/15/18.

Enouy, R. et al. 2015. "An implicit model for water rate setting within municipal utilities." *JAWWA* 107:9. http://dx.doi.org/10.5942/jawwa.2015.107.0122.

Farmer, L. 2016. "How oil states are dealing with sinking prices and revenue." *Governing.* February 4, 2016. http://www.governing.com/topics/finance/gov-oil-tax-revenue-states-finances.html.

GAO. 2015. *Financial Audit: U.S. Government's Fiscal Years 2014 and 2013 Consolidated Financial State-ments.* GAO, Washington, D.C. http://www.gao.gov/assets/670/668739.pdf. Accessed 2/2/17.

GFOA. 2016. *Debt Management Policy.* GFOA, Chicago, IL. http://gfoa.org/debt-management-policy. Ac-cessed 11/14/2015.

Gordon, E. et al. 2011. "Water works: rebuilding infrastructure, creating jobs, and greening the environment. Washington and Oakland: American Rivers, Economic Policy Institute, and Pacific Institute." http://www.pacinst.org/reports/water_works.pdf.

Gordon, R. and Slemrod, J. 2000. "Are 'real' responses to taxes simply income shifting between corporate and personal tax bases?" In: Slemrod, J., ed.: *Does Atlas Shrug? The Economic Consequences of Taxing the Rich*. Russell Sage Foundation, Harvard University Press, Cambridge, MA.

Grigg, N. 2016. *Integrated Water Resource Management: An Interdisciplinary Approach*. Springer, New York, NY.

Holland, J. 2015. "20 people now own as much wealth as half of all Americans." *The Nation*. https://www.thenation.com/article/20-people-now-own-as-much-wealth-as-half-of-all-americans/. Accessed 2/15/17.

Jansen, B. 2017. "Nearly 56,000 bridges called structurally deficient." *USA Today*. Published 1:20 a.m. ET, Feb. 15, 2017. https://www.usatoday.com/story/news/2017/02/15/deficient-bridges/97890324/.

Jarocki, W. 2003. *Cap Finance, Version C2.1 User Manual*, Environmental Finance Center, Boise State University, Boise, ID.

Kavanagh, S. and Williams, W. A. 2004. *Financial Policies: Design and Implementation*: Government Finance Officers Association Budgeting Series, Vol. 7. GFOA, Chicago, IL.

Kenny, J. F. et al. 2009. *Estimated use of water in the United States in 2005*. U.S. Geological Survey Circular 1344.

McNichol, E. and Waxman, S. 2017. "States faced revenue shortfalls in 2017 despite growing economy, policymakers can take steps to strengthen their tax systems and reserves." https://www.cbpp.org/research/state-budget-and-tax/states-faced-revenue-shortfalls-in-2017-despite-growing-economy.

Miranda, R. A. and Picur, R. D. 2000. "Benchmarking and measuring debt capacity." GFOA, Chicago, IL.

NCES. 2003. "Financial accounting for local and state school systems." https://nces.ed.gov/pubs2004/h2r2/ch_3.asp. Accessed 2/11/16.

NRC. 1997. *Safe Water from Every Tap Improving Water Service to Small Communities*. National Academy Press, National Research Council, Washington, D.C.

Ori, R. J. and Mantz, B. A. 2013. "Preparing for rate studies and bond financings: Is your utility ready?" *FWRJ*. 64(5), p. 6.

Pacific Institute. 2009. "Per capita water use in the U.S. drops." (For Immediate Release: October 28, 2009). http://pacinst.org/news/397/. Accessed 2/13/16.

Pew. 2015. "The American middle class is losing ground." Pew Research Center, Washington, D.C. http://www.pewsocialtrends.org/2015/12/09/the-american-middle-class-is-losing-ground/. Accessed 2/11/17.

Piketty, T. E., Hess, P., and Saez, E. 2004. "Income inequality in the United States, 1913–2002." file:///E:/CGN%206506%20Infra%20Mgmt/9964%20piketty-saezOUP04US%20(2015_03_08%2017_53_48%20UTC).pdf.

Rahill-Marier, B. and Lall, U. 2015. "America's water: An exploratory analysis of Municipal Water Survey Data." file:///E:/CGN%206506%20Infra%20Mgmt/9941%20Aquanauts_Study_Data%20(2015_03_08%2017_53_48%20UTC).pdf.

Reilly, T. E. et al. 2009. "Ground-water availability in the United States." USGS Circular 1323, USGS, Reston, VA.

Saez, E. 2013. "Striking it richer: the evolution of top incomes in the United States." (Updated with 2012 preliminary estimates) UC Berkeley. September 3, 2013, an updated version of "Striking it richer: the evolution of top incomes in the United States." Pathways Magazine, Stanford Center for the Study of Poverty and Inequality, Winter 2008, 6–7.

Setliff Law. 2018. "States' latest weapon in the struggle for revenue: Gross receipts taxes and what it could mean for your business." https://www.setlifflaw.com/news/2018/10/states-latest-weapon-in-the-struggle-for-revenue-gross-receipts-taxes-and-what-it-could-mean-for-your-business/. Accessed 12/15/18.

Slemrod, J. 1995. "Income creation or income shifting? Behavioral responses to the Tax Reform Act of 1986." *American Economic Review Papers and Proceedings*. 85(2), 175–180.

Stone, C. and Sherman, A. 2010. "Income gaps between very rich and everyone else more than tripled in last three decades." New Data Show, Center on Budget and Policy Priorities, Washington, D.C. http://www.cbpp.org/sites/default/files/atoms/files/6-25-10inc.pdf. Accessed 12/14/16.

U.S. Conference of Mayors. 2012. "USCM staff report on water and wastewater affordability." November 16, 2012. Washington, D.C.

USDA. 2014. "Rural American at a Glance, United States Department of Agriculture Economic Research Service." Economic Brief Number 26, November 2014. https://www.ers.usda.gov/webdocs/publications/eb26/49474_eb26.pdf.

Varela, L. 2016. Report of the Legislative Finance Committee to The Fifty-Second Legislature First Session Volume I. "Legislating for results: Policy and performance analysis." January 2015, For Fiscal Year 2016, New Mexico Legislative report, Santa Fe, NM. https://www.nmlegis.gov/lcs/lfc/lfcdocs/budget/2016RecommendVolI.pdf.

WRF. 2014. "Chapter 2: Assessing the revenue resilience of the industry's business model." Water Research Foundation, Alexandria, VA. file:///E:/CGN%206506%20Infra%20Mgmt/7%204366_Chapter%202%20josh.pdf.

12

OPERATING ENVIRONMENT

Every organization that provides infrastructure services requires an appropriate staff to operate, maintain, plan, engineer, and manage its assets (see Figure 12.1). Most local infrastructure organizations are heavy on operations, and it usually serves as the largest portion of any agency budget. Part and parcel to operations is some means to manage the organization; to direct the vision. Both tend to be internal to the organization and rely on the experience of the agency. Maintenance and capital construction tend to be related to operations and may or may not be included as a part of the operations budget. However, both tend to be allocated based on available revenues as opposed to long-term planning as discussed in the prior chapter. Non-operations and support services like purchasing, budget and finance, legal services, insurance, human resources, public information, technical information, and general management are normally internal to the organization, and because they are necessary for operations, may take priority over certain maintenance or capital expenses. Allocation of these non-operating costs should be prorated over the operating budgets or personnel. With respect to finance, human resources, and purchasing, none of these services should be deemed as frontline agencies—they provide services only within the organization and their mission should be internal only. They, like management, are responsible for acquiring the necessary resources for the other portions of the organization so that the line organizations (engineering, operations and maintenance)

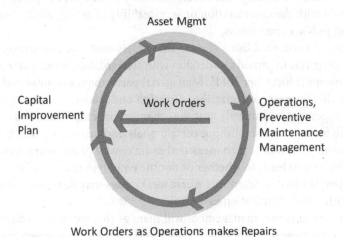

Figure 12.1 The infrastructure construction and maintenance process—this is a cyclical function starting with the capital improvement plan (CIP) process, adding the asset to the inventory, operating the assets, making repairs, and then ultimately reconstructing it.

can do their work. Each function is important to the success of the organization and failure of any part may cascade to other portions of the organization. Most design and construction is outsourced because the skill sets are less cost effective to retain in-house.

MANAGEMENT

Management has many responsibilities. There are seven major areas that management must focus on (Bloetscher 2011):

- *Operational planning*—where the failure of operations will lead to potentially significant impacts on the community and include maintenance, system monitoring (like water plant operations or traffic control), data acquisition, and resource utilization
- *Administrative*—including staffing, communication, performance monitoring, data analysis and reporting of traffic flow and delays, monitoring water quality, tracking work orders, or water volumes sold
- *Project planning*—scheduling, construction, project oversight, and public discussion
- *Finance*—revenue generation, accounting, and paying bills
- *Emergency planning*
- *Organizational planning*—human resources, knowledge retention, and training
- *Communications planning and strategies for media relations*

In order to facilitate the responsibilities of management, management needs a means to organize information, and then use this information to make decisions on the allocation of resources (human, financial, material, equipment) to implement those decisions and ensure desired results. This includes developing a hierarchy of authority to delegate responsibility and authority for decisions (which can be a huge problem if too much or too little delegation occurs) and setting expectations for supervisors. The success of those delegations will be measured through the accumulation of data in the field and from internal activities and budgets in order to monitor progress and revise decisions on how resources can be more effectively used. At the same time, it is management's responsibility to address the public information parameters associated with the deterioration or serviceability of assets, while meeting regulatory requirements and public expectations.

Astute management will establish a climate that unites staff and encourages a commitment to success and willingness to provide leadership with stewardship of the assets and with utility-customer relationships (Bloetscher 2011). Managerial performance is measured by overall project management effectiveness, organization, direction and leadership, and team performance. Performance management is ultimately how the organization, employees, and the managers will be evaluated. The focus is on establishing certain goals and criteria by which the employee is expected to perform. The work is then measured as successful or not when it comes to meeting those criteria, which often leads to whether or not the employee is successful in the job. Performance management is a tool to determine where weaknesses may lie versus strengths within the personnel and within the organization as a whole.

Since the expectation is that management will manage the system on a day-to-day basis effectively and ensure regulatory compliance, public and local officials expect that managers will resolve problems and implement policy as directed. To this end, the following are the questions that the three bond-rating agencies (noted in Chapter 11) try to answer to discern the utility's managerial capabilities (Bloetscher 2011):

- Are there institutionalized management systems for controlling cash, debt, and budgets?
- What is done to attract and maintain qualified managers and middle managers within the organization?
- Are appropriate and customized financial management goals and standards developed that take into consideration current and future public health, safety, and welfare obligations and/or meet economic sustainability goals?
- How are costs controlled that seek to identify the extent at which methods have been examined that might promote economies and efficiencies in providing services to the customers—not just whether rates, fees, and charges are low?

The financial controls and policies are not the only concern, but also the quality of management. Reliability and continuity are important elements of management, as well as exhibiting a willingness to make hard choices.

Likewise, management must address ongoing trends with technology, training needs, and workforce changes. The demographics of employees and their workplace attitudes continue to evolve. The skills of yesterday may not apply tomorrow. There is more education available and more information exchanged. Any organization must keep up with accelerating changes in technology and communication. Additionally, the old adage or expectation that *all employees will be treated equally* does not apply because it is not possible to treat everyone exactly the same. People respond to different stimuli. Some employees need rigid repetition to be successful and successful management requires addressing these changing trends. Table 12.1 outlines the differences in what employees want and what the industry needs—these are not aligned.

Table 12.1   Changes in employee expectations between what current employers want compared to the needs of the infrastructure system

Workforce Desire	Infrastructure System Needs
Recognition	Everyone integral part of team
Pay	Budget
Freedom with time	Rigid schedules
Flexibility	24/7
Technology	Not particularly, yet
Challenging	Repetitive work
Personal interaction	Size => economy of scale
Nontraditional benefits	Traditional benefits

Visioning

Management and policy-setting are not the same things and in most public-sector organizations, they are performed by different entities. Management directs the overall operations of an organization while a governing board sets policies that managers carry out. To set these policies, and therefore management direction, a series of goals and objectives needs to be set. Visioning and goal setting are performed among upper management and elected or appointed officials. However, the results are likely to be problematic if the entire organization is not involved because the entire organization must buy in. This takes a degree of leadership, and as will be discussed, management and leadership are very different things. Neither result from entitlement or position, just as respect is not automatically granted, although that may be difficult for some managers or local officials to accept.

The overall goals for the organization are set with respect to policy, finance, and vision. The policy piece means setting the parameters for operations. Policies involve anything from levels of service and usage of the service, to how service is provided and how it is paid for (the finance piece). Setting policies based on understanding and managing the system's assets requires important information related to the ongoing future needs of the system—paying for needed upgrades. However, the only method to develop that vision or future direction is through the planning process.

Planning is a process that is utilized to reach a vision of the organization as defined by the customers or the governing board, or to meet certain demands for service projected to be required in the future. Development of vision allows all members of the organization to work toward a common set of goals and outcomes. A previous paragraph noted that governing board responsibilities primarily include setting overall vision for the organization and tasking management to implement the vision. But what is a vision? It is supposed to be a concept of what the leaders and managers want their organization to be like in the future. A vision may also be a vehicle for change, and those developing the vision are outlining the change they want in the organization. What services are to be provided? What roads are needed? What bridges need to be expanded? What water sources are to be used? Is energy self-sufficiency a goal? How about wastewater reuse opportunities, incorporation of stormwater to sources waters, etc.? All possible ideas, but they only scratch the surface of the universe of opportunities that might exist. The key is change, which normally requires thinking outside the proverbial box. Change rarely comes from doing the same thing over and over. Change requires innovation. So as a result, by its very nature, the status quo is not leadership because no change is required. Managers who "don't rock the boat" may be excellent managers, but they are not leaders. Elected officials whose mantra is not to raise rates are not leaders either. Customers often are a great source for defining a vision. They will tell local officials what services they want.

With any infrastructure system, it is imperative to have ongoing planning and communication activities since many necessary improvements and programs take months or years to implement and/or complete. Planning the future of an infrastructure system's needs and communication of those needs to the public are critical components of the visioning process (see Figure 12.2). Planning should be undertaken on a regular basis by all enterprises in an effort to anticipate needs, clarify organizational goals, provide direction for the organization to pursue, and communicate each of these to the public. Without a short- and long-term plan to accomplish future needs, the agency will suffer errors in direction and build unnecessary or inadequate infrastructure. Also, the agency may pursue programs that later are found to provide the wrong information, level of service, or type of treatment (Bloetscher 2009). Planning can provide for a number of long-term benefits—but most important, a *vision* for the organization and its service delivery.

Complementary tasks of local officials include approving the annual budget, creating financial plans, developing strategic plans, approving capital improvement, and special programs. Typically, only the governing board can approve agreements, issue debt, or use reserve funds. The board is tasked with setting policies and procedures that govern operations—including such items as the billing and payment policies, rates, fees, charges and/or taxes, and other customer service policies. The board also provides the staff with standards of performance and accountability for daily operations. None of these represent leadership—they are management.

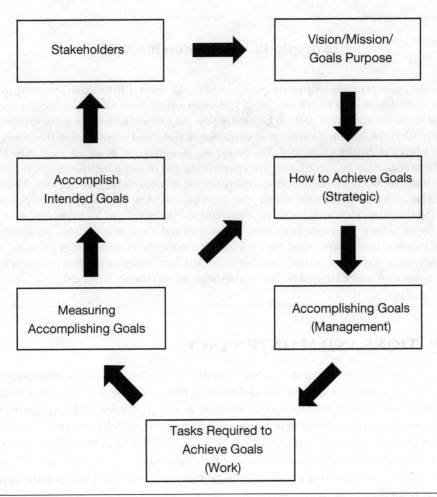

Figure 12.2 Planning for future needs of infrastructure systems and communicating those needs to the stakeholders (public) are critical components of the visioning process outlined in this flow chart. *Source*: Grigg 2016

Doing a good job of these tasks sets the organization on a path of organizational health. Organizational health is an important driver, especially for good employees. *Good* and *great* organizations are looking for people who strive to be great, who have the requisite skills, and who can provide perspective on their areas of responsibility (Collins 2002). Having differing opinions is good, assuming some consensus can be reached. Good or great organizational culture is created by:

- Developing consistent direction
- Establishing an internal way of life
- Determining internal effectiveness
- Setting expectations

This is where managers rise to achieve effectiveness in the organization or fail personally and organizationally. It is the latter issue that most supervisors do not understand—their responsibility is to the organization.

Example 12.1: A Leadership Void

A medium-sized, but relatively newly incorporated public sewer district called a meeting to talk about leadership to some elected officials. A leadership specialist was asked to attend (for free) by one of the commissioners in order to help initiate the discussion. He did. The public was present in force. They brought up the concept of developing a vision and expressed that they wanted to know where the district was headed. The public was encouraged to speak out about their ideas, which encouraged more conversation. The conversation was all very good and engaging until one of the board members informed everyone that vision statements were the job of the attorney and he would just have the attorney write one up. That did not go over nearly as well as that board member had hoped. He was abdicating his role in overseeing the utility and any leadership role he might have hoped to have. The public knew what they wanted and it was clearly change, something the board member clearly did not want. They wanted input and leadership. They wanted a vision. They got something very different. The conversation turned from engaging to unhappy immediately. The commission ignored the public. Years later, the public still has not forgotten.

OPERATIONS AND MAINTENANCE

Directly related to maintaining the system is system reliability. An old, inadequately funded maintenance program will increase the likelihood of failure despite the best efforts of staff. Infrastructure systems strive to make service available to all customers at all times and the regulations are designed with this goal in mind. Service disruptions, which can have serious political consequences to local elected and appointed officials, are designed to minimize impacts. However, the management of operations requires an understanding of maintenance and operating needs of the system, and having personnel, equipment, and funding in place to allow operations personnel to do their jobs.

The economic growth of the United States in the late 1990s was in many ways directly related to an improvement in productivity among American workers. The utilization of technology helped spur this growth because it made the collection of data easier, and therefore, better analysis of conditions could be undertaken and operations optimized. Sensors and computers are now used for water and wastewater treatment plants and pumps and for the power grid and traffic logistics. Operations and maintenance manuals provide information on the equipment, parts listings, and frequency recommendations for preventive maintenance, along with replacement programs that should be pursued by field crews and plant operators. Included in the manuals should be shop drawings and flow schematics that will help staff in the future when improvements are made. All of these data should be part of the operational asset management plan that should be part of every vision of an entity and every *toolbox* of management. Grigg (2016) outlines the basic concepts for asset management success in Table 12.2. Table 12.3 outlines the asset management goals.

Given tools and maintenance data, operations management still requires that there be a clear set of goals and objectives and an appropriate organizational structure to be in place, along with appropriate lines of communication. Good communications will facilitate maintained productivity within and between work groups because operations personnel have a variety of jobs to do on an ongoing basis, such as:

Table 12.2   Concepts for asset management success

Strategy	Action
Commitments	Asset management system in place
	Policy not to defer maintenance
	5-year capital plan
Planning for maintenance management system (MMS)	Interagency discussion
	Work order historian to denote frequency of issues
MMS	Emphasis on maintenance
	Schedule for preventive maintenance
Condition assessment	Condition asset program
Repair and replacement (R&R) program	Asset replacement schedules
	Condition monitoring
	Life-cycle decision making
	Require assets be designed to be maintained
Data	Inventory of data

Source: Grigg 2016

Table 12.3   Goals for asset management

Goals
Better customer service
Better preparation of capital plans
Cost control
Community support for capital improvements
Guide operations and maintenance practices
Regulatory compliance

- *Day-to-day jobs*—this will be things like testing water, monitoring chlorine residuals, operating treatment plants, daily log sheets, cleaning drains, street sweeping, replacing signs and traffic light bulb, and fixing potholes
- *Maintenance of facilities*—general cleanup; painting, striping, verifying that generators are working; back-up systems; emergency procedures; greasing of bearings; leak checks; repairs to concrete
- *Periodic jobs*—such as requisitioning and accepting the delivery of chemicals, parts, or supplies; pump repairs; pump tests; reprogramming traffic control devices; well or intake cleaning; and talking to supervisors
- *Reactive issues*—something went wrong and needs to be fixed

The crews normally meet at a central site where they receive work assignments, materials, and supplies to accomplish the day's work. They then leave for the job site, which may be some distance away. Having the appropriate supplies and materials is important to maximize field crew productivity. Field crews then spend time on the following activities:

- Excavation and routine repairs
- Preventive maintenance of pumps, drains, catch basins, and generators/mechanical equipment

- Emergency repairs
- Response to customer requests or inquiries

Management is in part responsible for productivity. Managers must ensure that field crews have materials and parts in their inventory to reduce delays in arriving to the job site in a timely manner. In addition, there needs to be effort and thought put into scheduling of work and workloads so that *down time* is minimized. The proper tools are critical. Productivity is an important aspect of maintenance—and better tools, procedures, and communication should translate to the ability to accomplish more work. Management should note that assessing productivity may concern employees about job losses without understanding that every task may be different from similar tasks performed elsewhere. Optimal crew sizes will vary by job and work—crew sizes that are too large or too small create inefficiency. Two-man crews are often all that is needed for most jobs; one-man crews are not recommended for safety reasons. In a high traffic area, more staffing is likely needed.

When looking at the type of people on the site from a labor perspective, there are a variety of tasks, and therefore, different job classes that should be expected (Bloetscher 2011). There will be some or all of the following: a supervisor or crew chief, laborers, drivers, equipment operators, and perhaps skilled labor, such as pipe-layers, concrete finishers, and electricians. Any job requires the right skills and equipment to be productive; it requires more than the day labor ditch digger and a shovel to lay pipe, concrete, sewers, or other general construction or repair work productively. The appropriate people with the right skills need to be in place. Good concrete finishers will save a significant amount of time and effort by ensuring that the concrete looks appropriate. Likewise, experienced pipe layers will do it correctly, so that there is a limited amount of follow up and repairs required due to construction deficiencies. Sewer lines require a different crew than a water main; water mains are not required to be on grade, whereas sewer mains must be on grade and require a crew that is used to dealing with lasers and grade lines. Likewise, roadway crews require persons who are familiar with grade lines and surface paving, but do not need excavation experience. Concrete and asphalt are different and require different crews.

The appropriate construction equipment operator is required for each. An equipment operator may be able to operate several pieces of equipment, but it is unlikely that he or she will be able to operate the whole range of equipment that is required for all jobs. Training and ongoing reviews are likely needed for pieces of equipment, such as backhoes and excavators, to ensure that the proper skills remain in place. For lesser-operated pieces of equipment, refresher classes may also be beneficial. Making sure that having the right tools to do the job is the province of management.

As a part of operations and maintenance, targets of work units for productivity purposes are required. In other words, if the meter reader is expected to read 200 meters per day, someone needs to track to see if he or she actually reads that many or if there are complicating factors (recall the earlier example where the meter reader had many other tasks and could not keep up). At the same time, managers and supervisors need to determine if the goals are overly easy. For example, in some agencies in the past, solid waste workers took eight hours to accomplish their jobs. Once they were converted to a system where they could go home when they were done with the work, workers were suddenly done in three or four hours. Clearly the managers and supervisors were not gauging whether or not the employees were productive during the time. The result is that much more work could get done, and in the organization where staffing is typically limited, the ability to do more work with existing staff is an important consideration.

Crew productivity is difficult to measure without tracking the work they perform so a mechanism must be in place to record field activities. In a perfect world, the infrastructure organization will have complete, detailed, and accurate system maps to locate all assets (with installation date and material). The organization would also have details on all repairs that were made and the conditions surrounding those repairs, reports on all new assets installed and conditions found during excavation, and a number assigned to each asset. It is surprising how few organizations actually have maps and locations of their assets.

Good record keeping should be started as soon as practical to build a base of knowledge that will improve the efficiency of repairs. Staff must be assigned to perform the record keeping, data entry, and mapping. System maps can be maintained on a geographic information system (GIS), Access databases, Excel spreadsheets, or other computer programs—simple is best. Roadway surface data, pothole repair history and location, repaving history, and other data should be included in a roadway layer. Whenever an asset is removed or repaired, information should be recorded about its location, conditions of the soil, and other relevant data, combined with a photograph. Photographs are always useful, especially digital photos that can be put into report files and stored online. This data can be compiled for later use when performing condition assessments, identifying areas that may potentially fail, or excessive wear areas. Some entities may wish to keep a small piece of pipe tagged with the date and location to help with long-term assessments.

A work order tracking system is required to accomplish this goal (see Chapter 7). Figure 12.3 outlines how work orders should be used to achieve goals that are used in the process of asset management. While many agencies have not implemented work order tracking systems, work orders are helpful in dealing with reluctant governing bodies and managers. A work order tracking system and photographs are hugely beneficial in dealing with the public. They demonstrate

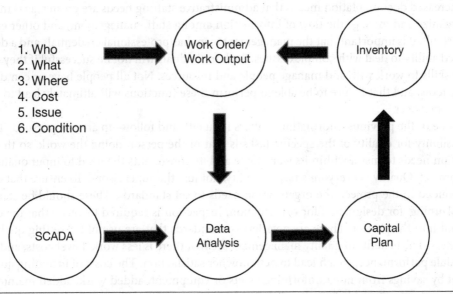

Figure 12.3 Flow chart of how work orders are the central operations tracking tool. Information comes in on an issue (who, what, where, etc.), and then the work orders are created. However, the work orders require input from inventory developed as a part of the asset process. Work order outputs and supervisory control and data acquisition (SCADA) are analyzed and the results are fed into the capital planning process.

that as infrastructure ages, it is normal for maintenance to increase. Tracking maintenance costs demonstrate this trend, but the trend can only be accomplished by tracking work orders and the investment of time, machines, and parts.

Work orders and record keeping help resolve another problem—the loss of institutional knowledge when people leave the organization. Given that infrastructure is designed to last 50 to 100 years, careers go by before replacement is needed. Loss of institutional knowledge about infrastructure systems continues to be a significant issue for most communities because it is so difficult to retrieve. Many assets can be covered up through paving and other construction activities and thus, may be lost. Connections or repairs that have been done on pipe or field modifications that have been made to solve a given problem are also likely to be lost because very few pictures have been taken of these repairs. This is a flaw in the current practice for engineering oversight of construction projects. Record drawings rarely include photographs, which could easily be inserted. The fact that institutional knowledge will be lost indicates that additional training will be required. Having employees who are getting set to retire train the incoming employees is important for knowledge transfer.

Parts inventories should be tied to the work order tracking program to provide some control over the amount of parts and supplies that are available and used in the field. This has the side benefit of reducing losses of inventory. All of these issues work together to gain information to help the asset through its life.

More recent developments include the incorporation of maintenance management systems through computer programming that attempt to optimize the time at which certain maintenance activities should take place in order to ensure the long-term reliability of infrastructure systems. Maintenance management systems are linked to the fiscal side through asset management practices, whereby asset management looks at the additions, subtractions, and depreciation of the physical assets with time.

Increased documentation means that administrative staffing needs are greater, as is the need to organize and manage the flow of information among staff, management, and other entities. As a result, it is important that the management team has professional credentials and a demonstrated ability to deal with complex issues. Such individuals will not be successful if they do not have skills to work with and manage people and resources. Not all people have or can develop these tools, and that failure to be able to perform these functions will ultimately lead to failure by the manager.

None of the previous information matters if quality and follow-up are not pursued. The responsibility for quality of the specific tasks is that of the person doing the work, so the organization needs to instill within its work force and its consultants the need to input quality into the project. Quality is everyone's responsibility, not just the supervisor's. To ensure that quality is included in the project, the organization needs to set standards. These should be standards for planning, for design, and for construction. Inspection is required to verify that the quality is built into the project. There are a variety of standards that are useful for setting quality parameters. The benefits of quality management are that there is less work, lower costs, and better schedule performance, which lead to better owner satisfaction. The costs of including quality is offset by savings from non-conforming assets or equipment, added work, added maintenance and operating costs, added repairs, shortened asset life, unreliability, and dealing with that work that does not meet the requirements. This may mean a little more cost up front, but what the organization should be looking at is life-cycle savings. The costs to ensure the necessary performance are up front but not significant in the total life cycle of the project (Chapter 13). Poor work quality costs the owner significantly over the life of the project and creates a much higher life-cycle cost than if the work had been constructed properly up front.

Monitoring the project ensures that specific results are obtained to ensure compliance with quality standards and identifying ways to eliminate causes of unsatisfactory results. This will include inspections, reports from inspections and planning, and quality control logs to indicate where or when corrective actions must be taken or have been taken. Such records will permit utility managers to make necessary decisions and recommendations on improvements and programs. Records must be accurate as to work performed, parts used, and findings. Record drawings and files should be updated continuously to reflect these differences. Daily logs and monthly summaries of work are useful to track field crew performance.

ENGINEERING

The goal of engineering is to evaluate, assess, and design infrastructure solutions—new, replacement, or major rehabilitation. Engineering services involve the design and analysis of infrastructure—condition, needs, and solutions to problems. Figure 12.4 outlines the process they use to design, construct, and then operate assets. Most public entities encounter the need to construct major facilities on an ongoing basis. With proper planning and anticipation of the needs for construction, the new capital projects can be constructed with minimum impact to operations. For the most part, large capital items are designed infrequently, which makes the justification for having in-house expertise difficult to make. For example, if a major treatment plant upgrade occurs every 10 years, the design engineers would be used less judiciously in between. The same goes for roads, bridges, rail, and many stormwater systems. Hence, having the ability of external expertise is useful and productive. It also can save time and money, and the

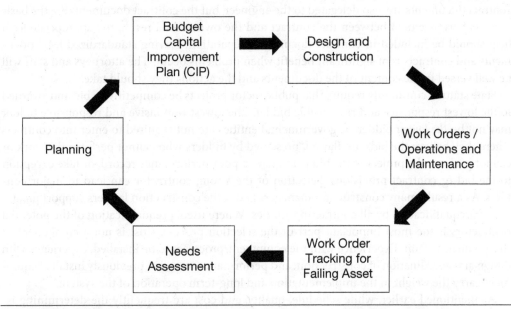

Figure 12.4 Flow chart for the asset development and operations process—the process is cyclical. The CIP results in construction, and then operations. The operations process creates work orders for maintenance, which feed into a needs assessment, and a plan for new capital.

staffing leaves when the project is complete. For capital project management, the staff is only needed during the project.

Engineers design how systems will be rehabilitated or replaced—repairs (more specifically, routine repairs) are the province of maintenance and operations staff. There are some means to improve the process. First, the internal staff will need to define the project fully. This may require external help. Next, there must be a procedure to select the external services. Standardizing forms and procedures that apply to all projects will facilitate the efficient and effective construction of capital projects to repair or replace assets. Included in these documents should be legal documents, contracts and change orders, technical specifications for water, stormwater, roadway, and sewer improvements, and various formats for elected official agenda items. The result is consistency and ease of administration for the capital projects.

The primary issues of importance when implementing any capital construction upgrade are cost, construction quality, and completion schedule. All of these issues are separate, but interrelated. Therefore, the entity must choose which objectives are most important since that choice may adversely affect the other two—i.e., increasing quality requirements may increase the schedule, whereas accelerating the schedule and maintaining quality will increase costs. While schedule, quality, and cost are frequently the determining issues, they are not the only issues of concern—control over the final product is also a key concern. Technological changes are rapidly making the operation of infrastructure systems more complicated and emphasize the need for greater expertise to deal with ongoing operations and maintenance for the future. Personnel must have the knowledge and abilities to deal with complicated electronics, machinery, and new treatment methods, and also understand regulations well enough to maintain compliance.

The development of appropriate construction specifications and contract documents is important to ensure that a construction project proceeds as smoothly as possible. The specifications, like the plans, are the responsibility of the design engineer. Too often, preparation of the contract documents are also delegated to the engineer, but the contract documents are the basis of a legal arrangement between the contract and the owner. As a result, the appropriate legal help should be included in contract document preparation. Having standardized legal documents and contracts is of significant benefit when problems occur. The attorneys and staff will be well versed in the content of the documents and the actions they should take.

State statutes commonly require that public sector projects be competitively bid and awarded to the lowest responsive and responsible bidder. The lowest responsive and responsible bidder may not be *the lowest* bidder, as governmental entities are not required to enter into contracts where they believe the bids are flawed, presented by bidders who cannot perform the work or cannot perform the project at the bid price, have a poor performance record, or take exception to the bid or contract provisions. Selection of the wrong contractor can lead to major conflicts. As a result, many construction managers and some construction lawyers support greater use of prequalification by all contracting entities. Where used, prequalification of the potential contractors is the most important part of the selection process. Cost is not a component of this prequalification. Experience with the equipment/processes to be installed, experience with design and coordination of construction, and performance record of previously installed equipment carry the weight in the implementation and long-term operation of the system.

As mentioned earlier, while schedule, quality, and cost are frequently the determining issues, they are not the only issues of concern—control over the final product is also a key concern. Also, staff often has specific ideas on the details of the project—the more details, the more control. Retaining control means more involvement in design and construction efforts, which may detract from other activities. Management needs to be cognizant of this tradeoff.

However, productive oversight can be lacking since public agencies often build a limited number of new facilities during any given employee's tenure, while outside consulting engineers may have specialists on staff that have built similar facilities many times in the past. By assuming a degree of control, staff of the infrastructure entity will reduce responsibility of the design team and the contractor because decisions the contractor might be forced to make will have already been made.

An infinite variety of issues can arise during construction, but generally construction projects involve observations of the contractor's work, progress reports and meetings, and the processing of pay requests. Other issues involve field changes and time extensions, which may lead to change orders. Defective plans and specifications, as a result of errors, omissions, or unforeseen field conditions, may involve additional changes.

Progress reports are developed to track progress on the project. Progress reports have the following parts:

- A review of the status of the project overall (from a macro-scale)
- A review of what has happened since the last progress report (in detail)
- A review of what will happen in the coming progress report period (in detail)
- An update of the schedule

In addition, the owner needs to know problems, foreseen and unforeseen, with the plan/project. The progress report needs to identify areas requiring input from the owner, as well as recommendations on how to resolve problems.

Project cash flow is important to consider. Contractors want to be paid monthly. However, the amounts the contractor desires will vary from month to month. To develop the cash flow for the project, the contractor needs to define the work load for each task and set up the schedule. Generally, cash flow for a project starts slowly, depending on whether the contractor can be paid for stored materials under the contract. The middle part of the project includes the majority of the installation, so money is drawn quickly. The final stages of the project include finishing work. This is generally slow work that has little return for the contractor. To ensure that the finishing work is actually completed, and all the record drawings and other deliverables are delivered by the contractor, most contracts permit an amount to be withheld each month as *retainage*. The final contract payment is the retainage for the project, paid upon final completion of the punch-list and receipt of all deliverables. Management should be cognizant of the time commitments of staff, which can become a more important issue than cost on projects that are not well defined, or where the contractor is difficult to deal with.

HUMAN RESOURCES

There are three support agencies that must be well managed for the organization to function and operate effectively: human resources, finance, and purchasing. Every organization must hire people to get the work done. A good human resource division is needed to be able to identify the people with the appropriate skill, based on the needs of line organizations. They will post job vacancies, collect résumés, perform background checks, check social media, and verify education and experience in order to whittle down the number of candidates and to improve the efficiency of review by line supervisors. Contacts in the community are important when it comes to finding good candidates to apply for the jobs. Setting up interviews and finalizing paperwork to hire candidates concludes the process.

As Collins (2002) notes in *Good to Great*, great organizations strive to place the best qualified people in the proper positions. The concept is that people who have the skills and the drive to do their work will do it well and will be happy. They will work as a team, demonstrating respect for one another while solving organizational issues. The key is making the organization great, not the individuals. The latter comes with recognition of the organization. Organizations that just fill positions are generally more hostile.

There are four barriers to filling positions. One is a highly competitive environment. If there are a lot of companies hiring for the same position, it will be difficult to secure the best people. Private entities have a better opportunity to deal with that problem in the short term because they're not required to go through elected bodies to increase salaries (something that's generally avoided during the fiscal year). The second issue is inadequate human resource support. If job descriptions are poor, recruiting is limited, or salaries are inappropriate, good employees will not apply for the job. Third, the lack of training opportunities, educational opportunities, or a skilled labor pool will limit options. The options that do exist may be expensive or require extensive training. Fourth, inadequate compensation will not appeal to better workers or those currently with jobs.

Once they are selected, new employees who meet all the criteria within the objectives of recruitment will need to be properly acclimated. Unfortunately, many organizations do not do a good job with the acclimation of employees; they send them out to do the job, and the employees succeed or fail on their own. This does not build a long-term support structure, nor does it indicate a long-term willingness by management to support the success of the employees. Instead, it is better to assign the employee to a veteran of the department so that they can learn the organizational goals, work guidelines, and objectives (i.e., mentoring).

A major issue associated with management is the organization and staffing to accomplish the tasks needed to accomplish the organization's mission. From an organization and staffing standpoint, matching organizational goals with operations defines how the work will actually proceed and by whom. In order to determine organization and staffing, certain issues need to be identified. First, there needs to be a manager and it needs to be determined what the manager's assets and skills should be. There may be assistants to the manager and their skills and assets need to be determined as well. An office or some sort of office to work from needs to be secured and then a team of employees should be hired to actually accomplish the tasks. The manager may be the most important of these positions because it is through the manager that the direction and coordination will occur. The manager may be the supervisor or vice versa. Supervisors will certainly have management skills.

Human resources responsibilities do not stop at hiring. Pay scales, review of pay grades, review and creation of positions, review and creation of job descriptions, and the like are ongoing tasks to ensure the organization is competitive and is a place of choice for employment. Excessive turnover of employees can often be rectified by the right human resource professional. Training and development of employee benefits are usually responsibilities of human resources. A good active human resource group can improve the productivity and perception of the organization by identifying and holding good, engaged employees.

FINANCE/BUDGETING

Finance is responsible for collecting revenues and paying for resources used by line agencies. Part of this responsibility involves methods about how collection occurs (monthly bills, tax bills, separate invoices) and cutting checks to pay for services. They may also be responsible for the creation of the budget. The budget is a plan based on the needs of the line agencies and should identify how funding will be secured (see Chapter 11). Support will include notices, public hearings, and creating the tracking system to report responsible status of the expenditures on a regular (typically monthly) basis.

Budgets are an annual plan for expenditures. Based on the budget, finance is responsible to track the revenues and expenditures. Ongoing status of expenses is a critical need of line organizations. Because infrastructure organizations normally are public entities, there are laws and rules that require judicious tracking of expenditures to ensure generally accepted accounting principles and reporting for financial conditions. Comprehensive annual financial reports (CAFRs), as discussed in Chapter 11, are required in most states. Finance is responsible for these annual audits.

Finance may be asked to create and track metrics for management. This is a potential area of conflict since those in finance are normally not overly familiar with operations. Creating metrics such as the number of breaks fixed per crew, catch basins cleaned per day, or roadway potholes fixed, may not relate to the actual difficulty in the field. The metrics should probably be defined by the line agencies and then tracked by management or finance.

A failure by the finance group to be effective, pay bills on a timely basis, or provide timely information will create significant conflict in the organization and diminish public perception. If bills are not paid, vendors and contractors will not perform. If tracking of expenses is poor, it will be difficult for line agencies to plan work, order parts, or hire contractors. It will also complicate the ability to meet emergency needs due to the reluctance of outside agencies to work with the organization based on poor contact with finance.

PURCHASING

Purchasing is the third critical support service agency to the organization. Failure in the purchasing area will create line failures that may create or exacerbate negative public perception. Table 12.4 outlines the steps to purchase items for most public infrastructure agencies. These rules are cumbersome. As a result, management has the responsibility to make purchasing a smooth, integrated function. The responsibility of the purchasing component of the organization is to acquire the materials, equipment, chemicals, and repair parts in order for line organizations to do their jobs. For most public sector organizations, there are statutes that create procedures for the acquisition of resources—and many organizations supplement these statutes with more comprehensive policies. In all cases, the goal is to spend the public's money in the most efficient manner by securing the best prices for resources, assuming these services meet some minimum quality requirement.

Efficient and timely acquisition of resources is critical. Failure to acquire resources will delay processes and force line organizations to make less-than-optimal decisions on repairs in the interim, meaning that the issue will likely need to be revisited later. The purchasing function is procedural, so it must rely on timely and complete specifications and design documents that are created by others in order to be successful. The procedural aspect to protect against theft, which may or may not be the responsibility of purchasing, will permit line organizations to focus on what they do best—getting the work done. A potential for conflict arises in two areas: (1) emergency repairs where the appropriate parts or skills might not be on hand, but for which there are immediate needs for repair and (2) on small projects that crop up regularly, but are generally unplanned. There are two ways to work though this. One is to have a series of blanket contracts

Table 12.4   Typical steps for purchasing items within public infrastructure agencies

1	Estimate of quantities
2	Obtain quotes from one or more parties[a]
3	Tally the quotes
4	Determine appropriate quantities and unit prices
5	Determine how much inventory can be stored
6	Get signoffs and authorizations[a]
7	Request purchase order
8	Place the order with a purchase order
9	Receive order and store it
10	Verify that quantities delivered match those ordered
11	Determine any defective material and reload it on truck
12	Adjust shipping receipt
13	Receive replacement parts for defects
14	Verify the count
15	Inspect replacement parts for defects
16	Receive invoice
17	Match invoice with delivery tickets
18	Process pay request
19	Pay invoice
20	Record payment in accounting system

Source: Grigg 2016
[a] Assume governing board approval is not required

bid ahead of time for repairs and parts. Parts inventories are important but must be tracked to avoid theft. The second is to create procedures to deal with the unplanned small items—how to acquire services and then obtain approvals after the fact since most public service acquisitions require governing board approval.

PRIVATIZATION

A recent study reported in *Governing* magazine is that privatization of services may increase income disparities within their communities (Farmer 2016). While the argument with privatization has been cost savings, improvements in efficiencies are rarely where the savings are. Private organizations must make profits for shareholders and do more with less—the only real means to accomplish this is to pay less with fewer benefits for workers (Farmer 2016). This is not a surprise. But, surprisingly, the same study found that not-for-profit organizations that perform services do not increase this disparity (Farmer 2016). However, there are reasons that certain services should be contracted to the private sector. Engineering is an example. Rarely do organizations continually design large projects like water plants or major highways, so having specialized staff for those types of projects is best contracted to outside experts. That does not mean that routine engineering like water main replacements or repaving projects that are done continuously should not be maintained in house.

DEALING WITH THE PUBLIC

Perception is reality in government. As a result, the activities of any infrastructure agency will be scrutinized constantly. Criticism will occur, often undeserved, but may be related to a lack of communication. There will be times when failures occur simply because pipes are old and they fail, underground conditions differ from what is anticipated, lightning strikes, severe weather occurs, corrosion breaks through metal, weathering cracks pavement, floods overwhelm the stormwater piping system, etc. Most of these occurrences cannot be predicted. Usually failures are small—blocked sewer lines, water main breaks that have people out of service for a few hours, potholes and other pavement issues that require a couple hours of inconvenience, collapsed or plugged storm drains, or pumps that short out. Minor issues are rarely remembered. Large concerns, mostly from natural incidents like weather, are remembered. Flooding from heavy rainstorms can flood homes and businesses and damage roadways. Damage to pavement can disrupt traffic and commerce. Negative events of this scale can affect public confidence and perception of the entity. This is the major reason why it is important that there is an ongoing public relations and communication effort with the public that maintains public confidence prior to a failure. When a failure occurs, the response to the public should be carefully planned and factual.

There are several issues that should be kept in mind when dealing with the public (especially in the age of social media):

- Attention spans are short (eight seconds—so make it count)
- Perception often matters as much or more than the facts
- The perception emerges before the facts
- Any failure is perceived as a negative regardless of the cause
- Publicity is what many people crave

Keeping these issues in mind when speaking to the public will help with public relations efforts and in maintaining public confidence if (Bloetscher 2011):

- The speaker is qualified to make the responses
- The answers are short and to the point—the press likes sound bites
- The speaker is positive and avoids industry jargon that makes people suspicious
- The speaker clearly explains how efforts are underway to fix the problem
- The speaker avoids negativity, promises, and discussions of costs

Of utmost importance—the speaker must be genuine—body language and the perception of the speaker is as important as the words being said. Of course, things are never this easy. When events happen, the infrastructure system managers will take the spotlight. When it goes smoothly, count on the fact that adequate information was in hand and that issues were discussed internally before the public presentation. That is why it is helpful if there is some planning for responsibilities of staff in the event of a failure, including the training of specific employees who are identified as having a potential role during a problem. A system to funnel information to a central point should be a standard protocol, regardless of who the speaker is. That will help boost confidence.

REFERENCES

Bloetscher, F. 2009. *Water Basics for Decision Makers: What Local Officials Need to Know about Water and Wastewater Systems.* AWWA, Denver, CO.

——. 2011. *Utility Management for Water and Wastewater Operators.* AWWA, Denver CO.

Collins, J. C. 2002. *Good to Great.* HarperCollins, Glasgow, UK.

Farmer, L. 2016. "Privatization may be worsening inequality." *Governing.* http://www.Governing.Com/Topics/Finance/Gov-Privatization-Inequality.Html. Accessed 10/10/16.

Grigg, N. 2016. *Integrated Water Resource Management: An Interdisciplinary Approach.* Springer, New York, NY.

13

LIFE-CYCLE COST ANALYSIS

Effective asset management programs include processes to optimize the life-cycle costs of owning, operating, and maintaining assets at an acceptable level of risk, while operating the system continuously at the designed level of service (RTI, no date). Life-cycle analysis permits the agency to invest in capital investments through a combination of optimizing installation costs of replacement infrastructure with existing operations and maintenance costs, without creating undue risk of failure. Note that the life cycle involves a number of steps, not just construction or operations (see Figure 13.1). Stadnycky (2010) notes that the goal is to establish a manageable program in terms of affordability and ease of implementation to aid in the renewal of infrastructure systems. A level of service (LOS) statement defines the way in which managers and operators want the system to perform over the long term (NMEFC 2006). The LOS must include standards for regulatory compliance and may include additional components such as quality, quantity, and reliability (Underwood Engineers 2014).

The life cycle of assets has four phases. The first phase is the planning phase. The planning phase deals with planning for service delivery. This phase will include input into the capital budget through information from the asset management plan. Various acquisition options should be considered during this phase. The second phase involves construction/acquisition—how

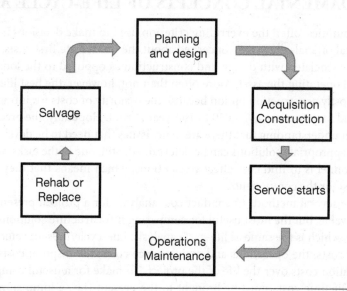

Figure 13.1 The life cycle of an asset involves a number of steps beyond just construction.

best to acquire the asset. The third phase is operations. During operations, maintenance will be required whether in the form of refurbishment, enhancement, rehabilitation, depreciation, or replacement. This phase includes activities of a capital and current nature. It also includes the vast majority of costs associated with owning the asset. Finally, the disposal phase deals with the timing of and disposal of the assets including the disposal costs and specific requirements for the assets (National Treasury 2004).

Ellison et al. (2014) notes that from an entirely economic perspective, infrastructure agencies are most likely to invest in asset condition assessments when the benefits of such an investment is likely to exceed the costs. From a comprehensive perspective, the costs and potential benefits of alternative solutions should be included. A cost-benefit analysis is a private sector model that can be difficult to apply for infrastructure because the benefits that accrue for infrastructure are often difficult to measure, e.g., the benefits resulting from avoidance of an accident or the avoidance of a water outage. So, most economic assessments of infrastructure focus only on tangible benefits, not the avoided economic or environmental costs that may exceed the value of the tangible benefits. The intangible benefits should be included, but because they are not, the true benefits are largely underestimated.

For high-consequence assets, performing a condition and vulnerability assessment is relatively easy to accomplish if the likelihood of failure is perceived to be high and the asset is located above ground. Little assessment actually may be done because repair or replacement would produce such an obvious benefit. In such situations, catastrophic failure and disruption is avoided (but not the avoided intangible costs, which are perceived to be *high*, as opposed to being actually *known*). Avoiding expensive, catastrophic failures can allow the agency to redirect resources to other assets because ongoing condition assessments might allow the agency to save money that might have been wasted in renewing an otherwise good asset. The concept of life-cycle analysis is often used to compare solutions or decide which improvements are best.

THE FUNDAMENTAL CONCEPTS OF LIFE-CYCLE ANALYSIS

For many communities, often the overriding criterion used to make decisions is cost. Unfortunately, many local officials are more concerned about the immediate first costs, which include the capital costs associated with design and construction as opposed to the long-term costs of maintaining and operating the asset. More often than not, however, the best life-cycle solution is not the least costly construction option because the majority of costs for infrastructure are in the operation and maintenance over a 30 to 100-year period. Hopefully, the preceding chapters have helped with understanding that there are many issues that need to be developed before the capital costs for appropriate solutions can be determined. Still, one of the tasks of engineers and operations personnel is to find cost-effective solutions, which means that they should look at the life-cycle cost for the components.

Engineers use several methods to conduct cost analyses for a project: present value, annual value, or future value, but the most useful for comparing infrastructure solutions is the concept of present value, which is the same as life-cycle analysis. Life-cycle/present value analyses compare the capital costs, the operations and maintenance costs, the disposal costs, and any suggested rehabilitation costs over the life of the project. To make for a useful comparison among alternatives, all lifetime costs must be brought to the present (i.e., a lump sum). All lifetimes must be the same. If one option has half the life expectancy of another option, the project for the shorter life expectancy must be assumed to occur twice (more on that later in Example 13.3). Rehabilitation should also be included if history suggests this can occur. For example, buildings typically undergo major renovations every 25–30 years. Rehabilitation must be programmed into the life-cycle analysis. With the correct analysis, the engineer can guide the client to the lowest overall life-cycle cost of a project, thereby reducing lifetime costs and improving performance and reliability of the asset.

DETERMINING LIFE OF ASSETS

The key to infrastructure management is that appropriate capital expenditures are made and properly maintained. In a time when there are limited dollars available from customers, spending large sums of money on capital infrastructure that may be of limited value in the future reflects poorly on management. The long term is evaluated in two ways. First is the useful life of the asset before it needs to be replaced. Second is the annual cost associated with maintaining the asset. Figure 13.2 shows two cash flow diagrams. The upper diagram is typical: initial cost plus the operation and maintenance (assumed to be constant here). The lower cash flow diagram reveals a lower initial cost, but the life expectancy of the asset is significantly less than the second, so it must be replaced sooner, and the maintenance costs are higher. So, the question is whether to

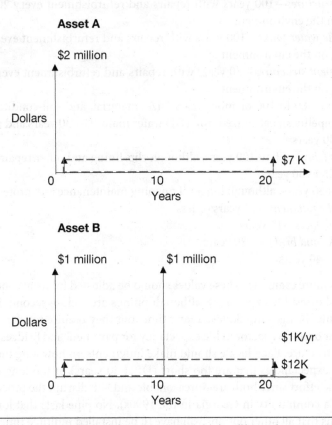

Figure 13.2 An asset with a lower initial cost but much higher life-cycle cost.

invest in the first option, which has a higher initial cost but the asset lasts nearly twice as long, or to invest in the less expensive option. Life-cycle analysis can show you the best option.

There is an incentive for utilizing cost to improve quality as a means to decrease the long-term asset costs. A present value analysis of the system would show that the annual cost of a higher quality item (longer lasting with minimal maintenance costs) has a lower life-cycle cost than cheaper items that need to be repaired more often or may need more frequent replacement.

So, the first thing that must be done is to determine how long an asset will last. This requires engineering judgment, knowledge of the area, knowledge of soils and traffic, and design standards. Many of these things are discussed as a part of developing a condition index. The condition index parameters should guide the owner and engineer to the appropriate life-cycle solutions. For example, ductile iron (DI) pipe is designed to last 100 years. However, in corrosive soils or in saltwater, the life will be reduced substantially unless measures are taken to address these issues, such as using other materials. C900 polyvinyl chloride (PVC) might be an acceptable solution because it does not have issues with corrosion. Its projected life is 80–100 years, which will be longer than the reduced performance expected of the DI pipe. Placing pipe in a plastic liner is a possible solution to make the pipe life the same, but adds to capital costs.

There are many thoughts on the life of assets. Here are some suggestions (Bloetscher 2011):

- *Concrete structures*—100 years with repairs and refurbishment every 20 to 30 years, depending on the environment
- *Steel potable water tanks*—100 years with repairs and refurbishment every 10 to 15 years, depending on the environment
- *Steel treatment structures*—50 years with repairs and refurbishment every 10 to 20 years, depending on the environment
- *Water mains*—80 to 100 or more years (size, material, and soil conditions may shorten the life of pipelines); galvanized iron (GI) water mains last 30 years and asbestos concrete pipes last 50 years
- *Sewer lines (clay)*—50 to 100 years, although slip-lining and other repairs will need to occur every 25 to 30 years
- *Manholes*—50 years, although longer if ongoing maintenance and protection is pursued
- *Mechanical equipment*—15 years or less
- *Roadway surfaces*—15 years
- *Bridge decks and bridges*—80 years
- *Road base*—40 years

Table 13.1 shows more examples. These values should be adjusted for local conditions. Roadway asphalt has the shortest life expectancy, although pumps are a close second. Buried pipe typically has a long life. Bridges should last a long time, but they require regular maintenance and inspections. Wear or use is a factor in life expectancy for pavement and bridges, but not so much for buried infrastructure. Each locale should make judgments on how long their infrastructure should last. Life expectancies that are too short (DI at 40 years) or too long (DI at 240 years) are harmful to the effort since both are unreasonable and will disrupt the process. The 240-year example is real (a community in Georgia in the 1990s). No pipe lasts that long, but assuming that it does means that all other options will have to be installed multiple times. As a result, the life-cycle analysis is skewed toward the DI pipe. There should be other reasons to formulate that decision, not one reliant on assumptions that are clearly unsubstantiated.

Table 13.1 Expected life of assets

Asset	Expected Life (Years)
Airports	up to 150
Runways	up to 50
Bridges (deck)	up to 50
Substructure	up to 125
Tunnels	up to 200
Ports/rail	up to 50–100
Dams	>100
Power, communication lines	>50
Nuclear power plants	50
Water lines	100
Sewer lines	100
Lift stations	50
Equipment	25
Treatment plants	>50
Equipment	25–50
Stormwater pipe	80

TIME VALUE OF MONEY

One problem with conducting life-cycle analyses is that the value of money changes with time. Everyone knows that things cost more today than they did last year, five years ago, or 20 years ago. This is called inflation. Inflation is an increase in the costs of goods to be purchased compared to those same costs in a prior period. This is reflected in the Consumer Price Index (CPI). Inflation creates a reduction in purchasing power—the same dollar will not purchase the same amount of goods over time since the value of money decreases with time. Investors will want to ensure their investments earn enough interest or return to exceed the inflation rate, much like people hope they get raises that are higher than inflation, otherwise they lose purchasing power. In most infrastructure situations, the cost of money is declining while needs are increasing, which creates an accelerated increase in budget needs. Inflation is sensitive to economic strength, consumer spending, borrowing rates, energy costs, and supply and demand. Also, it can be impacted by political turmoil, war, material shortages, and public perception. When performing a life-cycle analysis, it is important to understand how the time value of money alters the analysis. This is done by developing a cash flow diagram (every finance person should be able to do this) and by using the concepts of engineering economics. The cash flow diagram is the input to the economic analysis.

Cash Flow Diagrams

Engineering economic evaluations involve six common components (Bloetscher and Meeroff 2015; Blank and Tarquin 2014):

- *Effective interest rate (i)*—which refers to the rate of borrowed money, the rate of inflation, the rate of increase in operating costs, and the desired rate of return, which are all basically expressed in the form of interest rates. Higher interest rates mean higher borrowing costs. Higher inflation rates will increase operations costs geometrically. Combined interest rates may exist where there is inflation and growth happening at the same time. Interest rates should be assumed to compound (usually yearly for economic analysis).

- *Number of payment periods or number of payments (n)*—which refers to the length of time over which the analysis will be considered or the number of payments that will be made over a given period of time.
- *Present value (P)*—which refers to the cost of the asset or cash flow at the current time ($t = 0$). Capital costs are present value amounts.
- *Annuity or uniform payments (A)*—which refers to the periodic costs of the project converted to a uniform stream. Annuities can be used as annual amounts, but are often adapted for use with monthly, weekly, or daily payments by adjusting the number of payment periods (n). Debt service for capital components is an example of a uniform cash flow stream. For instance, some municipal bonds are issued for a period of 20 or 30 years but are compounded semi-annually. At a given interest rate, the annual debt service can be found from actuarial tables (more on this a little later). Annuities can be operations and maintenance (O&M), but O&M is rarely uniform, a fact that engineers need to keep in mind.
- *Future value (F)*—which refers to the value of an investment, such as real estate, that is expected to increase in value over time or the value of a series of payments such as a retirement account. Future value can be determined from P and i for a given term n. For instance, if a building is owned for 10 years, the future value can be estimated given P, i, and n. At a given interest rate and time, the future value of a series of payments can be found from actuarial tables.
- *Gradient value (G)*—which refers to the value of a cash flow that is increasing in constant increments over time. Operations and maintenance costs are annual amounts, but they are rarely uniform. Instead, they tend to inflate at a given percent each year, so a gradient of annual costs can also be used. Gradients may be useful in situations where arithmetic growth in a given number of customers each year (G) is occurring, but not on a percent-per-year basis (geometric growth). Many rate analysts mistakenly assume that growth is a given percent each year. This actually means growth accelerates with time, which is rarely the case. If growth is exponential, this would be treated more like an interest rate as opposed to a gradient. Instead, a linear or constant growth can be analyzed more appropriately using the G values in the actuarial tables.

Solving cost analysis problems will require knowing a minimum of four of these six key variables. Substantive discussion can be found in Bloetscher and Meeroff (2015), Blank and Tarquin (2014), or other engineering economics text.

There are three solutions to these six variables that are used to conduct analyses:

- *Present value or life-cycle analysis (LCA)*—brings all costs to the present for comparison, assuming the lives of the options is the same.
- *Annual value*—allows bankers to lend money at a given interest rate (higher than anticipated inflation) and then create a series of payments over a certain time frame. Local governments prefer that these are uniform (or near uniform) amounts. The uniform series accomplishes this. Debt service and borrowing is outlined in more detail in Chapter 11.
- *Future value*—is less useful for infrastructure. Future value converts cash flows to a dollar amount a number of years down the road. Since infrastructure is not acquired with

the intent of selling it, this has no real use for infrastructure analysis and will not be discussed further.

It is useful to create a *cash flow diagram* when undertaking a present value analysis. The cash flow diagram depicts the costs and revenues over a timeline. The horizontal line represents the timeline of the investment. For a series of annuities, each tick mark represents one payment period (n). For a future cost, an arrow is placed where that cost will occur. If disposed of, a future value line will represent any revenue returning to the owner (arrow the other way). Annual gradients are shown like a triangle or series of steps. Arrows can represent payments or revenues. A convention should be chosen—all payments should be represented as up arrows and all receipts should be represented as down arrows, or vice versa, as long as it remains consistent and understandable to the analyst. A simple cash flow diagram showing the current (present value) and the future value is shown in Figure 13.3.

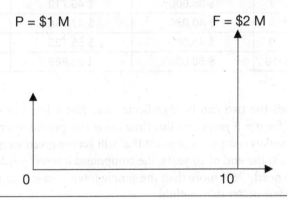

Figure 13.3   This simple cash flow diagram shows the current (present value) and the future value.

Interest Rates, Geometric Growth, and Inflation

Before proceeding further, a brief discussion of interest rates is needed. There are two types of interest rates: simple and compound. Simple interest rates are rarely used in commercial businesses, banks, the stock market, or virtually anywhere else. Simple interest is defined as follows:

$$\text{Simple interest} = \text{principal} \times \text{periods} \times \text{interest rate} = P \times n \times i$$

The concept would be as follows: If a principal of $P = \$10,000$ is borrowed from a friend for a period of $n = 5$ years, and they charged simple interest of $i = 10\%$ each year, the borrower would be responsible for paying $1,000 in interest each year. Over the five-year period, $5,000 in total interest would be paid. Few commercial entities use this method.

Businesses and banks typically will use compound interest rates for borrowing or other financial instruments. These rates are based on interest charged over time. In other words, interest is also paid on the accrued interest for each period (just like home mortgages, credit cards, or student loans). This concept forms the basis for the actuarial or interest tables located in Bloetscher and Meeroff (2015), Blank and Tarquin (2014), or at the following web address: http://global.oup.com/us/companion.websites/9780199778126/pdf/Appendix_C_CITables.pdf.

Compound interest is defined as follows:

$$\text{Compound Interest} = (\text{principal} + \text{accrued interest}) \times \text{interest rate}$$

Table 13.2 Comparison between the payback on $100,000 using simple interest, and comparing it to the use of compound at 5% per year interest rate

Principal	Interest Amount	Year	Amount of Interest Owed, Simple Interest	Amount of Interest Owed, Compound Interest	Different Using Compound Interest
100,000	5%	0			
		1	$ 5,000	$ 5,000	$ —
		2	$ 10,000	$ 10,250	$ 250
		3	$ 15,000	$ 15,763	$ 763
		4	$ 20,000	$ 21,551	$ 1,551
		5	$ 25,000	$ 27,628	$ 2,628
		6	$ 30,000	$ 34,010	$ 4,010
		7	$ 35,000	$ 40,710	$ 5,710
		8	$ 40,000	$ 47,746	$ 7,746
		9	$ 45,000	$ 55,133	$ 10,133
		10	$ 50,000	$ 62,889	$ 12,889

The difference between the two can be significant after just a few years (see Table 13.2). If $100,000 is borrowed for $n = 5$ years, but this time using compound interest, the difference in the amount of simple versus compound interest that will accrue given an interest rate of $r = 5\%$ per year is significant. At the end of 10 years, the compound interest would be $12,889 greater, which is equivalent to nearly 26% more than the simple interest case over the life of the loan. It is clear to see why lenders prefer this method.

The concept of annual compound interest rates was outlined previously, but everyone has seen advertisements for loans at lending institutions or on credit card offers that show a given annual percent interest rate, followed by some smaller-sized fine print with a larger value. The more easily read (bigger) number is the nominal annual interest rate (r). Most people do not know what the smaller number is, but are suspicious that the lender is not telling them something. That *something* is that interest rates for loans are rarely done on an annual basis. They compound monthly. Most lenders charge interest on a monthly basis because that is how often payments are made. As a result, interest is compounded monthly, not yearly. That number in the fine print is the equivalent annual percent interest rate (APR) or effective interest rate (i_a), assuming monthly compounding. And yes, it is always a larger number than the annual interest rate, but it is the actual result based on how loan payments are calculated. Essentially, interest can be compounded in any fashion and the conversion is relatively easy to do even without the tables. The formula for converting to virtually any compounding period is simple and is easily developed:

$$i_a = (1 + r \div m)^m - 1$$

Where:

i_a = effective interest rate per year in decimal form
r = nominal interest rate per year in decimal form
m = periods per year

So, for credit cards and mortgages, the conversion to monthly interest rates requires the formula to be:

$$i_a = (1 + r/12)^{12} - 1$$

As the number of payment periods increases, a limit is reached going to a natural log, as follows:

$$\lim_{m \to \infty} \left(1 + \frac{1}{m}\right)^m = e$$

Therefore,

$$i_a = e^r - 1$$

where r is the nominal interest rate. An example would help. Suppose a nominal interest rate of 4% per year is offered for a bond issue for replacement infrastructure. What would the effective interest rate be if the compounding period were monthly, weekly, daily, or continuous? Keep in mind that interest rates are actually decimals, so use the decimal form (4% = 0.04). The results would look like the following:

Biannual (typical for bond issues; $m = 2$):

$$i_a = (1 + r \div m)^m - 1 = (1 + 0.04 \div 2)^2 - 1 = 0.0404 \text{ (or 4.04\%)}$$

Monthly ($m = 12$): $i_a = (1 + r \div m)^m - 1 = (1 + 0.04 \div 12)^{12} - 1 = 0.04074$ (or 4.074%)

Weekly ($m = 52$): $i_a = (1 + r \div m)^m - 1 = (1 + 0.04 \div 52)^{52} - 1 = 0.04079$ (or 4.079%)

Daily ($m = 365$): $i_a = (1 + r \div m)^m - 1 = (1 + 0.04 \div 365)^{365} - 1 = 0.04081$ (or 4.081%)

Continuous ($m = \infty$): $i_a = e^r - 1 = e^{0.10} - 1 = 0.04081$ (or 4.081%)

From this analysis, it is understandable why monthly compounding is so popular. It generates more money for the lender for limited extra work. Trying to update all loans and accounts daily, or even weekly, creates only a marginal increase in revenues.

Inflation is also an interest rate. Most people are familiar with annual inflation rates. Since the 1920s, the United States has tracked inflation (see Figure 13.4). What is immediately obvious is that the rate of inflation decreased after the late 1970s when the Federal Reserve Bank made inflation control a monetary policy of the federal government. The Federal Reserve Bank and Congress tinkered with how the CPI was calculated in part to address perceived acceleration of costs for Social Security after 1980. The CPI is used by the U.S. Bureau of Labor Statistics to measure the average change over time in the prices paid by consumers for a common basket of goods and services. The CPI is based on prices of food, clothing, housing, fuel, transportation fares, charges for medical care, drugs, and other goods and services that people purchase for day-to-day living. The CPI is calculated by taking price changes for each item in the predetermined composition (weighted influence) of these goods and averaging them according to their importance. Prices are collected each month from 87 areas across the country—6,100 housing units and approximately 24,000 retail establishments, including department stores, supermarkets, hospitals, filling stations, and other types of stores and service establishments (Bloetscher and Meeroff 2015). There are a variety of other inflation indexes available. One is for construction—the Construction Cost Index published by *Engineering News Record*. Others exist for various industries. All inflation indices are complicated formulas that go well beyond the scope of this text.

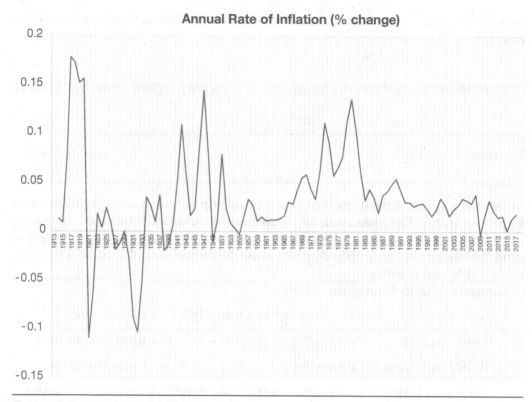

Figure 13.4   Inflation rate graph

Inflation is differentiated from compound interest rates (r and i_a) by using inflation (f) instead of i, and is usually accompanied by some growth or other interest value (which is why inflation is denoted with an f). Combining interest rates is slightly trickier. Let's assume a compounding rate for each. With inflation-adjusted interest rates, the new effective rate (i_a) is given as follows:

$$i_a = r + rf + f$$

For example, if the project was expected to increase O&M costs by 10% per year and inflation was 4% then:

$$i_a = r + rf + f = 0.10 + (0.10 \times 0.04) + 0.04 = 0.144 \text{ (or 14.4%)}$$

Now, what if an engineer is asked to project the expenses for a facility's power costs? In this case, power is related to product sales, which are expected to increase at 5% per year for the next several years (note this is actually the gradient g as opposed to i). If inflation is running at 3%, then project the coming year's cost of power as follows:

$$i_a = r + rf + f = 0.05 + (0.05 \times 0.03) + 0.03 = 0.0815 \text{ (or 8.15%)}$$

The 8.15% is a major difference from either the 3% for inflation or the 5% for growth, which may be the value the finance director wants to use.

Inflation works against investments. Assume a warehouse was purchased for $1,000,000. The real estate market is slack and the building only increases in value 1% per year. If inflation is 3% per year, what is happening to our investment? The answer is that it is effectively losing roughly

2% per year. Unfortunately, this is not uncommon in many older, Rust Belt cities. This creates problems in attracting investments and reduces available tax revenues to improve infrastructure.

Gradient Series Factor

Prior to applying the life-cycle analysis to examples, one other factor is worth discussing— gradients. Gradients come in two types: (1) a percent increase for each payment period (g) or (2) a constant increase for each payment period (G). The first is simply an interest rate like inflation and is treated the same way. Gradients, discussed previously, for increases in costs are a good example; the other is the constant increase, which occurs rarely, but can be useful for analysis (e.g., the cost of maintenance for a piece of equipment that increases $250 per year). One thing to note about gradients is that they are usually applied in conjunction with annuities. Most important, at the end of the first year of the gradient, the value of the gradient is zero. This confuses a lot of people.

Present Value Method

By far the most common economic analysis is present value analysis, in which all costs associated with the engineering alternative over its useful life are brought to current day dollars (present value). Essentially, this calculation determines the lifetime value of the investment if paid all at once in today's money (the present day life-cycle cost). In order to organize payments and receipts in a visual representation (a cash flow diagram) must be created. All costs should be included whether they are fixed annuities, future events (like planned rehabilitations), costs that inflate like power costs, or some combination of many different items. Note that care should be taken in assuming that all factors are interest rates since this creates an exponential relation that will become less accurate with time (shown in Figure 13.5). Present value analysis is useful for comparing options. The terminology used is as follows:

- Nothing needs to be done to convert present day costs into present value
- To convert an annuity to present value, use $P = A*(P/A, i, n)$ where the values in parentheses comes from standard actuarial interest tables

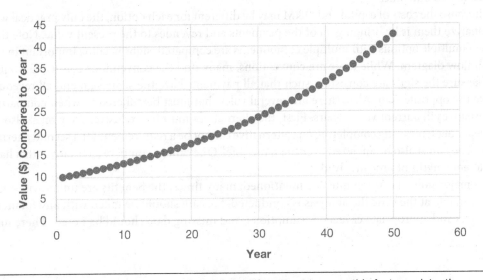

Figure 13.5 Exponential expense curve—it is rare for a cost to grow this fast consistently.

- To convert a future value to present value, use $P = F*(P/F, i, n)$ where the values in parentheses comes from standard actuarial interest tables
- To convert a gradient to present value, use $P = G*(P/G, i, n)$ where the values in parentheses comes from standard actuarial interest tables

The result is that all costs and revenues will align at the present or $t = 0$. The actual formulas for determining present value can be complicated. The potential for errors in calculating these complicated formulas is why the actuarial tables were developed—the calculations are already done for those who need to develop these analyses.

Comparing Examples—Evaluating Economic Alternatives

There are usually several options for resolving a challenge. The current situation, or the do-nothing alternative, is usually considered the base situation (and likely the least expensive option) for benefit-cost analysis, although it does not resolve the issue. Note that at this point, only acceptable alternatives—those that take into account the issues associated with the condition assessment, vulnerability, reliability, and risk—should be considered. It is assumed that engineers, operations personnel, and others involved in operating the facility will lead this effort, not finance or administrative personnel who may not fully understand the operational needs. The costs of these options will typically be developed by engineers based on experience with similar facilities or with similar materials. The O&M costs will be derived by engineers, operations personnel, and others involved in operating the facility. Changes in technology are always possible and it is useful to anticipate these types of changes and when they might occur. Solutions that would be obsolete as a result of technology changes in a short period of time should not be considered. Engineers can often identify technology that is becoming obsolete. Any option should be developed with an understanding about the needs for technology (advanced technology that creates information is not useful unless someone can analyze it). In many cases, certain benefits or efficiencies can be gained. Every system will be different, but engineers often have access to information that can be used to identify the moment when the need for rehabilitation will occur. These must be noted as well. Presenting bad alternatives leads to bad outcomes.

Because the costs of capital and O&M may be different for each option, the only practical way to analyze them is by bringing all of the payments and revenues to the present value. Note that when multiple options with multiple components are compared, signs become important in the cash flow diagram. While there are conventions, many find the conventions confusing, so just make sure the signs are consistent such that all payments have the same sign and all revenues have the opposite sign. Also, there are several rules that must be adhered to when comparing alternatives in current year dollars. First, the alternatives must be exclusive. They must also be independent; meaning completing one alternative cannot be a requirement for a second alternative. They must also be analyzed over the same "life." One cannot compare alternatives that have different lengths of time involved.

A major note of caution must be mentioned; many times, the benefits are unknown or not quantifiable at the time the analysis is conducted. So, care should be taken with any financial analysis and appropriate caveats, assumptions, and missing data should be noted. There may be future benefits or the alternative may encourage other actions that are difficult to decouple from the original decision. Water/sewer utilities, for example, are typically funded by the public because the benefit-cost analysis of extending pipelines into undeveloped areas will always show a benefit-to-cost ratio of less than one. This failure occurs because the timing for future development is unknown and there are relatively few customers at the time of construction to offset the initial costs. The benefit of extending piping may be to expand or stimulate development or discourage other utilities that might compete in the future, which would represent a significant detriment to the utility in the form of potential lost income. The same argument exists with extending roadways outside the central city; the initial demand may be lacking, but economic development (as discussed in Chapter 2) usually follows these urban roadway extensions. Some example applications may be helpful.

Suppose a stormwater utility must decide between a $100,000 pump with $15,000 annual operating costs (Pump A) and a $105,000 unit with $12,000 per year operating costs (Pump B). Which item is more cost-effective in the long term at 5% interest for 5 years? One way to handle this is to perform a present value analysis, which is conducted by bringing all anticipated costs of a project to the current time (see Figures 13.6 and 13.7 for the cash flow diagrams).

In this case, the present value for Pump A is calculated as follows:

$$PV_1 = \$100,000 + \$15,000(P/A, 5\%, 5) = \$100,000 + \$15,000(4.3295) = \pmb{\$164,843}$$

And the second option is calculated as follows:

$$PV_2 = \$105,000 + \$12,000(P/A, 5\%, 5) = \$105,000 + \$12,000(4.3295) = \pmb{\$156,954}$$

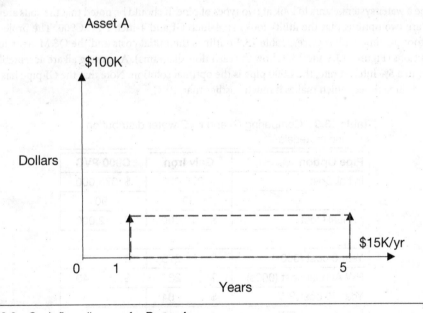

Figure 13.6 Cash flow diagram for Pump A.

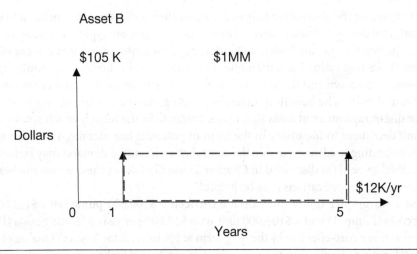

Figure 13.7 Cash flow diagram for Pump *B*.

Option 1, the cheaper initial cost, costs $7,889 more over the 5-year life of the pump. This approach can best be explained with a set of more realistic and complex examples.

Example 13.1: Pipe Comparisons

Assume a water system wants to look at two types of pipe. It should be noted that the soils are poor. There are two options that the utility looks at: 4-inch GI and 4-inch PVC C900. The project is a 5,000 foot pipeline with services. Table 13.3 outlines the initial costs and the O&M costs for the pipe options (Figures 13.8 and 13.9 show the cash flow diagrams). Assuming all are acceptable options, and a 5% interest rate, the C900 pipe is the optimal solution. Note that the GI pipe has to be replaced three times, which makes it much costlier than PVC.

Table 13.3 Comparing GI and PVC water distribution pipeline materials

Pipe Option	Galv Iron	C900 PVC
Initial Cost	$ 275,000	$ 375,000
Life	30	90
Annual Cost	$ 10,000	$ 2,000
Year 0 cost (000s)	$ 275	$ 375
PW Annual cost (000s)	$ 237	$ 40
Year 30 cost PW	$ 64	
Year 60 cost PW	$ 15	
	$ 590	$ 415

continued

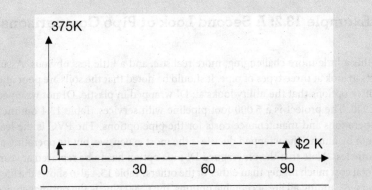

Figure 13.8 Cash flow diagram for PVC C900 pipe option with 90-year life.

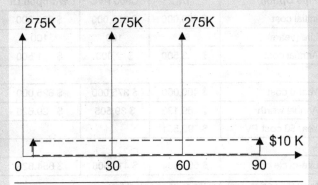

Figure 13.9 Cash flow diagram for galvanized pipe option noting that it will be replaced multiple times to allow for comparison to the PVC pipe.

Example 13.2: A Second Look at Pipe Comparisons

Let's make this a little more challenging, more realistic, and a little less obvious. Assume a water system wants to look at three types of pipe. It should be noted that the soils are poor and corrosive. There are three options that the utility looks at: DI wrapped in plastic, DI not wrapped in plastic, and PVC C900. The project is a 5,000 foot pipeline with services. Table 13.4 outlines the initial costs and operations and maintenance costs for the pipe options. The PVC is the least costly to install, but not by much. The wrapped ductile is the most expensive, but its operating and maintenance costs are less. Note that the unwrapped DI pipe option has to be replaced at year 50, so that makes the total cost much higher than either of the others. Table 13.4 also shows the life-cycle costs over 100 years. Assuming all are acceptable options, the C900 pipe is the optimal choice.

Table 13.4   Comparing three pipe options

Pipe Option	DI	C900 PVC	Wrapped DI
Initial cost	$ 400,000	$ 375,000	$ 625,000
Life (years)	50	100	100
Annual cost	$ 3,500	$ 2,000	$ 1,500
Year 0 cost	$ 400,000	$ 375,000	$ 625,000
Annual worth	$ 69,133	$ 39,505	$ 29,628
Year 50 cost PV	$ 92,551		
Life-cycle cost	$ 561,684	$ 414,505	$ 654,628

Note: PV = present value is the same as cost at time = 0

Example 13.3: Finding the Life-Cycle Cost of Equipment

Assume a public works department is doing a 30-year projection of capital costs. They realize that equipment must be included. They are going to buy a piece of equipment that has a service life of 10 years. They assume they will replace it in 10 years. The equipment costs $100,000. If the cost of maintenance is $5,000 the first five years, but then increases $1,000 each year after the fifth year until replacement, what is the present value cost of these equipment purchases?

The first task is a cash flow diagram. This is shown in Figure 13.10. In this case, there is a P, two Fs, three gradient triangles, and one annuity across all thirty years. To obtain the present

continued

value, create a series of present values for each of the cash flows. The problem would be solved as follows:

$P1 = \$100K$

The annuity:

$P2 = \$5K(P/A, i, n) = \$5K(P/A, 6, 30) = \$5K(13.7648) = \$68,824$

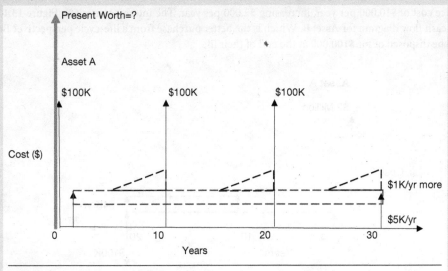

Figure 13.10 Cash flow diagram for Example 13.3.

The two future payments for replacement equipment must be brought back as follows:

$P3 = \$100K(P/F, 6, 10) = \$100K(0.5584) = \$55,840$
$P4 = \$100K(P/F, 6, 20) = \$100K(0.3118) = \$31,180$

The gradients occur at future years and must be brought to the present as follows, noting that the first year of the gradient is zero. That means that in years 5, 15, and 25, the gradient is zero. Hence, the present value of these gradients occurs at years 4, 14, and 24 respectively, but note they only occur over 6 years each (the first year, year 5, is a zero):

$PG1 = \$1K(P/G, 6, 6) = \$1K(11.4594) = \$11,459$
$PG2 = \$1K(P/G, 6, 6) = \$1K(11.4594) = \$11,459$
$PG3 = \$1K(P/G, 6, 6) = \$1K(11.4594) = \$11,459$

Each of these must be brought to the present:

$P5 = PG1(P/F, 6, 4) = \$11,459(0.7921) = \$9,077$
$P6 = PG1(P/F, 6, 14) = \$11,459(0.4423) = \$5,068$
$P7 = PG1(P/F, 6, 24) = \$11,459(0.246) = \$2,829$

The total present value of the equipment is the sum of the Ps and is $272,818.

Example 13.4: Comparing Two Purchase Options Using Life-Cycle Cost Analysis

Compare two equipment purchases. One will last 20 years, the other only 10. The initial cost for Asset A is $2 million. Figure 13.11 is the cash flow diagram for Asset A. The cost for Asset B is $1.3 million, hence why the manager wants to buy Asset B. The operating cost for Asset A is $7,000 per year for 10 years, then $11,000 per year for the remainder. Asset B has an operations and maintenance cost of $10,000 per year, increasing $2,000 per year. The interest rate is 6%. Figure 13.12 is the cash flow diagram for Asset B. Which is the better purchase from a life-cycle perspective? Both will be disposed of for $100,000 at the end of their life.

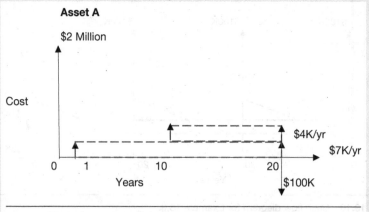

Figure 13.11 Cash flow diagram for Example 13.4, Asset A.

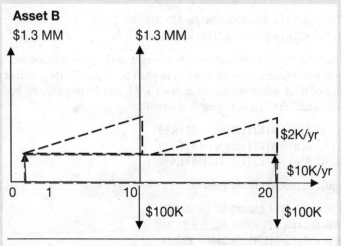

Figure 13.12 Cash flow diagram for Example 13.4, Asset B.

continued

For Asset A:

$P1 = \$2M$

$P2 = \$7K(P/A, i, n) = \$7K(P/A, 6, 20) = \$7K(11.4699) = \$80,289$

$P3 = \$4K(P/A, i, n) = \$4K(P/A, 6, 10) = \$4K(7.3601) = \$29,440$ but note this in the future

The future payments ($P3$ and the sale of the asset) must be brought back as follows:

$P4 = \$29,044K(P/F, 6, 10) = \$28,440(0.5584) = \$16,440$

$P5 = -\$100K(P/F, 6, 20) = -\$100K(0.3118) = -\$31,180$

The total present value/life cycle for Asset A is $P1 + P2 + P4 + P5 = \$2,065,549$.

For Asset B:

$P1 = \$1.3M$

$P2 = \$1.3M(P/F, 6, 10) = \$1.3M(0.5584) = \$725,920$

$P3 = \$10K(P/A, i, n) = \$10K(P/A, 6, 20) = \$10K(11.4699) = \$114,699$

The gradients occur out at future years and must be brought to a present value as follows, noting that the first year of the gradient is zero. Hence, the present value of these gradients occurs at years 0 and 10 years:

$PG1 = \$2K(P/G, 6, 10) = \$2K(29.6023) = \$59,204$

$PG2 = \$2K(P/G, 6, 10) = \$2K(29.6023) = \$59,204$

The future payments ($PG2$ and the sale of the asset) must be brought back as follows:

$P4 = PG2(P/F, 6, 10) = \$59,204(0.5584) = \$33,060$

$P5 = -\$100K(P/F, 6, 10) = -\$100K(0.5584) = -\$55,840$

$P6 = -\$100K(P/F, 6, 20) = -\$100K(0.3118) = -\$31,180$

The total present value/life cycle for Asset B = $\$2,145,863$. Asset A is the choice.

Payback Period

The payback period is the time required to earn back the amount invested in an asset from its net cash flows. Due to its ease of use, payback period is a common method used to express return on investment, though it is important to note that it does not always account for the time value of money. To calculate the payback period, simply divide the cash outlay (which is assumed to occur entirely at the beginning of the project) by the amount of net cash inflow generated by the project per year (which is assumed to be the same in every year).

Let's say a company builds a new factory. The annual revenues are $1 million per year in current dollars. Expenses are $800,000 per year. The positive net cash flow of $200,000 is the difference. Let's also say the new factory costs $1 million to construct. When will the breakeven point be? Simple: $1,000,000/$200,000 or 5 years if no interest rate is involved. If there is an interest rate, the time value of money will extend the payback.

Rate of Return Method

The concept of return on investment has to do with comparable interest rates. Take the earlier example of the warehouse; the rate of return on this investment was negative. No one wants to lose money on an investment. Most people have a goal of an interest rate they would like to see their investments earn. Let's assume the desired return on investment of an investor is 5% per year. That means that most government bonds, certificates of deposit, and commercial loans would not reach this value because in 2016 those interest rates were all below 5%. Instead, the investor might need to try the stock market, which has a historical return of about 7% per year. But note the stock market fluctuates; in 2008 it lost half its value. Thus, high returns on investments will increase risk. Bonds and CDs are low-risk investments—your money is safe, but the interest rates are lower as well. The return on investment is essentially an interest rate. While rarely used for infrastructure projects, the summary would be similar to the following example.

Example 13.5: Finding the Rate of Return Percentage on a Toll Bridge

Let's say your road division is looking at a new toll bridge across a river. This river crossing has been a major economic/transportation issue for years. As a result, it seems like a good toll option. The cost for the bridge is $100 million. It is expected that tolls will start immediately after it is completed. Your borrowing rate is 10%, which means the investment must have a higher rate of return than 10% in order for you to move forward with the project. The collections are anticipated at $10 million the first year, increasing by $2 million each year afterward for 10 years (see Figure 13.13). Find the rate of return (i) of the project and compare it to your borrowing rate, which is your minimum acceptable rate of return (MARR).

$P1 = \$100M$
$P2 = \$10M(P/A, i, 10)$
$P3 = \$2M(P/G, i, 10)$

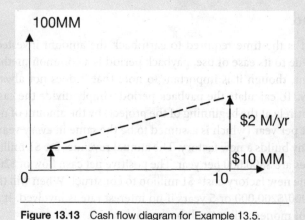

Figure 13.13   Cash flow diagram for Example 13.5.

continued

Set the values equal to one another and solve for i to get the two sides equal:

$$\$100M = \$10M(P/A, i, 10) + \$2M(P/G, i, 10)$$

Try 8% since that is below your MARR:

$$\$100M = \$10M(P/A, 8, 10) + \$2K(P/G, 8, 10) = \$10M(6.710) + \$2M(25.977)$$

$100M < 67.1M + \$53M$; $\$100M < \$120.1M$ which means at 8%, the numbers are off by $20.1M.

Let's try 10%, your MARR.

$$\$100M = \$10M(P/A, 10, 10) + \$2M(P/G, 10, 10) = \$10M(6.145) + \$2M(22.891)$$

$100MM < \$61.45M + \$45.8M$; $\$100M < \$107.25M$ which means at 10%, the numbers are still $7.25M off. This is closer but let's keep going.

Try 12%.

$$\$100M = \$10M(P/A, 12, 10) + \$2M(P/G, 12, 10) = \$10M(5.650) + \$2M(20.254)$$

$100M > \$56.5M + \$40.4M$; $\$100M > \$96.9M$. Now, we have gone under the $100M initial investment, which means the actual rate is now between 10% and 12%. Using a finance calculator, we can find the rate of return that makes these numbers equal is 11.9%. Since the 11.9% rate is above the MARR, this is an acceptable investment.

All engineers and public works professionals should be familiar with the principles of life-cycle analysis. Table 13.5 outlines questions that should be answered when making any repair or improvement. Honest answers will help guide officials to good solutions, rather than using good money to chase after solutions that cannot resolve an issue. This is important in making good choices with public funds and addressing risk and maintenance concerns. The cheapest first cost option is rarely the lower cost option in the long term.

Table 13.5 Life-cycle questions to ask

It is safe?
Is it structurally sound?
How has it been damaged?
Is any damage repairable at a reasonable cost?
Does it create delays/inconvenience?
Does it affect movement of goods and services?
Is it costly to maintain?
Can you get parts?
Is it obsolete?

REFERENCES

Blank, L. T. and Tarquin, A. J. 2014. *Basics of Engineering Economy, 2nd Ed.* McGraw-Hill Higher-Education, New York, NY.

Bloetscher, F. 2011. *Utility Management for Water and Wastewater Operators.* AWWA, Denver, CO.

Bloetscher, F. and Meeroff, D. M. 2015. *Practical Design Concepts for Capstone Engineering Design.* J. Ross Publishing, Plantation, FL.

Ellison, D. et al. 2014. "Answers to challenging infrastructure management questions." Report #4367. WRF, Denver, CO.

National Treasury. 2004. *Local Government Capital Asset Mangement Guidelines.* National Treasury, Pretoria, South Africa.

NMEFC. 2006. *ASSET Management: A Guide for Water and Wastewater Systems.* 2006 Edition. Environmental Finance Center, New Mexico Tech, Socorro, NM.

RTI. (no date). "A toolkit for municipal asset management." RTI International. https://community-wealth .org/sites/clone.community-wealth.org/files/downloads/tool-rti-asset-mgmt.pdf. Accessed 5/11/2018.

Stadnyckyj, M. 2010. "Condition assessment: Bridging the gap between pipeline investments and risk reduction." *Water Utility Infra Management*, E-Newsletter. May 2010. https://waterfm.com/condition -assessment-bridging-the-gap-between-pipeline-investments-and-risk-reduction/. Accessed 1/22/17.

Underwood Engineers. 2014. *Conway Village Fire District Water System Asset Management Plan*, Underwood Engineers, Concord, NH.

SECTION III

The Future, Based on Avoiding the Past

14

WHAT COULD POSSIBLY GO WRONG?

Throughout this book, there have been examples, good and bad, regarding infrastructure systems and associated protocols. In the public sector, the public health, safety, and welfare are designated as the priority by statutes, case law, and administrative rules. Therefore, when working for the public sector, the expectations and responsibilities to the collective health, safety, and welfare of the public cannot be avoided. Regardless of the infrastructure system, this is so because infrastructure by its nature serves society. Therefore, the stewards of these systems, those managing and operating them, inherit the responsibility of protecting the public's health, safety, and welfare. Welfare includes protecting their economic status.

While often the most talked about issue with public officials is cost (taxes, rates, fees, etc.), very few of the laws and rules affecting the public sector permit costs to be a means by which one makes decisions. The public health, safety, and welfare are paramount—cost is not. As a result, local officials must search for solutions that meet the health, safety, and welfare protection issues first, in a manner that can be managed cost-effectively. Given the condition of infrastructure in many communities, finding solutions to protect the public health, safety, and welfare as infrastructure ages, cannot be done using the current cost model. This is where grants have helped small and disadvantaged communities construct needed infrastructure, but the absence of that same funding in the future may imperil these same communities. When things fail, and they will, the first thing that the lawyers seek out is whether or not the stewardship of the system and the responsibilities to protect public health, safety, and welfare were met. Lawyers sue when they believe these mandates have not been met, when local officials have been poor stewards, or when negligence exists. Failure to protect the public health, safety, and welfare creates a lot of legal liability for someone.

All systems experience failure, but for the most part, the failures are minor, only last for a short time, and have no long-lasting effects—these are quickly forgotten. Far less frequently are the cases where things go wrong and there are big consequences—those are the ones people remember. The cryptosporidium incident in Milwaukee in 1993 is an example of where things went awry—that was 25 years ago, but no one has forgotten. There were severe repercussions where the decisions made by the operations staff at the time were with the best intent and testing did not indicate a problem. The Milwaukee incident brought a little-known protozoa to the forefront as a major potential water-quality issue. But, there was no knowledge that this issue might occur beforehand, so no preventive measures were in place. There is plenty of literature on Milwaukee and the extensive protocols and testing that were implemented *after the fact* to prevent such an incident from happening in the future. No one consciously did anything wrong in Milwaukee.

Of interest here are the incidents where there were poor decisions that led to the consequences. The majority of the time these incidents are cascading events, as was discussed in Chapter 9. One small incident creates more events, which have an even larger impact. Those will be the subject of this chapter. In addition, the focus will be on events where there exists some culpability on persons who were not optimal stewards of their infrastructure system. Often, decision making was poor, uninformed, or noncompliant with protecting the public health, safety, and welfare.

THE 2003 NORTHEAST BLACKOUT

While power infrastructure was not discussed at any length in this book, this power blackout is a perfect example of a cascade of failures that has had consequences that continue to this day. The power outage was triggered by high voltage lines touching unpruned vegetation outside of Cleveland, Ohio, and a bad computer alarm that did not respond to the trip (Minkel 2008). The software issues with the alarm were apparently known by operations at the power utility, but the issue was not addressed. Because of the alarm failure, the operators were unaware of or ignored an overloaded circuit when the problem happened. Over the next four hours, a progressive power outage encompassed the northeast. The full extent of the power outage manifested itself at 4:10 p.m. on August 14, 2003—affecting 55 million people in the United States and Canada (Minkel 2008). While many areas had some power back by midnight, others had their power out for up to a week in the middle of the summer.

Like roadway systems, the power grid across North America is interconnected. The reason for this is that power is deemed to be an essential service since much of the economy relies on access to adequate power at all times. The grid is interconnected so that if there is an increased demand in one part of the grid and an excess of power on another, the grid can automatically route power where it is needed. The Federal Energy Regulatory Commission (FERC) rules designate how power is regulated across the grid between competing power utilities and state lines. Computers deal with the routing; and where there are alarms, those alarms are supposed to alert operators of the need to reroute the grid. The issues that cause rerouting, in addition to demands, include downed power lines, damaged equipment, vegetation shorting lines, squirrels, weather, etc. Almost any potential impact to the grid infrastructure can create the immediate need to reroute power. Most communities and especially facilities like water and wastewater treatment plants, stormwater pumping stations, and certain transportation hubs are set up to have power from two or more directions to accomplish this goal.

The day of this event was very hot and the demand for air conditioning increased power needs, especially in densely populated communities. However, increased demand heats the wires, which causes the wires to sag more, creating the potential for touching vegetation. In such cases, the grid is supposed to automatically alarm and reroute power to shed the load and cool the wires. If the other lines do not have enough spare capacity to carry the extra current, the grid can overload circuits, causing the overload protection in those lines to trip, resulting in a cascading failure. The power stations are manned so that system operators can ensure that power supply and loads remain balanced and within safe operational limits in the event of computer malfunction (i.e., an alarm tripping). If a failure occurs, the operators seek more power from generators or other regions or they *shed load* (meaning to intentionally cut power or reduce voltage to a given area) until they can be sure that the worst remaining possible failure anywhere in the system will not cause a system collapse. Overloads can cause hard-to-repair, costly damage, so the affected device or segment is disconnected from the network if a serious imbalance is detected. In the 2003 blackout, the electrical demand plus the vegetation touching the sagging lines created an imbalance. However, neither the operators nor the computers were able to correct the load imbalances fast enough to prevent the system from collapsing.

According to the official analysis of the blackout prepared by the U.S. and Canadian governments, more than 508 generating units at 265 power plants shut down during the outage (Energy.gov 2004). In the minutes before the event, the New York Independent System Operator (NYISO)-managed power system was carrying 28,700 MW of load. At the height of the outage, the load had dropped to 5,716 MW, a loss of 80% (NYISO 2004). Non-power-related essential services remained in operation in some of these areas, although many backup generation systems failed. Hard-wired telephones remained in service, while cellphone and internet networks generally remained operational if they were using backup power. The same was true with television and radio stations.

Water and sewer systems remained in service only to the extent they had backup power. Those that did not, lost pressure, requiring boil-water notices. Detroit was one of those systems, which would have consequences for Flint, Michigan, ten years later. Four million customers of the Detroit water system in eight counties (including Genessee County—Flint) were under a boil-water advisory until August 18, four days after the initial outage. A total of 2.3 million businesses and customers were affected in some way in the Detroit area. Detroit Edison's entire power system went down. Cleveland's water system also lost water pressure and instituted a boil-water advisory. Cleveland and New York reported sewage overflows into waterways. Many gas stations were unable to pump fuel due to lack of electricity, not that the gas was really needed—all of the traffic lights, which had no backup power, were not working; traffic and trains were effectively gridlocked.

A joint federal task force was formed by the governments of Canada and the United States to oversee the investigation and report directly to Ottawa and Washington. In February 2004, the U.S.-Canada Power System Outage Task Force released their final report, placing the causes of the blackout into five groups (Energy.gov 2004):

1. FirstEnergy (FE) and its reliability council "failed to assess and understand the inadequacies of FE's system, particularly with respect to voltage instability and the vulnerability of the Cleveland-Akron area."
2. FirstEnergy "did not operate its system with appropriate voltage criteria."
3. FirstEnergy "did not recognize or understand the deteriorating condition of its system."
4. FirstEnergy "failed to manage adequately tree growth in its transmission rights of way."
5. Finally, there was the "failure of the interconnected grid's reliability to provide effective real-time diagnostic support."

The report stated that a generating plant in Eastlake, Ohio, a suburb of Cleveland, went offline amid a high electrical demand, putting a strain on high-voltage power lines (located in a distant rural setting), which later went out of service when they came in contact with overgrown trees (Energy.gov 2004). The cascading effect was to reroute the power to other transmission lines, which were also not able to bear the load, thereby tripping their relays (Energy.gov 2004). Once these multiple trips occurred, multiple generators suddenly lost parts of their loads, so they tripped to prevent damage. In other words, portions of the system worked properly (tripped), because other parts of the system were not properly maintained. Known deficiencies were not addressed. Usually cost savings or profits are associated with such decisions.

The blackout prompted the U.S. government to include reliability provisions in the Energy Policy Act of 2005. The standards of the North American Electric Reliability Corporation became mandatory for U.S. electricity providers (Minkel 2008). Homeland Security raised concerns about the potential for exposure and undetected, unauthorized access to vulnerable sites. The failure highlighted the ease with which the power grid could be taken down, and the difficulty in thwarting it. Cybersecurity measures were recommended to address this, but the grid still remains at risk.

HURRICANE KATRINA AND THE DIKES OF NEW ORLEANS

Many people remember watching the disaster that was Hurricane Katrina in New Orleans on television. The hurricane had glanced off of the southeast Florida coast a week earlier, leaving some damage in its wake, before pushing its way into the Gulf of Mexico, and at one time intensifying to a Category 5 hurricane. As the point of landfall became clearer, the concern was raised about the potential for disaster should the storm hit New Orleans.

The city of New Orleans is an old city, established by the French over 300 years ago along the Mississippi River. The city is at a low elevation; in fact, areas of the city are well below the water level of the Mississippi River, necessitating extensive dikes and pumping systems to keep the city dry. New Orleans contained about 480,000 people before the storm hit on August 29, 2005. In its aftermath, over 150,000 of those people did not return.

It was not as if New Orleans had not flooded previously. In 1953, a storm had overrun its dikes and new improvements were undertaken. In 1965, Congress withdrew the relief responsibility for hurricane protection from the state and assigned it to the Army Corps of Engineers in the Flood Control Act of 1965 (PL-89-298 1965). However, in the 40 years prior to Katrina, few gave much concern to these dikes and their condition. As Douglas Brinkley noted in a *Smithsonian* article, ". . . that the federally funded flood walls looked unsafe and poorly constructed never concerned me, because in America we built structures" (Brinkley 2015). However, the *Houston Chronicle* published a story in 2001 that predicted that a severe hurricane striking New Orleans, "would strand 250,000 people or more, and probably kill one in 10 left behind as the city drowned under 20 feet (6.1 m) of water" (Berger 2001). One-foot thick concrete walls could not hold back flood waters. As a result, once the levees started to break, 80% of the city was inundated. At least 50 breaches were noted. Pumps did not work, and many will recall the ugly floodwaters that left 1,500 dead. FEMA could not access the city. The Louisiana National Guard was stationed too close and had lost its gear in the floodwaters. The city was on its own in those early days after the storm.

In the aftermath of the storm, the investigations focused on two parties—the U.S. Army Corps of Engineers and the City of New Orleans Levee Board (Grunwald 2006). The way the levees are operated is that the Army Corps of Engineers transferred the ownership of the levees to local agencies that were then responsible to budget the funds to maintain the condition and operations of those assets. No one checked to ensure they actually did the job (Grunwald 2006). Funds were always an issue, as is the case in most communities, especially one with a significant disadvantaged population. In the case of New Orleans, the ownership, operation, and maintenance of the system was with the New Orleans Levee Board. Monies for the levee board appeared to have been spent on light maintenance, like mowing grass, not on structural maintenance. Visual observation by many before the storm indicated the lack of maintenance, but no one addressed the situation with the levee board. As a result, the levees were in poor condition before the storm due to neglect and a lack of inspection.

The U.S. Army Corps of Engineers was responsible for the design and construction of the flood protection measures for the city. Congress had directed the Corps to fix the levees in 1955 and 1965, but work was still not complete in 2005—50 years later. The Corps only designed its hurricane levees for a 200-year storm—which the Corps calculated as a Category 3 hurricane, even though Betsy, which hit the area in the 1950s, was Category 4. A Category 5 storm would be a 1:800-year storm. By 1976, the Corps built the levees for a 1:200 year storm, but the soils upon which they were built were wet, poor soils, with limited reinforcement. Roads fail when the base gets wet. So do levees. Pumps on certain canals were not installed. Additional charges included that the Corps built the levees to encourage development instead of safety, trying to maximize economic benefits when prioritizing where to build (Grunwald 2006).

Ultimately, a series of lawsuits were filed against all parties. Congress granted sovereign immunity to the Army Corps of Engineers under Section 3 of the Flood Control Act of 1928, which states "no liability of any kind would attach or rest upon the United States for any damage from or by floods or flood waters at any place, provided that if on any stretch of the banks of the Mississippi River it was impracticable to construct levees,"—33 U.S.C. §702c. That settled certain suits, but not all. The U.S. Fifth Circuit Court of Appeals upheld a lower court ruling finding the Army Corps of Engineers responsible for flooding a portion of the Lower Ninth Ward and St. Bernard Parish during Hurricane Katrina because of its failure to maintain the Mississippi River-Gulf Outlet (MRGO) canal (Schleifstein, 2012). The case was ultimately appealed because the Army Corps of Engineers claimed sovereign immunity (PL-89-298 1965). On September 24, 2012, with respect to the Katrina Canal Breach Litigation, the United States District Court for the Eastern District of Louisiana noted that with respect to the MRGO canal, the Corps reported studies recognizing that erosion from wave wash had widened the channel and that the north shore was close to being breached, thereby exposing development and inhabitants to the southwest to hurricane force winds and wave damage from Lake Borgne. They had proposed two erosion-control plans in a 1984 report, although dredging would have increased substantially. The Corps took the position that design modification was "not warranted under the cost-benefit ratio." They believed the continued need for dredging to maintain the channel's navigability was less expensive than the cost of providing foreshore protection. The Corps also refused to undertake the cost of foreshore protection unless there was local cost participation under the Water Resources Development Act, 33 U.S.C. They did not prioritize protecting the north shore because there was no levee to protect, and the primary mission of the Corps was to keep the channel navigable. In 1994, the Corps issued another report in which it still proceeded under the assumption that the cost of foreshore protection was greater than it proved to be. The appeals court, however, noted that sovereign immunity released the Corps only from damages caused by floodwaters released on account of flood-control activity or negligence therein. Some Katrina-related flooding was caused not by flood-control activity (or negligence therein) but by the MRGO, a navigational channel whose design, construction, and maintenance could not be characterized as flood-control activity. Therefore, the appeals court refused to apply sovereign immunity on behalf of the Corps for that flooding.

Newberry (2006) noted that "The USACE had tried for many years to obtain authorization to install floodgates at the north ends of the three drainage canals. . . Dysfunctional interaction between the local Levee Board (who were responsible for levees and floodwalls, etc.) and the local Water and Sewerage Board (who were responsible for pumping water from the city via the drainage canals) prevented the installation of these gates. As a result, many miles of the sides of these three canals had to be lined with levees and floodwalls."

Newberry (2006) also noted that "For many years, the Corps of Engineers has been subjected to extreme pressures at the federal and state levels to do more with less; to do their projects better, faster, and cheaper. The Corps knew of land subsidence problems, improved storm forecasting, etc., for years, but had not modified the design or designated repairs that the New Orleans Levee Board was to make."

In the aftermath, the news was not much better. Despite spending over $10 billion in an effort to rebuild the system, the Corps believes there's still a significant risk of flooding from major hurricanes or river floods that are greater than the design heights of the levees (Schleifstein 2012). In both cases, the levees were rated Class II or *urgent* (unsafe or potentially unsafe) (Schleifstein 2012).

ST. PETERSBURG/PINELLAS COUNTY SEWER OVERFLOWS

Sanitary sewer overflows are a bane of the existence of sanitary sewer systems. Overflows permit the surface and nearby surface waters to be exposed to untreated sewage, which can lead to negative health impacts for the public and the environment. As noted, prior to the development of modern wastewater treatment systems, waterborne illness caused by the exposure to wastewater was rampant leading to many deaths each year. The capacity, management, operations, and maintenance (CMOM) program discussed in Chapter 4 was created specifically to try to reduce the sanitary sewer overflow problem. Most of the rules were developed to minimize this exposure. However, that does not mean they can be fully eliminated if there is a large weather event, if the system was not constructed properly, or if the system was not maintained as it aged. Overflows occur if a line breaks or a gravity line gets plugged—these are usually small, limited events that impact few and are cleaned up promptly. Of greater concern are overflows that occur when it rains. Then, millions of gallons can be spilled. Overflowing sanitary sewers is a public health hazard, as well as a nuisance. Utilities can be fined if overflows occur regularly and may pay substantial amounts to fix private property as well.

St Petersburg, Florida, has operated a wastewater collection system since 1894 (City of St. Petersburg 2016). The system grew slowly through the 1920s as the Tampa Bay region grew into an import/export center. By 1933, the sewer system was about 25% of what it is today (City of St. Petersburg 2016). Most of the rest of the nearly 900 miles of gravity sewer pipe was constructed between 1950 and 1962 (City of St. Petersburg 2016). The pipe is primarily vitrified clay pipe with Orangeburg service lines. Both deteriorate with time as discussed in Chapter 4.

In June 2016, a rainstorm caused an overflow that spilled nearly 10 million gallons into Tampa Bay (Frago 2016). As a result, the city spent $400,000 to add three million gallons of emergency storage to its old Albert Whitted Wastewater Treatment Plant. The job was expected to take four months to do, but was done in two. However, as Category 1 Hurricane Hermine approached in late August 2016, city officials were still concerned that additional sewer overflows were likely to occur due to the potential rain and leaky service lines (Danielson, DiNatale, Solomon 2016; Sowers 2016).

When Hurricane Hermine hit Tampa Bay on September 2, 2016, estimates of between 30 and 40 million gallons of partially treated wastewater were reportedly discharged into the waters of the Tampa and Boca Ciega Bays, and into watersheds of all the bay area (Frago 2016). Millions of gallons of untreated sewage overflowed from the St. Petersburg sewer collection system into the bay (Deeson 2016). The overflows were so widespread that many local governments had difficulty calculating the amounts. The Florida Department of Environmental Protection and local legislative representatives were concerned about the impact to the communities from the pervasive overflow situation.

The overflows appeared to have had multiple causes, but two came to the forefront quickly. The first was that the city's wastewater treatment plants were running at capacity before the storm, and they had little ability to store the excess flows. One wastewater plant had been decommissioned several years earlier in a cost-saving mode that reduced redundancy on the system. 10Investigates (2016) reported that the consultant who prepared the report on the plant decommissioning recommended against it until plant capacity was expanded elsewhere, but that expansion was never funded (nor does it appear that funds were requested for the same). The risks associated with lack of storage and redundancy was noted in an engineering report concerning the decommissioning. The reason to close the plant had been obsolescence and cost to operate it. The report and its findings were never conveyed to elected officials or management. As a result, the mayor suspended two senior employees who were involved in that decision when the overflows occurred during Hurricane Hermine. The governor ordered an investigation and sources told the news media that there could be a criminal investigation into the sewage overflow after some discussion about the spill and the wastewater quality was reported (10Investigates 2016; Puente 2016).

The city staff attributed much of the overflow problem to the 80,000+ sewer laterals that serve private property. St. Petersburg has no ownership in these lines and therefore staff had concerns about their ability to address them. In addition, there is no data to pinpoint where private laterals may be playing a role in the overflows. While the city does have an ordinance that prohibits excessive discharges from private property, and therefore does have a mechanism to force homeowners to address impacts to the sewer system on private property (City of St. Petersburg 2016), much work would need to be undertaken to identify which properties would be required to make improvements and no monies were in the budget for this purpose (nor were any requested).

Laterals may be a concern, but the dynamics of the event suggest that capacity of the pipelines was insufficient and that pumping overwhelmed the pipe capacity. Excessive flows after a rainstorm indicate inflow—connections between the surface and the sewer lines (see Chapter 4) and not infiltration. Given the flow in the system, leaky manhole rings, connected stormwater lines, connections of gutters to sewers, and broken or removed cleanouts are likely to be far more significant than private laterals. The corrections are discussed in Chapter 4—inflow correction. The mayor, the new public utilities director, and the local legislators undertook an effort to start addressing the problem in the winter of 2017.

WALKERTON, ONTARIO

Since Dr. John Snow's 1854 discovery in London that drinking water could kill people by transmitting disease, water treatment and transmission infrastructure have mostly eliminated the transmission of water-borne diseases (Livernois 2002). The Walkerton experience serves to remind local officials that despite first-class facilities, much of the public takes for granted that their local government will operate and maintain these facilities to ensure drinking water safety. However, if the proper resources are not put in place to ensure that the infrastructure is operated and maintained properly, problems can occur and affect an unassuming customer base.

Walkerton is in a small Canadian community near Lake Huron. The community relies on groundwater for its water supply. The utility system was started in 1949 (WQHC 2005) with surficial wells constructed for the municipal water supply. In May 2000, the utility had three wells—Wells 5, 6, and 7—although only Well 5 was actually operational (WQHC 2005).

The first significant infrastructure failure of the new millennium started on May 18, 2000. Twenty children were absent from Mother Teresa School, two of which were admitted to the Owen Sound hospital in Walkerton with bloody diarrhea (O'Connor 2002a). Identification of the Walkerton outbreak was initiated by the early recognition of pediatric cases dealing with bloody diarrhea and severe abdominal cramps at the local hospital on May 19, 2000 (Bruce-Grey-Owen 2000). The next day there were widespread reports of gastrointestinal illness as reported by local doctors and hospitals. Many more students did not attend school (O'Connor 2002a). The region's Medical Health Office was first notified about the issue on May 20th, but later found out that local doctors had been treating patients with gastrointestinal illness symptoms since May 17th (CBC News 2010).

By May 21st, the Medical Health Office determined that the illnesses were caused by E. coli O157:H7 (Bruce-Grey-Owen 2000). A stool culture taken from one of the initial cases was reported on Saturday, May 20, to be presumptive positive for sorbitol negative E. coli (a marker for E. coli O157:H7), and a preliminary report early on May 21 identified the isolate as E. coli O157:H7 (Bruce-Grey-Owen 2000). A boil-water advisory was issued despite claims from the utility that the water was safe to drink (Bruce-Grey-Owen 2000). Testing of water samples from the distribution system on May 21 and of water from Well 5 on May 23 demonstrated significant contamination with coliform and E. coli bacteria.

Nearly half the population fell ill, and seven people died (CBC News 2010). A person infected with E. coli O157:H7 experiences intestinal disease lasting on average four days, but sometimes longer. After 24 hours, the person often experiences bloody diarrhea, and in some cases very severe abdominal pain. The illness usually resolves itself without treatment other than rehydration and electrolyte replacement, but persons with compromised immune systems, the elderly, and children under age five are at risk of death from the effects of these organisms.

The events that precipitated this incident were discovered over the course of the following summer. The provincial government assigned Justice Dennis R. O'Connor to lead the investigation. Justice O'Connor found that the Walkerton incident was caused by errors and omissions by various individuals, as well as by systemic failures at the local, regional, and provincial levels (Lindgren 2003). Cascading effects had led to the deaths. The contamination was ultimately due to manure that had been spread on a farm near Well 5. The owner of this farm followed proper practices. However, the well was located at the lower end of the field and the well seal was not properly maintained, so floodwaters contaminated the well. Justice O'Connor noted that the failure to inspect and require the well seal to be corrected resulted from shortcomings in the approvals and inspections programs of the Ministry of the Environment and the utility staff (O'Connor 2002a). In addition, the Walkerton Public Utilities Commission operators lacked the training and expertise necessary to identify both the vulnerability of Well 5 to surface contamination or the resulting need for continuous chlorine residual or turbidity monitoring. The operators did not monitor the chlorine residual and there was some concern that the data provided to the province may have been inaccurate. Justice O'Connor (2002a) noted that the operators had engaged in a host of improper operating practices, including failing to use adequate doses of chlorine, failing to monitor chlorine residuals daily, making false entries about residuals in daily operating records, misstating the locations at which microbiological samples were taken for many years, and that the operators knew that these practices were unacceptable and contrary to regulatory guidelines and directives. However, no action was taken by the utility managers of the city of Walkerton, and data on the concerns were not transmitted upward. The public utility committee's (PUC) general manager concealed information about a chlorinator being off-line and water quality test failures from water samples taken on May 15, 2000, by the local Health Unit. Had he disclosed either of these facts, the health unit would have issued a boil-water advisory on May 19 and many illnesses would have been avoided (O'Connor 2002a).

In addition, the provincial Ministry of the Environment should have, but did not, detect the problem nor was any effort undertaken to address the practices of the operators. He indicated that the provincial government's budget reductions, which led to the discontinuation of government laboratory testing services for municipalities in 1996, were partially to blame (O'Connor 2002a). Such independent testing might have uncovered the prior test failure and operational deficiencies (chlorine residuals), causing enforcement action and boil-water notifications. Justice O'Connor recommended a new regulation mandating that testing laboratories immediately and directly notify both the Ministry of the Environment and the Medical Officer of Health of any adverse results. Had the government done this, the boil-water advisory would likely have been issued by May 19, 2000, thereby preventing most of the illness and possibly all of the deaths (O'Connor 2002a). He also suggested that boil-water advisories should be more broadly disseminated because some residents of Walkerton did not become aware of the boil-water advisory on May 21, 2000.

Within the utility, the PUC commissioners were not made aware of the improper treatment and monitoring practices of the PUC operators. However, those who were commissioners had failed to properly respond to a Ministry of the Environment inspection report in 1998 that set out significant concerns about water quality and that identified several operating deficiencies at the PUC. These deficiencies included a number of measures to reduce exposure risk: (1) placing multiple barriers aimed at preventing contaminants from reaching consumers, (2) adopting a cautious approach to making decisions that affect drinking water safety, (3) ensuring that water providers apply sound quality management and operating systems, and (4) providing for effective provincial government regulation and oversight. The failure of the utility to provide for or to request funds to correct the 1998 problems was cited as a causal factor.

Justice O'Connor found that inadequate funds had been allocated by the community government for public utility oversight, which limited the ability of the city and its manager to oversee the utility, which may have either mitigated or altogether avoided the outbreak (WQHC 2005). The failure of the local government to allocate funds or to involve themselves in the affairs of the utility they oversaw was also a casual concern. There was a demonstrated need by the utility and regulatory agencies to ensure that multiple protection barriers for public health protection were needed. These were not provided. Environmental factors were not considered—a well at the low end (elevation-wise) of a farm field on which manure was applied would appear to be an obvious potential health issue, but it was not addressed by the utility, local government, or regulatory agencies. Justice O'Connor noted that groundwater sources must be managed with consideration for their surrounding environments, topography, and land uses (WQHC 2005).

O'Connor (2002) noted that in the aftermath, the community had widespread feelings of frustration, anger, and insecurity. The economic cost of the Walkerton tragedy was estimated to be more than $64.5 million, which is huge for a community of this size (Livernois 2002). Five years later, Walkerton's water system managers pleaded guilty to criminal charges, closing five years worth of investigation.

Walkerton was the first cascading failure in the new millennium. Money, regulatory and management oversight, and stewardship all failed the public in this instance. If responsibility had been taken by anyone at any point, the incident likely would not have happened.

ALAMOSA, COLORADO

Falco and Williams (2009) reported that as recently as the 1980s, disease outbreaks that were associated with drinking water were relatively common in Colorado, often occurring more than once a year. However, since 1990 they have been relatively uncommon due to added testing and disinfection of water systems. That leads us to Alamosa.

Alamosa is a city of 8,900 people in southern Colorado. Alamosa's water system distributes over 800,000 gallons of water per day through 50 miles of distribution pipe (Falco and Williams 2009). According to engineering reports, the system is composed primarily of polyvinyl chloride (PVC), asbestos cement (AC), and cast iron (CI) pipe. A small amount, about 1,800 linear feet, is high-density polyethylene (HDPE). The newer PVC pipe is generally less than 15 years old. The AC and CI pipe is typically older than 20 years, with some of it installed in the early part of the twentieth century (Falco and Williams 2009).

Alamosa relies on deep artesian wells for its water supply. The well permits were issued between 1963 and 1983. Records also appear to indicate that original well construction was as early as 1936 for the Cole Park well and 1956 for the Weber well (Falco and Williams 2009). Falco and Williams (2009) note that the aquifer was considered to be a protected groundwater source by the state. Waivers of disinfection requirements are granted on a case-by-case basis and are permanent unless the state's safe drinking water program has reason to withdraw the waiver. A waiver from the statewide requirement for disinfection was granted to Alamosa in 1974. As a result, the city's drinking water was not chlorinated for disinfection. Not chlorinating the water saves money for the city by reducing chemical costs and monitoring requirements. The city was deemed to be in compliance with all health-based drinking water standards with the exception of the arsenic standard, which required a treatment plant be built years earlier.

During the second week of March 2008, multiple residents of Alamosa fell ill with severe gastrointestinal symptoms, including diarrhea and abdominal cramps (Marler 2015). On March 17, 2008, the Colorado Department of Public Health and Environment (CDPHE) notified the Centers for Disease Control and Prevention (CDC) that 18 Alamosa residents who had fallen ill in the month of March 2008 had tested positive for *Salmonella typhimurium* and that 19 others were suspected to be suffering from gastrointestinal symptoms of the same organism. A series of tests on randomly sampled water from the city's water system confirmed that the potable water being distributed was heavily contaminated by *Salmonella typhimurium*—thus, the source of illness in the developing outbreak (Marler 2015). Colorado's chief medical officer and the CDPHE issued an order on March 19, 2008, requiring the City of Alamosa to advise its residents to drink only bottled water (Marler 2015). As of that date, at least 79 people were believed to have been infected by the heavily contaminated public water supply; and that number appeared to be rising. Ultimately, the outbreak resulted in 442 reported illnesses—122 of which were laboratory-confirmed—and one death (Falco and Williams 2009). Epidemiological estimates suggest that up to 1,300 people may have been ill (Falco and Williams 2009).

Based on a review of records and discussion with city personnel, the city did not maintain a program of routine flushing, disinfecting, or removing sediment from their tanks as part of routine operations. In fact, according to Falco and Williams (2009), at the time of the outbreak, the city did not have an active distribution-system maintenance and flushing program or a cross-connection control program. All the distribution crews did was respond to line breaks.

It was later noted that the Weber reservoir was in poor condition and had been poorly maintained for a number of years. It was suggested that the contamination may have entered this reservoir though snowmelt, small animals, or fecal contamination. The actual means of transmission was never fully defined. The treated, high-quality ground water was stored in a reservoir that was not sealed, thereby negating the protection provided by the aquifer.

On October 10, 2008, the CDPHE signed an Enforcement Order on Consent (*consent agreement*) for violations related to the Salmonella outbreak. The requirements of the consent agreement were discussed between the CDPHE and the city and were agreed upon in advance. The consent agreement clarified outstanding requirements and ongoing expectations as they related to the Salmonella outbreak and Alamosa's operation of its public water system. Requirements included: (1) continuous chlorination of the well sources, (2) ongoing monitoring of water quality in the distribution system, (3) cooperation with the CDPHE on the investigation into the cause of the outbreak, and (4) submittal of other technical and administrative documentation (Falco and Williams 2009).

A lawsuit was filed against Alamosa claiming that the city was negligent in maintaining its water facilities and had sold tainted water to its customers. On September 16, 2010, it was announced that the City of Alamosa, through its insurance carrier, Travelers, would pay $360,000 to 29 Alamosa residents who became ill because of the salmonella outbreak in 2008 (Marler 2015).

This is another cascading event, although much more limited than Walkerton. Here the city declined to chlorinate the system (which would have addressed the issue immediately), did not maintain or flush their distribution system, and had not properly maintained the reservoir (Falco and Williams 2009). Cost was an issue, although whether utility staff had asked for monies to fix the reservoir was unclear. Testing for salmonella was neither required nor undertaken. The waiver does not appear to have been reviewed after 1974.

FLINT, MICHIGAN, 2014–2017

Then, there is the most recent failure—Flint, Michigan. There was a lot of coverage in the news about the troubles in Flint throughout 2016. However, if one reads between the lines, there are two issues. First, this was not new—it was several years old, going back to when the city's water plant came back online in May 2014. Second, this was a political/financial issue that became a public health issue. In fact, the political/financial goals appear to have been so overwhelming that the public health aspects were scarcely considered.

Flint's first water plant was constructed in 1917. The source of water was the Flint River. This plant was abandoned in 1952 when the city's second plant was constructed (note the original 1917 plant is still located—in dilapidated condition—on the east side of the current plant). Because of declining water quality in the Flint River, the city had made plans to build a pipeline from Lake Huron to Flint in 1962, but a real estate scandal caused the city commission to abandon the pipeline project in 1964. Instead, the city entered into a contract to buy water from the City of Detroit (whose source is Lake Huron). Flint stopped treating its water in 1967 when a pipeline from Detroit was completed (Flint Water Advisory Task Force 2016). Starting in 1967, the City of Flint purchased almost 100 millions of gallons per day (MGD) from Detroit, but this amount had diminished to under 14 MGD by 2013 (Flint Water Advisory Task Force 2016) as industry left and the population diminished.

While the City of Flint had purchased water for years from Detroit, the 60-year-old Flint water treatment plant had been maintained as a backup to the Detroit system, operating approximately 20 days over those 60 years at 11 MGD. In 2010, Detroit declared bankruptcy. Flint declared bankruptcy about the same time. Between 2011 and 2015, Flint's finances were controlled by a series of receivers/emergency managers appointed by the governor—as were

Detroit's (different receivers, however). Both receivers were told by the governor to reduce costs (the finance/business decisions—Flint Water Advisory Task Force 2016). Public health was not a priority.

In 2010, the City of Flint joined the Karegnondi Water Authority (KWA), which consisted of a group of local communities that decided to support and fund construction of a raw water pipeline from Lake Huron. Flint was one of the communities and the concept was for Lake Huron water to be delivered to the old Flint water plant for treatment. Since Flint was in bankruptcy at the time, final approval of the deal was delayed until the state and the emergency manager approved of the agreement (Flint Water Advisory Task Force 2016). While discussions were ongoing for several years thereafter, the *Detroit Free Press* reported a seven to one vote in favor of the KWA project by Flint's elected officials in March 25, 2013 (City of Flint 2016). On April 16, 2013, Emergency Manager Ed Kurtz signed the contract effectively purchasing 18 MGD of capacity from KWA (City of Flint 2016). The cost of the new KWA pipeline was estimated to be $272 million with Flint's portion estimated at $81 million (TYJT 2013).

An engineer's report noted that a Genesee County Drain Commissioner stated that one of the main reasons for pursuing the KWA supply was the perceived lack of reliability of the Detroit supply, given the 2003 power blackout that left Flint without water for several days (City of Detroit 2013). Another issue was that Flint had no say in the rate increases issued by Detroit. Given the likelihood of increased prices due to Detroit's bankruptcy, this may also have been a factor.

The City of Detroit objected due to loss of revenues at a time when a receiver was trying to stabilize the city's finances (in conjunction with the state treasurer, City of Detroit 2013). In February 2013, the engineering consulting firm of Tucker, Young, Jackson, Tull, Inc. (TYJT 2013), at the request of the state treasurer, performed an analysis of the water supply options being considered by the City of Flint (City of Detroit 2013). They noted that the raw water transmission system had a proposed 60 MGD capacity and sized to deliver a maximum of 18 MGD to the Flint water treatment plant with an average day supply of 12 MGD. The term of the KWA contract for Flint was 40 years (TYJT 2013). They indicated that improvements would also be required at the Flint water treatment plant to treat the lake water because the plant was designed to treat the Flint River water (TYJT 2013; City of Flint 2016). The preliminary investigation evaluated the cost associated with the required improvements to the plant, plus the costs for annual operation and maintenance including labor, utilities, chemicals, and residual management. They indicated that the pipeline cost was likely too low, that Flint's financial obligation to the pipeline could be $25 million higher than projected, and that there was less redundancy in the KWA pipeline than in Detroit's system (City of Detroit 2013). In 2013, the City of Detroit made a final offer to convince Flint to stay on Detroit water with certain concessions. Flint declined the final Detroit offer. Immediately after Flint declined the offer, Detroit gave Flint notice that their long-standing water agreement would terminate in twelve months, meaning that Flint's water agreement with Detroit would end on April 17, 2014. However, construction of KWA was not expected to be completed until the end of 2016 (TYJT 2013; City of Flint 2016).

On June 23, 2013, the engineering firm of Lockwood, Andrews, and Newnam was hired to design and oversee a $7 million renovation/upgrade to the old Flint River plant to allow the filters to treat Flint River water (Flint Water Advisory Task Force 2016). The project was designed to take water from the Flint River for a period of time until the Lake Huron water pipeline was completed (Fonger 2015). Flows were designed for 16 MGD. Lime softening, sand filters, and disinfection were in place, but water quality testing was suggested to have been lacking.

The change in water quality and treatment created other water quality challenges that re-sulted in water quality violations. Like most older, northern cities, the water distribution system was almost 100 years old. Many of the service lines from the CI water mains (with lead joints) to customer's homes were constructed with lead goosenecks and copper and lead service lines. Lead piping existed within many houses. The potential impact of lead in children engendered the lead and copper rule of the late 1980s.

In the early 1990s, water systems addressed this rule by testing and by corrosion control ad-ditives. The concept was that on the first draw of water in the morning, the lead concentration should not exceed 0.015 mg/L and copper should not exceed 1.3 mg/L. Depending on the size of the utility, sampling was to be undertaken twice at a random set of houses with the number of samples dependent on the size of the system. The sampling was required to be performed twice, six months apart (note routine sampling has occurred since then to ensure compliance). Resi-dents were instructed on how to take the samples, and results were submitted to the appropriate regulatory agencies. If the system came up *positive* for either compound, the utility was required to make adjustments to the treatment process. Ideally, water leaving the plant would have a slightly negative Langlier saturation index (LSI) and would tend to slightly deposit on pipes. Coupon tests could be conducted to demonstrate this actually occurred—as they age, the pipes develop a scale that helps prevent leaching. Most utilities tested various products in 1989/1990. Detroit did this and there were no problems.

Flint did not do this testing in 2013/2014 (Flint Water Advisory Task Force 2016), which was a major oversight since the water quality in the Flint River water is different than the water qual-ity coming from Detroit. Salinity, total organic carbon (TOC), pH, and overall aggressiveness of the water (as measured by how easily it might dissolve the pipe) were significantly greater. The variable flow rates in the river, with upstream agriculture, industry, and a high potential for contamination, make it nowhere near as easy to treat as the cold water from Lake Huron. Accommodations were not made to address the problem and the state found that no poly-phosphates were added to protect the pipes against corrosion, as directed by the State. Veolia, a contract operator, reported that the operations needed changes and that the operators needed training (Fonger 2015). Additional facilities were needed to address quality concerns (including granular activated carbon filter media).

As a result, Flint appears to have sent corrosive water into the piping system, which dis-solved the scale that had developed over the years from the Detroit water, exposing raw metal, and thus creating the leaching issue. Volunteer teams led by Virginia Tech researchers reported that at least a quarter of Flint households had levels of lead above the federal level of 15 ppb, and as high as 13,200 ppb (Lazarus 2017). Aging CI pipe compounded the situation, leading to aesthetic issues, including taste, odor, and discoloration that result from aggressive water (brown water). Once the city started receiving violations, public interest and scrutiny of the drinking water system intensified.

Operator Glasgow noted that there was pressure from local officials to start the water sup-ply from the Flint River on April 16, 2014 (Flint Water Advisory Task Force 2016). The City of Flint began using the Flint River as a water source on April 29, 2014, knowing that treatment would need to be closely watched since the Michigan Department of Environmental Quality, in partnership with the U.S. Geological Survey and the City of Flint Utilities Department, had conducted a source water assessment and determined there was a very high susceptibility of potential contamination because of a very high susceptibility to potential contaminant sources (Cooper 2004).

In May and August of 2014, total trihalomethane (TTHM) samples violated the drinking water standards (Flint Water Advisory Task Force 2016). This means two things—TOC was present in the water and too much chlorine was being added to disinfect and probably reduce color caused by the TOC. Of interest, Flint's plant is a lime-softening plant. Lime softening does not remove TOC and is not generally used for surface water treatment. Filtration is not very effective in removing TOC either. High concentrations of TOC must be removed with granular activated carbon, ion exchange, or membrane treatment. The Flint plant had none of these, so the carbon stayed in the water.

To address the TTHM issue, chlorine appears to have been reduced as the TTHM issue was in compliance by the next sampling event in November 2014. However, in the interim, new violations included coliform and E. coli in August and September of 2014 (Flint Water Advisory Task Force 2016), an indication of inadequate disinfection. That meant residents had to boil their water and there was much public outcry.

Experts say Flint's lead problems could have been held in check if the city had added phosphates to the water, as Detroit had done for years, but even when phosphates were added, it corroded at 16 times the rate of the Detroit water system (Wisely and Erb 2015). The tests showed Flint River water without added phosphates corroded the lead at 19 times the rate of Detroit water (Ganim and Tran 2016). Once the switch was made, the state began testing for lead and copper, as is required by the United States Environmental Protection Agency (EPA). The first round of testing completed in December 2014 showed lead levels of six parts per billion. The second round, completed in June 2015, showed they had almost doubled to 11 parts per billion (Wisely and Erb 2015). The EPA requires a remediation plan when levels reach 15 parts per billion.

At Flint's General Motors (GM) engine plant, workers there began noticing rust spots on newly machined parts (Wisely and Erb 2015). Citing high chloride levels in the water, GM advised the city that it could no longer use the water being delivered to its engine plant, terminated the connection, and began drawing water from neighboring Flint Township (a system separate from the City of Flint). The City of Flint approved letting GM switch to water from Flint Township, but didn't change its water treatment procedures (Wisely and Erb 2015).

Veolia, the contract operator mentioned earlier, was asked to help with the TTHM issues (Fonger 2015). They recommended installing granular activated carbon filters and providing extensive training for staff members who were not familiar with operating this type of plant (Fonger 2015a). The city commission reportedly asked the receiver to switch back to Detroit water, but that request was initially rebuffed. Finally, in October 2015, the water supply was switched back to Detroit and the city started adding additional zinc orthophosphate in December 2015 to facilitate the buildup of the phosphate scale eroded from the pipes by the Flint River water. But that meant the pipes that were once stable, then destabilized, were now destabilized again by the switch back. It will take some time for the scale to rebuild and to lower lead levels, leaving the residents of Flint at risk because of a business/finance/political decision that had not considered public health impacts. It also indicates that the chemistry profile and sampling prior to conversion and startup was not fully performed to identify the potential for the lead leaching to occur.

The city estimates that there are at least 4,000 known lead service lines in Flint based on city records and a study performed by Rowe Engineering. However, for another 13,000 service lines, the city can't say for sure whether they are lead, galvanized steel, or copper—but history indicates that many of those are likely to be made of lead. It costs approximately $6,000 per home to replace the pipe. Changing these out would take time even with a service line replacement project as discussed in Chapter 3. And, that does not count the lead plumbing in the house which could be many thousands more, a major concern for residents. Premise plumbing is a major cost and legal issue for residents and the city.

Just when things appeared to improve in the media, a hospital in Flint reported in January 2016 that low levels of Legionnella bacteria had been discovered in the water system and that 10 people had died and another 77 to 85 were affected way back in 2014. Shigella was also reported (McFarland and Anderson 2016). The source is unclear, but Legionella does occur when chlorine residuals are reduced (Bloetscher and Plummer 2011). The lawsuits have begun, but where does the problem lie? First, it is clear that public health was not the primary driver for the decisions. Second, treating water is not as simple as cost managers think. Decision makers need to understand that water quality, pipe quality and stabilization, and potential problems with new water sources are incredibly important factors to consider. Their initial decision making appears to have been flawed in Flint.

Former Mayor Dayne Walling and former Emergency Manager Darnell Earley blamed the high lead levels in the city's water on state and federal officials (Shepardson 2016). Note that Earley was later indicted along with a number of others for his role in the decision making. Residents, the former mayor, the current mayor, Congressman Kildee, and city workers all blame the governor's office and the state Department of Environmental Quality (Ganim and Tran 2016). The state's focus on balancing the city's books and choosing low cost over human consequences appears to have created more expensive public problems (Shepardson 2016). Where the lawsuits and criminal indictments will go remains to be seen.

The guidance from the consultants or other water managers is unclear. If the due diligence by the engineers was not undertaken as to what the water quality impacts would be due to a change in waters, then the engineering design effort appears to have been flawed. If the engineer recommended this work but it was denied by the receivers, then it creates another flawed business decision that fails the public health test.

Clearly, there were issues with operations. Stating that state phosphates were used when they were not appears to be operational malfeasance. Walkerton also had operational issues; a major concern when public health is at risk. Veolia came to a similar conclusion (Fonger 2015).

The state has received its share of blame in the press. The regulatory staff was reduced as the state trimmed its budget, but it is unclear if there were sufficient resources to ensure oversight of water quality. That would then involve the governor and legislature. This is similar to the Walkerton example where the lack of provincial resources to monitor water quality was an issue and lack of oversight compounded local issues. Multiple state employees were indicted.

So many confounding problems, but what is clear is that Flint is an example of why public utilities should be operated with public health at the forefront and not by cost or politics. Neither cost nor politics protect the public health. The operations personnel should get the funds needed to protect the public health and the politicians should work with the staff to achieve this goal.

John Young is the former receiver in Birmingham and was employed to help the City of Flint. He was asked a question in a meeting at the American Water Works Association headquarters in Denver: "When will the water emergency be over in Flint?" He said the answer is unclear (Young 2016). Residents were provided water filters, but those filters may not be properly maintained. So, now what? Public trust is missing which makes fixing the problem difficult. And given the percentage of disadvantaged people in the community, how do they solve the revenue problem?

The KWA pipeline was completed in late 2017 and started delivering water to Genesee County shortly thereafter. Flint is not part of the raw water distribution system. Detroit's water system was sold to a newly created public entity called the Great Lakes Water Authority, which continues to supply Flint. Trust remains an issue. Young suggested that the Flint incident is un-likely to be replicated due to the cascade of bad decisions made there. But, that is what was said of Walkerton so public officials must be vigilant.

HOW TO PREDICT THE NEXT FLINT

Flint's deteriorated water system is a result of financial stress, but the community has a lot of pov-erty and high water bills, so they can't pay for improvements. They are not alone. Utilities all over the country have increasing incidents of breaks and age-related problems. So, the real question then is: who are the at-risk utilities? Who is the next Flint? The answer to the first resides in the measures that might identify the future *Flint*. These could be determinations like those used for asset condition assessment—age of the system, materials used, etc., as discussed in Chapter 10. Also, community characteristics like economic activity trends, income, poverty rate, unemploy-ment rate, utility size, reserves, utility rates, history of rate increases, etc., should be included (see Chapter 12). Are there work orders to track consequences? If not, how can that data be gener-ated? Could these be developed into a means to evaluate risk? If so, who would use it and how would the high-risk cases be addressed? Lenders have means to evaluate this using many of these same measures, but from a risk-of-event perspective, this method has not been applied.

CONCLUSIONS

There are many opportunities for impacts on the public from the failure of infrastructure sys-tems, as indicated by these case studies. There are many more examples of failures that can hap-pen for any infrastructure system. As seen in the 2003 blackout, one system's failure can affect other systems, and even lead to poor decisions on those systems. These examples were chosen because they are recent and highlight several issues that should be of concern to local officials. First, the desire to cut costs was present in every one of these incidents. The desire to cut costs appears to have trumped the obligation to protect the public health, safety, and welfare, in all cases. In several cases, regulatory oversight—the bane of existence in some political circles—was absent or underfunded. Regulations are created because there was a problem. Regulations are rarely enacted if there is not a problem first. Money is required to enforce the regulations to reduce the potential for repeats. Oversight might have helped prevent the impacts (for the blackout, the regulatory oversight impacts were never very clear). Issues with a lack of resources, training, or supervision for staff were present.

Local elected officials shoulder part of the blame in most of these cases—in part because they control the money and in part because they control the environment within which work gets done in the organization. Legal action resulted in all cases. A cascade of failures is also com-mon for all. All of these incidents indicate a need for a paradigm shift in public funding and operation. That means a new perspective on leading these organizations. That will be the topic of Chapter 16.

REFERENCES

Berger, E. 2001. "Keeping its head above water: New Orleans faces doomsday scenario." *Houston Chroni-cle*. December 1, 2001.

Bloetscher, F. and Plummer, J. D. 2011. "Evaluating the significance of certain pharmaceuticals and emerg-ing pathogens in raw water supplies." *Environmental Practice* 13. (9).

Bloetscher, F. et al. 2016. "Assessing potential impacts of sea level rise on public health and vulnerable populations in Southeast Florida and providing a framework to improve outcomes." *Sustainability.* 2016, 8(4), 315; doi.10.3390/su8040315.

Brinkley, D. 2015. "The broken promise of the levees that failed New Orleans." *Smithsonian Magazine.* September 2015. http://www.smithsonianmag.com/smithsonian-institution/broken-promise-levees-failed-new-orleans-180956326/#dgtHH72yWi5Tcler. Accessed 12/13/16.

Bruce-Grey-Owen Sound Health Unit. 2000. "The investigative report of the Walkerton outbreak of waterborne gastroenteritis May—June, 2000." Bruce-Grey-Owen Sound Health Unit, Toronto, ON.

CBC News. 2010. "Inside Walkerton: Canada's worst-ever E. coli contamination." CBC News Posted, May 10, 2010. CBC/Radio-Canada.

City of Detroit. 2013. "Water war undermines Flint-DWSD relations." 4/1/2013. file:///E:/CGN%20 6506%20Infra%20Mgmt/9932%20pr2013-04-01_water_war_undermines_flint-dwsd_relations.pdf.

City of Flint. 2016. "State of emergency declared in the City of Flint." City of Flint, MI. https://www.cityof flint.com/state-of-emergency/. Accessed 3/15/16.

City of St. Petersburg. 2016. "Wastewater Collection & Maintenance, City of St. Petersburg." St. Petersburg, FL.

Cooper, J. 2004. "A biological assessment of the Flint River and selected tributaries in Lapeer, Genesee, Oakland, and Saginaw Counties, Michigan." June 30–August 2003. Surface Water Quality Assessment Section, Water Bureau, Michigan Department of Environmental Quality, MI/DEQ/WD-03/114.

Danielson, R., DiNatale, S., and Solomon, J. 2016. "Tampa, St. Petersburg officials say sewage overflows less likely in this storm." Times Staff Writers Published: August 30, *Tampa Bay Times*, Aug. 30, 2016. http://www.tampabay.com/news/weather/tampa-st-petersburg-officials-say-sewage-overflows-less-likely-in-this/2291515. Accessed 12/12/16.

Deeson, M. 2016. "Top St. Pete water officials suspended over sewage crisis." WTSP 6:36 p.m. EST. September 21, 2016. http://www.wtsp.com/news/investigations/top-st-pete-water-officials-suspended-over-sewage-crisis/323104234. Accessed 12/12/16.

Energy.gov. 2004. "Final report on the August 14, 2003 blackout in the United States and Canada: Causes and recommendations." U.S.-Canada Power System Outage Task Force. https://energy.gov/sites/prod/files/oeprod/DocumentsandMedia/BlackoutFinal-Web.pdf. Accessed 12/12/16.

Falco, R. and Williams, S. I. 2009. "Waterborne salmonella outbreak in Alamosa, Colorado, March and April 2008—Outbreak identification, response, and investigation." Colorado Department of Public Health and Environment, Denver, CO.

Fischetti, M. 2001. "Drowning New Orleans." *Scientific American.* October, 2001.

Flint Water Advisory Task Force. 2016. "Final Report." http://www.michigan.gov/documents/snyder/FWATF_FINAL_REPORT_21March2016_517805_7.pdf. Accessed October, 22, 2016.

Fonger, R. 2015. "Flint water consulting team gets $900 an hour for advice." Report due next week 2/28/15. file:///E:/CGN%206506%20Infra%20Mgmt/9933%20Flint%20water%20consulting%20team%20gets%20$900%20an%20hour%20for%20advice;%20Report%20due%20next%20week%20_%20MLive.pdf.

Fonger, R. 2015a. "Flint water consultant's final report recommends $3 million in changes." 3/18/15. file:///E:/CGN%206506%20Infra%20Mgmt/9934%20Flint%20water%20consultant's%20final%20report%20recommends%20$3%20million%20in%20changes%20_%20MLive.pdf.

Frago, C. 2016. "Tampa Bay's sewage mess: 29 million gallons spilled into the bay and rising." *Tampa Bay Times.* Tuesday, September 6, 2016. 8:18pm.

Ganim, S. and Tran, L. 2016. "How tap water became toxic in Flint, Michigan." updated 10:53 a.m. ET. Wed. January 13, 2016. https://Www.Michigan.Gov/Documents/Snyder/FWATF_FINAL_REPORT_21March2016_517805_7.Pdf.

Grunwald, M. 2006. "The Army Corps of Engineers is the real culprit behind New Orleans' devastation." *GRIST*, on Aug. 30, 2006. http://grist.org/article/grunwald/. Accessed 12/12/16.

Lazarus, O. 2017. "In Flint, Michigan, a crisis over lead levels in tap water." *Public Radio International.* Minneapolis, MN. https://www.pri.org/stories/2016-01-07/flint-michigan-crisis-over-lead-levels-tap-water. Accessed 1/20/17.

Lindgren, R. D. 2003. "In the wake of the Walkerton tragedy: The Top 10 Questions." *National Symposium on Water Law (Canadian Bar Association, Vancouver, March 28 & 29th, 2003, Canadian Environmental Law Association, Toronto, Ontario.* Publication No. 440 ISBN No. 1-894158-83-0.

Livernois, J. 2002. "The economic costs of the Walkerton water crisis." Walkerton Inquiry Commissioned Paper 14.

Marler, W. 2015. "Salmonella Litigation A resource for Salmonella Outbreak Legal Cases link: Alamosa, Colorado, Municipal Water System Salmonella Outbreak Litigation." Marler-Clark law firm blog. http://www.salmonellalitigation.com/salmonella_caseupdates/view/alamosa_colorado_municipal _water_system_salmonella_outbreak_litigation. Accessed 11/11/16.

McFarland, P. H. with Anderson, M. 2016. "Crisis in Flint underscores a national problem." *ENR Online*. 10/12/16. http://www.enr.com/articles/40518-crisis-in-flint-underscores-a-national-drinking-water -quality-problem?v=preview. Accessed 11/12/16.

Minkel, J. R. 2008. "The 2003 Northeast Blackout—Five Years Later; Tougher regulatory measures are in place, but we're still a long way from a "smart" power grid." August 13, 2008. https://www .scientificamerican.com/article/2003-blackout-five-years-later/. Accessed 11/14/16.

Murphy, V. 2005. "Fixing New Orleans' thin grey line." *BBC News*. October 4, 2005.

Newberry, B. 2006. "Risk and safety in engineering: Lessons from Hurricane Katrina.", Mechanical Engineering, Baylor University, Waco, TX. http://www.baylor.edu/content/services/document .php?id=42035. Accessed 12/2/16/.

New York Independent System Operator (NYISD). "2004 Interim Report on the August 14, 2003, Black-out." Accessed 12/12/16.

O'Connor, D. R. 2002. "Part I: Report Walkerton Disease Outbreak." *Pipeline*. Vol. 17, No. 2, p. 1. http:// www.ohd.hr.state.or.us/dwpVol.17.

O'Connor, D. R. 2002a. "Part II: A Summary Report of the Walkerton Inquiry: The Events of May 2000 and Related Issues." Ontario Ministry of the Attorney General, Queen's Printer for Ontario. Toronto, ON.

PL-89-298. 1965. "Flood Control Act: Public Law 89-298, 89th Congress, S. 2300." https://www.fws.gov/ habitatconservation/Omnibus/R&HA1965.pdf.

Puente, M. 2016. "St. Petersburg sewage spill stories remain at odds." *Tampa Bay Times*, Monday, September 26, 2016. 10:26 p.m. http://www.tampabay.com/news/environment/water/st-petersburg-sewage -spill-stories-remain-at-odds/2295353. Accessed 12/12/16.

Schleifstein, M. 2012. "Hurricane Katrina flood ruling upheld by federal appeals court." NOLA.com. *The Times-Picayune*, March 03, 2012, 8:00 a.m. http://www.nola.com/katrina/index.ssf/2012/03/ hurricane_katrina_flood_ruling.html. Accessed 12/10/16.

Shepardson, D. 2016. "Two former Flint officials blame state and feds for water crisis." *Reuters*. http:// www.reuters.com/article/us-michigan-water-idUSKCN0WG2N0. Accessed 11/22/16.

Sowers, L. 2016. "St. Pete monitors sewage system for overflow." Posted August 31, 2016. http://www .fox13news.com/news/local-news/198222152-story. Accessed 12/12/16.

10Investigates. 2016. http://www.wtsp.com/news/investigations/10-investigates-allegations-st-pete-sewage -result-of-ignoring-consultant/319475312. Accessed 12/12/16.

Tucker, Young, Jackson, and Tull, Inc. 2013. "City Of Flint Water Supply Assessment State Of Michigan Contract No. 271N3200089." Tucker, Young, Jackson, and Tull, Inc. Detroit, MI.

Wisely, J. and Erb, R. 2015. "Chemical testing could have predicted Flint's water crisis." *Detroit Free Press*. October 11, 2015. http://www.freep.com/story/news/local/michigan/2015/10/10/missed -opportunities-flint-water-crisis/73688428/. Accessed December 20, 2016.

WQHC. 2005. "Walkerton—Five years after lessons learned in the aftermath of Canada's worst E. coli contamination." Health, Water Quality, and Health Council of the American Chemistry Council. WQHC.

Young, J. 2016. Presentation in person; 10/19/2016 at AWWA Council Summit, Denver Westin, Denver CO.

15

SUSTAINABILITY OF INFRASTRUCTURE: LOOKING AT THE LONG-TERM FUTURE TRENDS

The history of civilization was written as a result of infrastructure systems. The future will be no different. We currently see our infrastructure systems stressed with the combined effects of increasing populations and age, but there are a number of emerging trends that will complicate things further. This chapter is dedicated to discussing these trends. The focus will be in the United States, but the concepts are similar—and perhaps exacerbated overseas. The first major issue is the limitation of water supplies in many places. Rural electrification efforts were great as a means to help rural communities develop. Access to power brought pumps to places that had limited surface water but had groundwater sources that had not yet been tapped. Unfortunately, today we know that much of that groundwater does not recharge at nearly the same rate as it is being pumped out, leading to a sustainability issue. Reaching sustainability may require alternative water supplies, like direct potable reuse (DPR) or indirect potable reuse (IPR) of wastewater and stormwater. Those will create major public education challenges for communities.

Climate change, politically acceptable or not in any given community, is a reality. One problem with climate change is that the effects manifest themselves differently in different communities. Coastal areas have issues with sea level rise (SLR) today. Inland areas may see more or less rainfall, warmer weather, or altered weather patterns, which will impact the needs for water, particularly in regard to agricultural practices.

To cope with certain climate issues, some communities adopt green building practices, which have applications for infrastructure, (although green building does not apply to unoccupied buildings). Affordability is an issue discussed in Chapter 11 (and not repeated here), but will remain a hot topic given that income disparities continue to increase. The impact of driverless cars, trucks, and trains (as discussed in Chapter 6) is a longer term future issue, but vehicles, maintenance of roadways, and power are clearly related.

WATER SUPPLIES

Water supply is often conducted based on the concept of *basins* or catchment areas, defined as *the area where all rainfall that falls on the ground in that vicinity will drain to one exit point*. For example, an area surrounded by hills or mountains will have all rainfall stay within that basin or catchment area, exiting via one river or stream. There are a variety of users competing for water resources in a given basin, and each basin has unique characteristics:

- Agriculture
- Ecosystems
- Urban demands
- Industrial demands
- Cooling water for power plants

Estimates indicate that from 1950 to 1980, demands for water increased steadily across all of those sectors, with 1980 being the peak water use year. With rural electrification after World War II, much of the increase in water use has been via groundwater, especially in dry areas that never had a lot of water on the surface and as a result never really developed much before then. The total amount of fresh groundwater pumped each day in the United States is approximately 83 billion gallons per day (Hutson et al 2004), which is about 8% of the estimated one trillion gallons per day of natural recharge to the Nation's groundwater systems (Nace 1960)—except that, as noted, the most common withdrawal sites are arid states, which are among those least likely to recharge. For example, in parts of the western Great Basin, the aquifers have dropped hundreds of feet since electricity arrived. With an average of 13–18 inches per year of rainfall and high evaporation rates throughout the summer, little of this water has the potential to recharge the aquifer (Bloetscher and Muniz 2008).

The Great Basin is not alone. Throughout the West/Southwest, Plains states, upper Midwest (Wisconsin, Minnesota, Iowa), and Southeast (South Carolina, North Carolina), over-pumping exists because groundwater has been the obvious choice for water supply after rural electrification in the mid-20th century. This is because groundwater is supposedly a drought-proof problem. But, is it? Groundwater recharge is affected by precipitation (recharge), actual evapotranspiration (less recharge), topography, land use, soil type, land cover, aquifer transmissivity, vegetation characteristics, and contributions to recharge along active stream channels (Herrera-Pantoja and Hiscock 2008). Terrestrial changes impact the rainfall retention and intensity, quantity, and timing (Marshall et al. 2003), while reducing infiltration to recharge groundwater. In rural areas, increased evapotranspiration (ET) is observed in areas with large-scale irrigation, which then alters regional precipitation patterns (Moore and Rojstaczer 2002; Scanlon et al. 2005). Evidence also indicates that deforestation increases runoff, while decreasing the time of runoff and the amount of time available for infiltration. As a result, determining groundwater availability involves more than calculating the volume of groundwater within any given aquifer: it requires a consideration of recharge, water quality, interconnectedness with the hydrologic system, and ecosystem/user demands.

Since 1980, withdrawals have declined slightly (Reilly et al. 2008). Solley, Pierce, and Perlman (1998) estimates that during 1995, total surface-water withdrawals were 324,000 Mgal/d, which is about the same as during 1990, and total groundwater withdrawals were 77,500 Mgal/d, or 4% less than during 1990 (DOE/NETL 2004). Kenny et al. (2009) estimate that about 410 Bgal/d were withdrawn in 2005 in the United States for water use. This total is slightly less than the estimate for 2000, and about 5% less than total withdrawals in the peak year of 1980 (Kenny et al. 2009). Despite the leveling off of water use, the fact that water supplies appear to be decreasing in many basins creates a concern about the sustainability of these groundwater supplies.

The concept of *sustainable water supplies* became popular in the late 2000s as a forum to discuss water supply limitations and seek solutions. A number of organizations, like the American Water Works Association, have committees and studies focused on sustainable water supplies or ancillary issues. The key component in planning the utilization of water supplies is to determine how the hydrologic cycle provides water to the service area (e.g., recharge basin), in what quantities, and with what reliability. From a hydrologic perspective, the term *sustainable yield* is the amount of water that can be withdrawn from a water source at rates that are less than the recharge potential and that do not deteriorate the source or basin (note that this is actively changing in many locales—much of the northwest is losing the snowpack that has stored water in glaciers). But withdrawals less than the sustainable yield is not what happens in many basins and competition for water throughout the United States creates further challenges with supplies and quality.

The largest user in the United States is agriculture (40%), followed by power (39%) (Lisk, Greenberg, and Bloetscher 2012). Withdrawals for irrigation have remained constant due to water conservation efforts, despite irrigated acreage increasing from 25 million acres in 1950 to 58 million acres in 1980; but at the same time, agriculture has relied increasingly on groundwater. Five states—California, Texas, Nebraska, Arkansas, and Idaho—account for half of the irrigation withdrawals for agriculture in the United States. Most of these states are relatively arid, and most of that is groundwater. Water levels in these areas are of increasing concern.

While agriculture is the largest water user in the United States, the built environment (urban and suburban centers) accounts for only 12.7% of the water demands (Bloetscher and Muniz 2012). Urban demands are more concentrated, although since most urban areas are very small, their water can come from much smaller sources than large agri-businesses. Power demands are generally large (for cooling purposes), and the need for power is mostly confined to areas near urban centers. As a result, water supplies for power are primarily withdrawn from large volume sources, such as lakes, rivers, oceans, and in a few cases, underground aquifers (Cifreno 2009). By contrast, agricultural needs are spatially extensive (NSTC 2007).

In addition to water use, each potential user contaminates and/or reduces the volume of the receiving waters they use. The changes in water quality resulting from each user are of interest because it can limit downstream water supply options. Agricultural users consume the water for plants, although runoff can contaminate surficial water bodies during rain events with nutrients, pesticides, and herbicides, which can make downstream water bodies uninhabitable for native species and difficult to treat for water supply purposes (see Walkerton example from the previous chapter). Power entities create thermal pollution and evaporative losses (10%) that can make downstream water bodies uninhabitable for native species and also creates a treatment challenge for potable purposes. Urban runoff is rarely treated, carrying nutrients and particulates into nearby waters—another environmental and water supply challenge. Wastewater plants discharge treated wastewater that can also contribute nutrients to the streams. Each creates downstream challenges for potable water supply users and amplifies the importance of examining the future adequacy of the nation's water supply (Brown 1999). As shown in Figure 15.1, one person's waste may be the downstream user's water supply and therefore, care should be undertaken to understand the impacts of discharges to avoid the next Flint.

The need for more water for urban and agricultural uses drives more competition for limited supplies in stressed basins. In all cases, as withdrawals to off-stream users increase, more water is consumed, leaving less water in streams and increasing contaminant concentrations. Changes in weather patterns appear to have decreased stream flows at the same time as additional in-stream demands increase (natural system demands), and droughts have become more frequent. During the 2007 drought in the southeast United States, water levels were so low that power production at some power plants had to be stopped or reduced (Kimmel and Veil 2009). The Tennessee Valley Authority (TVA) Gallatin Fossil Plant was not permitted to discharge water

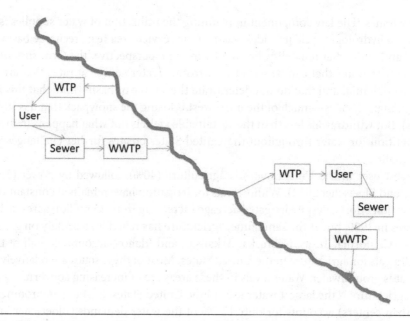

Figure 15.1 Typical stream condition—one community's wastewater can be another's water supply.

used for cooling back into the Cumberland River due to thermal pollution (water > 90°F) (Kimmel and Veil 2009). Nuclear and coal-fired plants within the TVA system were forced to shut down (e.g., the Browns Ferry facility in August 2007) and curtail operations at others. A similar situation occurred in August 2006 along the Mississippi River (Exelon Quad Cities Illinois plant). Other plants in Illinois and some in Minnesota were also affected (Union of Concerned Scientists 2007).

The objective of effective resource utilization is equivalent to the goal of sustainable project design (Virjee and Gaskin 2005), whereby the institutional, social, and informational mechanisms to keep in check the feedback loops that cause exponential population growth and natural capital (water supply) depletion (Meadows and Randers 2004). But, while researchers can define a comprehensible concept of sustainability, practitioners emphasize feasibility, which factors against many of the tenets needed to reach sustainability of the ecosystem (Starkl and Brunner, 2004). Scanlon et al. (2005) identified the need fully to optimize management of water resources, but this poses a challenge when water rights and permits are involved.

A United States Geological Survey (USGS) report (#1323 by Reilly et al. 2008) provides a little insight on these challenges and the impact they may have on the future. Figure 15.2 shows the difference between rainfall and ET. The large tan area is where the ET rate is higher than the rainfall, meaning no net rainfall for crops and other purposes. It also indicates where groundwater is likely to be used, as confirmed by aerial views that show extensive irrigation or *crop circles* in many of these areas (Figure 15.3). Figure 15.4 shows that many of these areas never have surplus amounts of water, so high groundwater use is not a sustainable practice because recharge occurs at a very small rate. Confirming these conclusions is Figure 15.5, which shows the amount of water available for recharge throughout the United States. Most areas have very little water available for recharge. Figure 15.6 combines regional water-level declines and local water-level declines for changes over the last 40 years throughout the United States. The tan regions indicate areas in excess of 500 square miles that have water-level decline in excess of

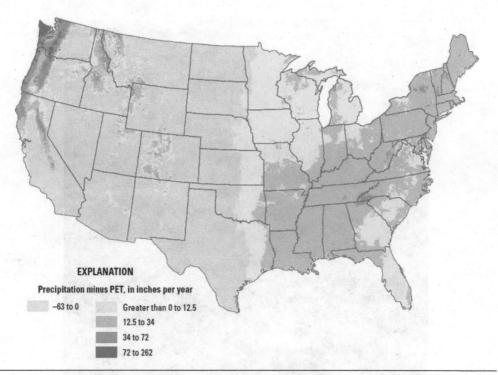

EXPLANATION

Precipitation minus PET, in inches per year

−63 to 0	Greater than 0 to 12.5
	12.5 to 34
	34 to 72
	72 to 262

Figure 15.2   Difference between average annual precipitation and potential ET rates. *Source*: Reilly et al. 2008

40 feet in at least one confined aquifer since predevelopment, or in excess of 25 feet of decline in unconfined aquifers since predevelopment. Blue dots are wells in the USGS National Water Information System database where the measured water-level difference over time is equal to or greater than 40 feet. These match up with the areas noted in previous maps; areas where there are already indications that water supplies are insufficient to provide the full needs of the community (Reilly et al. 2008; Bloetscher, Meeroff, and Heimlich 2009; Bloetscher and Muniz 2012). The lowering of water levels that USGS and state agencies see is an indication that recharge is generally over estimated, giving a false picture of water availability. If an aquifer declines year after year, it is not drought—it is being mined. The continued withdrawal of water from that source appears to be a permanent loss of the resource in the long term.

Reilly et al. (2008) found that the loss of groundwater supplies in many areas could be catastrophic, affecting the economic viability of communities and potentially disrupting lives and ecological viability. By 2025, Ciferno (2009) suggests the most vulnerable areas for water shortages are fast growing areas: Charlotte, North Carolina; Chicago, Illinois; Queens, New York; Atlanta, Georgia; Dallas, Texas; Houston, Texas; San Antonio, Texas; and San Francisco, California. Immediately behind these areas are Denver, Colorado; Las Vegas, Nevada; St Paul, Minnesota; and Portland, Oregon. Some areas have more options than others, but drilling deeper is not a long-term solution. Deeper waters tend to have poorer water quality as a result of having been in contact with the rock formation longer and dissolving the minerals in the rock into the water. If wells are drilled deeper, then harder water is withdrawn and additional power will be required to further treat limited, lower quality supplies. Therefore, while some deep aquifers may be prolific, the quality of water obtained from a well may not be desirable or even usable for drinking water without substantial amounts of treatment. In addition, deeper aquifers are

Figure 15.3 Aerial view of *crop circles* that are irrigated areas from groundwater pumpage can be seen in the arid western parts of the U.S.

generally confined and therefore do not recharge significantly locally. In many cases there is no deeper solution, so lowering groundwater levels is a not-to-distant future problem, especially for small communities. Groundwater has been a small utility solution, so lowering levels of groundwater puts many of the over 40,000 groundwater systems, serving nearly 100 million Americans, at risk; and in a similar position as the eastern Carolinas found themselves during the mid-1990s.

To address the issues, USGS suggests the need for a nationwide effort to organize available information on changes in groundwater storage, similar to what was done for the High Plains aquifer (Reilly et al. 2008), because when looking at the average water demands by area and the total water withdrawals across the United States, the highest withdrawals per square mile are in the east, where populations are largest (Bloetscher 2012; Ciferno 2009; Elcock 2009). However, over the next 20 years, water demands are expected to shift south and west—and per capita demands are much higher in the west than in the east (Elcock 2009; Bloetscher 2012).

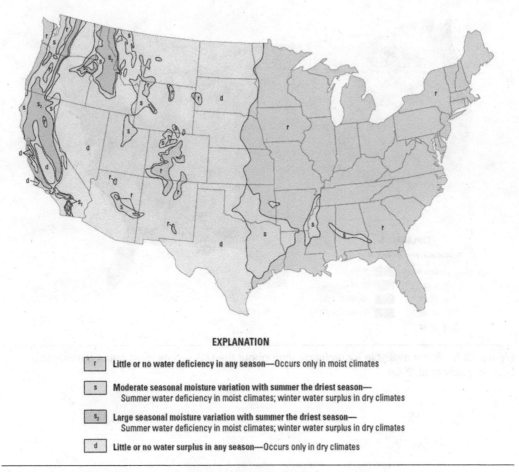

EXPLANATION

r   Little or no water deficiency in any season—Occurs only in moist climates

s   Moderate seasonal moisture variation with summer the driest season—
     Summer water deficiency in moist climates; winter water surplus in dry climates

s₂   Large seasonal moisture variation with summer the driest season—
     Summer water deficiency in moist climates; winter water surplus in dry climates

d   Little or no water surplus in any season—Occurs only in dry climates

Figure 15.4   Water deficit areas. *Source*: Reilly et al. 2008

Population migrations to the south and west will be problematic since the west receives the least amount of rainfall of any area of the United States (see Figure 3.4). Some of these areas are also forecasted to continue to have decreased rain in light of climatic changes. These shortages could be brought about or exacerbated by increased rates of water consumption and withdraws along with forced power plant shutdowns due to lack of water in others (Sovacool and Sovacool 2009). Confounding the issue further is the U.S. Census Bureau (2004) projection that the national population will increase from 282 million people in 2000 to 364 million by 2030 and 420 million by 2050 which means 50% more water will be needed, primarily in these water-limited areas.

Changing water supplies is a high-cost item. There have been a couple of recent examples of *extreme measures* that will become more common. Wichita Falls and Big Spring, Texas, went the DPR route (wastewater piped to the water plant for treatment) due to drought that caused their surface reservoirs to diminish to very low levels. California has multiple IPR projects where highly treated wastewater is injected into aquifers and recovered later (Orange County Groundwater Recharge project being the biggest). Nearly a dozen utilities in Florida are looking at other options as well (more on this shortly).

What is the solution for agricultural operations and utilities where groundwater is quickly diminishing? When can we start the dialogue to manage groundwater resources better in the

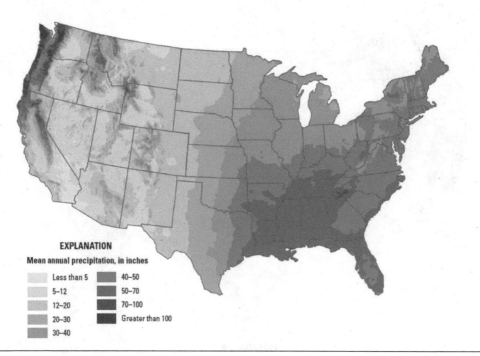

EXPLANATION
Mean annual precipitation, in inches

Less than 5	40–50
5–12	50–70
12–20	70–100
20–30	Greater than 100
30–40	

Figure 15.5   Water available for recharge throughout the U.S.—note most areas are very low. *Source*: Reilly et al. 2008

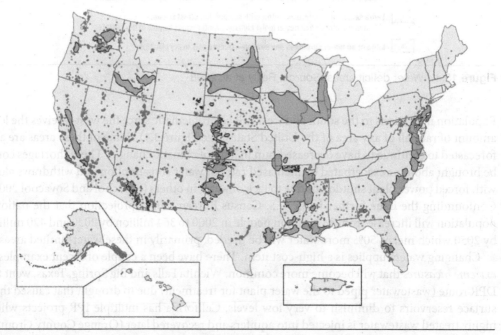

Figure 15.6   Water-level declines—red regions indicate areas in excess of 500 square miles that have water-level decline in excess of 40 feet in at least one confined aquifer since pre-development, or in excess of 25 feet of decline in unconfined aquifers since pre-development. Blue dots are wells in the USGS National Water Information System database where the measured water-level difference over time is equal to or greater than 40 feet. *Source*: Reilly et al. 2008

United States without all the legal and political constraints that currently work against protecting our nation's groundwater supplies? Clearly, that won't make everyone happy and may make a lot of people very unhappy, but it is better to make those decisions now than in 20 or 30 years when the groundwater runs out. Current policies of many elected officials and developers allow the continuation of building to attract more people and business. This is in contrast to the fact that water supplies in most basins are relatively finite or fixed, which means that inevitably, the supply will be exceeded by local demands, the opposite of *sustainability* from a water resource perspective. For many places, groundwater should probably be the backup plan only, not the primary source.

The good news is that across the United States, the regulatory discussions are about managing groundwater supplies. There are 20-year plans (which many think is the long-term perspective), 50-year plans, and 100-year plans; and no doubt a myriad of others. The concept of managing groundwater seems reasonable, but in most cases, the concept of managing aquifers for finite periods is associated with the need or desire by local and state officials to develop a certain region and obtain the necessary water to meet development projections. Sustainable groundwater will be a 21st-century issue for both water utilities and agriculture.

WATER-POWER CONFLICT

Confounding the development and water allocation issues is the increase in power demands that development may require, which further impacts water supply availability, particularly surface waters. Many societies never get started because they lack infrastructure, but even with the infrastructure in place, they require power to grow. Initial power was simply wood burning or efforts by people or beasts. Eventually, wind and coal were used for power. None however, were sufficient to meet the demands required for economies to grow exponentially. That changed in the 19th and 20th centuries when the use of steam, generators, gasoline, nuclear, and other power solutions were employed at efficiencies that make them economically viable. However, modern power-generating equipment requires water for cooling, creating an interdependency between water and energy infrastructure and the potential for conflict over water resources. At the nexus of water and energy exists a host of societal issues: policy and regulatory debates, environmental concerns (local and global), technological challenges, and economic impacts that must be balanced or optimized to permit ongoing economic development for all (NETL 2008).

The total U.S. power consumption continues to climb as a result of population increases. The U.S. Department of Energy (DOE) estimates that total electricity consumption will grow at an annual rate of 1.3% per year, or from 3,821 billion kilowatt-hour (kWh) in 2006 to 5,149 billion kWh by 2030 (DOE 2008), and will double nationally by 2050 (Sovacool and Sovacool 2009). At the same time, according to the U.S. Energy Information Administration (EIA), energy demands per person have decreased nearly 15% since 1970 (EIA 2010), while the actual cost for power has decreased about 50%, making energy cheaper than any time in the past (EPA 2013). The changes in climate patterns interrupt historical projections of water availability, impacting cooling water reliability. The reliability is a risk issue. This begs the question: is the precipitation consistent or are there significant fluctuations that disrupt ongoing basin development (Molak 1997)?

Each kWh of electricity requires on average about 25 gallons of water to produce (Hoffman, Feeley, and Carney et al. 2005). Therefore, we may use almost three times as much water turning on lights and running appliances as we do taking showers and watering lawns (Hoffman, Feeley, and Carney 2005). Fossil fuel thermoelectric generation represents the largest segment of U.S.

electricity production. Thermoelectric power plants—coal, oil, natural gas, and nuclear-fueled power generators using a steam turbine based on the Rankine thermodynamic cycles—require significant quantities of water for generating electrical energy as a result of the need for cooling the prime movers (Ciferno 2009).

For the foreseeable future, thermoelectric power plants will continue to require copious amounts of water for cooling purposes. Thermoelectric facilities generally can be used to produce power as needed, according to consumer demand and fuel supply. This responsiveness to demand makes electricity from these thermoelectric facilities particularly attractive (EPRI 2003). It is no surprise that most thermoelectric cooling systems lie on the rivers and lakes of the East, the Great Lakes region, the Southeast, and the Mississippi/Missouri River valleys. Consistent, perennial streamflow is not available in much of the West. The Department of Energy's National Energy Technology Laboratory (Shuster 2008) projects that the Southeast, Florida, and the Rocky Mountain states will have the greatest demands for thermo-cooling systems—and all are water limited.

Water use efficiency in thermoelectric power generation plants can be enhanced by upgrading these plants with replacing evaporative cooling towers with closed-loop dry cooling systems (DOE 2006). Dry cooling systems use high flow rates of ambient air for exhaust steam cooling, thus reducing evaporative losses (Feely 2008), but it reduces the efficiency of the plant. Another solution to reducing thermoelectric water use/losses is air cooling, where the turbine exhaust steam flows through tubes of an air-cooled condenser (ACC). It is here where the steam is cooled directly via conductive heat transfer using a high flow rate of ambient air that is blown by fans across the outside surface of the tubes (DOE/NETL 2004). But air cooling costs about 30% of their capacity, which means burning more fossil fuels. Another water-saving option is combined cycle power plants. In these plants, steam produced from generators is used to power another turbine making the system more water efficient. The DOE (2006) suggested that natural-gas-fired combined-cycle gas turbines use half as much water as typical coal-fired plants, which may facilitate increased utilization of natural gas as a fuel source.

The DOE calculated that more than 50% of all domestically available energy resources were in the form of wind and solar (Sovacool and Sovacool 2009). Wind power has been used since the 1870s, but has not been commercially viable for large-scale use until recently. Wind power has increased substantially since 2000, but remains a minor player in the power generation scheme, well behind biofuels and hydroelectric power (EIA 2010). According to a report from the National Renewable Energy Laboratory (NREL), there are over one million windmills in the United States, Argentina, and Australia alone (Argaw 2003). Electrical wind pumps are twice as efficient as traditional windmills and are often a cost-effective alternative to traditional power supplies (Argaw 2003; Mead et al. 2009), but sustained winds are required for efficiency. Table 15.1 outlines the typical water demand for power production by type of generating facility. Table 15.2 shows that renewable fuels—wind and solar—are much more efficient users of water.

But there are some potential solutions for the water-power nexus. Water and wastewater utilities are among the largest power users in a community. Many innovative and emerging equipment and operations-related energy conservation methods (ECMs) have been used successfully and applied to save energy and costs.

On-site generation of renewable power (solar or wind) is an option for water utilities but may not be embraced by power utilities. A membrane/reverse osmosis water treatment plant might be able to recover 10% of its power by using more efficient membranes and optimizing pumps (lower pressure). Energy recovery turbines might increase this further. Wastewater

Table 15.1 Summary of water needs to develop power plant fuel

Fuel/Process	Water Need	Water Use (MGD) Water/MWh	Gallons of Water/ Gallon of Fuel
Oil/gas refining	Refining	20 to 70	1.5
Oil/gas extraction	Extraction	6 to 10	1.5
Oil/shale	Refining	3 to 30	2
Oil sands	Extraction	50 to 150	3
Biofuels—ethanol	Growing fuel stock	10,000 to 100,000	1,000
Biodiesel process	Growing fuel stock	15 to 20	1
Biofuel—soy	Growing fuel stock	50,000 to 200,000	6,500
Biomass conversion	Growing fuel stock	50 to 350	4

Source: Brown 1999

Table 15.2 Summary of water demands by power plant type

Power Plant Technology	Cooling Demand MG/MWh	Other Use or Consumption
Coal fired	0.05	0.0005
Coal—gas	0.0002	0.0003
Natural gas—open loop	0.02	0.001
Natural gas—closed loop		
Nuclear—open loop	0.06	minimal
Nuclear—closed loop	0.001	0.0001
Geothermal	0.02	0.02
Wind	0.00075	0.00075
Solar PV	0	minimal

Source: Shuster 2008

plants create biosolids that can effectively be converted to methane gas through digesters, so a wastewater treatment plant can create a portion of its energy demands. Larger wastewater plants have pursued this technology for years. It has been found that adding fats, oils, and grease from private grease trap haulers can dramatically increase gas generation. The methane gas can then be burned in turbines and fuel cells. Micro-turbines provide a cost-effective opportunity for some utilities because they can create a similar amount of power and require minimal maintenance. Micro-turbines require no major repair parts and the technology is well developed, with recoveries exceeding 80%. A premanufactured fuel cell module could operate continuously to produce power usage from plant methane. Fuel cells can use methane, but require higher capital and higher maintenance costs. Both require *cleaning* the methane of impurities, but in many cases the gas needs only limited cleaning to be efficiently burned. It should be noted that methane has 22 times the greenhouse gas effects of carbon dioxide—but, methane is a useful fuel if it can be captured.

Research shows that recovery of methane becomes cost effective for wastewater treatment plants above 5 MGD (Lisk, Greenberg, and Bloetscher 2012). The incineration of biosolids with electricity generation is an effective energy recovery option that uses multiple hearth and fluidized bed furnaces. Incineration can be used to power a steam cycle power plant. Wastewater utilities can now strategically replace incineration with advanced energy recovery technologies like gasification and pyrolysis, which offer the prospect of greater energy recovery than incineration. Energy credits and grants for local governments wishing to pursue this type of power generation may be available.

Finally, both water and wastewater treatment plants usually have large land areas and/or large tanks. Lakes and reservoirs are usually off limits to the public. These sites can make good sites for solar farms and wind turbines, if in the right locations. Co-op agreements with power companies may be fruitful for both the water/wastewater utilities and the power company since the distance to the users will be less than, say, a solar farm in Montana, if you live on the east or west coast.

DIRECT/INDIRECT POTABLE REUSE

The state of Florida is the third most populous state in the United States, passing New York in 2015. Florida's population exceeds 20 million. U.S. Census population in 2010 for Southeast Florida was approximately 5.5 million and is expected to increase to approximately 7.4 million by the year 2030 (Broward-by-the-Numbers 2004). Considering the projected population growth in the area, it can be anticipated that water demand will increase approximately 45%—to 1,110 MGD over the same years, assuming an average water use of 150 gallons per capita per day (Whitcomb 2006). Water supply is a serious issue for Southeast Florida as a result of weather patterns, climate variations, and the lack of topography. While the area receives nearly 60 inches of rainfall per year, these water resources in South Florida occur in distinct wet and dry seasons. The dry season occurs concurrently with the period of highest population, while the wet season rainfall cannot be stored due to a lack of topographic relief.

The Biscayne aquifer is the only source of freshwater in this area. The Floridan aquifer is a source of brackish water 1,000 feet below the surface and does not recharge locally. As a result of it being a brackish water source, the costs for operations in the Floridan aquifer are significantly higher, and the sustainability of drawing large amounts of water from a confined aquifer is questionable. Desalination is far more costly due to power demands, which creates concerns regarding the ability of the area to meet future water demand while maintaining the stability of the natural systems. In addition, wastewater production is also expected to increase, and by 2025 the ocean outfalls that are used to discharge treated wastewater off-shore are to be discontinued, creating an impaired water source that has no obvious disposal option (Hudkins and Fox 2006). Large-scale reuse is not feasible for Southeast Florida utilities due to difficulties with the volume of reuse generated, the lack of open space, high chloride levels, low elevation, and the small size of lots.

As a result, seven projects that have looked at indirect potable recharge have taken place; among them are Miami-Dade County, Plantation, Sunrise, Davie, and Pembroke Pines. In each case, there is wastewater to dispose of and limitations on raw water supplies, so the communities have considered the use of reclaimed wastewater to offset raw water supply needs (Bloetscher et al. 2012). The intent of each project was to inject the treated wastewater into the Biscayne aquifer for recovery downstream in the future. The wastewater treatment concept includes the use of reverse osmosis membranes, ultraviolet disinfection, and advanced oxidation processes downstream of activated sludge and microfiltration. Treating for recharge standards requires membranes to treat the water to acceptable standards. The standard bearer for such recharge projects is the Orange County Groundwater Recharge project (old Water Factory 21) in Orange County, California. Orange County has been operating a groundwater recharge facility using storm and wastewater for over 40 years. A 1996 epidemiological study on the health impacts of recharging the aquifer with reclaimed wastewater using the treatment protocol found no measurable differences in the incidence of diseases between Orange County and the Los Angeles basin where the water supply is not recharged with reclaimed water (Sloss et al. 1996).

The micro-filtration—reverse osmosis—ultraviolet/peroxide process works. For example, the city of Pembroke Pines' process demonstrated compliance with the Broward County (Florida) code limitation of 0.01 mg/L for phosphorous, while also demonstrating a true 3-log removal of emerging contaminants (Bloetscher et al. 2014). However, one of the problems is that the permeate leaves the process grossly lacking stability or hardness, because with respect to minerals, virtually everything in the water is removed by the reverse osmosis membranes. Injecting this corrosive water into a limestone formation creates significant potential for damage to the rock. An innovative concept of using kiln dust and limestone filters to permit the water to pick up hardness and alkalinity resolved the concern (Bloetscher et al. 2013), but there are other concerns with respect to reliability of the process and the *yuck* factor. To date, none of the projects have moved forward, partly because of the 2008 recession, and partly because issuing permits for the water supply is currently outside the rules of the regional water management districts or the Florida Department of Environmental Protection.

These South Florida projects have been considering IPR. What about direct reuse? While several central Florida utilities are looking at it, two places in Texas have actually done it. In Texas, the Colorado River Municipal Water District (CRMWD) serves a total of 250,000 consumers in Odessa, Big Spring, Snyder, and Midland. The area incurred a serious drought that resulted in the reservoir levels dropping to 2% of full. Several issues complicated efforts: (1) they had been looking for water supplies as far back as 2002, (2) there was no place for surface reservoirs—and the high ET limits this option, and (3) most of the fresh groundwater had already been developed. So, in 2013 they completed a 2 MGD, $14 million DPR plant to convert wastewater to drinking water standards. The process uses microfiltration, reverse osmosis, and ultraviolet disinfection. The treated wastewater went to the head of the water plant, demonstrating that the technology to convert wastewater to potable water is clearly available. That leaves monitoring and public perception as keys. Monitoring requires surrogates and redundancy—the multiple barrier approach. Technology will create more monitoring tools, but what needs to be monitored is only somewhat understood. The CRMWD did this by going through a water plant after the wastewater plant—something that is likely going to be the procedure going forward because the processes are fundamentally different and remove different constituents. Wichita Falls, Texas, did a similar project for the same reasons.

Technology is feasible. Public perception is more of a challenge. Overcoming the *yuck* factor took some ongoing customer interaction for the CRMWD, but in west Texas, the need for water, and the limited availability of the same, overwhelmed the perception—a situation that may not exist in most areas, but which will increase with time. The Texas examples indicate that the reality is that DPR is likely to receive more interest and less pushback in those dry, western areas with high population growth and limited supplies, as opposed to wet areas that appear to have more water, like Florida. But, expect to see more application of IPR and DPR. Figure 15.7 shows areas in the United States that utilize or have piloted direct potable water projects. This is a solution for land-locked areas with limited water supplies and the desire to grow. It also may be for areas where regulations, manuals of practice, and consumer acceptance are starting to grow. It is a potential future solution for water-limited areas.

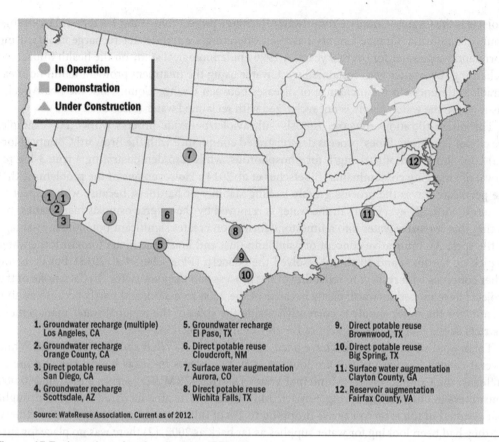

1. Groundwater recharge (multiple)
   Los Angeles, CA
2. Groundwater recharge
   Orange County, CA
3. Direct potable reuse
   San Diego, CA
4. Groundwater recharge
   Scottsdale, AZ

5. Groundwater recharge
   El Paso, TX
6. Direct potable reuse
   Cloudcroft, NM
7. Surface water augmentation
   Aurora, CO
8. Direct potable reuse
   Wichita Falls, TX

9. Direct potable reuse
   Brownwood, TX
10. Direct potable reuse
    Big Spring, TX
11. Surface water augmentation
    Clayton County, GA
12. Reservoir augmentation
    Fairfax County, VA

Source: WateReuse Association. Current as of 2012.

Figure 15.7   Location of active direct or IPR projects

CHANGING CLIMATE

The most contentious future trend is climate. While much of the focus of scientists has been on the causes of the correlation between the changes in greenhouse gas concentrations in the atmosphere and temperature changes, the focus for infrastructure managers should be on how potential impacts should be addressed in light of long-term infrastructure investments and on water supply planning. Infrastructure managers need to plan and must account for potential changes in local situations such as more intense rain, less overall precipitation, more energy demands, storm and drainage pattern changes, roadway base impacts, wastewater flooding, water limitations/drought, etc.

First, let's discuss the literature with respect to observed impacts. There is strong evidence that global climate change is impacting the global water cycle, natural environments, the built environment, infrastructure systems, and global water resources (IPCC 2013; UNEP 2009; Karl and Melillo 2009; Heimlich et al. 2009; Bloetscher and Romah 2015). Examples are rising global average air and ocean temperatures, increased and earlier snow and ice melt, shorter subtropical rainy seasons, shifted seasons, SLR, and greater variations in temperature and precipitation (IPCC 2013; Freas et al. 2008; Marshall et al. 2003; Bloetscher et al. 2010). The National Oceanographic and Atmospheric Administration (NOAA) and the Intergovernmental Panel on Climate Change (IPCC 2013) predictions are that by 2100, the warming will be on the order of 2–3°C and will increase SLR by up to three feet. Significant changes in precipitation patterns are expected in many areas, creating increased drought frequency and duration in the Plains states, the Southeast, and the Southwestern United States (Backlund 2008; Grinstead 2009; Rahmstorf 2007). Accompanying these drivers are potential changes in storm frequency and intensity, desertification, population migration, ocean acidification, and coastal flooding (IPCC 2013). Impacts to water supplies will be from recharge aquifers and discharges to streams, rivers, and wetlands (Adriaens et al. 2003; Scanlon et al. 2005; Marshall et al. 2003; Salmun and Molod 2006). Land cover and land use changes will impact timing and quality of water, leading to changes in the timing and volume of streamflow (Richey and Costa-Cabral 2006) and flooding (Moore and Rojstaczer 2002; Scanlon et al. 2005). The American Meteorological Society published a supplement to its official 2013 bulletin titled *Explaining Extreme Events from a Climate Perspective*, compiling empirical data which indicated an annual maximum 24-hour precipitation increase of 3.3% on average. This increase is consistent with a 5% to 7% median increase in precipitation per 1°C rise in global mean temperature as simulated by global climate modeling (Stott 2015; Herring et al. 2015). Evidence from multiple studies suggests observed changes in several parameters related to the intensity and frequency of extreme events (Herring et al. 2015). According to the IPCC (2013), available data also suggests that a warming climate has contributed to intensification of storm activity and heavy precipitation (Stocker 2014). Similarly, upward trends in monthly averages and daily temperature ranges have been observed in several regions (Stott 2015; Herring et al. 2015).

Having viewed the literature, what changes are already visible? Changes have already begun to affect precipitation and temperature patterns across the United States. The new National Climate Assessment (Chapin et al. 2014) says that temperatures are warmer; there will be more droughts, less rainfall, less available water, more intense storms, and SLR. Some of these phenomena that are already occurring include flooded streets in coastal areas like South Florida, severe weather (Colorado, New Jersey), droughts and fires (California), melting permafrost, glacial melt, and changes in the arctic air currents that may be affecting Northeastern and Midwestern winter storm frequency and timing. Coastal areas and northern areas may be the most impacted. For example, over the past 60 years, Alaska has warmed more than twice as rapidly as the rest of the United States, with state-wide average annual air temperature increasing by 3°F (Chapin et al. 2014) and average winter temperature by 6°F (Stewart et al. 2013; Bieniek et al. 2014; Wendler, Chen, and Moore 2012; CCSP 2008). Most of Alaska's glaciers have seen significant retreat in the past 30 years as a result of rising land temperatures (Jacob et al. 2012; Larsen et al. 2007; Arendt et al. 2002, 2009), a trend that the Rocky Mountains, Pacific Northwest, and California have also seen. The loss of glaciers results in changes to surface water runoff timing and volume. Earlier melt provides earlier surges in rivers and streams while lessening snowpack will result in less available water to those streams during the late summer and fall, an issue seen in California and the Pacific Northwest already. That means roads, bridges, and stormwater systems may require altered design assumptions and different maintenance schedules.

Rising temperatures have begun to melt the permafrost as much as 50 meters below the surface in Alaska (Markham 2015; Jorgenson et al. 2008; Osterkamp and Romanovsky 1999; Romanovsky, et al. 2012). Permafrost melt causes the soil to heave and subside—creating a spongy surface condition that releases large amounts of methane in to the atmosphere. Increases in evaporation are expected to reduce water availability in Alaska (Hinzman et al. 2005) and other areas, but lessen the extent of wetland areas. Thunderstorm activity resulting from an increase in ET (Kasischke et al. 2010) will increase forest fires and airborne particulates (Hu et al. 2008; Rosen 2015; NAS 2015; Mack et al. 2011; Balshi et al. 2008; Chapin et al. 2014). The changes may increase the risk of extreme events such as hurricanes, floods, heatwaves, droughts, and wildfires, which are known to damage infrastructure, disrupt medical services delivery, and trigger public health emergencies. Climate change has the potential to create a serious public health threat that affects human health outcomes and disease patterns (Bloetscher et al. 2016). Although preventive and adaptive strategies for climate change will help lessen negative health impacts, human health will continue to be affected from present climate change conditions (Hess, McDowell, and Luber 2012; Kjellstrom and McMichael 2013; Portier et al. 2010). It is expected that climate change will both aggravate existing human health risks and conditions and create new ones, while the health impacts will vary and have both direct and indirect effects (Kjellstrom and McMichael 2013). Populations with combined health, socioeconomic, and place-based vulnerabilities will be most affected (Portier et al. 2010). The health impacts will be felt to different degrees depending on the action taken to adapt and on location (Hanson 2011; Parkinson 2009; Doherty and Clayton, 2011; Rose et al. 2001).

With respect to a given community's infrastructure and way of life, some evaluation of the impacts from changing climatic conditions is required, so a realistic and practical plan can be developed because we expect our infrastructure, factories, houses, and economies to last a long time. These impacts can be temperature, water availability, power demands, agricultural crops, wetter weather, more intense storms, and heave. Climate change impacts are felt globally but some areas and populations are recognized as being particularly vulnerable (Heimlich et al. 2009; Bloetscher, Meeroff, and Heimlich 2009; Pachauri and Reisinger 2007). In some places, the impacts may be minimal. For example, Detroit is forecasted to be the least impacted city in the United States—southeast Florida communities, the most.

Sea Level Rise (SLR)

The southeast Florida region, with its low-lying coasts, subtropical climate, porous geology, and distinctive hydrology, is one of the world's most vulnerable areas for SLR. Due to these unique conditions, SLR is the principal long-term, permanent impact of climate change for the region, threatening both its natural systems and densely populated and highly diverse built environment (Heimlich et al. 2009; SEFRCC 2011). The region constitutes one-third of the state's total population, one-third of the state's economy, and is among the highest rates of projected population growth (U.S. Census 2012), so the impact on the state and national economy will be significant. It is a microcosm for the world since nearly half the world's population lives near the ocean.

During the past 140 years, a consistent increase in sea level has been observed (Bloetscher 2012). Using the Key West tidal data, the U.S. Army Corps of Engineers projected SLR. Results suggested that sea level in southeast Florida will rise one foot from the 2000 baseline by 2040, and could rise two feet by 2060 (USACE 2013). SLR is a major concern for coastal areas because nearly half the U.S. population lives within 50 miles of the coast, involving most major commercial, leisure, and import/export enterprises.

While much of the focus has been on the immediate coastal communities due to the direct threat of SLR, Bloetscher et al. (2012), Bloetscher and Romah (2015), and Romah (2012) noted that groundwater levels in southeast Florida are intrinsically linked to sea level; thus, while coastal populations are particularly at risk due to erosion, inundation, and storm surge, interior populations are also susceptible to rising water tables and extended periods of inundation. Chang et al. (2011) describes an overall *lifting process* by which there is a 1:1 ratio in water table elevation that correlated to sea-level rise. Higher groundwater levels mean reduced aquifer storage, thereby lessening the capacity of soil to absorb precipitation and floodwaters, increasing the risk of groundwater flooding (Bloetscher and Romah 2015; Romah 2012; Bolter 2014). Because of the associated loss of soil storage capacity due to SLR, more intense storms will overwhelm the current stormwater infrastructure. Projections indicate the potential for severe damage to southeast Florida's energy systems, transportation infrastructure, water infrastructure, agricultural lands, and the Everglades ecosystem (Zhang 2011; Karl and Melillo 2009). Due to flooding issues, the state is expecting flood insurance premiums to increase 25% or more.

Much of the current work on SLR adaptation focuses on understanding the physical and economic vulnerability of infrastructure, as well as on developing adaptation strategies for the natural and built environments (Zhang 2011; Hanson et al. 2011; Parkinson 2009; SEFRCC 2011; Tebaldi, Strauss, and Zervas 2012; Titus and Richman 2001; Weiss, Overpeck, and Strauss 2011). Long-term decisions that consider this systems approach are essential as local governments and businesses examine long-term viability, particularly in respect to relocating populations while maintaining growth. Property values are also dependent upon the maintenance of transportation and utilities, including wastewater treatment and water supply. The insurance industry, which has traditionally been focused on a one-year vision of risk, is beginning to discuss long-term risks. As properties rebuild in risk-prone areas, there will be an impact on how bankers look at lending practices and a potential likelihood that property values will decrease. This is a game changer for local officials because it will adversely impact property values and as a result, tax revenues. This is why it is in the community's interest to develop a planning framework to understand, adapt, and protect vulnerable infrastructure for the long term.

Mapping of Sea Level Rise

SLR will impact most coastal communities to some degree. Depending on topography and how the regional water management system is set up, the impact could be significant. As a result, it is useful to examine the communities most at risk—especially in southeast Florida. Figure 15.8 is a summary of high-tide data from 2008 to 2013, showing that in southeast Florida, the king tides occur annually, in October (Bloetscher and Romah 2015). Storms may alter this pattern slightly, but storms are atypical events. A level of service (LOS) would indicate how often it is acceptable for street flooding to occur in a community on an annual basis. The effects of SLR on the LOS should be used to update the mapping in terms of demonstrating LOS changes. For example, a 1% flooding frequency translates to four flood days per year. Using this concept, SLR vulnerability maps for Miami Dade and Broward Counties are shown in Figure 15.9 based on the current, one-, two-, and three-foot of SLR scenarios. The red areas are those at risk for repetitive flooding and where efforts should be put in place to analyze efforts to reduce impacts (so infrastructure may already be present). Roads are the first thing to flood in all of these areas, so transportation infrastructure is the first to be at risk (along with sanitary sewers). Related to the transportation infrastructure is the related effectiveness of flood control and stormwater drainage systems for the transportation corridors.

Road integrity relies on adequate drainage. The Florida Department of Transportation (FDOT) reports that SLR will cause increased water table levels, which could compound the risk of flooding in low-lying areas, depending on the regional water management system setup for that area (FDOT 2012). According to the North American Vertical Datum of 1988 (NAVD88), road bases below five feet NAVD88 would become saturated under this scenario, likely causing premature base failure, a problem discussed in Chapter 6. Frequent flooding of roadways would likely damage pavements (FDOT 2012), especially since many local roads do not meet FDOT standards. Some solutions exist; for example, as water levels rise, wellpoint systems could be installed for

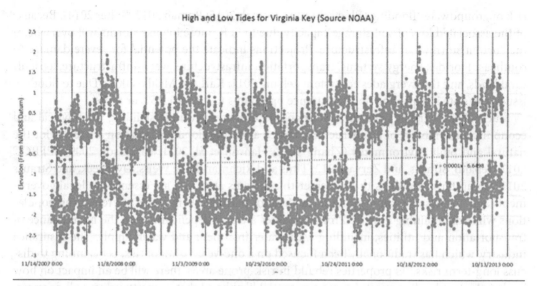

Figure 15.8   Six years of high tides at the Virginia Key Tidal Station (2007–2013)—the highest tides each year occur in mid-Fall (October). Note that the overall trend (green dotted line) is upward. *Source*: NOAA

more permanent drainage. However, wellpoint water is usually turbid—containing sand, other particles, and contaminants from runoff, which requires an off-site discharge zone that is often lacking. Treatment areas for removal of particulates and sand will also be required, requiring additional area for discharge purposes. Wellpoint pump stations need to be regularly spaced so a series of pump stations might be needed for every mile of roadway. Since wellpoints do not function in flood conditions, additional drainage measures must be taken to address wellpoint failure during heavy rainfall events. The costs for such systems could exceed $1 million per lane mile based on data gathered by the FDOT (2012) project team and therefore, may not be an economical solution.

Several Florida municipalities rely heavily on exfiltration trenches or French drains. These systems work because the perforated piping is located above the water table. From a practical standpoint, they cease to function if they are located below the water table. Stormwater gravity wells are a useful option where saltwater underlies the surface. However, as the sea level rises, the potential for pumping will be altered. Wells of this type generally cost about $150,000 each for a 24-inch diameter well. They also require splitter boxes and filters to remove solids, regular inspections, and regular maintenance—thereby increasing transportation system budgets.

For low-lying areas, elevated roads may be an option. However, this option comes with two significant issues: roadway elevations and impacts on adjacent properties. Roadways are designed for a 50- to 100-year service life. As a result, transportation agencies should design roadway bases to be above the mean high water table of five feet. Such roads might need surface elevations at or above eight feet NAVD88, well above many of today's low-lying roads and in many cases above the finished floors of adjacent properties, thereby creating a potential stormwater runoff concern. It is good practice to have road crowns 18 inches below first floors of buildings, whenever possible. That is not the case in many areas today. Raising roadways is

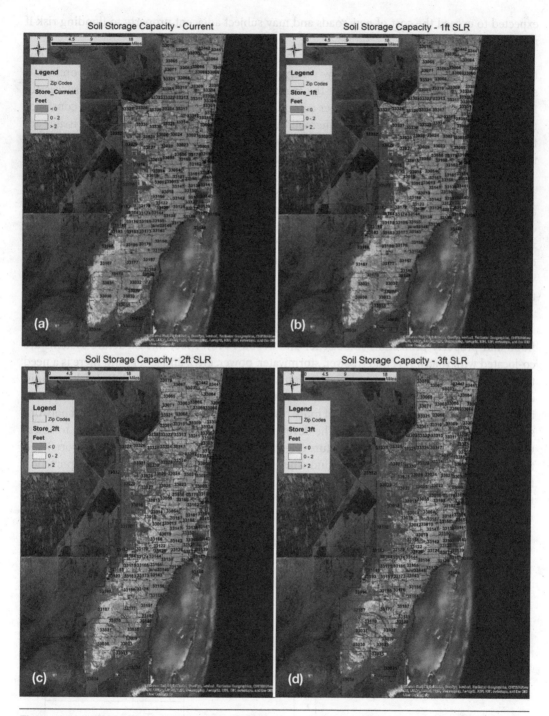

Figure 15.9   Miami-Dade and Broward Counties—vulnerability at 0, 1, 2, and 3 foot SLR at 99 percentile groundwater/tidal elevations (ignoring current infrastructure). *Source: Sustainibility*, Bloetscher et al. 2016 published by MDPI

expected to exceed the cost of new roads and may subject adjacent property to flooding risk if not also raised.

Based on findings of the vulnerable areas, the development of scenarios whereby toolbox options are utilized to address community vulnerability should be pursued. Figure 15.10 outlines a simplified flow chart used as a basis for the evaluation with the goal being to identify successful flood mitigation strategies used by other communities that face similar drainage and construction problems based on identified vulnerabilities and cost effectiveness. These two issues are then combined to develop a framework to evaluate the impacts of climate change on infrastructure and urban development (since they are intrinsically intertwined).

The strength of this framework lies in the proposed holistic and incremental approach to addressing climate impacts, which entails understanding of combined social and health vulnerabilities in the context of higher exposure of the physical infrastructure to hazards. As such, it combines physical vulnerability with health indicators and social evaluation criteria and conveys the notion that a plan is not a fixed document, but rather a process that evolves with the changing environment. The final two steps occur at regular intervals by the community level with associated adjustments made to the initial plans for improvements to various infrastructure systems.

The final task is for a community to develop a toolbox of strategies that could be used to improve the regional resiliency to SLR. Table 15.3 outlines hard infrastructure solutions for flood and property protection (Bloetscher and Wood 2016). Most infrastructure systems are co-located with roadways (water, sewer, stormwater, power, etc.), and as a result, there is a need to prioritize where funds are spent. Catastrophic flooding would be expected during heavy rain events because of reduced capacity of the drainage system. The vulnerability of transportation infrastructure will require the design of more resistant and adaptive infrastructure and network systems. This would, in turn, involve the development of new performance measures to assess the ability of transportation infrastructure (e.g., roadways, bridges, rail, sea ports, airports) and to enhance resilience standards and guidelines for design and construction of transportation facilities. Specifically, considerations must include retrofitting, material protective measures, rehabilitation, and in some cases, the relocation of a facility to accommodate SLR impacts. Since

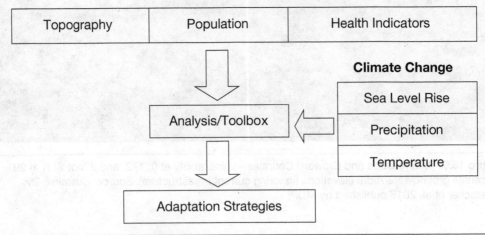

Figure 15.10   Analytical framework for toolbox development.

Table 15.3 Hard infrastructure improvements (SFWMD = South Florida Water Management District)

Implementation Strategy	Benefits	Cost	Barriers to Implementation	Point When Action May Need to Be Abandoned
Exfiltration trenches	Excess water drains to aquifer, some treatment provided	$250/ft.	Significant damage to roadways for installation, maintenance needed, clogging issues reduce benefits	If groundwater table is above exfiltration piping, the exfiltration efficiency diminishes quickly
Infiltration trenches	Excess water gathered from soil and drained to pump stations, creating storage capacity of soil to store runoff, soil treatment	$250/ft. plus pump station	Significant damage to roadways for installation, maintenance needed, clogging issues, costs for pump station	Complete inundation means pumps run constantly and pump the same water over and over
Install stormwater pumping stations in low-lying areas to reduce stormwater flooding (requires studies to identify appropriate areas, sites, and priority levels)	Removes water from streets, reduces flooding	Starts at $1.5 to 5 million each, number unclear without more study	Permits, maintenance cost, land acquisition, discharge quality	When full area served is inundated (>3–5 ft. SLR)
Added dry retention	Removes water from streets, reduces flooding	$200K/ac.	Land availability, maintenance of pond, discharge location	When full area served is inundated
Armoring the sewer system (G7 program)	Keeps stormwater out of sanitary sewer system and reduces potential for disease spread from sewage overflows; major public health solution	$500/manhole	Limited expense beyond capital cost	None
Central sewer installation in on-site sewer treatment and disposal system (OSTDS—septic tank) areas	Public health benefit of reducing discharges to lawns, canals, and groundwater from septic tanks	$15,000 per household	Cost, assessments against property owners	None

Continued

Implementation Strategy	Benefits	Cost	Barriers to Implementation	Point When Action May Need to Be Abandoned
Raise roadways	Keeps traffic above floodwaters	$2–4 million/lane mile	Runoff, cost, utility relocation	When full area served is inundated
Class V gravity wells	Means to drain neighborhoods	$250K ea.	Needs baffle box, limited flow volume (1 MGD)	When full area served is inundated
Class I injection wells	Means to drain neighborhoods, 15 MGD capacity	$6 million	Needs baffle box	When full area served is inundated
Bioswales	Means to drain neighborhoods, provides treatment of water	$0.5 million/mi.	Land area, flow volume, maintenance	When full area served is inundated
Raise sea walls	Protects property	$.1–1 million/lot	Private property rights, neighbors	N/A
Relocate wellfields westward/horizontal wells	$20 million assuming locations can be permitted in Biscayne aquifer	$20 million assuming locations can be permitted in Biscayne aquifer	Cost, concern over saltwater intrusion east and west, inundation of wellfields, permitting by SFWMD	When well is inundated
Salinity/lock structures	Keeps sea out, reduces saltwater intrusion	Up to $10 million, may require ancillary stormwater pumping stations at $2–5 million each	SFWMD, western residents, private property rights arguments	N/A – solution to retard sea encroachment and saltwater intrusion
Regional relocation of locks to pump stations	Creates regional system to use coastal ridge to protect inland property, keeps saltwater out	$200 million ea.	SFWMD, western residents, private property rights arguments	N/A – solution to retard sea encroachment and protect property which can exist at levels below sea level
Pump to Everglades via regional system	Huge volume of water can be removed from urban area and used to recharge Biscayne aquifer	Unknown	Water quality	When full area served is inundated
Pump to tide	Huge volume of water can be removed from urban area	Unknown	Water quality to reefs, sea grasses, etc.	When full area served is inundated

they are related, groundwater is, similarly, expected to have a significant impact on flooding in these low-lying areas as a result of the loss of soil storage capacity; yet, this continues to not be the focus of many planning efforts. Note that if SLR continues to increase, at some point there may be areas that are no longer able to remain dry. That creates a tough decision for government officials, but one that should be made if the welfare of the public is really being taken into account.

A number of strategies can be considered for improving water supplies, although the applicability will vary from one location to the other. Table 15.4 summarizes tools that can be used to help protect water resources from the impacts of climate change, which would in turn protect public health via drinking water supplies. Table 15.5 outlines efforts to address social issues. At the center of these planning efforts should also be the provision for an adequate drainage system, designed to accommodate increased volumes of water. This provision will be critical in protecting the roadway base. Because these systems will not be viable as sea levels rise, future stormwater systems should be designed like sanitary sewers with tight piping, with minimal allowances for infiltration, and adequately sized pumping stations that permit discharge points and means for associated treatment of the stormwater. Discharges of stormwater into water bodies may threaten vital seagrasses and reefs, so some effort will be required to determine the level of treatment needed to protect the ecosystem in the face of excessive water levels.

Table 15.4   Tools for protection of water resources from climate change impacts

Water Resource Adaptation Alternatives
Water conservation   • Reducing requirements for additional treatment capacity and development of alternative water supplies (AWS)
Reducing the impact of SLR on existing water sources   • Hydrodynamic barriers: aquifer injection/infiltration trenches to counteract saltwater intrusion using treated wastewater   • Horizontal wells   • Salinity structures and locks control advance of saltwater intrusion   • Relocation of wellfields when saltwater intrusion or other threats render wellfield operations impractical
Gaining access to alternative water resources   • Desalination of brackish waters   • Regional alternative water supplies   • Capture and storage of stormwater in reservoirs and impoundments   • Aquifer storage and recovery (ASR)
Wastewater reclaim and reuse   • Irrigation to conserve water and recharge aquifer   • Industrial use and for cooling water   • Indirect aquifer recharge for potable water
Stormwater management   • Reengineering canal systems, control structures, and pumping

Adapted from Bloetscher et al. 2015

Table 15.5   Soft infrastructure improvements

Implementation Strategy	Benefits	Cost	Barriers to Implementation	Point When Action May Need to Be Abandoned
Increase access to health care	Improved health care access should reduce impacts	Unknown	Cost, ongoing operations	When full area served is inundated
Reduce potential for forced migration	Lessens risk of socially vulnerable people moving out of vulnerable areas	Unknown	Pressure from developers, rental properties at risk	When full area served is inundated
Redevelopment control ordinances and policies	Reduces competition for land by removing land from redevelopment	Unknown	Pressure from developers, rental properties at risk, property rights issues	When full area served is inundated
Assessments for hard infrastructure	Provides funding to support social efforts	See Table 15.3	Public resistance or public support	N/A
Public acquisition of at-risk property	Reduces potential for migration to vulnerable property by taking property out of circulation	Various land regulatory tools: land lease, outright purchase, condemnation; may provide short-term income	Public resistance or public support	N/A
Vaccinations	Reduces risk	N/A	Public resistance or public support	N/A
Risk communication	Improves communication to residents about their vulnerability	Unknown	Public awareness	N/A
Outreach	Improves communication to residents about vulnerability	Unknown	Public awareness	N/A

Adapted from Bloetscher et al. 2015

Extreme Events

Previous research has documented the escalating cost of extreme weather events to society (Lott and Ross 2006; Brody et al. 2007, Smith and Katz 2013, Smith and Matthews 2015). Lott and Ross (2006) quantified the damage cost of 58 weather-related events exceeding $1 billion in losses in the United States between 1980 and 2003 (all monetary losses were converted in 2002 dollars). The study found that the combined effect of hurricanes, tropical storms, and nontropical floods accounted for approximately 45% of losses or $155 billion, whereas droughts and heat waves alone caused losses in the amount of $144 billion (Smith and Katz 2013). Empirical data also suggest that the frequency of occurrence of billion-dollar events each year, as well as the annual aggregate losses from such events, show an upward trend (Smith and Katz 2013). Similar trends have been observed in Europe where estimated losses over the past 20 years accrued to 400% of those accumulated over the preceding 20-year period (Funari, Manganelli, and Sinisi 2012). Over the same period, the estimated losses from weather-related natural disasters (particularly floods) amounted to $270 billion (Funari, Manganelli, and Sinisi 2012).

Although there is a growing body of literature on economic losses from extreme weather events, relatively little attention has been paid to the health costs of disasters. Knowlton et al. (2011) estimated that the cumulative health costs of six recent extreme-weather events in the United States exceeded $14 billion. According to the International Disaster Database EM-DAT developed by the Center for Research on the Epidemiology of Disasters (Catholic University of Louvain, Belgium), nearly 38 million people in Europe were in need of medical assistance, shelter, food assistance, and sanitation following a major natural disaster over the past 20 years (Funari, Manganelli, and Sinisi 2012). Against the backdrop of these findings, studies also suggest that the impact of climate change and climate variability cannot be understood independently from other global processes including patterns of urban development, population growth, poverty, and adequate access to health care services (Lott and Ross 2006; Haines et al. 2006; Brody et al. 2007).

Planning for extreme events should be a part of all infrastructure agencies planning modes. As discussed in Chapter 8, the less likely the risk, the more impact it is likely to have. However, no one can plan for there to be only minimal damage from the most extreme storms. That would bankrupt government. As a solution, many local officials adopt a level of service, say for stormwater a 1:100-year event—or for roadways, a 1:10-year event. The storm event sets limits for elevation. SLR may cause this number to change with time, but as infrastructure ages and replacement is needed, future conditions may need to be considered. Each agency should plan accordingly.

BUILDING GREEN

About 10 years ago, the American Society of Civil Engineers (ASCE) updated its code of ethics to include the concept of sustainability. In the field of engineering, the concept of sustainability refers to designing and managing to fully contribute to the objectives of society, now and in the future, while maintaining the ecological, environmental, and economic integrity of the system. Buildings, which usually have lifespans measured in decades, are important to sustainability, and as such, are looked at as opportunities for developing sustainably. It is less clear for more traditional infrastructure systems like transportation and water. However, that potential exists even when traditional building measures do not apply, since infrastructure systems have a considerable impact on the natural environment and human health (EPA 2010).

Green or high-performance building is the practice of creating structures by using processes that are environmentally responsible and resource efficient throughout a building's life cycle— from site to design, construction, operation, maintenance, renovation, and deconstruction (EPA 2010). High-performance building standards expand and complement the conventional designs to include factors related to: economy, utility, durability, sustainability, and comfort. At the same time, green building practices are designed to reduce the overall impact of the built environment on human health and to use natural resources more responsibly by more efficiently using energy, water, and other resources, while protecting occupant health and improving employee productivity (Bloetscher and Meeroff 2015). Despite the economic issues post-2008, it is expected that green building will support two million U.S. jobs and pump over $30 million per year into the American economy (Booz Allen Hamilton 2009). Governments have taken the lead in building green, often requiring new buildings to meet certain certifications, but the commercial sector is growing (Bloetscher and Meeroff 2015). Data indicates that the initial construction cost of buildings certified by Leadership in Energy and Environmental Design (LEED) are sometimes only minimally more than traditional building practices. A case study done by the U.S. Green Building Council (USGBC) showed that the average premium for a LEED-certified silver building was around 1.9% per square foot more than a conventional building. The premium for gold is 2.2% and 6.8% for platinum (Kubba 2010). These numbers are averaged from all LEED-registered projects, so the data is limited, but demonstrates that in many cases, it may cost only marginally more to deliver a LEED-certified project, which improves the value of the building and lowers operating costs (Kubba 2010). The experience with the Dania Beach, Florida nanofiltration water treatment plant indicated the premium was under 3% to achieve LEED-Gold certification compared to standard construction (the Dania Beach Nanofiltration Plant was the first, and only, nanofiltration plant in the world to be certified Gold). It is noted that the Florida State Revolving Fund programs prioritize green infrastructure, so why don't more people pursue them?

An issue that arises with green building is determining how to evaluate its performance. There are several agencies that have created a means to accomplish this goal. The main organizations to focus on include the International Organization for Standardization (ISO) (for ISO 14001), the EPA, the USGBC, the Green Building Certification Institute (GBCI) and more specifically for infrastructure systems, Envision®.

International Organization for Standardization (ISO)

ISO, located in Geneva, Switzerland, has established nearly 20,000 standards to verify that materials, products, processes, and services are fit for their purpose (ISO 2011). Some of these standards are used by communities and organizations in an attempt to reduce their environmental footprint. They apply primarily to management processes. The ISO 14000 family of standards addresses environmental management to minimize harmful effects on the environment and to achieve continual improvements in environmental performance (ISO 2011). The ISO 14001 standard focuses on the development of an environmental management system (EMS). The ISO 14001:2004 provides the requirements for an EMS (ISO 2011), enabling an organization to identify and control the environmental impact of its activities, products, or services; to improve its continual performance; and to implement a systematic approach to setting better objectives and attainable goals (ISO 2011). ISO 14001 provides a framework for an EMS for any type of industry's strategic approach to environmental policy, plans, and actions. The main requirements of an EMS include a commitment to pollution prevention, continuous improvement, and compliance with any environmental or other applicable standards and regulations (EPA 2011).

ISO 14001 provides an audit process that demonstrates that the organization is compliant as long as the systemic, policy, planning, operational, checking, and review requirements are met as opposed to specific levels of performance or improvements (ISO 14001:2002). The organization develops the framework that allows their own EMS to be environmentally responsible. Incumbent in the process is to identify and set objectives for the organization's environmental impact from an operational standpoint and requires that all environmental policies developed must be documented. The legal and operational requirements are designed to ensure ongoing training and environmental awareness throughout the organization.

The internal objectives provide assurance to management that it is in control of the organizational processes and activities that have an impact on the environment (ISO 2011). The external objectives achieved by ISO 14001:2014 provide assurances about environmental issues to external stakeholders, including regulatory agencies, that the corporation complies with regulations and supports the organization's claims through communication about its own environmental policies, plans, and actions.

EPA Programs

The EPA has created two programs that evaluate key aspects of high-performance building: ENERGY STAR® and WaterSense®. Both have some impact in the infrastructure industry since lights, sensors, and water equipment fall under the purview of these standards. The EPA created the ENERGY STAR system as a voluntary labeling program to help promote energy efficient products to reduce greenhouse gas emissions. The labeling started off with computers, expanded to include office equipment, and then expanded further to include most household appliances, lighting, and many other electronics devices in residential, commercial, and industrial buildings. ENERGY STAR has partnerships with over 20,000 private and public sector organizations to provide the technical information and tools necessary to obtain the most energy efficient products and provide the best management practices.

U.S. Green Building Council

The USGBC is a Washington, D.C., based, nonprofit organization whose mission is to transform the way buildings and communities are designed, built, and operated in order to enable an environmentally and socially responsible, healthy, and prosperous environment that improves the quality of life (USGBC 2011). The goal of the USGBC is to influence the design of new buildings and communities through energy efficiency, water efficiency, sustainable materials, indoor air quality, and innovations in design. Verification of a high performance building is accomplished through the use of a checklist developed by USGBC members, known as the LEED certification checklist, that undergoes continual updates. Buildings and communities that are designed to be LEED certified are not only more efficient than conventional buildings, but the USGBC suggests that their carbon footprint will be greatly diminished as well (USGBC 2011a). LEED-certified buildings are designed to certain standards and requirements that will ensure the building's greenhouse gas emissions are reduced, operating costs are lower, value is increased, waste sent to landfills is reduced, energy and water demands are decreased, and occupant health and comfort is improved, while also qualifying for tax rebates, zoning allowances, and other incentives (USGBC 2011, Kubba 2010).

LEED certification provides independent, third-party verification through the GBCI. The GBCI was established in 2008 as a separately incorporated entity with the support of the USGBC, specifically to confirm that a given building was designed and built using strategies aimed at achieving the goals of the USGBC. Although non-occupied buildings do not qualify to be certified, that does not mean that the principles should not be pursued. Things like using recycled steel and concrete, reducing construction waste, reducing carbon footprints, and pursuing site sustainability are goals. LEED points are awarded to achieve one of four levels of certification: LEED Certified, Silver, Gold, and Platinum.

Triple Bottom Line (*Inception 2012*)

A concept for all design professionals to understand is the *triple bottom line*, which promotes environmental, economic, and social improvements in the approach to projects. The goal is to consider and then optimize the three. The concept is to foster social interaction and community connectivity—worthy goals of infrastructure systems—to provide a healthy life for building users and their surrounding communities. The environmental imperative takes into account the way humans and industries are negatively affecting the planet. The economic imperative provides evidence that sustainable construction practices will be considered best practices for builders within the foreseeable future (Montoya 2010). The last factor to be considered is the social imperative. Many of the projects that are built consider only the economic and environmental impacts on the building's users and owners. These projects also affect the overall community and society in which they are constructed. Social responsibility considers and promotes interaction and cultural enrichment for building users and nearby communities. All projects should consider the basic needs of people—such as happiness, health, safety, freedom, sustainability, and other positive attributes that lead to more environmentally sustainable lifestyles (Montoya 2010).

Envision®

Envision is an asset management product designed for infrastructure systems developed by the Institute for Sustainable Infrastructure (ISI). ISI is based in Washington, D.C., and was created through a partnership of the ASCE, American Public Works Association, and the American Council of Engineering Companies with a group of Harvard professors and students. The ISI mission is to help communities build more sustainable infrastructure projects (ISI website) using a rating system for all civil infrastructure. The goal is to provide guidance to agencies as a holistic framework for evaluation infrastructure condition and setting priorities. Envision is the tool they use. The process includes 60 criteria and is available at no cost via computer (certification does have a cost associated with it). Criteria include technical consideration, environmental issues, processes, leadership, resource allocation, climate, risk, and issues associated with the life cycle of the asset. The process can be completed in-house or via consultants. Members of a team are certified by ISI for the process. Life-cycle costing is part of Envision (it is not for LEED).

Several major infrastructure agencies have used Envision. The Los Angeles County Board of Supervisors directed staff to adopt the use of Envision for county infrastructure projects, to reach out and engage with the construction and small business community to assist them with implementation, and to coordinate with cities and public agencies within the county that may be interested in adopting the Envision framework (City of Los Angeles 2016). This is a recent example of the use of Envision as a means to help with infrastructure. There are many communities—both large and small—that have utilized this tool to help with green concepts or sustainability.

TRAINING AND WORKFORCE

The current economic conditions have created limitations within agency budgets with respect to training. Most agencies are looking for low-cost, high-impact options. Physical contact is often limited as a result of budget limitations, which makes the mission of providing more effective training even more difficult. Many look to online opportunities for training to save costs for travel, hotels, food, and other incidentals. Major, cutting-edge companies spend 2–3% of gross income on training, but the infrastructure agencies spend under 0.2%. There is also major competition for the same training dollars, so quality matters. At the same time, the industry needs more training opportunities that are oriented to current issues.

There is a need to identify the appropriate avenues, sources of material, delivery methods, and audiences for training of staff. No one solution fits all circumstances. Face-to-face learning is often preferred by those being trained—in part because half the learning involves conversations with other attendees. That is why conferences are so effective—managers get together and talk about things they are working on. Supervisors often overlook the benefits of the "learning-from-other-students" portion of training. One only needs to observe a college class and note that the in-class students generally perform a grade higher than the distance learners.

Training should be a long-term vision. Organizations should be thinking about where they want to be five and 10 years out. There is a need to identify a means to interface with other entities to accomplish education goals. There needs to be a discussion of the quality and cost effectiveness of training, which will allow agencies to argue for larger training budgets. Training is often seen as a luxury item and is easily cut by those who are not part of the industry or by managers who figure they are training people to go elsewhere. Both are incorrect; and that perception needs to change. Cheap training is rarely better—or even useful. Any class can provide information that is either wrong, dated, or both, which is not of any value. The agency benefits when something is learned from proper training. Getting employees trained benefits the effectiveness and efficiency of the workforce. In times of scarcity, having had the proper training means being able to do more. With a large portion of the infrastructure workforce able to retire in 10 years, the effort to educate new people is critical. This argument needs to be made to decision makers.

EMERGING TOPICS

There are many concepts that we cannot even conceive that will impact the future. However, there are three current questions that could have significant potential impact. The first is: *What does the driverless car mean for the future?* Think about the old Arnold Schwarzenegger movie (recently remade) *Running Man*; recall the part where the lead actor just jumped into a jitney that was going down the street. This issue was discussed in Chapter 6.

The next issue connects with the first: *Will robots be doing all of our repetitive jobs?* If so, what does that mean for all the people doing those jobs now, like truckers? Most repetitive jobs do not require a large number of skills, and these jobs will be eliminated in favor of technology, just as assembly line workers have been displaced by robots. That will create a new wave of displaced workers. Training becomes an issue because it is doubtful that those who lost jobs have the skills for the new jobs created. We should also ask, does the $15 per hour minimum wage accelerate this transition? How do either affect the water industry? Meter readers might be replaced with automatic meter-reader systems. Field crews need a lot more training and education to use technology. Work orders are the tip of the technology iceberg. There is a change coming to the industry.

A third issue addresses materials. Lower pH readings in coastal waters will negatively impact concrete and accelerate the deterioration of reinforcing steel that holds much of the load. What solutions will be developed? Solutions for pipe and protection of infrastructure systems are of obvious concern as well. Materials and construction techniques will change. The industry must embrace it.

REFERENCES

Adriaens, P. et al. 2003. "Intelligent infrastructure for sustainable potable water: A roundtable for emerging transnational research and technology development needs." *Biotechnology Advances*. Vol. 22, pp. 119–134.

Arendt, A. A. et al. 2002. "Rapid wastage of Alaska glaciers and their contribution to rising sea level. Science." 297, 382–386, doi:10.1126/science.1072497.

Arendt, A. A., Luthcke, S. B., and Hock, R. 2009. "Glacier changes in Alaska: Can mass-balance models explain GRACE mascon trends?" *Annals of Glaciology*. 50, 148–54. doi:10.3189/172756409787769753.

Argaw, N. 2003. "Renewable energy in water and wastewater treatment applications." National Renewable Energy Laboratory (NREL). Golden, Colorado. http://www.nrel.gov/docs/fy03osti/30383.pdf.

Backlund, P. 2008. "Synthesis and assessment product 4.3: The effects of climate change on agriculture, land resources, water resources, and biodiversity in the United States." May 27, 2008. Available online: http://www.climatescience.gov/Library/sap/sap4-3/finalreport/default.htm#EntireReport. Accessed on 2/22/12.

Balshi, M. S., McGuire, D. A., Duffy, P., Flannigan, M., Walsh, J., and Melillo J. "Assessing the response of area burned to changing climate in western boreal North America using a Multivariate Adaptive Regression Splines (MARS) approach." *Global Change Biology*. 15, 578–600. doi:10.1111/j.1365-2486.2008.01679.x.

Bieniek, P. A. et al. 2014: "Using climate divisions to analyze variations and trends in Alaska temperature and precipitation." *Journal of Climate*. In press. doi:10.1175/JCLI-D-13-00342.1.

Bloetscher, F. 2012. "Protecting people, infrastructure, economies, and ecosystem assets: Water management and adaptation in the face of climate change." *Journal of Water*. Water 2012, 4; doi:10.3390/w40x000x.

Bloetscher, F. and Meeroff, D. M. 2015. *Practical Design Concepts for Capstone Engineering Design*, J. Ross Publishing, Plantation, FL.

Bloetscher, F. and Muniz, A. 2008. "Water supply in South Florida—The new limitations." *Sustainable Water Sources (Reno) Conference Proceedings*, AWWA, Denver, CO.

Bloetscher, F. and Muniz, A. 2012. "Where is the power to treat all the water? potential utility driven solutions to the coming power-water conflict." *Florida Water Resource Journal*. Vol. 64, No. 3, pp. 32–46.

Bloetscher, F. and Romah, T. 2015. "Tools for assessing sea level rise vulnerability." *Journal of Water and Climate Change*. Vol. 6, No. 2, pp. 181–190. © IWA Publishing 2015. doi:10.2166/wcc.2014.045.

Bloetscher, F. and Wood, M., 2016. "Assessing the impacts of sea level rise using existing data." *Journal of Geoscience and Environment Protection*. 4, 159–183.

Bloetscher, F., Meeroff, D. E., and Heimlich, B. N. 2009. "Southeast Florida's resilient water resources." Florida Atlantic University, Boca Raton, FL. http://www.ces.fau.edu/files/projects/climate_change/SE_Florida_Resilient_Water_Resources.pdf. Accessed 2/1/12.

Bloetscher, F. et al. 2010. "Improving the resilience of a municipal water utility against the likely impacts of climate change: A Case Study: City of Pompano Beach." Florida Water Utility: City of Pompano Beach, FL. *Journal American Water Works Association*, 102:11, pp. 36–46. November 2010. http://www.awwa.org/files/secure/index.cfm?FileID=186789.

Bloetscher, F. et al. 2012. "Identification of physical transportation infrastructure vulnerable to sea level rise." *Journal of Sustainability*. Vol. 5, No. 12.

Bloetscher, F. et al. 2013. "Use of lime, limestone and kiln dust to resolve saturation issues with reverse osmosis treated wastewater for injection." *Journal of Water Reuse and Desalination*. 3(3), pp. 277–290. doi:10.2166/wrd.2013.09.

Bloetscher, F. et al. 2014. "Comparing contaminant removal costs for aquifer recharge with wastewater with water supply benefits." *JAWRA*. Vol. 50, Issue 2, pp. 324–333.

Bloetscher, F. et al. 2016. "Assessing potential impacts of sea level rise on public health and vulnerable populations in Southeast Florida and providing a framework to improve outcomes." *Sustainability*. 8(4), 315.

Bolter, K. P. 2014. "Perceived risk versus actual risk to sea-level rise: A Case Study in Broward County." Florida Atlantic University, Boca Raton, FL.

Booz Allen Hamilton. 2009. "Green Jobs Study, prepared for U.S. Green Building Council." Booz Allen Hamilton, McLean, VA. http://www.usgbc.org/Docs/Archive/General/Docs6435.pdf.

Brody, S. D. et al. 2007. "The rising costs of floods: Examining the impact of planning and development decisions on property damage in Florida." *Journal of the American Planning Association*. 73(3), 330–345.

Broward-by-the-Numbers. 2004. "The southeast Florida MSA. Broward County Planning Services Division." www.broward.org/planningservices/bbtn20.pdf. Accessed 10/16/06.

Brown, Thomas C. 1999. "Past and future freshwater use in the United States." Gen Tech. Rep. RMRS-DTR-39. Fort Collins, CO: U.S. Department of Agriculture, Forest Service, Rocky Mountain Research Station. p. 47.

CCSP. 2008. "Weather and climate extremes in a changing climate—regions of focus—North America, Hawaii, Caribbean, and U.S. Pacific Islands." A Report by the U.S. Climate Change Science Program and the Subcommittee on Global Change Research. Vol. 3.3 Karl, T. R. et al., eds. Department of Commerce, NOAA's National Climatic Data Center. p. 164.

Chang, S. W. et al. 2011. "Does sea-level rise have an impact on saltwater intrusion?" *Advances in Water Resources*, 34(10), 1283–1291.

Chapin, F. S. et al. 2014. Ch. 22: "Alaska." *Climate Change Impacts in the United States: The Third National Climate Assessment*. Melillo, J. M., Richmond, T. C., and Yohe, G. W. eds. U.S. Global Change Research Program. 514536. doi:10.7930/J00Z7150.

Ciferno, J. 2009. "Use of non-traditional water for power plant applications: an overview of DOE/NETL R&D efforts." DOE/NETL, 311/040609.

City of Los Angeles. 2016. "L.A. selected as an Envision America City." https://www.lacity.org/blog/la-selected-envision-america-city. Accessed 6/2/16.

DOE. 2006. "Report to Congress on the interdependency of energy and water." United States Department of Energy. Washington, D.C. http://www.sandia.gov/energy-water/congress_report.htm.

———. 2008. "Fossil energy: how coal gasification power plants work." Department of Energy. p. 6. http://www.fossil.energy.gov/programs/powersystems/gasification/howgasificationworks.html.

———. 2014. "The water-energy nexus: Challenges and opportunities." Dept. of Energy, Washington, D.C.

DOE/EIA. 2008. "Annual energy outlook 2008 with projections to 2030." February, 2008.

DOE/NETL. 2004. "Estimating freshwater needs to meet 2025 electricity generating capacity forecasts." June 2004.

DOE and NARUC. 2005. "Liquified natural gas: Understanding the facts." United States Department of Energy and National Association of Regulatory Utility Commissioners. DOE/FE-0489.

Doherty, T. J. and Clayton, S. 2011. "The psychological impacts of global climate change." *Am Psychol*. 66(4), 265.

EIA. 2008. "Annual Energy Review, 2007." Energy Information Administration, Washington, D.C. http://www.eia.doe.gov/aer/.

EIA. 2010. "Annual Energy Review, 2009." DOE/EIA-038 (2009), Department of Energy, Washington, D.C. www.eia.gov/aer/.

Elcock, Deborah. 2009. "Energy production and other uses.", *Groundwater Protection Council Annual Forum Proceedings—Salt Lake City*. GWPC, Oklahoma City, OK.

Environmental Information Administration. 2008. EIA Annual Energy Outlook.

EnvisionGroup. 2009. "Use Envision® to create more sustainable infrastructure." Institute for Sustainable Infrastructure, Washington, D.C. https://sustainableinfrastructure.org/.

EPA. 2006. "Wastewater management fact sheet: Energy conservation." EPA-832-F-06-024. EPA, Office of Water, Washington, D.C.

———. 2009. "State clean energy and climate program." *Clean Energy Lead by Example Guide: Strategies, Resources, and Action Steps for State Programs.* Prepared by Joanna Pratt and Joe Donahue, Stratus Consulting, Inc.

———. 2010. "Green building home." From: http://www.epa.gov/greenbuilding/pubs/components.htm. Accessed 11/12/12.

———. 2011. "Frequent questions about environmental management systems." EPA, Washington, D.C. website. https://www.epa.gov/ems/frequent-questions-about-environmental-managment-systems. Accessed 2/2/12.

———. 2012. "History of ENERGY STAR." EPA, Washington, D.C. website. http://www.energystar.gov/index.cfm?c=about.ab_history. Accessed 2/1/12.

———. 2012a. "What is WaterSense?" U.S. Environmental Protection Agency. http://www.epa.gov/WaterSense/about_us/what_is_ws.html. Accessed 2/1/12.

———. 2013. "Water & energy efficiency in water and wastewater facilities." EPA, Region 9. file:///E:/CGN%206506%20Infra%20Mgmt/9974%20Water%20&%20Energy%20Efficiency%20in%20Water%20and%20Wastewater%20Facilities_%20Benefits%20and%20Challenges%20_%20Region%209_%20Sustainable%20Water%20Infrastructure%20_%20US%20EPA.pdf.

———. 2013a. "Climate change adaptation for state and local governments overcoming the uncertainty barrier to adaptation." Webcast U.S. EPA State and Local Climate and Energy Program. April 17, 2013. EPA, Washington, D.C.

EPRI. 2003. "A survey of water use and sustainability in the United States with a focus on power generation." Topical Report, Nov. 2003.

Feely, T. et al. 2008. Water: "A critical resource in thermoelectric power industry." *Energy.* 33, 1–11.

FDOT. 2006. *Service Development Plan.* Chapter 2. Florida Department of Transportation, Tallahassee, FL.

FDOT. 2012. BDK79 977-01 "Development of a methodology for the assessment of sea level rise impacts on Florida's transportation modes and infrastructure." Summary [PDF—470 KB], Final Report. [PDF—13,326 KB].

FL DEP. 2002. "Implementing regional water supply plans: Is progress being made?" Florida Department of Environmental Protection, Tallahassee, FL. May 2002. Accessed April 8, 2006 at www.dep.state.fl.us/water/waterpolicy/docs/rwsp_2002.pdf.

Freas, K. et al. 2008. "Incorporating climate change in water planning." *J. Am. Water Works Ass.* 100, 93–99.

Funari, E., Manganelli, M., and Sinisi, L. 2012. "Impact of climate change on waterborne diseases." *Ann Ist Super Sanità,* 2012, 48(4), 473–487, doi:10.4415/ANN_12_04_13.

GBCI. 2011. "Current certification fees." *Green Building Certification Institute,* @http://www.gbci.org/main-nav/building-certification/resources/fees/current.aspx. Accessed 1/31/12.

Grinsted, A. J. 2009. "Reconstructing sea level from Paleo and projected temperatures 200 to 2100 AD." *Clim. Dyn.* 34, 461–472.

Haines, A. et al. 2006. "Climate change and human health: Impacts, vulnerability and public health." *Pub Health.* 120, 585–596.

Hanson, S. et al. 2011. "Global ranking of port cities with high exposure to climate extremes." *Clim. Chang.* 104, 89–111.

Heimlich, B. N. et al. 2009. "Southeast Florida's resilient water resources: Adaptation to sea level rise and other climate change impacts." Florida Atlantic University, Boca Raton, FL. 2009. Available online: http://www.ces.fau.edu/files/projects/ climate_change/SE_Florida_Resilient_Water_Resources.pdf. Accessed on 3/26/12.

Herrera-Pantoja, M. and K. M. Hiscock. 2008. "The effects of climate change on potential groundwater recharge in Great Britain," *Hydrol. Process.* 22, pp. 73–86.

Herring, S. C., Hoerling, M. P., Kossin, J. P., Peterson, T. C., and Stott, P. A., eds. 2015. Explaining Extreme Events of 2014 from a Climate Perspective. *Bull. Amer. Meteor. Soc.,* 96 (12), S1–S172.

Hess, J. J., McDowell, J. Z., and Luber, G. 2012. "Integrating climate change adaptation into public health practice: Using adaptive management to increase adaptive capacity and build resilience." *Environ Health Perspect.* 120(2), p. 171.

Hinzman, L. D. et al. 2005. "Evidence and implications of recent climate change in Northern Alaska and other Arctic regions." *Climatic Change*. 72, 251–298. doi:10.1007/s10584-005-5352-2.

Hoffmann, J., Feeley, T., and Carney, B. 2005. U.S. Department of Energy/National Energy Technology Laboratory, DOE/NETL's power plant water management R&D program—responding to emerging issues. 8th Electric Utilities Environmental Conference, Tucson, AZ.

Hu, F. et al. 2010. "Tundra burning in Alaska: Linkages to climatic change and sea ice retreat." *Journal of Geophysical Research*. 115, G04002. doi:10.1029/2009jg001270.

Hu, Y. et al. 2008. "Simulation of air quality impacts from prescribed fires on an urban area." *Environ. Sci. Technol*. 42, 3676–3682.

Hudkins, J. M. and Fox, J. D. 2006. "Water conservation methods at the source: A review of membrane treatment minimization strategies." *Florida Water Resource Journal*. Vol. 58, No. 8, 34–35.

Hutson, S. S. et al. 2004. "Estimated use of water in the United States in 2000." U.S. Geological Survey Circular 1268, p. 46.

Intergovernmental Panel on Climate Change (IPCC). 2013. *Climate Change 2013: The Physical Science Basis; Working Group I Contribution to the Fifth Assessment Report of the Intergovernmental Panel on Climate Change*. Cambridge University Press, New York, NY.

ISO. 2011. ISO 14000 environmental management. *International Organization for Standarization*. http://www.iso.org/iso/iso_catalogue/management_and_leadership_standards/environmental_management.htm. Accessed, 1/2212.

ISO. 2015. *14001 Environmental Management System (EMS). International Organization for Standarization*, Geneva, Switzerland.

Jacob, T. et al. 2012. "Recent contributions of glaciers and ice caps to sea level rise." *Nature*. 482, 514–518. doi:10.1038/nature10847. Available online at: http://www.nature.com/doifinder/10.1038/nature 10847.

Jorgenson, T. et al. 2008. "Permafrost characteristics of Alaska." *Extended Abstracts of the Ninth International Conference on Permafrost, June 29-July 3, 2008*. D. L. Kane, and K. M. Hinkel, eds., University of Alaska Fairbanks, 121–123. Available online at http://permafrost.gi.alaska.edu/sites/ default/files/AlaskaPermafrostMap_Front_Dec2008_Jorgenson_ etal_2008.pdf.

Karl, T. R. and Melillo, J. M. 2009. "U.S. global change research program, global climate change impacts in the United States." Peterson, T. C., eds.; Cambridge University Press, Cambridge, UK. Available online: http://www.globalchange.gov/publications/reports/scientificassessments/us-impacts/full-report. Accessed on 3/26/12.

Kasischke, E. S. et al. 2010. "Alaska's changing fire regime—implications for the vulnerability of its boreal forests." *Canadian Journal of Forest Research*. 40, 1313–1324. doi:10.1139/X10-098.

Kats, G. 2003. "The costs and financial benefits of green buildings: A report to california's sustainable building task force." http://evanmills.lbl.gov/pubs/pdf/green_buildings.pdf. Accessed 12/13/12.

Kaushik, S. 2012. "Corporate conceptions of triple bottom line reporting: An empirical analysis into the signs and symbols driving this fashionable framework." *Social Responsibility Journal*. Vol. 8, Issue: 3, pp. 312–326. doi: 10.1108/17471111211247901.

Kenny, J. F. et al. 2000. "Estimated use of water in the United States in 2005." U.S. Geological Survey Circular 1344, USGS, Washington, D.C.

Kimmel, T. and Veil, J. A. 2009. "Impact of drought on U.S. stream electric power plant cooling water intakes and related water resource management issues." DOE/NETL 2009/1364. DPE, Washington, D.C.

Kjellstrom, T. and McMichael, A. 2013. "Climate change threats to population health and well-being: The imperative of protective solutions that will last." *Glob Health Action*. 10, 3402.

Knowlton, K. et al. 2011. "Six climate change-related events in the United States accounted for about $14 billion in lost lives and health costs." *Health Aff* (Millwood). 30, 2167–2176.

Kubba, S. 2010. *LEED Practices, Certification, and Accreditation Handbook*, Butterworth-Heinemann/Elsevier, Burlington, MA.

Larsen, C. F. et al. 2007. "Glacier changes in southeast Alaska and northwest British Columbia and contribution to sea level rise." *Journal of Geophysical Research*. 112, F01007. doi:10.1029/2006JF000586.

Lisk, Bryan, Greenberg, Ely, and Bloetscher, Frederick. 2012. "Implementing renewable energy at water utilities web report #4424." Water Research Foundation, Denver, CO.

Lott, N. and Ross, T. 2006. "Tracking and evaluating U.S. billion dollar weather disasters, 1980–2005." The 86th annual meeting of the American Meteorological Society, Atlanta, Georgia, January 29–February 2.

Mack, M. C. et al. 2011. "Carbon loss from an unprecedented Arctic tundra wildfire." *Nature*, 475, 489–492. doi:10.1038/nature10283.

Markham, A. 2015. "Six ways climate change in Alaska will affect you." *The Equation Blog/Union of Concerned Scientists.* Accessed 9/1/15.

Marshall, C. H. et al. 2003. "The impact of anthropogenic land—Cover change on the Florida Peninsula sea breezes and warm season sensible weather." *Mon. Wea. Rev.* 132, 28–52.

Mead, S. P. et al. 2009. "A water/energy best practices guide for rural Arizona's water & wastewater systems." Northern Arizona University, Flagstaff, AZ. http://www.waterenergy.nau.edu/docs/Best_Practices_Guide_2009.pdf. Accessed 3/31/2012.

Meadows, D. and Randers, J. 2004. *Limits to Growth: The 30 year Update.* Chelsea Green Publishing Co., White River Junction, VT.

Molak, V. 1997. *Fundamentals of Risk Analysis and Management.* Lewis Publishers, Boca Raton, FL.

Montoya, M. 2010. *Green Building Fundamentals.* Pearson Education, Inc., Upper Saddle River, NJ.

Moore, N. and Rojstaczer, S. 2002. "Irrigation's influence on precipitation: Texas High Plains, U.S.A." *Geophysical Research Letters*, 29, 2-1–2-4.

Nace, R. L. 1960. "Water management, agriculture, and ground-water supplies." *U.S. Geological Survey Circular 415*, 12.

NAS. 2015. "Arctic Matters." http://dels.nas.edu/resources/static-assets/materials-based-on-reports/booklets/ArcticMatters.pdf. Accessed 2/29/2016.

NCA. 2014. globalchange.gov/highlights/regions/Alaska. Accessed 2/29/16.

NETL. 2008. "2008 Update." Office of Systems Analyses and Planning, DOE/NETL-400/2008/1339.

NSTC (National Science and Technology Council). 2007. "A strategy for federal science and technology to support water availability and quality in the United States." Committee of Environment and Natural Resources Washington, D.C., p. 35.

Osterkamp, T. E. and Romanovsky, V. E. 1999. "Evidence for warming and thawing of discontinuous permafrost in Alaska." Permafrost and Periglacial Processes, 10, 17–37. doi:10.1002/(SICI)1099-1530(199901/03)10:13.0.CO;2-4. Available online: http://onlinelibrary.wiley.com/doi/10.1002/(SICI)1099-1530(199901/03)10:1%3C17::AID-PPP303%3E3.0.CO;2-4/pdf] 67.

Pachauri, R. and Reisinger, A. eds. 2007. "Climate Change 2007." *Synthesis Report. Contribution of Working Groups I, II and III to the fourth assessment report to the Fourth Assessment Report of the Intergovernmental Panel on Climate Change.* IPCC, Geneva, Switzerland.

Parkinson, R. W. 2009. "Adapting to rising sea level: A Florida perspective." In *Sustainability 2009: The Next Horizon; In Proceedings of the AIP Conference, Melbourne, FL, USA, March 3–4, 2009.* Available online: https://411.fit.edu/sustainability/documents/FINAL%20-%20With%20Cover%2010-1-09.pdf. Accessed on 3/28/16.

Portier, C. et al. 2010. "A human health perspective on climate change: A report outlining the research needs on the human health effects of climate change." Research Triangle Park, NC: Environmental Health Perspectives/National Institute of Environmental Health Sciences. doi:10.1289/ehp.1002272. URL: www.niehs.nih.gov/climatereport. Accessed 10/23/14.

Rahmstorf, S. 2007. "A semi-empirical approach to projecting future sea-level rise." *Science* 2007, 315, 368–370.

Reilly, T. E. et al. 2008. "Ground-water availability in the United States." USGS Circular 1323, USGS, Reston, VA.

Richey, J. E. and Costa-Cabral, M. 2006. "Floods, droughts, and the human transformation of the Mekong River basin." *Eos Trans. AGU*, 87(52), Fall Meet. Suppl., Abstract.

Romah, T. 2012. "Advanced methods in sea level rise vulnerability assessment." Master's Thesis. Florida Atlantic University, Boca Raton, FL.

Romanovsky, V. E. et al. 2012. "[The Arctic] Permafrost [in "State of the Climate in 2011"]." Bulletin of the American Meteorological Society, 93, S137-S138. doi:10.1175/2012BAMSStateoftheClimate.1. Available online at http://www1.ncdc.noaa.gov/pub/data/cmb/bams-sotc/ climate-assessment-2011 -lo-rez.pdf.

Rose, J. B. et al. 2001. "Climate variability and change in the United States: Potential impacts on water and foodborne diseases caused by microbiologic agents." Environ Health Perspect. 109(Suppl 2), 211.

Rosen, Y. 2015. "Arctic Study: Warmer Arctic means faster mosquito growth, spelling hazard for caribou." August 29, 2015. https://www.adn.com/arctic/article/warmer-arctic-weather-means-faster -mosquito-growth-and-more-ill-timed-pests 2015/09/15/. Accessed 2/14/16.

Salmun, H. and Molod, A. 2006. "Progress in modeling the impact of land cover change on the global climate." Progress in Physical Geography. 30, 6, pp. 737–749.

Scanlon, B. R. et al. 2005. "Impact of land use and land cover change on groundwater recharge and quality in the southwestern US." Global Change Biology. 11, 1577–1593.

SEFRCC. 2011. "Southeast Florida Regional Climate Change Compact (SFRCC) Inundation Mapping and Vulnerability Assessment Work Group." Analysis of the Vulnerability of Southeast Florida to Sea Level Rise. http://www.ces.fau.edu/files/projects/climate_change/SE_Florida_SLR_Vulnerability _May2011_0.pdf.

Shuster, E. 2008. Estimating Freshwater Needs to Meet Future Thermoelectric Generation Requirements, 2008 Update, NETL, Pittsburgh, PA.

Sloss, E. M. et al. 1996. Groundwater Recharge with Reclaimed Water: An Epidemiologic Assessment in Los Angeles County. 1987–1991. RAND Corp., 139.

Smith A. and Katz, R. 2013. US Billion Dollar Weather and Climate Disasters: Data Sources, Trends, Accuracy and Bias, Natural Hazards, Volume 67, Issue 2, pp. 387–410.

Smith, A. B. and Matthews, J. L. 2015. "Quantifying uncertainty and variable sensitivity within the U.S. billion-dollar weather and climate disaster cost estimates." Natural Hazards 77(3), 1829–51.

Solley, W. B., Pierce, R. R., and Perlman, H. A. 1998. "Estimated use of water in the United States in 1995." United States Geological Survey.

Southeast Florida Regional Climate Compact (SFRCCC). 2012. "Analysis of the vulnerability of Southeast Florida to sea-level rise." Available online: http://www.southeastfloridaclimatecompact.org/wp -content/uploads/2014/09/regional-climate-action-plan-final-ada-compliant.pdf. Accessed 3/29/16.

Sovacool, B. and Sovacool, K. 2009. "Identifying future electricity water tradeoffs in the United States." Energy Policy. 37, 7(7): 2763–73.

Starkl, M. and Brunner, N. 2004. "Feasibility versus sustainability in urban water management," Journal of Environmental Management, Vol. 71, pp. 245–260.

Stewart, B. C. et al. 2013. "Regional climate trends and scenarios for the U.S. National Climate Assessment: Part 7. Climate of Alaska." NOAA Technical Report NESDIS 142–7. p. 60.

Stocker, T. F. 2014. "Climate Change 2013: The Physical Science Basis." Cambridge University Press, p. 1535.

Stott, P. A. 2015. "Weather risks in a warming world." Nat. Climate Change, 5, 517–518. doi:10.1038/ nclimate2640.

Tebaldi, C., Strauss, B. H., and Zervas, C. E. 2012. "Modelling sea level rise impacts on storm surges along U.S. coasts." Environ. Res. Lett. 7. Article 1.

Titus, J. G. and Richman, C. 2001. "Maps of lands vulnerable to sea level rise: Modeled elevations along the U.S. Atlantic and Gulf coasts." Clim. Res. 2001, 18, 205–228.

United States Census Bureau, 2004. "Projected population of the United States, 2000–2050." Census Bureau, Washington, D.C. Available at /http://www.census.gov/population/ www/projections/ usinterimproj/natprojtab01a.pdf.

United States Census Bureau. 2012. "State and county quickfacts: Florida." URL: http://quickfacts.census .gov/qfd/states/12000.html. Accessed 03/15/2015.

USACE. 2013. "Incorporating sea level change into civil works programs." ER 1100-2-8162. http://www .publications.usace.army.mil/Portals/76/Publications/EngineerRegulations/ER_1100-2- 8162.pdf.

USGBC. 2011. "Buildings and climate change." USGBC, http://www.documents.dgs.ca.gov/dgs/pio/facts/ LA%20workshop/climate.pdf. Accessed 2/22/12.

USGBC. 2011a. "What LEED is." *U.S. Green Building Council.* http://www.usgbc.org/DisplayPage .aspx?CMSPageID=1988. Accessed 1/22/12.

USGS. "Ground-water depletion across the nation." U.S. Geological Survey Fact Sheet 103-03. November 2003.

Union of Concerned Scientists. 2007. "Freshwater use by U.S. power plants: electricity's thirst for a precious resource." A Report of the Energy and Water in a Warming World Initiative, UCS, Cambridge, MA.

UNEP. 2009. "Climate Change Science Compendium 2009." New York: United Nations Environmental Programme. United Nations, New York, NY.

U.S. National Energy Technology Laboratory and the Department of Energy. 2006. "Energy Demands on Water Resources: Report to Congress on the Interdependency of Energy and Water." DOE, Washington, D.C.

Virjee, K. and Gaskin, S. 2005. "Fuzzy cost recovery in planning for sustainable water supply systems in developing countries." *Energy*, Vol. 30, pp. 1329–1341.

Weiss, J. L., Overpeck, J. T., and Strauss, B. 2011. "Implications of recent sea level rise science for low-elevation areas in coastal cities of the conterminous USA." *Clim. Chang.* 105, 635–645.

Wendler, G., Chen, L., and Moore, B. 2012. "The first decade of the new century: A cooling trend for most of Alaska." *The Open Atmospheric Science Journal*, 6, pp. 111–116. doi:10.2174/1874282301206010111.

Whitcomb, J. B. 2006. "Florida water rates evaluation of single-family homes." South Florida Water Management District. www.swfwmd.state.fl.us/documents/reports/water_rate_report.pdf. Accessed 10/16/06.

World Business Council for Sustainable Development. 2008. "Energy efficiency in buildings." World Business Council for Sustainable Development, Geneva, Switzerland. https://www.c2es.org/docUploads/EEBSummaryReportFINAL.pdf. Accessed 2/2/12.

Zhang, K. 2011. "Analysis of nonlinear inundation from sea-level rise using LIDAR data: A case study for South Florida." *Clim. Chang.* 106, 537–565.

16

LEADERSHIP

Ultimately, the solution to many of the infrastructure issues that are faced by infrastructure professionals today will have more to do with creating local leadership than it will be associated with state or federal political decisions. Many infrastructure issues are ultimately local, and local entities have made significant investments to benefit their citizens. Local leadership will be required to protect those assets. It is therefore important to discuss leadership, which is distinctly different from management; a distinction that is often misunderstood by the public. In addition, leadership often manifests itself at levels other than the top of an organization.

Leadership is both a short- and a long-term concern. For the long term, what must be answered is how the decisions made today will impact the course of the organization. One thing many people do not understand is that while we live in the moment, it is how people view our actions afterward that often defines leadership. In other words, it is easy to see leadership after the fact but very difficult during the event. Public officials that oversee public works agencies need to ask themselves *if their tenure has added value to the system*—whether that includes better roads and bridges, improved stormwater system reliability, less ongoing flooding, improved treatment plant capacity and quality, better public health protection, or improved reliability of the system. This is a legacy leadership issue that should be answered by elected and appointed officials alike, as well as the employees of the organization. More difficult is how to measure the value added, since monetary value is not the only means to add value. Keep in mind, no one remembers the guy who did not raise rates or taxes nor the person who deferred investments. People tend to remember those who implemented a vision, navigated through a crisis, or left behind a legacy that demonstrated that the system was better after the fact. Likewise, people are likely to recall those who did not budget to replace the infrastructure that failed, negatively impacting their communities.

It is hard to define leadership because it comes in many forms and is often specific to the approach to a situation. A quarterback who is a great leader on the field might not be the best choice to lead the reorganization of a major corporation. Both positions require leadership, but the skill sets required for each position are situational. Because no one can define the skill set for every situation, the tendency is to look at examples of people who are leaders or who have exhibited leadership in the past and try to draw from their experience—what made him/her a leader? An analogy might be ethics. We know when we don't see it—just like leadership. The reverse is a little more of a challenge, but the two concepts are related. One cannot be a leader if he or she is perceived to be unethical. It offends the sensibilities of most people and they will not follow that person.

Another conundrum with leadership is actually trying to define what it is when it happens. It was not clear that Abraham Lincoln would be remembered as one of the greatest leaders in

American history while he was president. If you read accounts of his presidency, the early years are marked with indecision and backtracking before he got it right. Most of that is forgotten in lieu of the ultimate results. Results matter and conveying the message is required.

There are many things that are required for leaders. They must bring value to the organization, as well as skills and knowledge. Leaders should be able to clearly communicate their vision or risk losing their position. Then, they need people to buy into that vision or they cannot lead. When provided a challenge, it is how the leader attacks it and how they marshal resources that determine success or failure. Leaders will have confidence in their abilities and strive to make everyone better in the organization. That means understanding that if people are put in positions to fail, they will. It is a failure of both management and leadership to keep putting people in these situations. Many athletic coaches know this; others found out the hard way when they didn't follow this concept and lost their job. Leaders may exist at every level of the organization. The challenge is seeking them out and putting them in position to succeed.

Leadership can be evaluated by those who are successful in getting others to follow them. They will also seek to bring in people to fill those gaps. That means leaders will hire the best people they can, without worrying about whether they are vying for some future position. In some cases, the strategy must be adjusted based on the skill set of the people on hand. Again, a couple of sports examples are instructive. Legendary NFL coach Don Shula was successful with the Baltimore Colts when he had a prolific passer in Johnny Unitas. He was even more successful in adjusting to a run-oriented game with the Miami Dolphins when he had Larry Csonka, Jim Kiick, and Mercury Morris in the backfield leading the team to the only perfect season in the NFL and two Super Bowl championships (three trips). He adjusted again when those players moved on and he drafted Dan Marino. The Dolphins became a pass-oriented team to match up with Marino's skill set. Shula adjusted how his team operated based on how the personnel could be successful. He also hired excellent coaches to help implement his vision.

Legendary NBA coach and general manager Pat Riley did the same when he went from coaching the Los Angeles Lakers (high scoring run-and-gun offense), to the New York Knicks— a plodding defensive team, to a mix with the Miami Heat based on players (both ways). Leaders will set a vision and adapt strategies to make the people they lead successful. Leaders tend to know their own limitations and the limitations of their staff—and play to their strengths. Those who follow the Miami Heat know there is a *Heat* way—and players must adjust to that *way*.

If all this sounds like sports, it should. In the NFL, the skill sets of players are similar, but it is the mental aspects—the vision, the ability to work together, the willingness to do the little things that do not always get noticed, along with the ability to exploit someone else's weaknesses by using your strengths—that wins games. You need talent, but you need leadership to be successful. Examples of teams with lots of great players that never win are abundant. Teams change coaches and players, trying to find that right mix.

When the team finds the right mix, success follows. Lincoln found this during the Civil War. He spent time with the troops. He communicated his vision to them, expressed his appreciation for their efforts; he supported them and they were enthusiastic supporters in return. His generals, well, were another matter. So, Lincoln kept changing generals until he found Ulysses S. Grant—a general who would fight and end the war, because his vision of the end game was to win.

Many of the issues facing society today need real leadership to create a long-term solution. For example, the public understands the long-term need for solutions to high medical costs since this is discussed routinely in the media. The public asked for a solution in the 2008 election— President Obama proposed one and Congress approved it, although it was not perfect. Trying to fix it has been the challenge since, which is why the public sees little leadership in Congress. The need to fix the infrastructure that made our economy strong should be a priority as well, but many things are a higher priority in Congress. Unfortunately, that priority is often lower at the local and state level, as well. Public infrastructure officials are going to have to deal with tough issues like rebuilding deteriorating infrastructure, sea level rise, climate change, stressed water supplies, energy demands, and a more demanding electorate. Future leaders will have to raise funds to pay for these needs and will likely have to recommend increasing taxes, along with water and wastewater fees. But, will they have the skills to encourage decision makers to move forward to address the needs of the system while understanding that there are political repercussions in doing so? Leadership involves understanding that there is a limit to the inexpensive and temporary solutions, while at the same time there is a grander vision of the future that involves a group of allied people.

Think about the city of Los Angeles. The only reason large numbers of people can live in Los Angeles is because of the aqueducts that were started back in the 1920s by William Mulholland under the guidance of Mayor Fred Eaton. The vision was to grow the city, but the limitation was the water supply. The aqueducts sparked water wars (think of the old Jack Nicholson movie *Chinatown*) that developed through the 1930s. The Hetch Hetchy reservoir was established as San Francisco's water supply over 100 miles east of the city back in 1913. The reservoir system continues to supply San Francisco today. Denver Water acquired and/or constructed reservoirs and tunnels to the west side of the Rockies for water supplies prior to 1940, realizing that the water supplies that were needed in order to sustain growth in the Denver area were not available east of the Rockies. Pinellas County, Florida, and Orange County, California, started projects to reuse treated wastewater for irrigation of private yards and for aquifer recharge in the 1970s to sustain their supplies. Sustainability of water supplies, management of water sources (including wastewater and stormwater as a part of an integrated program), and sustaining the financial and infrastructure condition of the utility are the long-term priorities. Leaders in New York, San Francisco, and Chicago put transit and train systems into place before the ridership existed to help with crowded roads and excessive commuting time. These forward-looking decisions provided an option that encouraged people to change their behaviors by riding transit.

Difficulties can occur when conditions change. Using one of the previous examples: while Denver Water built tunnels and reservoirs to transfer water from the west side of the Rockies to the east, other Denver area communities did not. Those communities relied on groundwater that was far less costly to obtain than transfers across the Rockies via tunnels. This worked until the population in the Denver area started to explode, exceeding the capacity of the groundwater. Buying from Denver Water worked, until their transfer capability was nearing capacity. A 100-year management plan was developed and approved by the State Legislature in 1985 to allow water to be withdrawn from the Denver Basin, despite very limited recharge. This is not to say that the plan for management was not a good leadership start (certainly it is an improvement over doing nothing), but what happens in 70 years? It is generally assumed that officials will come up with a solution to extend the life of the aquifer, but when will that occur and who will lead that charge? What will the political backlash be when the initial rumblings begin? The good news is that the major users are utilities that have resources to pay for treatment, aquifer storage, indirect potable reuse, direct potable reuse, and a host of other potential options—but not every basin is so lucky. If the major users are agriculture or ecosystems, who pays that bill? Or small communities? If the answer is no one, what happens to the industry? The jobs? Communities? People? Proper decisions regarding infrastructure must be made for the long-term benefit of society.

So, in reality, leadership is defined by what is left behind not by the current condition; it's how we change our thinking and actions to adapt to the changed conditions. Look back at the great water projects of the 20th century—the Hoover Dam, the channels carrying water to Los Angeles from the Colorado River and central California that allowed southern California to develop, or the numerous dams across the west that permitted crops to grow in arid regions. Look at the vision of the highway system that allowed for improved transportation and commerce, the development of suburban communities, and connections with rural communities. San Francisco's transit system, along with those of New York and Chicago, are examples of a vision 100 years ago. You can search out who led those projects. That is their legacy. Those that come afterward reap the rewards created from the efforts of those leaders.

Leadership requires creativity. A book that relates to this need and is especially apropos to engineers is *The Cult of the Mouse* by Henry Caroselli (2003), where the author encourages creativity above profits in the workplace. Creativity is what will provide innovative solutions that change how people live. It is also where the patents and economic opportunities exist in the private sector—America rose to greatness in the 20th century in large part because of automobiles, airplanes, energy, and computers, which made many things possible. Innovation is what made Disney great, as Caroselli, a former Disney executive, writes in his book. Computers became commonplace in the latter part of the century, a technology used in all fields of public works today. In fact, computers have made the industry so much more efficient that costs have not climbed as fast as they might have, which is why cable television is usually more expensive than water and sewer.

Creativity and vision can be found in the public sector. The city of Dania Beach built the world's first LEED Gold water plant. That took vision on the part of the utility director, in league with a cooperative team of consultants, contractors, and students. When young people get the opportunity to be creative on projects, they often don't know the adage: "that's not the way we do it." Creativity, innovation, and the *can-do* mentality are part of leadership, but not all of it. There is that ability to set a vision and the ability to convince decision makers of the wisdom of that vision, as Caroselli noted. Selling innovation is often the hard part because that's where the costs are. One can be ridiculed by those maintaining the status quo when recommending change; which discourages innovation and change. As a result, many ideas are just lost in the shuffle because they never receive a voice.

How to find leaders in the public sector and what skills are required to be a leader are the big questions. If picking leaders was easy, all organizations would be successful—as opposed to being fodder for Scott Adams' Dilbert® comic series. If leadership skills were easily defined, there would be a lot more schools trying to teach leadership and they would create generations of leaders—but they don't do either well. The public often looks at elected officials for leadership, but one could spend pages discussing the fallacy of that argument. No offense intended here, but can one really say that Millard Fillmore, Andrew Johnson, Franklin Pierce, James Buchanan, or Warren Harding were great leaders? They rank in *U.S. News*'s 10 worst presidents of all time. And our perception is generally the same (assuming one knows enough U.S. history to know these characters). Was it their fault? It is hard to tell, but circumstances were not in their favor. They certainly did not lead the nation from difficulties.

Situations also matter. Ulysses Grant and Zachary Taylor were great leaders on the battlefield, but not as president. Why? Different skill sets. Their best skills were not transferrable to the presidency. Lincoln is recognized as a great leader, but was never a field general. History treats our Revolutionary forefathers, and Presidents Franklin D. Roosevelt, Teddy Roosevelt, and John F. Kennedy, in a similarly positive light because they were instruments of change. Our crazy, radical forefathers had the audacity in 1776 to think that average people could actually govern themselves when the whole world knew that was impossible. We have no conception today what a crazy idea that was in the 18th century, as we take it for granted today. Teddy Roosevelt changed how we viewed open spaces by altering the status quo. Franklin Roosevelt navigated us through the Depression, built much of the 1930s infrastructure using federal funds, navigated us through most of World War II, implemented social security and a host of other programs that served Americans for the balance of the 20th century. John F. Kennedy negotiated us through parts of the Cold War while giving us a vision of a future that was tragically cut short by his assassination—but many still hold out hope for that future. It is often the change that people wrought that made them leaders.

Likewise, we have business leaders, but mostly they are making money for their stockholders; few are making a big difference to the public today. The latter is why we all know Steve Jobs, Bill Gates, Paul Allen, Mark Zuckerberg, Henry Ford, Thomas Edison, and Harvey Firestone—they all made a difference in our lives and how we live. But, who is the CEO of Goldman Sachs? That's the problem—that person is not a leader because the impact on your life is missing.

It is what leaders envision that causes others to buy into their vision and cooperate toward achieving their goals. It is how they approach a challenge and coalesce resources to resolve it, regardless how big the issue may be. That is why if one asks, he or she can find out in who are the *go-to* people in an organization—the ones they rely on and follow (a recent but still unpublished survey of the water industry found that only 10% of respondents thought that leadership existed at the executive management level compared to the over 40% at the field and professional level). Those are the true leaders. They often outlast the managers, especially if the positional leadership does not tap into their skill set. Leaders set a direction and then support their staff, which does the work. It is how they guide, as opposed to directing employees. It is how they share accolades with the staff and accept the blame for failure, as opposed to the opposite.

Infrastructure is a public good and the difference between it as a public good versus a private good is worth discussing since the leadership differences are stark. One often hears officials argue how "we should be running government like a business." However, this appears to be an oversimplified argument that ignores true differences in the objectives of the public and private sector. The two sectors are different—the following example may be instructive. Assume a person is in charge of Ford Motor Company and he makes only two vehicles—the F150 pickup (largest selling vehicle in the United States) that has a high profit margin and a passenger vehicle that does not have a high profit margin and does not sell nearly as well. If he determines that his revenues are likely to decrease as a result of the economy, where does he make cuts? Theoretically, cutting costs and reducing production of the passenger vehicle should maintain or improve Ford's profit margin. So, that manager looks like a brilliant leader in the business world.

But, the public and private sectors are intrinsically different, and while there are commonalities, the inability to directly measure impacts on the public sector make private sector applications suspect in many situations. Curtailing services (cutting the passenger vehicle) may have unanticipated consequences. Simplistic solutions that are commonly offered up mean that these leaders do not adequately understand what their products are, nor which ones are priorities. Consequently, they abdicate their decision making for simplistic solutions that seem fair.

So let's assume our executive above gets hired to run a city because of his success at Ford. The city, of course, has a revenue shortfall, so what does he do? This is far more difficult. The city has police, fire, parks and recreation, planning, etc., so where does he make cuts? Fire? Police? Parks? None of them are profit centers; they are all services, the value of which cannot easily be measured. He could evaluate the risk of higher losses if he cuts the fire department, but that creates other issues. There is a distinct difference in the metrics between the sectors. So, he cuts all services the same amount—sharing the pain because there is no means to measure the impact of success of cutting costs. Every government employee recognizes this method to reduce the budget. So, how would that have worked at Ford? Well, cutting back on the F150 and the passenger vehicle the same percent would likely make the overall situation worse, not better. A Ford executive making that type of decision would be roundly criticized and likely dismissed, but would that same person be viewed as a successful manager in the public sector? It happens all the time—share the pain—but in reality, this is a cop-out and so he deserves to be fired as Ford would do. And, ditto for the other officials who go along with such simplistic decision making.

Leadership is needed to help make better decisions. Leaders need to set a vision, and this is perhaps the biggest barrier to public sector leadership. One would think that applying private sector business principles would help with the vision process, but it does not because the terms for officials are comparatively short. Infrastructure requires long-term decisions—often at least 100 years. No private firm thinks 100 years out. Stockholder demands on the private sector are short-term profits, which has hurt the long-term vision of both public and private sectors. Private sector payback is three to five years, seven at the most. Most public infrastructure may not pay back for 5, 10, or 20 years. Government takes that risk. So, the question is: "Where are the leaders and are we so afraid of change that we cannot tolerate leadership?" That will be the question going forward. Can we lead?

REFERENCE

Caroselli, M. H. 2003. *The Cult of the Mouse.* Ten Speed Press, Berkeley, CA.

17

CONCLUSIONS

The economy of the United States and much of the developed world was built on advanced (for their time) infrastructure systems constructed by governments with a vision to the future. As society has evolved, the need for infrastructure systems has evolved with it. For example, the lack of water severely inhibits many third-world nations. Even when they have water, it is unsafe to drink or use. In the United States at the turn of the 20th century, 1:100,000 people *died* each summer from typhoid; just typhoid, not all the other waterborne diseases. Many people were sick—and the population was much smaller. Talk about reduced productivity. Now, we have advanced water systems, disinfection practices that protect people and pipes, and few get sick from contaminated water. Those who do, become headlines. No one wants to be a headline (Flint!). Productivity is up—but we expect good water and not to see the pipes.

Sewer is an even better example. People just want to flush the toilet and see that it's gone. But, the equipment, treatment, and materials may be even more complex than the water system. Few people get sick from sewage because of the systems we have built. Now, think about third-world examples. Or think of conditions seen in documentaries, the news, or movies. Being in sewage is not a great place to be. Even the manhole-thriving cockroaches agree.

Stormwater is probably the laggard here, in part because changes in development patterns have overwhelmed the old systems. Miami Beach experienced this when redevelopment replaced small houses on permeable lots with large houses with mostly impermeable property. Meanwhile, roads and bridges have received a lot of funding—with much remaining to do (see the bridge that collapsed on I75 in Cincinnati in 2016 or the I35W bridge collapse in Minneapolis in 2007). Most states fund transportation projects at a magnitude greater than water and sewer projects because they see water and sewer as a local issue.

Infrastructure needs such as water, sewage, and stormwater conveyance systems provide the benefit of a sanitary living environment. Transportation systems have changed from trails or gravel roads into massive multi-lane highways and rail, air, and sea transport that allow for easier movement of goods from one place to another. Transportation connects the world through economic activity. Gas and electricity provide means for heating and cooking. One of the newest infrastructure utilities is widespread mass communication systems, such as cellular networks, cable systems, and internet infrastructure. These systems, when properly operating, allow for improved productivity and a better quality of life for all users.

We now face a changing condition in the 21st century and the question is: who will take the 21st-century infrastructure leadership mantle? How will it change our viewpoint to protect our resources? Will that leadership persuade residents to back infrastructure renewal, i.e., can we sell it? Local public officials often fail on the marketing end, especially in dealing with infrastructure issues. Everyone knows that infrastructure is in poor condition and that billions, perhaps trillions are needed to upgrade the system to serve society's needs. It's a huge number, but who takes ownership of that number? So much is hidden and park projects are far more glamorous, so guess which gets funded? Parks—at least until a failure occurs. We also face the need for more infrastructure, so making useful assumptions about increases in demands, prices, inflation rates, etc., are key to useful projections and long-term sustainability. Building too much or too little capacity, for example, can have disastrous consequences (to the ratepayers on the former, to the local economy on the latter). Long-term sustainability of these infrastructure systems is important and requires vision well beyond our working years.

The public and private sectors are different. That does not mean one cannot apply some concepts of the business world into the public sector, but we need to be careful which ones are chosen. For the most part, the public sector does those things that the private sector deems to be less (unprofitable). Clearly, everyone needs water, but if you can't get people to pay for it, you can't make a business out of it. Enter government, which has the ability to lien and condemn property for the failure to be connected. Or take fire service. Fire service in New York was once a private affair. You paid and the fire company would respond. If your house caught fire and you had not paid, then what? No one fights the fire and you lose everything. This was illustrated nicely in the movie *Gangs of New York* and was the catalyst for creating the New York City Fire Department in the 1860s. It simply is not acceptable to provide some people with service, but not all, because of the risk to everyone. Vaccinations are the same way. They are much easier to implement by the government. Historically, that is exactly what has happened.

Fortunately, most of our infrastructure systems are operating well and were made with good materials that require limited maintenance. Unfortunately, that has led to these same infra-structure systems often being completely neglected by their users. This is the main benefit of infrastructure: it fulfills the need for a particular purpose (or utility) so completely that its users can benefit from its function without concern for the complexities of its operation. This allows users to take the system for granted because their need is fulfilled, so they can focus on other activities in life. However, a danger presents itself when the expectation of that infrastructure is taken for granted.

When consumers use a service regularly, it is only remembered when it fails to provide the expected service. Service interruptions or degradations (such as roadwork, brief electrical inter-ruptions, pipe repairs) due to maintenance will be looked at unfavorably by the customers. The reality with infrastructure management in the United States is that those who do those jobs have been so successful at delivering the various services, they are largely forgotten until something goes wrong—and then they are roundly criticized for the failures. It is difficult to be perfect, yet most infrastructure systems are nearly perfect when delivering to their customers. As noted in Chapter 14, it is the odd cascade of errors that lead to failures, and people remember those. An increase in a monthly bill or annual tax assessment can bring up negative feelings about the per-formance of an infrastructure organization, since this is the only feedback that many customers receive regarding that entity. The job today is both managing those assets to ensure their long-term sustainability and leading people to the upgrades and renewals to ensure the infrastructure systems can continue to be managed successfully by meeting service expectations.

Asset management is a tool that should be used by all governmental agencies—large and small alike. It is essential to have an asset management plan and to implement it. Work order tracking is a primary component of any asset management plan and is essential for provid-ing useful data in order to make decisions. Implementing work order systems will take leader-ship because change will be required in most organizations. Providing transparency and public information on the functions and benefits of an infrastructure system are important by reminding end users about the complexity of operations and how maintenance operations benefit the user by helping preserve the value of the system, thereby reducing the long-term capital and operational costs. This should result in lower costs and increased long-term reliability for the user. A properly maintained infrastructure allows for the continued advancement of society.

Stewardship, leadership, communications, and ethics are key to local officials, stakeholders, and operations and engineering staff. Communication, consensus, and ethics are what builds trust between boards and management. All are needed to show the public the leadership needed to trust the organization, which is especially important during a crisis or when rate increases are needed. Responsible stewardship includes:

- Clear delineation of expectations of staff with regard to hours, work, and duty station
- Clear fiscal policies, purchasing requirements, and authority for purchases
- Transparent bidding and procurement standards
- Clear policies with regard to use of equipment
- Clear delineation between staff, management, and local officials with appropriate liaisons

To resolve the long-term infrastructure needs for society, all parties must take their stewardship responsibility seriously. There should be ongoing measures of how the system has improved from a condition and reliability perspective. How has service improved? Where is the system headed? What needs are there to make the system more efficient or provide better services? Has the system value increased with time?

Whether the infrastructure system is water, sewage, stormwater, power, transportation, or any other system, society relies on it to function as intended 100% of the time. Failure is not an option for a first-world country. To maintain society's place in the global economy and to maintain society's standard of living, infrastructure must be maintained to provide optimal service. It also takes leadership. This is the 21st-century leadership that needs to be shown today.

INDEX